AQUATIC ORGANIC MATTER FLUORESCENCE

This is the first comprehensive text on the theory and practice of aquatic organic matter fluorescence analysis, written by the experts who pioneered the research area. This book covers the topic in the broadest possible terms, providing a common reference for making measurements that are comparable across disciplines, and allowing consistent interpretation of data and results. The book includes the fundamental physics and chemistry of organic matter fluorescence, as well as the effects of environmental factors. All aspects of sample handling, data processing, and the operation of both field and laboratory instrumentation are included, providing the practical advice required for successful fluorescence analyses. Advanced methods for data interpretation and modeling, including parallel factor analysis, are also discussed. The book will be of interest to those establishing field, laboratory, or industrial applications of fluorescence, including advanced students and researchers in environmental chemistry, marine science, environmental geosciences, environmental engineering, soil science, and physical geography.

PAULA G. COBLE is Professor in the College of Marine Science at the University of South Florida and has been involved in research on the fluorescence of natural organic matter for twenty years.

JAMIE LEAD is Professor of Environmental Nanoscience and Risk and Director of the Environmental Nanoscience and Risk SmartState Center at the University of South Carolina and Adjunct Professor of Environmental Nanoscience and Director of the Facility for Environmental Nanoscience Analysis and Characterisation at the University of Birmingham, UK.

ANDY BAKER is Director of the Connected Waters Initiative Research Centre at the University of New South Wales and Research Program Leader in the National Centre for Groundwater Research and Training, Australia.

DARREN M. REYNOLDS leads a research group within the Centre for Research in Biosciences and sits on the Scientific Advisory Board for the Institute of Bio-Sensing Technology at the University of the West of England, Bristol.

ROBERT G. M. SPENCER is Assistant Scientist at the Woods Hole Research Center, Falmouth, Massachusetts. He is an Earth system scientist whose research encompasses both aquatic geochemistry and biogeochemistry, focused predominantly on the carbon and nitrogen cycles.

CAMBRIDGE ENVIRONMENTAL CHEMISTRY SERIES

Series editors:
P. G. C. Campbell, Institut National de la Recherche Scientifique, Université de Québec, Canada
R. M. Harrison, School of Chemistry, University of Birmingham, UK
S. J. de Mora, Plymouth Marine Laboratory, Plymouth, UK

This wide-ranging series aims to cover all areas of environmental chemistry, placing emphasis on both basic scientific and pollution-oriented aspects. A central core of textbooks, suitable for those taking courses in environmental sciences, ecology, and chemistry, provides comprehensive coverage at the undergraduate and first-year postgraduate level of atmospheric science, chemical sedimentology, freshwater chemistry, marine chemistry, and soil chemistry. At a more advanced level, the series contains topical accounts of current research interest.

All books available in the series:
P. Brimblecombe *Air Composition and Chemistry, Second Edition*
A. C. Chamberlain *Radioactive Aerosols*
M. Cresser, K. Killham, and A. Edwards *Soil Chemistry and its Applications*
A. Edwards and M. Cresser *Acidification of Freshwaters*
R. M. Harrison and S. J. de Mora *Introductory Chemistry for the Environmental Sciences, Second Edition*
T. D. Jickells and J. E. Rae *Biogeochemistry of Intertidal Sediments*
S. J. de Mora *Tributyltin: Case Study of an Environmental Contaminant*
S. J. de Mora, S. Demers, and M. Vernet *The Effect of UV Radiation in the Marine Environment*
S. Roy, E. S. Egeland, G. Johnsen, and C. A. Llewellyn *Phytoplankton Pigments: Characterization, Chemotaxonomy and Applications in Oceanography*
E. Tipping *Cation Binding by Humic Substances*
D. A. Wright and P. Welbourn *Environmental Toxicology*
S. Roy, C. A. Llewellyn, E. S. Egeland, and G. Johnson *Phytoplankton Pigments*
P. G. Coble, J. Lead, A. Baker, D. M. Reynolds, and R. G. M. Spencer *Aquatic Organic Matter Fluorescence*

AQUATIC ORGANIC MATTER FLUORESCENCE

Edited by

PAULA G. COBLE
University of South Florida

JAMIE LEAD
University of South Carolina

ANDY BAKER
University of New South Wales, Sydney

DARREN M. REYNOLDS
University of the West of England, Bristol

ROBERT G. M. SPENCER
Woods Hole Research Center

CAMBRIDGE
UNIVERSITY PRESS

CAMBRIDGE
UNIVERSITY PRESS

32 Avenue of the Americas, New York, NY 10013-2473, USA

Cambridge University Press is part of the University of Cambridge.

It furthers the University's mission by disseminating knowledge in the pursuit of education, learning, and research at the highest international levels of excellence.

www.cambridge.org
Information on this title: www.cambridge.org/9780521764612

First published 2014

Printed in the United States of America

A catalog record for this publication is available from the British Library.

Library of Congress Cataloging in Publication data
Aquatic organic matter fluorescence / [edited by] Paula Coble, Jamie Lead, Andy Baker.
pages cm. – (Cambridge environmental chemistry series)
Includes bibliographical references and index.
ISBN 978-0-521-76461-2 (hardback)
1. Chemical oceanography. 2. Water chemistry. 3. Biogeochemistry. 4. Seawater – Organic compound content. 5. Bioluminescence. I. Coble, Paula.
GC116.A78 2014
572′.435809162–dc23 2013038167

ISBN 978-0-521-76461-2 Hardback

Contents

Contributors

George Aiken
U.S. Geological Survey, Water Resources Discipline, National Research Program, Boulder, Colorado, US

Andy Baker
Connected Waters Initiative Research Centre, University of New South Wales, Sydney, Australia

Thomas J. Boyd
Chemistry Division, U.S. Naval Research Laboratory, Washington, DC, US

Rasmus Bro
University of Copenhagen, Faculty of Life Sciences, Department of Food Science, Frederiksberg C, Denmark

Robert F. Chen
University of Massachusetts, Boston, US

Paula G. Coble
College of Marine Science, University of South Florida, St. Petersburg, US

Robyn N. Conmy
U.S. Environmental Protection Agency, National Risk Management Research Laboratory, Cincinnati, Ohio, US

Rose M. Cory
Department of Earth & Environmental Sciences, University of Michigan, US

Carlos E. Del Castillo
Johns Hopkins University, Baltimore, Maryland, US

Rossana Del Vecchio
Earth System Science Interdisciplinary Center, University of Maryland, College Park, US

Bryan D. Downing
U.S. Geological Survey, Sacramento, California, US

Rachel S. Gabor
Environmental Studies and Institute of Arctic and Alpine Research, University of Colorado, Boulder, US

John R. Gilchrist
Gilden Photonics Ltd., Clydebank, UK

Diane M. McKnight
Department of Civil, Environmental, and Architectural Engineering and Institute of Arctic and Alpine Research, University of Colorado, Boulder, US

Matthew P. Miller
U.S. Geological Survey, Utah Water Science Center, Moab, US

Kathleen R. Murphy
Water Environment Technology, Department of Civil and Environmental Engineering, Chalmers University of Technology, Gothenburg, Sweden

Christopher L. Osburn
Department of Marine, Earth, and Atmospheric Sciences, North Carolina State University, Raleigh, US

Darren M. Reynolds
Centre for Research in Biosciences, University of the West of England, Bristol, UK

Robert G. M. Spencer
Woods Hole Research Center, Falmouth, Massachusetts, US

Colin A. Stedmon
National Institute for Aquatic Resources, Technical University of Denmark, Kavalergården, Denmark

About the Editors

PAULA G. COBLE is Professor in the College of Marine Science at the University of South Florida and has been involved in research on the fluorescence of natural organic matter for twenty years. She is one of the original researchers to use excitation–emission matrix spectroscopy (EEMS) for characterization of aquatic colored dissolved organic matter (CDOM) and has published several reviews on the topic. Coble's research interests have focused on cycling of CDOM in marine and estuarine systems, development and deployment of in situ fluorometers, and using fluorescence to trace organic matter cycling and water mass mixing. Most recently, Coble has been advancing application of the EEMS technique to the study of oil spills in the marine environment, including distinguishing between the effects of physical and chemical dispersion.

JAMIE LEAD is Professor of Environmental Nanoscience and Risk and Director of the Environmental Nanoscience and Risk SmartState Center at the University of South Carolina and Adjunct Professor of Environmental Nanoscience and Director of the Facility for Environmental Nanoscience Analysis and Characterisation, University of Birmingham, United Kingdom. His research aims to understand both natural and manufactured nanoparticles and their interactions, as well as their fate and impact in the environment, and encompasses work from synthesis and characterization to risk and regulation.

ANDY BAKER is Director of the Connected Waters Initiative Research Centre at the University of New South Wales and Research Program Leader in the National Centre for Groundwater Research and Training, Australia. He is a researcher in the contrasting fields of past climate change, hydrogeology, organic matter characterization, and isotope geochemistry. Common themes include an interdisciplinary scientific approach between the subject areas of geography, Earth science, environmental science, and engineering. Baker has published more than one hundred fifty internationally refereed papers, including papers in high-citation-rate journals such as *Nature* and *Science*, and has been awarded and successfully managed more than sixty research grant applications. He was awarded a Phillip Leverhulme Prize in 2003 and a Durham University Institute for Advanced Studies Fellowship in 2009. Baker's role in aquatic organic matter fluorescence commenced in the 1990s with his investigation of organic matter preserved in cave stalagmites. This led to research on organic matter fluorescence properties of karst groundwaters, UK rivers, and, finally, engineered systems such as drinking and waste water treatment systems.

DARREN M. REYNOLDS leads a research group within the Centre for Research in Biosciences and sits on the Scientific Advisory Board for the Institute of Bio-Sensing Technology at the University of the West of England, Bristol. He is a multidisciplinary scientist concerned with the development of technology platforms for applications in the fields of environment/agri-food and health. Reynolds's research interests include the development of optical technologies and techniques for environmental and biological sensing and bio-prospecting applications. He is currently involved in the development and use of synthetic biology (bioluminescent reporter bacteria) for the discovery and characterization of bacteriophage and to assess their potential use as a bio-control/antimicrobial in the health and agri-food sectors. He is developing, in collaboration with industrial partners, in situ fluorescence sensors to help further understand the role of dissolved organic matter in underpinning the microbial processes in aquatic systems though time.

ROBERT G. M. SPENCER is Assistant Scientist at the Woods Hole Research Center, Falmouth, Massachusetts. He is an Earth system scientist whose research encompasses both aquatic geochemistry and biogeochemistry, focused predominantly on the carbon and nitrogen cycles. The majority of his research centers on the interface between the hydrosphere and biosphere over environmental time scales within natural systems. Spencer is a highly interdisciplinary scientist and uses a suite of analytical techniques to examine the influence of physical, chemical, and biological processes – especially watershed hydrology, ecosystem processing, global climate change, and land use change – on the carbon and nitrogen cycles within aquatic and terrestrial environments. A major focus of Spencer's current research is examining controls on the export, processing, and fate of organic matter in the critical zone from the tropics to the Arctic, from soils and glaciers through rivers and estuaries and into the ocean. Spencer has utilized organic matter fluorescence extensively in this research for delineating sources and degradation histories of organic matter, as well as to improve resolution of components of the organic matter pool (e.g., dissolved organic carbon and biomarkers).

Preface

This volume is a response to the explosion of interest in the use of fluorescence spectroscopy to analyze organic matter in the aquatic environment and the realization that a book was needed that combined relevant information on fluorescence principles, laboratory and field methodologies, and data handling and interpretation. The editorial and author teams are drawn from a wide range of disciplines, which reflects the multidisciplinary interest in aquatic organic matter fluorescence. Despite more than thirty years of interest in the topics, until recently, research was undertaken primarily within disciplines, without the benefits of a genuine interdisciplinary approach. We hope that this volume goes some way to address this. It is intended as a core text for anyone starting to undertake research into aquatic organic matter fluorescence, with a potential readership in the fields of environmental chemistry, marine science, environmental geosciences, environmental engineering, soil science, and physical geography. The first three chapters provide an overview to the field of aquatic organic matter fluorescence: Chapter 1 outlines the principles of fluorescence, Chapter 2 synthesizes the relationship between fluorescence and dissolved organic matter chemistry, and Chapter 3 provides a multidisciplinary overview of the history and current understanding of aquatic organic matter fluorescence. As such, we anticipate that these chapters will be of broad interest to students and researchers interested in optical methods in environmental science and the environmental chemistry of organic matter. The next three chapters focus on instrumentation and sampling: Chapter 4 details sampling protocols essential for successful fluorescence analyses, and Chapters 5 and 6 provide insights into laboratory (bench) and field (in situ) methodologies. These chapters will be of particular interest to anyone intending to establish field, laboratory, or industrial applications of fluorescence, including undergraduate project students, PhD and postdoctoral researchers, and research and development officers. Chapters 7 and 8 investigate environmental effects on aquatic organic matter fluorescence, detailing the chemical and biological reprocessing of organic matter in natural and engineered systems. These chapters are relevant to anyone interested in the nature of organic matter transformations in the natural and engineered environments, from transformations within river systems through to chemical properties relevant to water treatment processes. Finally, Chapters 9 and 10 summarize methods used in the analysis and interpretation of fluorescence data, focusing on the use of indices and multivariate statistical and modeling approaches.

As with all research textbooks published in a rapidly moving field, our challenge is to produce a volume that is relevant both today and in the years to come. The future for research into aquatic organic matter fluorescence is exciting, as research opportunities develop with improvements in technology. At the time of writing, routine fluorescence analyses are still three-way analyses (excitation, emission, intensity), with the use of xenon light sources, diffraction gratings, and photomultiplier tubes. However, the use of ultraviolet light-emitting diodes (LEDs) as light sources, and charge-coupled devices (CCDs) as detectors, is increasing. The former permit a decrease in instrument power output and size, leading to increased portability and more diverse in situ applications. The latter increases analysis speed and the possibility of four-way analyses (excitation, emission, intensity, time) and novel experimentation. The recent development of turnkey instrumentation, with integrated absorbance measurements and multivariate modeling software, makes aquatic organic matter fluorescence measurements easier for the non-expert than they have ever been before. We feel this latter development makes the need for this book more relevant than ever.

This volume is an outcome of a series of research initiatives led by the editors over the last decade. In part, it is a result of the UK initiative Fluoronet, a knowledge transfer network funded by the Natural Environment Research Council between 2006 and 2009 and led by Andy Baker and Jamie Lead when they both worked at the University of Birmingham, United Kingdom. Numerous training courses and workshops run by Fluoronet members, including Darren M. Reynolds and Robert G. M. Spencer, have helped guide the requirements of the contents of this book. At the same time, scientific sessions on organic matter characterization at the American Geophysical Union Fall Meetings in 2006 and 2007 started to bring together research teams from diverse disciplines. Informal gatherings, and then formal meetings, occurred around these sessions. These led to the idea of the need for both a specialist conference, to compare methodologies and establish protocols between different disciplinary groups, as well as for a multidisciplinary research textbook. The resulting meeting was an American Geophysical Union Chapman Conference on Organic Matter Fluorescence, proposed and convened by Paula G. Coble and Andy Baker at the University of Birmingham, United Kingdom, in 2008. Several years later, we are proud that this edited volume has been produced by Cambridge University Press.

Part I

Introduction

1

The Principles of Fluorescence

DARREN M. REYNOLDS

1.1 Luminescence

Although this chapter is concerned primarily with the fluorescence process it is important to understand that fluorescence is one of several phenomena that give rise to the collective term *luminescence*. Broadly speaking, luminescence is defined as the emission of light by a substance, where the emitted light cannot be attributed to incandescence, that is, thermal radiation. In the case of fluorescence, light is emitted from a substance after it has been irradiated, usually with visible or ultraviolet light. There are other important kinds of luminescences and some of these are described in the following paragraphs.

Phosphorescence is a light emission that often lasts minutes or sometimes hours after irradiation. The slow release of light characterized by phosphorescence is the result of the storage of energy in *metastable* states and is often thermally activated. In this instance "metastable" refers to the local stability of an energy state with respect to surrounding systems that exhibit a different energy state. *Photoluminescence* is a more general term that encompasses both fluorescence and phosphorescence.

Chemiluminescence is light emitted during cold chemical reactions whereas *bioluminescence* is essentially chemiluminescence from living organisms. *Thermoluminescence* is a type of phosphorescence but one that occurs at elevated temperatures. Thermoluminescence is not related to incandescence, as thermal excitation is used to initiate the release of energy from another source.

Electroluminescence is light emission initiated by electric influences. For example, in cathodoluminescence, the emission of light is initiated by excitation with an electron beam. *Radioluminescence* is caused by excitation with nuclear radiation or X-rays, whereas *triboluminescence* occurs when certain materials are mechanically altered, such as when fractured or polished.

Although several investigators reported luminescence phenomena during the 17th and 18th centuries, it was British scientist Sir George G. Stokes (1852) who first described "fluorescence" in his reference to the light-emitting properties of the mineral fluorite (fluorspar). It was not until the discovery of the electron by Sir Joseph John Thomson (1897a,b) and the discovery of the quantized nature of matter by Planck (1900, 1902) and light by Einstein (1905) that our comprehension of matter and energy was spectacularly revolutionized, and indeed it is the current theory of quantum mechanics that underpins our understanding of the fluorescence process.

1.2 The Relevance of Quantum Mechanics and Electronic Theory

Wave–particle duality is a central concept for our current understanding of modern quantum mechanics (Anastopoulos, 2008). The fact that particles and matter exhibit both wave and particle-like properties helps us to explain their behavior at the quantum scale. To appreciate *how* light interacts with matter, it is important first to consider the nature of light and the role of matter in terms of electronic structure. Unfortunately, an in-depth discussion pertaining to quantum theory and the magnificent discoveries throughout the history of science is beyond the scope of this chapter. However, for readers to gain an insight into *how* light can interact with matter in ways that result in the emission of light it is necessary first to consider the nature of light and how matter is organized in terms of electronic structure. Although there have been monumental discoveries over the ages, all of which have contributed to our understanding of the universe, for simplification this chapter focuses attention toward scientific discoveries achieved throughout late 19th and the 20th centuries.

1.2.1 Wave–Particle Duality and Quantization of Energy and Matter

During the early 19th century, atoms were the smallest particles known, and were believed to be indestructible and indeed indivisible, as such the knowledge of subatomic particles and their role in energy transfer processes in light–matter interactions were unknown. Many of the early advancements in electromagnetic theory were achieved owing to curiosity surrounding the phenomena of magnetism, electricity, and light.

1.2.1.1 Subatomic Particles

In 1838, Michael Faraday passed an electric current through a glass tube containing rarefied air (partially emptied). Faraday observed an arc of light emanating from the negative electrode (cathode) almost reaching the positive electrode (anode). These so-called cathode rays, what we now know to be electron beams, were the subject of great interest (Faraday 1844; Dahl, 1997). Shortly after the work of Faraday, in 1839, the French physicist Edmund Becquerel, who was fascinated by the properties of light, observed that certain materials produced electricity (the emission of electrons) when exposed to sunlight (Becquerel 1839). In 1857, German physicist Heinrich Geissler repeated Faraday's experiment but this time he was able to evacuate more air from specially designed glass tubes (10^{-3} atmospheres) using an improved pump. Geissler found that, instead of an arc, the light glow filled the tube completely (Dahl, 1997). James Clerks Maxwell's work regarding the nature of electromagnetic fields paved the way for a greater understanding of the nature of light, and between 1862 and 1864 Maxwell demonstrated that electric and magnetic fields propagated through space, in wave forms, at the speed of light. From this, Maxwell deduced (Maxwell's equations) that electricity, magnetism and light were all manifestations of the same phenomenon, and in 1865 proposed a unifying theory of electromagnetism through his publication "A dynamical theory of the electromagnetic field." In 1876, the German physicist Eugen Goldstein coined the phrase *cathode rays* when he demonstrated that the glow from the cathode cast a shadow (Hedenus, 2002). The English scientist Sir William Crookes developed the first cathode ray tube during the 1870s with high vacuums. Using these tubes he was able to demonstrate that luminescence rays appearing within the tube actually carried energy from the cathode to the anode. Crookes also deflected these cathode rays using magnetism and showed that the cathode beam behaved as if it were negatively charged. In 1879, he proposed that these observations could be explained by a fourth state of matter in which negatively charged molecules were projected at high velocities from the cathode. Crookes termed this proposed fourth state "radiant matter" (Crookes, 1879; Eliezer & Eliezer, 2001).

Edmund Becquerel's work and the conversion of light into electricity were of great interest to the German physicist Heinrich Hertz. In 1887, before the discovery of the electron, Hertz performed experiments demonstrating that an electric spark across an air gap between two electrodes is more easily emitted when ultraviolet light is shone on the cathode. Finally, in 1897, in his experiments investigating how gases at low temperatures conducted electricity, J. J. Thomson proved that cathode rays were composed of negatively charged *particles* that we now know as electrons, and that these particles were much lighter than the smallest ions known at that time, hydrogen. These observations (Thomson, 1897a,b), along with the accidental discovery of radioactivity from studies performed on natural fluorescing minerals by Antoine Henri Becquerel (1896) at the same time, provided evidence that atoms are not indestructible and that they are composed of *subatomic* particles. Thomson realized that because many atoms appeared to be electrically, other "positively charged" subatomic particles must also exist within the atom. It was in 1903 that Thomson postulated that individual atoms were spheres of "uniform positive electrification," scattered with electrons rather like "currants" in a bun.

1.2.1.2 Quantized Matter and Energy

Before the discovery of the electron and the possibility of further subatomic particles, matter was known to have mass, chemical and electromagnetic properties. In the main, the aspects of matter that gave rise to many of the observed chemical and electrical properties were still largely unknown during the late 19th century. It was generally accepted that the arrangement of matter involved the presence of tiny oscillating particles (invisible to the naked eye) and that it was these oscillating properties that gave rise to observed chemical and physical properties. In 1894 Wilhelm Wien used theories about heat, and also Maxwell's electromagnetic theories, to account for the relationship between wavelength distribution and radiated heat energy from a theoretical body of matter that absorbed *all* radiation (black-body). In 1896 Wilhelm Wien performed experiments designed to understand the spectral radiance of electromagnetic radiation from a black body in thermodynamic equilibrium (within a cavity). Wien presented his laws of thermal radiation work in a Nobel lecture given in December 1911 (Wien, 1911). Wien's Law accurately predicts the behavior of black-body radiation at high frequencies (short wavelengths) but fails to predict accurately the behavior of black-body radiation at low frequencies (longer wavelengths). Max Planck, who is considered by many to be the founder of quantum mechanics, discovered that the intensity of electromagnetic radiation emitted by a black body is dependent on *both* the frequency of the radiation (the color of light) and the temperature of the emitting body. Planck (1900) stated that the energy of the charged oscillators in a black body must be *quantized* and that electromagnetic energy can be emitted only in a quantized form. This is to say that the energy (E) can only ever be a multiple of an elementary unit given by the equation

$$E = h\nu \tag{1.1}$$

where h is Planck's constant, and ν (the Greek letter nu) is the frequency of the oscillator. This later became known as the Planck postulate. The assumption that electromagnetic radiation (light) is quantized allowed Planck to derive a mathematical formula that could be applied to the *entire* electromagnetic spectrum, unlike Wien's Law, which was true only for short wavelengths (UV-Vis). At the time, Planck believed that the quantization of energy applied only to the tiny oscillators related to matter under investigation and made no assumption that light itself is quantized. Planck's concern was one of solving the mathematical problem highlighted earlier by Wien rather than proposing a fundamental change in the understanding of the world. Despite this, Planck's postulate was to help transform our understanding of the world and universe in which we exist.

The photoelectric effect is the phenomenon whereby electrons are emitted from material, such as metals, nonmetals, liquids, and gases as a direct consequence of their absorption of energy The achievements of Hertz in observing the photoelectric effect were very important as it paved the way for Johann Elster and Hans Geistel to pioneer the reliable production of photoelectric devices at the turn of the 20th century. These photoelectric devices could accurately measure the intensity of light far beyond the capability of the human eye. In 1902 before the discovery of the electron, Philipp Eduard Anton von Lenard observed that the energy of individual emitted particles from a cathode ray increased with the frequency of the light rather than the intensity of light (Philipp Lenard – Biography). At the time this postulate was in direct conflict with James Clerk Maxwell's electromagnetic wave theory, which predicts that the energy of the electromagnetic wave would be proportional to the intensity of the radiation as opposed to frequency. In 1905, Albert Einstein described light as being composed of discrete quanta (what we now know as photons), rather than as a continuous wave of energy. Using Max Planck's theory of black-body radiation, Einstein theorized that the energy in each quantum of light was equal to the frequency multiplied by a constant (later named Planck's constant). Therefore a photon above a threshold frequency has the required energy to eject a single electron. This work led to the theory of unity, which took into account that both electromagnetic waves and subatomic particles possessed properties both of particles and electromagnetic waves, the so-called wave–particle duality (Einstein, 1905).

In 1903, just before Einstein's theory of unity, Thomson postulated that individual atoms were spheres of "uniform positive electrification," scattered with electrons rather like "currants" in a bun. Thomson also realized that because many atoms appeared to be electrically neutral, other "positively charged" subatomic particles must also exist within the atom. Shortly after this proposed model of Thomson, in 1910 Lord Ernest Rutherford and his researchers led to the proposition that an atoms mass must be concentrated at its center, that is, the nucleus (Rutherford, 1911). Much of Rutherford's work was complemented by the Danish physicist Neils Bohr, who, in 1913, proposed that electrons exist in *quantized* states. Bohr's physical model postulated that the energy of these quantized states was determined by the angular momentum (motion through space) of the electron's orbit about the nucleus. Quantized states do not vary "continuously" but rather in permitted quantum leaps, that is, between precise values. Furthermore, electrons were free to leap between these states, or orbits, by the emission or absorption of photons at discrete frequencies. Bohr used the notion of quantized orbits to account for the emitted spectral lines of hydrogen atoms. Although momentous in our understanding of physics, Bohr's model failed to predict the observed relative intensities of spectral lines, and more importantly the spectra of more complex atoms with fine and hyperfine structure. Despite the shortfalls of Bohr's theory, which was constrained to the simplest known atom, hydrogen, the notion that an atom is a dense nucleus of positive charge surrounded by lower-mass orbiting electrons was an established idea by 1914.

Bohr's initial model (Bohr, 1922) helped scientists advance our understanding of chemical bonding between atoms and better understand the quantum state. In 1916, American scientist Gilbert Newton Lewis proposed the idea of the covalent chemical bond, in which the bond between two atoms is maintained by a pair of "shared" electrons. The work of Lewis was elaborated further in 1919 by the American chemist Irving Langmuir. Langmuir suggested that all electrons were distributed in consecutive spherical "shells" of equal thickness. Langmuir further divided these shells into a number of cells each containing one pair of electrons. Using this model Langmuir was able to explain the chemical properties of all elements in the periodic table according to the periodic law, which states that the chemical properties of the elements are periodic functions of their atomic numbers.

In 1923, Walter Heitler and Fitz London fully explained electron-pair formation and chemical bonding in terms of quantum mechanics (Heitler and London, 1927). In the same year the French physicist Louise de Broglie proposed that wave–particle duality applied not only to photons, but also to electrons and every other subatomic physical system; this work was published in his PhD thesis in 1924. Austrian physicist Wolfgang Pauli (1925) observed that the shell-like structure of the atom could be explained by a set of four parameters that define every quantum energy state, as long as each state was inhabited by no more than a single electron.

These parameters are:

- Principle quantum number, n. In Bohr's model this number largely determines the energy level and the average distance of an electron from the nucleus.

- Magnetic quantum number, l, which represents the orbital angular momentum and describes the number of possible angular momentum states.
- Azimuthal quantum number, m, where azimuthal represents the angular measurement in a spherical coordinate system.
- Spin quantum number, s. This number represents the intrinsic angular momentum.

It is important to note that matters are further complicated because for each principle quantum number value (n) there are $n - 1$ values for l. In addition, when s is taken into consideration for any quantum value n then there are a total of $2n^2$ states of the same energy possible. This prohibition against more than one electron occupying the same quantum energy state became known as the Pauli exclusion principle (Pauli, 1925, 1926; Massimi, 2005).

1.2.1.3 Copenhagen Interpretation

The work of Louise de Broglie in 1923 linked wavelength, frequency, and momentum, and de Broglie formulated the theory that any moving subatomic particle or object had an associated wave. This theory saw the birth of wave mechanics (*mécanique ondulatoire*), a mathematical unifying of the physics of energy (wave) and matter (particle). In 1925 an explanation of the spin quantum number (the fourth parameter), which had been shown to have two distinct possible values, was provided by the Dutch physicists George Uhlenbeck and Abraham Goudsmit when they suggested that an electron, in addition to the angular momentum of its orbit, could possess an intrinsic angular momentum. This property became known as spin and explained the previously mysterious splitting of spectral lines observed with a high-resolution spectrograph; this phenomenon is known as fine structure splitting.

In Copenhagen between 1925 and 1927, in an attempt to overcome the physical constraints and limitations of his theories, Bohr collaborated with the German physicists Werner Heisenberg and Max Born and the Austrian physicist Erwin Schrödinger to develop the use of abstract mathematical and theoretical formulations instead of physical empirical experiments. This was an important shift in scientific thinking, the main thrust of which was to explain the observations of everyday life and observation through mathematics, the so-called 'matrix mechanics (Born et al., 1925; Born and Jordan, 1925; Heisenberg, 1925). These models utilized matrices (rectangular array of numbers) to describe properties such as momentum, energy, and position as opposed to ordinary numbers. In 1927, Heisenberg published the uncertainty principle. The Heisenberg uncertainty principle gives an insight into the nature of the quantum system itself and states that it is impossible to simultaneously know the momentum and position of a quantum object (e.g., electron) with perfect accuracy. Furthermore, Heisenberg continued to show that the more precisely one property is measured, the less precisely the other *can* be measured. The very act of observing a particle at any one point in time and space will change the behavior of that particle within the quantum system. Therefore the uncertainty principle is not concerned about the limitations of scientists or measurement techniques, but is a mere description of the nature of the quantum system itself. Consequently, it is not possible to know the values of all of the properties of the system at the same time and therefore the unknown properties must be described in terms of probability. Erwin Schrödinger used de Broglie's concepts on wave mechanics to describe the time dependence of a physical quantum state. Schrödinger's attempt to describe how a quantum state changes over time assumed that because all matter has wave-like properties, then all physical quantum states could be explained using wave functions. Initially there was much debate concerning what the wave function (ψ) of the equation was. It is now generally accepted that a wave function is a probability distribution (Born interpretation). The Schrödinger equation is used extensively in modern quantum mechanics to discover the allowed energy levels of quantum mechanical systems (e.g., atoms, molecules, and transistors). Schrödinger (1926a,b) is seen by many as the most significant contributor to the wave theory of matter.

These attempts by Bohr, Heisenberg, Born, and Schrödinger to interpret experimental observations through mathematical formulations became known as the Copenhagen interpretation.

The principles of the Copenhagen interpretation state that

- All quantum systems can be completely described by wave functions.
- The description of nature is probabilistic.
- Matter has wave–particle duality and experiments can determine only if matter is behaving either as a particle or as a wave.
- It is not possible to know the values of all of the properties of any system at the same time. Therefore, the unknown properties can be described only in terms of probability (Heisenberg's uncertainty principle).

1.2.2 Chemical Bonding and Molecular Orbitals

The interactions between electrons and orbitals within atoms ultimately lead to chemical bonding and the formation of molecules, and it is these interactions that are mostly responsible for the absorption and light-emitting properties of molecules. Because this book is principally concerned with the properties of dissolved organic fluorophores, it is necessary to focus our attention on the nature of chemical bonding and molecular orbitals. The theory of covalent bonding, as proposed by Gilbert Lewis in 1916, states that a covalent bond involves the sharing of two electrons between two atoms. However, this theory predated the theory of quantum mechanics, and currently there are two basic models that have been developed to explain how electrons are shared by atoms, the valence bond (VB) theory and molecular orbital (MO) theory (Hückel, 1930, 1931, 1932; Pauling, 1931, 1940). Both of these theories introduce wave functions from quantum mechanical theory. The following sections discuss the nature of bonding albeit in a limited way. Useful underpinning reading can be found from most modern chemistry textbooks (Atkins, 2007; Atkins et al., 2009; Brady, 2011).

According to the Bohr theory (1922), all electrons in the same orbit (shell) have the same energy. However, we now know that with the exception of electrons in the first orbit that this is not the case. Therefore the energy orbitals within atoms (**s, p**, **d**, and **f**) also possess sublevels (principle quantum number) of energy. Owing to the wave–particle duality of electrons it is impossible to pinpoint their exact position; instead we can only consider the probability of finding an electron in a region of space. The probability (expressed as the volume around a nucleus in which an electron is 90% probable to be found) is referred to as the atomic orbital. The orbital of **s** electrons is spherical (**s** orbital). There are three different **p** orbitals ($\mathbf{p}_x$, $\mathbf{p}_y$, and $\mathbf{p}_z$), which have equal energies but different directions in space. These are often described as dumbbell orbitals. For electrons at higher energies, **d** and **f** orbits become available. These orbitals are more complex and numerous than the orbits observed for **s** and **p** electrons.

The distribution of electronic charge throughout the axis of a chemical bond is important. In covalent bonding the region where the value of an orbital wave function (ψ) equals zero (or is very low), defines a region of space within the system where there is zero electron density. This is known as a nodal plane, and quantum theory indicates that molecular orbitals with identical symmetries mix, and the wave functions for **s** + **s** and $\mathbf{p}_z$ + $\mathbf{p}_z$ become blended. The extent of this mixing (or blending) depends on the relative energies of the molecular orbitals involved and is extremely important for determining the number of nodal planes and distribution of energy within molecular bonding orbitals. This mixing of wave functions is termed resonance. Typically, molecules exist as a number of atoms bonded together via covalent bonding, and the collective arrangement of these atoms is such that the overall molecular structure is electrically neutral. Within this structure, all outermost electrons of the atoms involved are paired with other electrons, either in bonds or lone pairs. These outer electrons are termed *valence electrons* and are very influential in determining how atoms interact with each other (reactivity).

Lewis's original theory could not take into account the shape adopted by molecules. Gillespie and Nyholm (1957) developed the currently accepted modern theory of chemical bond formation (MO and VB theories), which uses the valence-shell electron pair repulsion model (VSEPR) to account for molecular structure (Gillespie, 1970). VSEPR states that molecular shape is caused by repulsions between electron pairs in the valence shell.

1.2.2.1 Sigma Bonds (σ Bonds)

Sigma bonds (σ bonds) are the strongest type of covalent chemical bond and are perhaps best illustrated in simple diatomic molecules such as H_2, F_2, Cl_2, Br_2, and I_2. Sigma bonding in diatomic molecules is always symmetrical with respect to the rotation about the bond axis (nucleus to nucleus). Therefore common σ bonds can be represented as **s** + **s**, $\mathbf{p}_z$ + $\mathbf{p}_z$, **s** + $\mathbf{p}_z$, and $\mathbf{d}_z{}^2$ + $\mathbf{d}_z{}^2$ (where z is defined as the axis of the bond). In σ covalent bonding the two "shared" electrons can either originate from the same atom, in which case the σ bond is covalent, or from each atom, where the σ bond is termed a coordinate covalent bond. For homo diatomic molecules, bonding σ orbitals have no nodal planes between the bonded atoms, whereas in the case of hetero diatomic atoms forming a covalent bond (where one atom is more electronegative than the other) the electron pair will spend more time closer to that atom. This is termed a polar covalent bond.

1.2.2.2 Pi Bonds (π Bonds)

Pi bonds (π bonds) occur where an orbital occurs in two regions above and below the bond axis between the two atom nuclei. Pi bonding is a much weaker interaction than sigma bonding and involves only electrons from **p** or **d** orbitals, never **s** orbitals. A pair of atomic nuclei may be connected by pi bonds only if a σ bond also exists between them. The forces involved in σ bonding are much stronger than for π bonding, which is explained by less overlap between the component **p**-orbitals due to their parallel orientation. The distribution of electronic charge through π bonding is concentrated outside of the bond axis, and therefore π electrons are more able to move between the atoms. This mobility of π electrons means that in certain situations, where multiple atoms are connected by a series of σ bonds along with the correct geometry of **p** and **d** orbitals, a system of delocalized π bonds can be formed that spreads over many atoms. The interaction of one **p**-orbital with another across an intervening sigma bond can also lead to conjugation. In reality, this manifests itself by the formation of a molecular entity whose structure may be represented as a system of alternating single and multiple bonds where the delocalized σ electrons do not belong to a single bond or atom, but rather to a group of atoms or the entire molecule. As long as each contiguous atom in a chain has an available **p**-orbital, the system can be considered to be conjugated. Examples of π and σ bonding are shown in Figure 1.1.

1.2.2.3 Antibonding Orbitals

If we consider a linear homonuclear diatomic molecule such as H_2 we can represent the interaction of both 1s molecular orbitals (wave function) associated with each hydrogen atom as

$$\Psi = \Psi_{A1s} \pm \Psi_{B1s} \tag{1.2}$$

This equation tells us that an electron can be found with equal probability in either orbital A or B. More explicitly, there are two possible wave functions for this arrangement and these are shown in Eqs. (1.3) and (1.4):

$$\Psi = \Psi_{A1s} + \Psi_{B1s} \tag{1.3}$$

or

$$\Psi = \Psi_{A1s} - \Psi_{B1s} \tag{1.4}$$

Equation (1.3) represents a sigma orbital (1σ), just like the ones described in Section 1.2.2.1 and shown in Figure 1.2. Both of these electron orbitals behave like waves, and therefore, according to wave theory, can interact constructively in the internuclear region. This constructive interference intensifies the amplitude of the wavefunction, and this in turn, increases the probability of finding the electron between the two nuclei. Any electron located between the two hydrogen nuclei will interact strongly with both of them. As a result, the overall

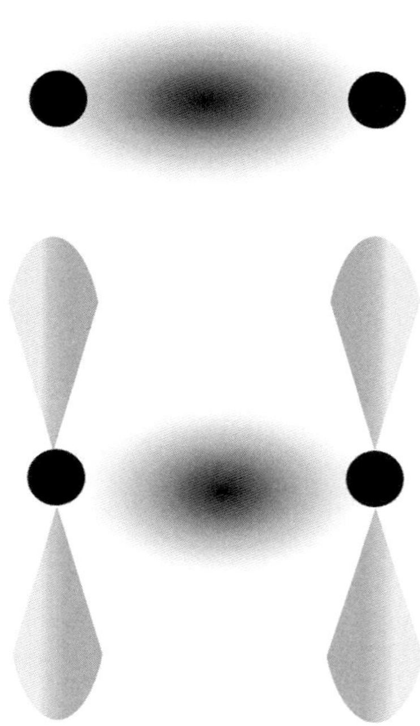

Figure 1.1. Illustration of localization of electron density in sigma (σ) and pi (π) bonding between two atoms. The σ bonding between two atoms is shown at the top while π and σ bonding between two atoms is illustrated underneath.

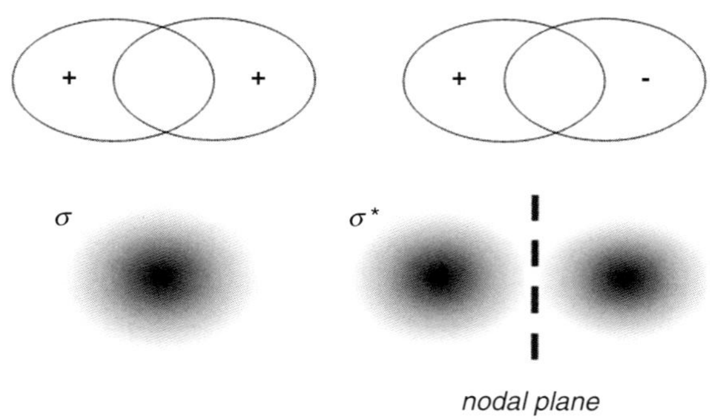

Figure 1.2. Overlapping molecular orbitals interfering constructively (same signs) and destructively (different signs). Constructive interference results in a (σ) molecular bonding orbital while destructive interference produces an antibonding sigma (σ*) molecular bonding orbital.

energy of the molecule is lower than that of the separate atoms because interaction of the electron within a single atom is much stronger when there is only one nucleus.

Let us now consider the second sigma orbital (2σ), shown in Eq. (1.4) and again in Figure 1.2. This wavefunction has symmetry identical to that of the 1σ orbital, described in Eq. (1.3). Schrödinger's equation calculates that this wavefunction has a higher energy than the 1σ orbital, and also higher than either of the individual atomic orbitals. This can be explained in terms of destructive interference of the two orbitals. The point in space which is equidistant from each nuclei and intersects the internuclear axis has a wavefunction equal to zero. This is known as the nodal plane and is referred to briefly in Section 1.2.2. Both *opposing* orbitals cancel each other on this plane as result of destructive interference. This 2σ bonding is known as an antibonding orbital and is denoted as 2σ *. This orbital excludes the electron from the internuclear region and relocates it to outside the region of bonding. The net result of this electron relocation is that the orbital is repulsive, pulling the nuclei apart. This is the main reason why the antibonding molecular orbital 2σ * exhibits higher energy than the 1σ molecular orbital.

A molecular orbital exhibits "antibonding" properties when the electron density between the two nuclei concerned is lower than would otherwise be predicted if there were no bonding interactions at all. In many systems molecules may exhibit molecular orbitals that change state from antibonding to bonding or vice versa. This is entirely dependent on the atoms/conjugated system involved, and the nature of the bonding is relative to the atoms concerned. In more conjugated systems involving molecules with several atoms, for example, benzene, a particular molecular orbital may be bonding with respect to some adjacent pairs of atoms and antibonding with respect to other pairs. In this instance the ratio of bonding to antibonding molecular orbitals becomes important. For example, if the number of *bonding interactions* outnumbers the *antibonding interactions*, the molecular orbital in question is deemed to be *bonding* and vice versa. For benzene each carbon atom contributes only one electron to the delocalized π-system of benzene, and because there are only six π-electrons only the three lowest-energy (bonding) molecular orbitals are filled.

1.2.2.4 Nonbonded Electrons

Valence electrons that are not used for bonding must be paired and are known as "lone pair electrons" or "nonbonding electrons" (*n* electrons). Almost all atoms have paired electrons in the valence shell. Although much attention is given to unpaired electrons in atoms, nonbonding electrons are important in determining the geometry of a molecule along with unpaired electrons using the VSEPR theory. Lone pair electrons exhibit higher energies than σ or π pairs because their repulsive forces are greater. In a σ bond a bonding electron pair lies farther away from the central atom than does a lone pair. Therefore, if the overall geometry of a molecule has two sets of possible positions, but each position has a different degree of repulsion, then lone pairs will occupy the position that has less repulsion. Although paired electrons are not specifically involved, in bonding interactions, they can contribute to the spectral features of a molecule and therefore must be considered.

1.3 Understanding the Fluorescence Process

British scientist Sir George G. Stokes first described "fluorescence" in 1852 after the blue-white fluorescent mineral fluorite (fluorspar). Stokes is perhaps better remembered for discovery of the observed differences (in wavelength or frequency) in positions of the band maxima of the absorption and emission spectra of the same electronic transition, the so-called Stokes Shift. There are three fundamental processes that are implicit in the emission of light by a molecule (Lakowicz, 2006): *excitation* of the molecule (i.e., absorption of an appropriate photon), *vibrational relaxation* (nonradiative decay), and finally the *emission of light* (radiative decay). All of these processes occur on timescales that are separated by several orders of magnitude (see Table 1.2). Excitation of a molecule by an incoming photon happens instantaneously (femtoseconds, $10E^{-15}$), while the vibrational relaxation of an electron in an excited state to the lowest energy level is slower, usually occurring over picoseconds (10^{-12}). The final process, light emission, almost always occurs at longer wavelengths and the return of the molecule back to the ground state occurs over nanoseconds (10^{-9} seconds).

1.3.1 Electronic Transitions

The interaction between a molecule and optical radiation manifests itself in various radiative and nonradiative processes as shown in Figure 1.3. Many of the colors that we experience in everyday life (flowers, green vegetation, synthetic dyes, etc.) are the result of the transition of an electron from one electron orbital into another. Sigma electrons require high energies found only at very short wavelengths in the vacuum UV (100 nm–200 nm) if they are to be promoted to an available molecular orbital at a higher energy level. They are of little relevance to most fluorescence techniques, which are usually concerned with wavelengths between 200 nm and 1000 nm. Pi electrons are less tightly bound to the nuclei than σ electrons. As a result, the energies required for ionization and electronic transition are lower than for σ electrons but still mainly in the vacuum UV or the middle UV (200 nm–300 nm). However, for delocalized π electrons in conjugated systems the energies required for electronic transitions are much lower and easily obtained in the near UV (300 nm–400 nm). For extensively delocalized systems or super-delocalized systems electronic transitions can occur at energies ranging from the near UV to the near infrared (200 nm–1500 nm). The preference of a particular process over another is dictated by the energy of the "exciting" field, that is, the wavelength of the light, the configuration of the molecule(s), and its environment.

1.3.1.1 Spin Multiplicity

The electronic state of a molecule determines both the distribution of negative charge and its overall geometry. All molecules exhibit different electronic states (illustrated as S_0, S_1, and S_2 in Figure 1.3), depending on the total electron energy and the symmetry of various electron spin states. Each electronic state contains a number of vibrational and rotational energy levels associated with the atomic nuclei and bonding orbitals. As discussed in Section 1.2.1.2, electrons are described within a particular electronic state by their spin quantum number (s), with values $ms = +½$ or $-½$. In Figure 1.4 the different spin states are shown in terms of their orientation, spin-down or spin-up, with the direction of the arrows

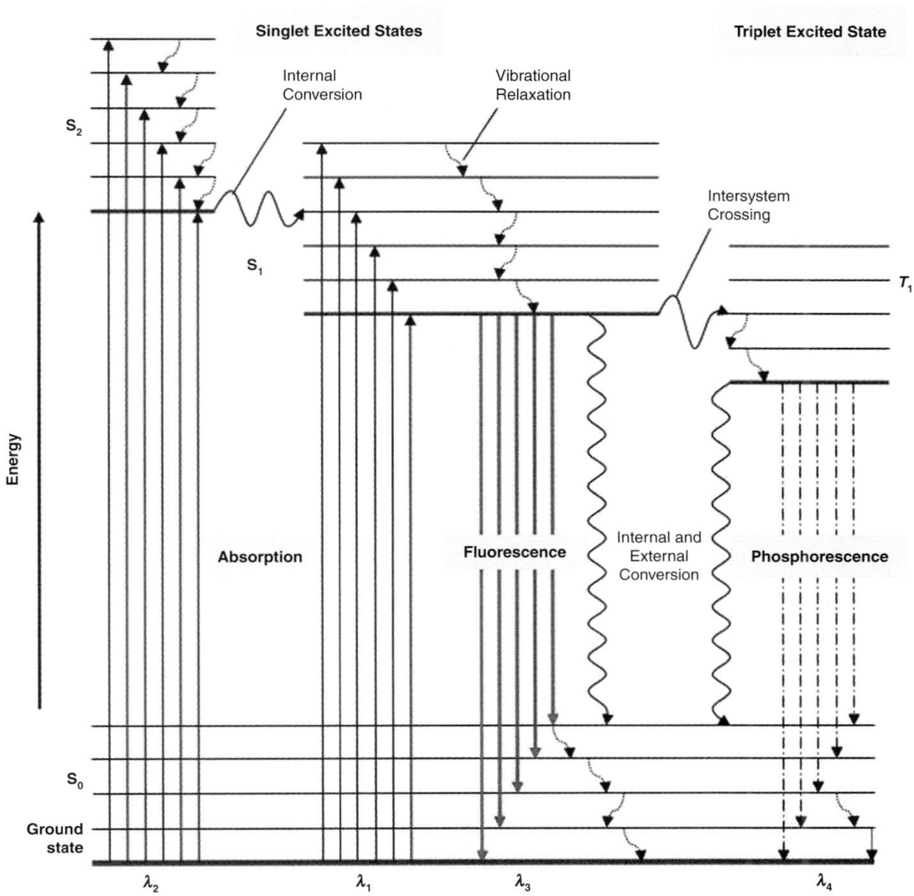

Figure 1.3. A Jablonski diagram for a photoluminescent system showing deactivation pathways from an excited state. The lowest vibrational energy level for each state is indicated by the blue lines and the fluorescence emission is highlighted in red.

representing the values of +½ or –½. For a specified electronic state the total spin quantum number (s) is given by the vector addition (angular momentum) of the individual spin quantum numbers. Therefore the spin multiplicity of an electronic state is defined as $2s + 1$.

If a molecule has two electrons in its highest occupied molecular orbital in the ground state then these have opposing spins of +½ or –½. Hence $s = 0$ and the multiplicity ($2s + 1$) of this specific state is 1; this is referred to as a singlet state (Figure 1.4a). In an excited state, the electrons will either remain opposing (Figure 1.4b) where $s = 0$ or else adopt parallel spins (Figure 1.4c) where $s = 1$, giving a multiplicity of 3. This is known as a *triplet state*. It is important to understand the *forbidden* nature of this state and as such triplet states are not populated via absorption but instead by intersystem crossing from an excited

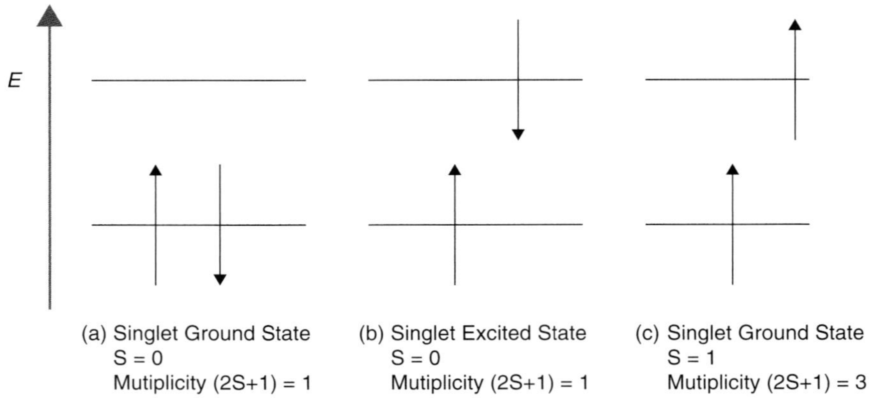

Figure 1.4. The relationship between spin orientation and multiplicity for electrons in singlet and triplet states. The direction of the arrows indicates the orientation of each electron or spin quantum number (ms) with values of either +½ or –½.

singlet state (Lakowicz, 2006). This is illustrated in Figure 1.3 and explained further in Section 1.3.2.2.

In generic terms a molecular energy state is the sum of the electronic, vibrational, rotational, nuclear, and translational components within a molecule. The energy required to change the distribution of an electron within in a molecular orbital is on the order of several electron volts. As a consequence, photons emitted or absorbed when such changes occur lie in the visible and ultraviolet regions (Table 1.1). In some cases the relocation of an electron may be so extensive that it results in the breaking of a bond and the dissociation of the molecule, as is the case in many photochemical reactions. For most organic molecules the lowest energy state, that is, the ground state, contains electrons that are spin-paired (*electronic singlet* state). Under standard temperature and pressure most molecules have only enough intrinsic energy to exist in the lowest vibrational level of the ground state. Therefore, excitation for these molecules originates from this vibrational level of the ground state. Before we examine the fluorescence process in detail it is important first to examine the interaction of energy and electronic states more closely.

1.3.1.2 Absorption

A prerequisite to fluorescence is absorption, so it is first important to understand some of the processes that occur during absorption. When a molecule *absorbs* radiation it's energy is increased (Lakowicz, 2006, chapter 2). This increase energy corresponds to the energy of the absorbed photon and can be expressed by Eq. (1.5):

$$E = h\nu = \frac{hc}{\lambda} \tag{1.5}$$

where h is Planck's constant, nu (ν) and lambda (λ) are the frequency and the wavelength of the radiation respectively, and c is the velocity of light. The change in energy of a molecule

Table 1.1. *Color, typical corresponding wavelengths, frequency, and energy of light*

Color	λ (nm)	5 (10^{-14} Hz)	E (eV)	E (kJ mol^{-1})
Infrared	>1000	<300	<1.24	<120
Red	700	428	1.77	171
Orange	620	484	2.00	193
Yellow	580	517	2.14	206
Green	530	566	2.34	226
Blue	470	638	2.64	254
Violet	420	714	2.95	285
Near ultraviolet	300	1000	4.15	400
Far ultraviolet	<200	>1500	>6.20	>598

Note: Italics represent the visible spectrum.

may manifest itself electronically, vibrationally, or rotationally depending on the energy of the incident photon. Due to the quantization of molecular energy levels, electronic excitation of a particular species occur only if E corresponds to the difference in energy between the ground electronic state and an electronically excited state of the absorber, which is illustrated in Figure 1.5.

The absorption by a molecule of a photon of light, with energy E, equal to the difference between energy states, can promote the transition of an electron from the ground electronic state to a vibrationally and electronically excited state. This is illustrated in Figure 1.3. The amount of incident radiation absorbed is proportional to the number of molecules (molecular concentration) in the path (photons^{-1} cm^{-2}), and is expressed in Beer and Lambert's Law (Eq. 1.6):

$$I_t = I_0 \ exp^{-\in cl} \qquad (1.6)$$

where I_t is the transmitted light intensity, I_o is the incident light intensity, $\in$ is the molar absorptivity (property of the absorbing species), c = the concentration of absorbing species, and l = the path length of the sample. According to Beer and Lambert's Law, the optical density of a solution should remain uniform as long as the product of the concentration and the path length is constant. However, the molar absorptivity often varies appreciably with the concentration of the solute and possible causes for this are molecular association of the solute at high concentrations; ionization of the solute in the case of acids, bases, and salts; and fluorescence of the solute.

1.3.1.3 Franck–Condon Principle

When molecules absorb energy that results in an electronic transition, molecular vibrations are always observed. In the electronic ground state of a molecule, the locations of nuclei in space are as a result of the Coulombic forces (electrostatic interaction between electrically charged particles such as electrons) acting on them. During an electronic transition

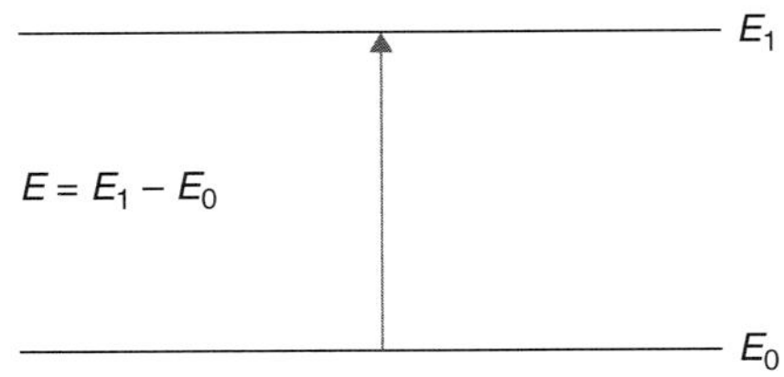

Figure 1.5. Absorption of radiation with energy E. The red arrow indicates an electronic transition of an electron from the ground electronic state (with energy E_0) to an excited electronic state with energy E_1.

electrons migrate to different parts of the molecule. This migration of electrons means that the Coulombic forces acting on the nuclei are subject to change. In response to changes in the Coulombic forces acting on the nuclei, molecules *vibrate*. Therefore, a proportion of the energy required to stimulate electronic transitions within a molecule is also used to excite *vibrations* of the absorbing molecule. This phenomenon explains why we don't usually observe pure electronic absorption lines. In fact, most observed absorption spectra are a composite of many discrete vibrational lines, and these discrete lines are representative of the vibrational structure of an electronic transition. In the case of gases, such vibrational structure can be resolved; however, in liquid or solid states, it is more common to experience featureless, broad-band absorption spectra. The vibrational structure of electronic transitions can be explained by the Franck–Condon principle (Lakowicz, 2006) and the concept of *vertical transitions* (Figure 1.5).

The Franck–Condon principle is perhaps best understood by first explaining that nuclei are massive in comparison to electrons; therefore an electronic transition takes place faster than the nuclei can actually respond to. As a result, an electronic transition is most likely to occur without changes in the positions of the nuclei in the molecular entity and its environment. In an electronic transition, changes in electron density are rapid as is the shift in the distribution of charge around the nuclei. These changes impact on the nuclei as new forces are created. The consequence of this is that nuclei begin to vibrate back and forth from their initial electronic ground state to a final electronic ground state. This change between electronic ground states represents a swinging (back and forth) in the molecular potential energy. This characteristic is termed *vertical transition* in accordance with the Franck–Condon principle. In practice the electronically excited molecule may manifest itself in one of several excited vibrational states; therefore absorption occurs at several different frequencies. With the exception of some gases, these transitions merge together to give a broad featureless band of absorption (Figure 1.6).

1.3.2 Nonradiative Decay

Any molecule in an excited state will return to the ground state, emitting the absorbed energy via a number of processes. The main mechanisms by which this occurs can be

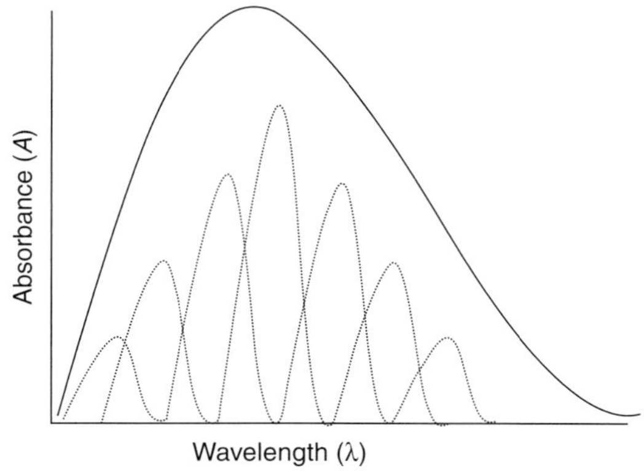

Figure 1.6. How superimposed vibrational structure merge together to give a broad unresolved electronic absorption band.

categorized into nonradiative and radiative decay (light emitting) processes. In this section we consider some of the mechanisms that lead to nonradiative decay.

1.3.2.1 Vibrational Relaxation

Vibrational relaxation (VR) is a radiationless process in which a molecule in an excited vibrational level returns to a lower vibrational energy level in the same electronic state (Lakowicz, 2006). Vibrational relaxation occurs by the transfer of energy from the excited vibrating molecule to the surrounding environment such as colliding into nearby solvent molecules. This energy transfer is very efficient and occurs over an average lifetime of picoseconds ($<10^{-12}$ s). This energy transfer process is very quick and much shorter than the average lifetime of an electronically excited state, which is on the order of nanoseconds. Molecules that are excited to different vibrational energy levels within the same excited electronic state very quickly return to the lowest vibrational energy level of that excited state. This is the main reason why, for most solvated molecules, fluorescence emission occurs only from the lowest vibrational energy level of an excited state. Therefore, fluorescence emission occurs from the level (or state) which is thermally populated according to the Boltzmann distribution. This is especially true for fluorophores in a condensed phase where their proximity to solvent or coordinating molecules is close. Hence, appreciable fluorescence yields can be obtained only from the *lowest* excited state of a given multiplicity; consequently the fluorescence emission of a fluorophore is independent of the excitation wavelength provided it is in a condensed phase. This concept is referred to as Kasha's rule and this concept is important in helping to understand the significance of excitation and emission spectra to investigate the vibrational structure of both the excited and ground states. Excitation spectra can show the variation in the emission intensity as a function of excitation wavelength, and this contains vibrational structural information relevant to the excited state. Emission spectra, on the other hand, are obtained by monitoring

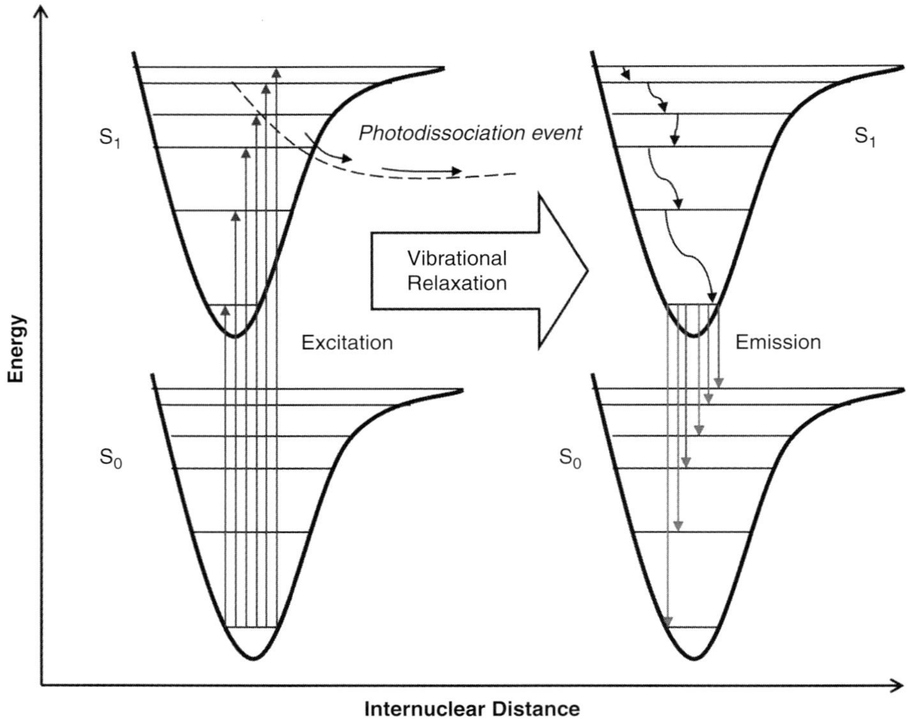

Figure 1.7. Electronic transitions during excitation and emission of a fluorophore. This illustration shows emission occurring from the lowest vibronic level of the excited state due to the high efficiency of vibrational relaxation.

the fluorescence emission intensity as a function of the emission wavelength and this provides information pertaining to ground state vibrational levels. This concept is illustrated in Figure 1.7.

1.3.2.2 Internal Conversion and Intersystem Crossing

Internal conversion (IC) is a form of radiationless deexcitation (relaxation), whereby a molecule that is in the ground vibrational level of an excited electronic state migrates *directly* to a higher vibrational energy level of a lower energy electronic state with the same spin multiplicity (Lakowicz, 2006) as shown in Figure 1.8a. Through a combination of internal conversion and vibrational relaxation the energy of the electronically excited state is given off to vibrational modes of the molecule. Intersystem crossing (ISC) is also a radiationless relaxation process whereby a molecule in the ground vibrational energy level of an excited electronic state passes directly into a high vibrational energy level of a lower energy electronic state (Lakowicz, 2006). However, unlike in internal conversion, the high vibrational energy level of the lower energy electronic state has a different spin

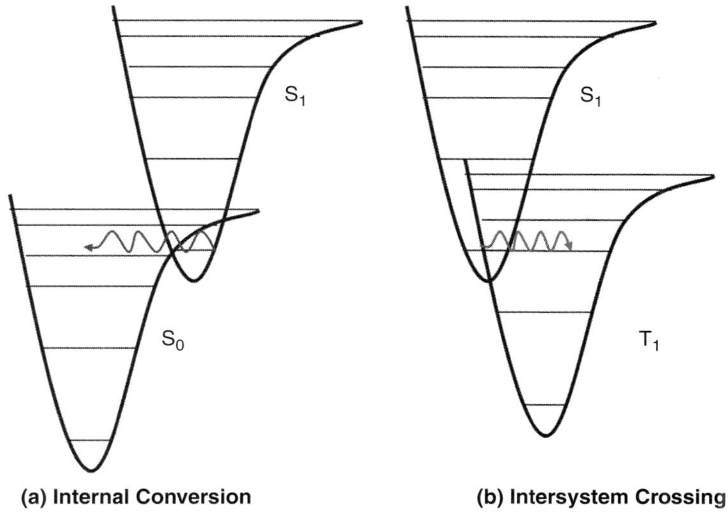

Figure 1.8. Internal conversion and intersystem crossing between different vibrational levels of the same energy (isoenergetic).

multiplicity (Figure 1.8b). Other processes that result in the loss of excitation energy as heat instead of as emitted light include external conversion. This is a related form of radiationless relaxation in which excess energy is transferred to a solvent molecule or another component in the sample matrix via molecular collisions, and is sometimes referred to as collisional quenching. This process is always present to some extent in solution samples and ultimately the excitation energy is transformed into heat.

1.3.3 Radiative Decay

The process by which an electronically excited molecule undergoes relaxation by the emission of a photon of radiation (light) is termed *radiative decay* (Lakowicz, 2006). In photoluminescence the two most relevant forms of radiative decay are fluorescence and phosphorescence. Both of these processes are driven exclusively by the multiplicities of the states from which the light-emitting transition occurs. For example, phosphorescence occurs when an electron in an excited triplet state relaxes to the singlet ground state (dashed vertical lines in Figure 1.1). A molecule may undergo intersystem crossing from an excited singlet state to a triplet excited state. Once in the lowest vibrational energy level of an excited triplet electronic state phosphorescence occurs when the molecule relaxes to the ground state (Figures 1.1 and 1.6b). Therefore, for the phosphorescence process, different states of multiplicities are required. In comparison, the fluorescence process involves a radiative transition between electronic states of the same spin multiplicity. For example, when in an excited single state, a molecule may relax to a singlet ground state by emitting

Table 1.2. *Time scales for radiative and nonradiative processes*

Transition	Process	Rate constant	Timescale (seconds)
S(0) => S(1) or S(*n*)	Absorption (excitation)	Instantaneous	10^{-15}
S(n) => S(1)	Internal conversion	k(IC)	10^{-14} to 10^{-10}
S(1) => S(1)	Vibrational relaxation	k(vr)	10^{-12} to 10^{-10}
S(1) => S(0)	Fluorescence	k(f)	10^{-9} to 10^{-7}
S(1) => T(1)	Intersystem crossing	k(pT)	10^{-10} to 10^{-8}
S(1) => S(0)	Nonradiative relaxation Quenching	k(nr), k(q)	10^{-7} to 10^{-5}
T(1) => S(0)	Phosphorescence	k(p)	10^{-3} to 100
T(1) => S(0)	Nonradiative relaxation Quenching	kNR, k(qT)	10^{-3} to 100

a photon of light (see solid vertical lines in Figure 1.1). If the excitation energy were to be removed then the lifetime of the fluorescence would be short, typically ranging from picoseconds to nanoseconds. In contrast, phosphorescence lifetimes are much longer in nature and range from millisecond to seconds (Table 1.2). These radiative decay characteristics are due to the transition between states of different spin multiplicities.

1.3.4 Fluorescence

Intersystem crossing competes with fluorescence for deactivation of the lowest excited singlet state. Intersystem crossing occurs from the lowest excited singlet state to the lowest excited "triplet" state and involves a change in spin angular momentum. Because intersystem crossing violates the law of conservation of angular momentum it is approximately a million times less probable (slower) than a typical singlet-singlet vibrational process such as internal conversion. Intersystem crossing is of a rate comparable to that of fluorescence and therefore competes for deactivation of the lowest excited singlet state. Molecules that populate the lowest excited triplet state undergo vibrational relaxation to the lowest vibrational level of the lowest excited triplet state. The return to the singlet ground state can occur either nonradiatively by triplet–singlet intersystem crossing, or by the emission of light. The latter process represents a "forbidden" transition, more so than fluorescence, and is characterized by a very long duration (10^{-14}–^{1}0s). The radiative transition from an upper electronic state to a lower electronic state of different spin is called *phosphorescence*.

For certain molecules, the return from the lowest excited singlet state by internal conversion or vibrational relaxation is improbable and is therefore termed "*forbidden*." The return to the ground electronic state for these molecules involves the emission of ultraviolet or visible radiation whose frequency is governed by the quantized difference between the lowest excited singlet state and the ground electronic state. Fluorescence occurs when a molecule in the lowest vibrational energy level of an excited electronic state returns to a lower energy electronic state by emitting a photon. Because molecules usually return to their ground state by the fastest mechanism, fluorescence is observed only if it is a more efficient means of relaxation than the combination of internal conversion and vibrational relaxation. These nonradiative processes compete with those that lead to radiative decay. Contrary to Planck's frequency equation (Eq. 1.1) fluorescence emission occurs over a range of wavelengths that correspond to several vibrational transitions, within a single electronic transition, that take place in the ground state. In Figure 1.3 the straight vertical arrows pointing downward represent fluorescence and phosphorescence, processes that involve the release of a photon of energy. The other deactivation steps, as indicated by wavy arrows, represent the radiationless processes, which compete with fluorescence.

1.3.4.1 Stokes Shift

The change in energy during a fluorescence transition is less than that for the absorption process. As a result, the fluorescence emission spectrum is almost always shifted to a longer wavelength than the respective absorption spectrum (Figure 1.9). This is due mostly to the high efficiency of vibrational relaxation of excited electron to the lowest vibrational level in an excited state. This observed shift between the absorption and spectrum spectra is called the *Stokes shift*. The Stokes shift is defined as the difference between the maximum of the first absorption band and the maximum of the fluorescence spectrum (Figure 1.9). Fluorescence may return the molecule to any of the vibrational energy levels in the ground electronic state, and therefore fluorescence emission occurs over a range of wavelengths.

1.3.4.2 Fluorescence Decay Kinetics

The decay of the excited state obeys first-order kinetic; therefore an excited population of fluorophores, [M*], decays according to

$$-\frac{d[M^*]}{dt} = k_F[M^*] \tag{1.7}$$

where the rate constant, k_F, is the sum of the rate constants for all radiative and nonradiative decay processes ($k_F = k_R + \Sigma\, k_{NR}$). Following integration of Eq. (1.7), the concentration of the excited state as a function of time can be represented as

$$[M^*](t) = [M^*]_0 \exp(-k_F t) \tag{1.8}$$

where $[M^*]_0$ is the initial concentration of M* at $t = 0$. The parameter monitored during fluorescence lifetime experiments is the fluorescence intensity, I, which is the rate of the emission of photons and is related to the excited state concentration by

$$I_F(t) = k_R[M^*](t) \tag{1.9}$$

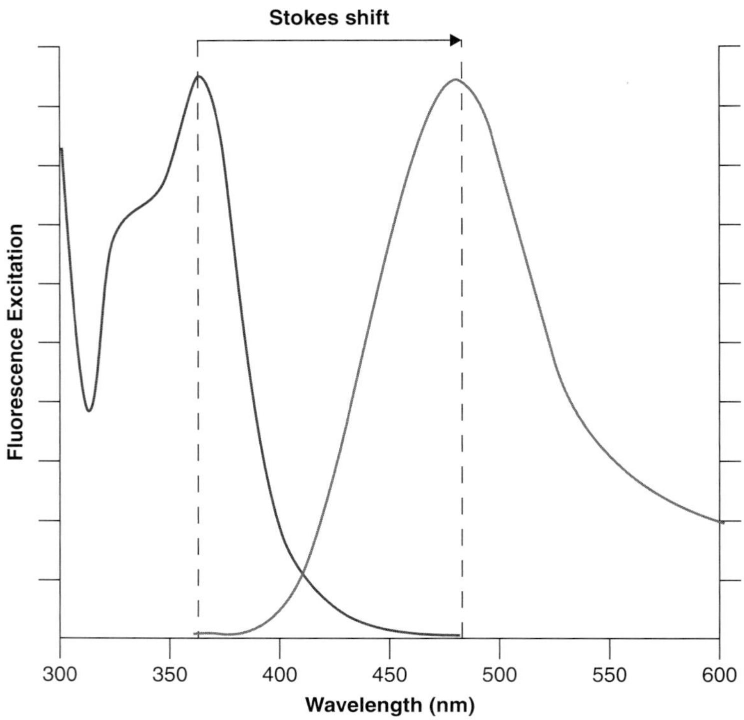

Figure 1.9. Stokes shift observed between the excitation (right) and emission (left) spectra.

Substituting for [M*] from Eq. (1.8) gives

$$I_F(t) = k_R[M^*]_0 \exp(-k_F t) \tag{1.10}$$

Substituting for $[M^*]_0$ from $I_F(0) = k_R[M^*]_0$, where $I_F(0)$ is the intensity at the time of the initial excitation, into Eq. (1.10) gives

$$I_F(t) = I_F(0)\exp(-k_F t) \tag{1.11}$$

Therefore the intensity decays exponentially after the initial excitation pulse. The fluorescence lifetime of the excited state, τ_F, can be represented as $\tau_F = (1/k_F)$. The fluorescence lifetime of a molecule is defined as the time taken for the excited state population to fall to 1/e of that initially excited. Equation (1.11) can then be rewritten as

$$I_F(t) = I_F(0)e^{-t/\tau_F} \qquad (1.11)$$

Equation (1.11) relates the measured parameter of intensity to the fluorescence lifetime, enabling its calculation experimentally. Possible methods for the determination of the fluorescence lifetimes include time correlated single-photon counting (TCSPC) spectroscopy and fluorescence lifetime imaging microscopy (FLIM), which spatially resolves the fluorescence lifetime of a sample (Lakowicz, 2006, chapters 4 and 17). Such techniques are not discussed here but additional information can be found in the references at the end of this chapter.

1.3.4.3 Fluorescence Efficiency (Quantum Yield)

Several processes compete with fluorescence for the deactivation of the lowest excited singlet state; therefore the intensity of fluorescence (I_f), is obtained by multiplying I_o (Eq. 1.2) by the fraction of excited molecules that "actually" fluoresce as shown in Eq. (1.3):

$$I_f = \Phi_F I_0 \tag{1.12}$$

Φ_f is called the quantum yield of fluorescence or the fluorescence efficiency.

The quantum yield can range from 1, when every molecule in an excited state undergoes fluorescence, to 0, where no fluorescence takes place. The fluorescence quantum yield can be related to the fluorescence lifetime by

$$\Phi_F = \frac{k_R}{k_F} = k_R \tau_F \tag{1.13}$$

Therefore the fluorescence lifetime, τ_F, is a measure of the fluorescence quantum yield, Φ_F. This indicates that the rate constant for the radiative decay processes k_R, is constant for a particular fluorophore as it is an intrinsic electronic property of the molecule. Consequently the fluorescence lifetime is influenced by changes in the nonradiative decay pathways. For example, a subsequent increase in nonradiative decay rates will reduce the fluorescence lifetime. As such, fluorescence lifetimes are extremely sensitive to the molecular environment surrounding the fluorophore. Therefore, this makes fluorescence lifetime measurement of individual fluorophores present in complex aquatic samples difficult to interpret.

The relationship between phosphorescence intensity (I_P) and analyte concentration is similar to that for fluorescence (Eq. 1.7).

$$I_P = \Phi_{ST}\Phi_P I_0 (1 - 10^{-\epsilon cl}) \tag{1.14}$$

In Eq. (1.14), I_0, ϵc, and l are as defined in Beer and Lambert's Law (Eq. 1.6), Φ_{ST} is the quantum yield of intersystem crossing, and ΦP is the quantum yield of phosphorescence. The former represents the fraction of excited molecules that undergo intersystem crossing from the lowest excited singlet state to the lowest triplet state, and the latter is the fraction of excited molecules having undergone intersystem crossing that are actually deactivated by phosphorescence.

1.3.4.4 Fluorescence Quenching

The fluorescence intensities and/or fluorescent lifetimes observed from luminescent species can be reduced or eliminated by interactions with other chemical species that increase the probability of deexcitation through nonradiative pathways. This phenomenon is known as quenching and can be either static or dynamic. Static quenching occurs when the potentially fluorescing chromophore, in the ground state, reacts with the quenching species to form a nonfluorescent species. The efficiency of this quenching is dictated by the rate of formation of the nonfluorescent species and the concentration of the quencher (Lakowicz, 2006, chapter 9). In dynamic quenching, the interaction with the quenching species is during the lifetime of the excited state of the potentially fluorescing species. Dynamic quenching is also known as collisional quenching and its efficiency depends on the viscosity of the solution, the lifetime of the excited state, τ_0, of the fluorophore, and the concentration of the quencher [Q]. This is summarized in the Stern–Volmer equation.

$$\frac{\Phi}{\Phi_0} = \frac{1}{1 + k_q \tau_0 [Q]} \tag{1.15}$$

where k_q is the rate constant for encounters between quencher and potentially luminescing species and Φ_0 and Φ are the quantum yields of luminescence in the absence or presence of concentration of the quencher [Q] respectively.

Perhaps the most well known collisional quencher is molecular oxygen, which is known to quench the vast majority of all known fluorophores. With this is mind it is generally accepted that observed fluorescence intensities from natural aquatic systems is quenched to some degree.

1.3.4.5 Influence of Molecular Structure on Fluorescence

High fluorescence yields are observed in highly conjugated "rigid" molecules. Therefore conjugated systems form the basis for many chromophores and chromophores are defined as a chemical group in which the electronic transition responsible for a given spectral band is approximately localized. Such chromophores are often present in naturally occurring organic compounds, and involve conjugated ring systems such as C=O and N=N in addition to conjugated C–C bonds (Lakowicz, 2006, chapter 3). Conjugation can also occur through carboxyl groups and are also important in conjugated systems. In these systems the double bond of the carbonyl group (–C=O) is adjacent to the single bond attaching the hydroxyl group (–OH) to the carbon atom. This functional group is of importance in many naturally occurring organic acids. The restricted rotational and vibration freedom of such molecules ensures that the energy gap between the lowest singlet state and the ground electronic state is too large for deactivation via internal conversion. However, if aromatic hydrocarbons have freely rotating substituents or lengthy side chains, then the fluorescence efficiency is greatly reduced due to the increase in rotational and vibrational freedom which subsequently increases the probability of internal conversion.

The molecular structure can have a profound effect on the wavelengths of fluorescence emission. The absorption properties of many biological pigments reflect the extent of conjugation; more specifically, when an electron in a conjugated system absorbs a photon of light at a suitable wavelength then it can be promoted to a higher energy level. Transitions like these are typically a π electron to a π^* electron. In addition to this nonbonding electrons can also be promoted (n to π^*). Such substances are said to be extensively delocalized. Conjugated systems of fewer than eight conjugated double bonds absorb exclusively in the ultraviolet region. Each additional double bond results in the conjugated system absorbing light at longer wavelengths (lower energy). Equation (1.2) states that the smaller the energy gap between the ground and lowest electronic excited state, then the longer the wavelengths of fluorescence. This is practically demonstrated by the fluorescence emission maxima for benzene, naphthalene, and anthracene, which are 262 nm, 320 nm, and 379 nm, respectively. With respect to biological pigments, beta-carotene has a long conjugated hydrocarbon chain and therefore a complex conjugated electron system, resulting in a strong orange color. Pigments utilizing conjugation in this way range from yellow to red. Blue and green (photosynthetic) pigments are less reliant on this type of conjugated electron system.

Molecules or compounds that undergo electronic transitions resulting in fluorescence are known as fluorescent probes, fluorochromes, or dyes. When such substances are conjugated to larger molecular structures such as nucleic acids, lignins, organic acids, and proteins they are termed fluorophores. Fluorophores can be further divided into two broad classes, "intrinsic" and "extrinsic" fluorophores. Intrinsic fluorophores, occur naturally and include aromatic amino acids, porphyrins, organic acids, and green fluorescent protein. Extrinsic fluorophores include synthetic dyes (Rhodamines) or modified biochemicals (fluorescent labeling) to produce fluorescence with known or specific properties.

1.3.4.6 The Effect of pH

The effect of pH on the fluorescence of many molecules is well known and understood. The influence of pH is derived from the dissociation or protonation of acidic and basic functional groups associated with the aromatic constituents of the fluorophores. Protonation or dissociation can alter the chemical nature of the fluorophore in such a way that the rates of nonradiative processes competing with fluorescence are increased or decreased. Protonation and dissociation can result in a shift in the fluorescence emission caused by alteration of the relative separation of the ground electronic states of the reacting molecules. For example, the protonation of electron withdrawing groups (carboxyl) results in shifts to longer wavelengths of fluorescence, while the protonation of electron-donating groups (amino groups) produce shifts to shorter wavelengths.

1.3.4.7 Effects of Solvents on Fluorescence Emission

The relative fluorescence intensity and spectral peak position of a molecule will vary in different solvents. The interactions between solvent and the solute molecules are largely electrostatic, and it is usually the differences between the electrostatic stabilization energies of ground and excited states that govern the fluorescence properties of a molecule in a given solvent. During fluorescence, changes in the π electron distribution causes changes in the dipolar and hydrogen-bonding properties of the solute (Lakowicz, 2006, chapter 6). If the solute is more polar in the excited state than in the ground state then the fluorescence will occur at longer wavelengths if the fluorophore is present in a polar solvent, as opposed to a nonpolar solvent. This is due to the stabilization of the excited state, by the more polar solvent. Because photoluminescence originates from an excited state, fluorescence emission occurs at longer wavelengths the more interactive the solvent is, that is, the stronger the hydrogen bonding or polarity.

1.3.4.8 The Heavy Atom Effect

The interactions between certain ions and conjugated fluorescing ligands are known to affect fluorescence in several ways. The fluorescence of a ligand may be somewhat enhanced or quenched depending on the influence that the ion has on the nonradiative processes competing with luminescence. In the case of the heaviest non-transition metal ions, for example Hg (II) and Bi (II), static quenching of fluorescence and sometimes phosphorescence, often results from a "heavy atom effect." This term is used to describe the influence of heavy atom substitution on spin-forbidden transitions. It is usually assumed that the dominant influence of the heavy atom is to enhance spin-orbit coupling, and enhances the probability of radiationless processes such as intersystem crossing. The fluorescence of luminescing ligands is usually quenched by complexation of the ligand with main group transition metal ions. This is believed to occur by way of a process known as paramagnetic quenching, in which the unpaired electrons of the metal ion interact initially with the π electrons of the ligand, thereby producing a pathway for intersystem crossing from the directly excited singlet state of the ligand to states of higher multiplicity introduced by interactions with the metal ion.

1.3.4.9 Fluorescence Spectra

Traditionally a fluorescence spectrum is either a plot of luminescence intensity at a fixed excitation wavelength as a function of emission wavelength (an emission spectrum) or a plot of luminescence intensity at a single emission wavelength as a function of excitation wavelength (an excitation spectrum).

The fluorescence of a complex mixture of fluorophores can be represented as a two-dimensional matrix of fluorescence intensity as a function of both excitation and emission wavelength. Cross sections of this excitation–emission matrix (EEM) at fixed excitation wavelengths and at fixed emission wavelengths are, respectively, standard emission and excitation spectra. The distribution of emission intensities among the emission wavelengths, that is, $I(\lambda_{em})$, may be expressed as

$$I(\lambda_{em}) = K\eta(\lambda_{em})I_0(\lambda_{ex})\{1-\exp(-\alpha(\lambda_{ex})cl)\} \tag{1.16}$$

where K is a constant that accounts for wavelength-interdependent experimental parameters, $\eta(\lambda_{em})$ is the quantum yield of fluorescence at λ_{ex} and λ_{em}. The intensity of the incident radiation is defined as I_0 (λ_{ex}), and $\alpha(\lambda_{ex})$ represents the absorption cross section at λ_{ex}.

Finally, c and l represent the concentration of the relevant fluorescent species and the sampling path length respectively.

Both the path length and absorptivity of the species/sample is important:

$$\alpha(\lambda_{ex})cl << 1; I(\lambda_{em}) \cong K\eta(\lambda_{em})I_0(\lambda_{ex})\alpha(\lambda_{ex})cl \tag{1.17}$$

In excitation and emission matrices the ratio of the emission intensity (λ_{em}) to the excitation intensity (λ_{ex}) is plotted. The relationship between λ_{em} and λ_{ex} can be expressed as

$$\frac{I(\lambda_{em})}{I(\lambda_{ex})} = Kcl\eta(\lambda_{em})\alpha(\lambda_{em} - \Delta\lambda) \tag{1.18}$$

From this it is clear that the intensity distribution will have pronounced peaks when the absorption maxima and the fluorescence quantum yield overlap. When a sample has a well defined absorption maximum and a well defined quantum yield maximum, then the most spectral structures and features are most likely to be observed when the $\Delta\lambda$ is equal to the difference in wavelength of the emission and excitation maxima (Stoke's shift).

1.3.4.10 Scattering of Radiation

The most probable interaction between a molecule and optical radiation is Rayleigh and Mie scattering. Both Rayleigh and Mie represent the major types of elastic scattering (negligible energy transfer) and unlike fluorescence emission there is no wavelength shift in this process. Rayleigh scatter involves the scattering of light by a spherical volume or entity with an uneven refractive index. This volume or entity could manifest itself as a colloid, particle, bubble, droplet, or even a density fluctuation. The size of the scattering species must, according to Lord Rayleigh's model be much smaller ($1/\lambda^4$ relation) than the wavelength of the incident light. For entities and volumes equal to or about the same size as the wavelength of incident light then Mie scattering (named after the German physicist Gustav Mie) occurs. From a practical aspect both scattering process occur in virtually all fluorescence applications, especially in highly heterogeneous samples such as natural and waste waters. Although Rayleigh–Mie scattering is an important natural phenomenon the scattered radiation does not bear any *identifiable signature* of the scattering species. Therefore in the application of fluorescence spectroscopy the reduction of the Rayleigh–Mie scattered is an important consideration.

Another interaction that takes place when light interacts with molecules is Raman scattering. In this process scattering takes place at discrete wavelengths governed by the size *and* symmetry of the molecule. For a nonlinear molecule of N atoms, there will be different possible scattering wavelengths or modes (3N – 6). Some of these modes may give overtones and combination modes, thereby creating a series of modes for each molecular species. The fundamental modes correspond to the energy levels of the vibrational levels and the wavelength of a particular Raman mode, λ_r, is given by

$$\frac{1}{\lambda_i} - \frac{1}{\lambda_r} = \Delta\nu \tag{1.19}$$

where λ_i is the wavelength of the incident radiation and $\Delta\nu$ is the frequency of the mode, which corresponds to the energy difference between the initial and final vibrational levels of the scattering process.

The intensity of a mode depends on the symmetry of the corresponding vibrational motion and the wavelength of the incident radiation. For most vibrational modes, the intensity of Raman scattering varies as the inverse fourth power of the Raman wavelength (α λ_r^{-4}). For some modes the wavelength dependence becomes much stronger when the excitation approaches an absorption band of the molecule (resonance excitation). Raman scattering is an inelastic (involving energy transfer) interaction process, having efficiencies of up to three to four orders of magnitude weaker than Rayleigh scattering. This is an important consideration because to retrieve a Raman signal from a species at a concentration in the region of parts per million of a host of species, an optical filter having a Rayleigh line rejection factor of 10^{-9}–10^{-10} would be required.

In water, Raman scattering is less probable (approximately one tenth) than elastic scattering (Rayleigh–Mie). Raman scattering occurs when molecules are irradiated with incident photons. The scattering molecule immediately scatters photons that are different in energy to the incident photon. Because this energy difference requires energy transfer, Raman scattering is said to be inelastic in nature as opposed to Rayleigh scattering, which is *elastic*. The difference in energy between the incident and scattered photons (either positive or negative) corresponds to the energy difference between two energy levels of the molecule. For a given pair of energy levels, the energy difference can be observed as a constant frequency difference, for example, 3400 cm^{-1} for the OH stretch vibrational mode in water, or a variable wavelength difference between the incident and scattered photons. Because Raman scattering follows a λ^{-4} relation law, there is a greater production of water Raman photons at shorter wavelengths.

Understanding the Raman scattering phenomena with respect to water molecules is important as water Raman scattered signals have been used extensively to normalize fluorescence signals. Although inherently a weak interaction, water Raman signals are commonly observed in the fluorescence spectra of aquatic samples given the high number of water molecules present per fluorophore. In addition, because irradiated water molecules will scatter photons probabilistically then water Raman signals observed in fluorescence spectra of aquatic samples can be used as an internal standard that allows the correction for differential absorption (inner filter) effects caused by the sample.

1.3.4.11 Normalization of Fluorescence Intensities

When a volume of water is probed with an incident beam of light the fluorescence emission is related to the number of the fluorophores present within the beam. As such the observed fluorescence emission is subject to large changes due to variations in the penetration of this incident light. These changes are produced by the optical attenuation properties of the sample caused by variations in the concentration of either the fluorophore(s) under investigation or other substances such as suspended particles and dissolved organics. The net effect of this interference is that the fluorescence signal no longer varies simply with the concentration of the fluorophore(s), but is also strongly dependent on changes in the optical attenuation coefficients of the sample, at the excitation and fluorescing wavelengths. This problem can be largely overcome by monitoring the water Raman signal and fluorescence emission. The relatively intense scattered signal from the O–H vibrational stretching mode of liquid water gives a good indication of the probed sample volume. By taking the ratio of the intensities for the fluorescence signal to that of the Raman signal, fluorescence emission data can be obtained that are independent of changes in the optical attenuation at the excitation wavelength and the water Raman signal wavelengths. For example, when a water sample is excited at 350 nm, water Raman signals occurs at 397 nm. The water Raman signal at 397 nm will be superimposed on a broad band of fluorescence due to the presence of dissolved organic matter (DOM). In samples with high levels of DOM the broad band fluorescence may be high and therefore it is necessary to separate the water Raman signal from the fluorescence spectra. This can be achieved by taking measurements on either side of the Raman band and calculating the Raman component by linear interpolation. The correction of fluorescence emission intensities using water Raman signals accounts for the attenuation of the incident light at the excitation wavelength and also for the attenuation at wavelengths corresponding to the water Raman line. There is still a problem of signal attenuation at other wavelengths and if this attenuation is significant then separate corrections have to be developed and applied to compensate for this differential absorption.

Mathematically, the attenuation of the exciting wavelength can be almost entirely compensated for by normalizing to the water Raman signal (I_R). However, this accounts only for the wavelength pertaining to the incident radiation and does not take into account the attenuation caused by differential absorption effects at other wavelengths. Therefore, to compensate for attenuation effects of "observed" measured Raman and fluorescence intensities (I'_R and I'_F respectively) which occur at specified wavelengths (λ_r and λ_f) then the transmittance (T) at all associated wavelengths and the respective path lengths must be considered such that

$$I_R = \frac{I'_R}{T_{\lambda_r}} \tag{1.20}$$

and

$$I_F = \frac{I'_F}{T_{\lambda_f}} \tag{1.21}$$

References

Anastopoulos, C. (2008). *Particle or Wave: The Evolution of the Concept of Matter in Modern Physics*. Princeton, NJ: Princeton University Press.

Atkins, P. (2001). *Elements of Physical Chemistry*. 3rd ed. Oxford: Oxford University Press.

Atkins, P., de Paula, J., and Friedman, R. (2009). *Quanta, Matter, and Change: A Molecular Approach to Physical Chemistry*. Oxford: Oxford University Press.

Becquerel, A.E. (1839). Mémoire sur les effets électriques produits sous l'influence des rayons solaires. *C. Rendus*, **9**, 561–567.

Becquerel, H. (1896). Sur les radiations émises par phosphorescence. *C. Rendus*, **122**, 420–421.

Bohr, N. (1913). On the constitution of atoms and molecules, Part I. *Philos. Mag.*, **26**, 1–24.

Bohr, N. (1922). The structure of the atom, *Nobel Lecture, December 11*. Retrieved from http://nobelprize.org/nobel_prizes/physics/laureates/1922/bohr-lecture.pdf (Accessed July 11, 2011).

Born, M. and Jordan, P. (1925). *Zur Quantenmechanik*, *Z. Physik*, **34**, 858–888. (1925). The paper was received on September 27, 1925. [English translation in: B. L. van der Waerden, editor, *Sources of Quantum Mechanics* (Dover Publications, 1968) (English title: *On Quantum Mechanics*).] This is the second paper in the famous trilogy that launched the matrix mechanics formulation of quantum mechanics.

Born, M., Heisenberg, W., and Jordan, P. (1925). *Zur Quantenmechanik II*, *Z. Physik*, 35, 557–615. The paper was received on November 16, 1925. [English translation in: B.L. van der Waerden, editor, *Sources of Quantum Mechanics* (Dover Publications, 1968).] This is the third paper in the famous trilogy that launched the matrix mechanics formulation of quantum mechanics.

Brady, J.E., Jespersen, N.D., and Hyslop, A. (2011). *Chemistry*, 6th ed., International Student Version. New York: John Wiley & Sons.

Crookes, W. (1879). On radiant matter 1. *Nature*, **20**, 436–440 (4 September 1879).

Dahl, P.F. (1997). *Flash of the Cathode Rays: A History of J. J. Thomson's Electron*. Bristol and Philadelphia: Institute of Physics Publishing.

de Broglie, L. (1923). Ondes et quanta (Waves and quanta). *C. Rendes*, 177, 507.

de Broglie, L. (1924). *Recherches sur la théorie des quanta* (Researches on the quantum theory), Ph.D. thesis, University of Paris.

Einstein, A. (1905). Über einen die Erzeugung und Verwandlung des Lichtes betreffenden heuristischen Gesichtspunkt. *Ann. Physik*, **17**(6), 132–148.

Eliezer, S. and Eliezer, Y. (2001). Highlights to plasma. *In The Fourth State of Matter, An Introduction to Plasma Science*, 2nd ed. (p. 155). Bristol, UK: Institute of Physics Publishing.

Faraday, M. (1844). *Experimental Researches in Electricity*. **2**. London: Richard and John Edward Taylor Publishers.

Gillespie, R.J. (1970). The electron-pair repulsion model for molecular geometry. *J. Chem. Educ.*, **47**(1), 18–23.

Gillespie R.J. and Nyholm, R.S. (1957). Inorganic stereochemistry. *Q. Rev. Chem. Soc.*, **11**, 339–380.

Hedenus, M. (2002). Eugen Goldstein and his laboratory work at Berlin Observatory. *Astron. Nachr.*, **323**, 567–569.

Heisenberg, W. (1925). Über quantentheoretische Umdeutung kinematischer und mechanischer Beziehungen, *Z. Physik*, **33**, 879–893 The paper was received on 29 July 1925. [English translation in B.L. van der Waerden (Ed.), *Sources of Quantum Mechanics* (Dover Publications, 1968) (English title: *Quantum-Theoretical Re-interpretation of*

Kinematic and Mechanical Relations).] This is the first paper in the famous trilogy that launched the matrix mechanics formulation of quantum mechanics.

Heisenberg, W. (1927). Über den anschaulichen Inhalt der quantentheoretischen Kinematik und Mechanik (The actual content of quantum theoretical kinematics and mechanics). *Z. Physik*, **43**, 172.

Heitler. W. and London, F. (1927). Wechselwirkung neutraler Atome und homöopolare Bindung nach der Quantenmechanik, *Z. Physik*, **44**, 455–472.

Hückel, E. (1930). Zur Quantentheorie der Doppelbindung (Quantum theory of double linkings). *Z. Physik*, **60**, 423–456.

Hückel, E. (1931). Quantum-theoretical contributions to the benzene problem. The electron configuration of benzene and related compounds. *Z. Physik*, **70**, 204–286.

Hückel, E. (1932). Quantum theoretical contributions to the problem of aromatic and non-saturated compounds. *Z. Physik*, **76**, 628–648.

Hertz, H.R. (1887). Ueber einen Einfluss des ultravioletten Lichtes auf die electrische Entladung, *Ann. Physik*, **267**(8), 983–1000.

Lakowicz, J.R. (2006). *Principles of Fluorescence Spectroscopy*, 3rd ed. New York: Springer Science+Business Media.

Langmuir, I. (1919). The arrangement of electrons in atoms and molecules. *J. Am. Chem. Soc.*, **41** (6), 868–934.

Lewis, G.N. (1916). The atom and the molecule. *J. Am. Chem. Soc.*, **38**, (4). Retrieved from http://osulibrary.oregonstate.edu/specialcollections/coll/pauling/bond/papers/corr216.3-lewispub-19160400.html.

Massimi, M. (2005). The origins of the exclusion principle: an extremely natural prescriptive rule. In *Pauli's Exclusion Principle, The Origin and Validation of a Scientific Principle* (pp. 35–73). Cambridge: Cambridge University Press.

Maxwell, J.C. (1865). A dynamical theory of the electromagnetic field. *Philos. Trans. R. Soc. London*, **155**, 459–512.

Minkin, V.I. (1999). Glossary of terms used in theoretical organic chemistry *IUPAC Pure Appl. Chem.*, **71**(10), 1919–1981.

Pauli, W. (1925). Über den Zusammenhang des Abschlussesf der Elektronengruppen im Atom mit der Komplexstruktur der Spektren (On the connection between the completion of electron groups in an atom with the complex structure of spectra). *Z. Physik*, **31**, 765. [Pauli exclusion principle].

Pauli, W. (1926). Quantum theory. In *Handbuch der Physik*, Vol. **23** [226] (pp. 22–29). Berlin: Verlag von Julius Springer,

Pauling, L. (1931). The nature of the chemical bond. Application of results obtained from the quantum mechanics and from a theory of paramagnetic susceptibility to the structure of molecules. *J. Am. Chem. Soc.*, **53**, 1367–1400.

Pauling, L. (1940). *The Nature of the Chemical Bond and the Structure of Molecules and Crystals: An Introduction to Modern Structural Chemistry*. Ithaca, NY, and London: Cornell University Press; H. Milford, Oxford University Press.

Philipp Lenard – Biography. Nobelprize.org. Retrieved from http://nobelprize.org/nobel_prizes/physics/laureates/1905/lenard-bio.html (Accessed August 14, 2011).

Planck, M. (1900). Entropy and temperature of radiant heat. *Ann. Physik*, **1**(4), 719–737.

Planck, M. (1901). On the law of distribution of energy in the normal spectrum. *Ann. Physik*, **4**, 553–563.

Rutherford, E. (1911). The scattering of alpha and beta particles by matter and the structure of the atom. *Philos. Mag.*, **21**, 668–669.

Schrödinger, E. (1926a). Quantizierung als Eigenwertproblem (Erste Mitteilung) (Quantization as a Problem of Proper Values. Part I.). *Ann. Physik*, **79**, 361 (1926). [Schrödinger equation]

Schrödinger, E. (1926b). Über das Verhältnis der Heisenberg Born Jordanischen Quantenmechanik zu der meinen (On the relation between the quantum mechanics of Heisenberg, Born, and Jordan, and that of Schrödinger). *Ann. Physik*, **79**, 734 (1926). [Equivalence of Heisenberg and Schrödinger formulations of quantum mechanics

Stokes, G.G. (1852). On the change of refrangibility of light. *Philos. Trans. R. Soc. London*, **142**, 463–562.

Thomson, J.J. (1897a). Cathode rays, *Philos. Mag.*, **44**, 293.

Thomson, J.J. (1897b). Cathode rays. *Electrician*, **39**, 104.

Thomson, J.J. (1903). *Conduction of Electricity Through Gases*, reprinted by Wexford College Press, 2005.

Uhlenbeck, G.E. and Goudsmit, S. (1925). Ersetzung der Hypothese vom unmechanischen Zwang durch eine Forderung bezüglich des inneren Verhaltens jedes einzelnen Elecktrons, *Naturwissenschaften*, **13**, 953–954.

"Wolfgang Pauli – Nobel Lecture." Nobelprize.org. Retrieved from http://nobelprize.org/nobel_prizes/physics/laureates/1945/pauli-lecture.html (Accessed July 28, 2011).

Wien, W. (1911). Nobel Lecture, 11 December 1911: On the laws of thermal radiation. In *Nobel Lectures, Physics 1901–1921*. Singapore: World Scientific.

2

Fluorescence and Dissolved Organic Matter: A Chemist's Perspective

GEORGE AIKEN

2.1 Introduction

The importance of dissolved organic matter (DOM) in aquatic systems has been clearly recognized during the past 30 years. The compounds that comprise DOM in aqueous systems often control ecological processes by influencing pH, serving as substrates for microbially mediated reactions (Tranvik, 1998; Findlay, 2003), controlling the depth of the photic zone (Wetzel, 2001), and influencing the availability of nutrients (Qualls and Richardson, 2003). DOM also exerts strong chemical controls on geochemical (Hoch et al., 2000; Waples et al., 2005) and photochemical (Moran and Covert, 2003; Stubbins et al., 2008) reactions, and interacts strongly with trace metals (Perdue, 1998; Haitzer et al., 2002) and organic pollutants (Chin, 2003), enhancing their apparent solubility and transport. In addition, DOM is a constituent of concern in drinking water supplies through the formation of disinfection by-products during the treatment process (Singer, 1994; Kraus et al., 2008), and is an important class of compounds comprising wastewaters (Baker, 2001; Westerhoff et al., 2001).

The study of the nature and environmental significance of organic matter in natural waters is hindered by its inherent chemical complexity, which poses a number of analytical problems (Aiken and Leenheer, 1993). A continuing need, therefore, is the development of analytical approaches that provide relevant data defining both its composition and, thus, reactivity. DOM optical properties, such as ultraviolet (UV)-visible spectroscopy, the spectral slope parameter, specific UV absorbance ($SUVA_{254}$), and fluorescence spectroscopy, have been shown to be useful in a number of disciplines in the water sciences for studying and monitoring both the concentration and nature of DOM in aquatic systems (Weishaar et al., 2003; Helms et al., 2008; Spencer et al., 2009). Utilizing optical data is an attractive approach for studying DOM because data collection is easy and straightforward, the data provide information about both the concentration and composition of DOM (Weishaar et al., 2003; Spencer et al., 2009), and detector systems can be employed for a variety of process-based studies and separation techniques to study DOM composition. In addition, optical data can be obtained in situ, allowing for the collection of high-frequency environmental data in real time that can be used to understand better the source influences and processing occurring within the system on the chemistry and export of DOM (Downing et al., 2009; Saraceno et al., 2009).

Although not a new technique (e.g., Hartley, 1893) for studying naturally occurring organic compounds, fluorescence spectroscopy is increasingly being used to study DOM in a range of aquatic systems including freshwater, marine waters, and wastewaters. It has been applied both to quantify DOM concentrations (Saraceno et al., 2009) and as a tool to "fingerprint" DOM composition in almost all aquatic environments (Green and Blough, 1994; Cory and McKnight, 2005; Spencer et al., 2007a; Larsen et al., 2010). It is an approach that is rapidly developing with respect to in situ probes, allowing for the collection of environmental data in real time (Downing et al., 2009; see Chapter 6 by Conmy et al., this volume). In addition, DOM fluorescence has been found to be useful in forensic applications, such as tracing sources of ship ballast water (Hall and Kenny, 2007; Murphy et al., 2006), is critical for remote sensing applications (Vodacek, 1989; Vodacek et al., 1995; Siegel et al., 2005), and has been applied to the determination of physical properties such as the electrostatic properties (Green et al., 1992) and diffusion coefficients (Lead et al., 2000) of humic substances. Excellent reviews (Blough and Del Vecchio, 2002; Coble, 2007; Hudson et al., 2007; Henderson et al., 2009; Fellman et al., 2010) provide thorough descriptions of the application of fluorescence analyses to the quantification and characterization of DOM across the range of systems and disciplines.

On the surface, fluorescence is an attractive method for studying DOM because data collection is easy and straightforward; it provides information about DOM composition; and, when properly calibrated with DOM measurements, can be used as a proxy for DOM concentration (e.g., Pellerin et al., 2011). For many practitioners in the water sciences, an underlying assumption in applying fluorescence spectroscopy to the characterization of DOM is that the compounds comprising DOM behave similarly to pure components in solution. In this context, changes in parameters such as intensity, peak width, fluorescence efficiency, and wavelengths of maximum intensity are often interpreted as variability in the provenance of DOM chemical components, as well as the production and removal of components via biological activity of bacteria and plankton. However, the ease of data collection and the potentially powerful applications of fluorescence spectroscopy to monitor compositional changes in DOM belie the inherent complexity of the method.

The measurement and comparability of fluorescence signals is nontrivial owing to many complicating factors. Some of these are nonchemical, such as instrumental inefficiencies and variability. Others are chemical in nature, such as inner filter effects and the dependency of measured data on environmental conditions (e.g., pH, temperature, redox status). In addition, fluorophores are very sensitive to chemical interactions. Indeed, the vast majority of papers in the biochemical and chemical literature using fluorescence make use of this sensitivity to chemical conditions to study structure and reactivity of fluorophores of interest (Lakowicz, 2006). Common approaches for interpreting the environmental and ecological significance of DOM fluorescence, however, often fail to adequately address the nonlinear behavior of fluorophores in natural samples. A goal of this chapter is to examine the potential influences of these chemical factors on the fluorescence properties of DOM. In addition, a cursory review of the large body of literature on the fluorescence of natural product compounds of interest in the study of DOM is presented, with an emphasis on the influences of chemical reactions on their fluorescence behavior.

2.2 Theory

Fluorescence describes the phenomenon wherein a molecule that has been excited to a higher electronic state by the absorption of UV-visible light returns to the ground state by the direct emission of light. Molecular structure is the primary determinant for both absorption and fluorescence of organic molecules. Both methods are used to provide structural information when studying individual molecules; however, the amount of structural information that can be obtained on natural organic matter samples is limited by the chemical complexities of these samples. Here a description of these phenomena is provided as a basis for understanding the controls on DOM fluorescence and the limitations associated with interpretation of fluorescence data. This topic is treated in a more thorough fashion in a number of excellent references (e.g., Skoog and West, 1982; Schulmann, 1985; Del Vecchio and Blough, 2004; Lakowicz, 2006).

2.2.1 Absorption

The absorption of light in the UV-visible range involves the excitation of electrons associated with chemical bonds from the ground electronic state (bonding orbital) to an excited electronic state (antibonding orbital). This process is sensitive to chemical structure, and, those structures that absorb light are referred to as chromophores. The wavelengths at which light can be absorbed by an organic molecule are determined by the differences in energy between bonding and antibonding orbitals. For many electronic structures, such as those in alkanes and carbohydrates, absorption occurs only at wavelengths that are shorter (higher energy) than those wavelengths available with most spectrophotometers. The energies associated with π-bonds, such as those in alkenes, aromatic molecules, and in some organic molecules containing heteroatoms, fall in the practical UV-visible range (e.g., 190–780 nm). Therefore, UV-visible spectroscopy of organic molecules largely deals with the absorption of light by conjugated systems (Silverstein et al., 1974). As molecules become more conjugated, the energy differences between bonding and antibonding orbitals decrease and molecules can absorb light at longer wavelengths (less energy), even into the visible portion of the spectrum. The UV spectrum of a molecule, therefore, indicates the presence of specific bonding arrangements within the molecule. In the case of absorption in the near UV and visible portions of the spectrum, conjugated systems, such as those in aromatic molecules, generally have the greatest absorptivities.

An advantage of the structural selectivity of UV-visible absorption is that characteristic features or bonding arrangements may be recognized in molecules of varying complexity (Silverstein et al., 1974). When measured in the near UV, many of the bonds present in a complex molecule or complex mixtures of molecules are transparent to UV radiation. Increased structural complexity, therefore, does not necessarily result in increased spectral

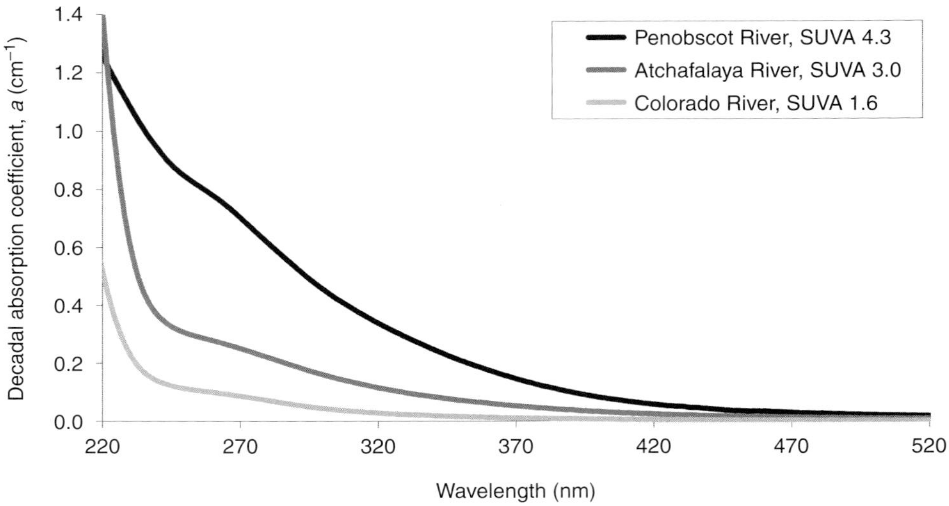

Figure 2.1. Dissolved organic matter UV-Visible absorption spectra and specific UV absorbance data for the Penobscot River, Atchafalaya River and the Colorado River.

complexity. Spectra obtained for complex mixtures of molecules, such as DOM, are generally considered to represent the average of individual compounds that comprise the mixture (Miller, 1994), although recent studies have shown that intramolecular interactions also strongly influence DOM UV-visible spectra, especially at longer wavelengths (Del Vecchio and Blough, 2004; Boyle et al., 2009 Ma et al., 2010). UV-visible spectra for water samples are generally featureless, with absorptivity increasing toward shorter wavelengths (Figure 2.1). Absorbance measurements are primarily detecting the presence of aromatic compounds, and, for analyses of aquatic humic substances, UV absorbance measurements at a wavelength of 254 nm have been shown to strongly correlate with aromatic carbon contents (Weishaar et al., 2003).

It is worth noting that there is some confusion in the water sciences literature concerning units for expressing absorbance data. Marine chemists often express absorption coefficients (α) in Naperian units, whereas the freshwater and wastewater communities usually express decadal absorption coefficients (a). The Naperian system is based on natural logarithms whereas the decadal system is based on common logarithms and the two coefficients are related by a factor of 2.303. According to the International Union of Pure and Applied Chemistry (IUPAC; Braslavsky, 2006), the Naperian absorption coefficient is given by:

$$\alpha(\lambda) = 2.303A(\lambda)/l \qquad (2.1)$$

where $A(\lambda)$ is the decadal absorbance and l is the cell pathlength in meters (Green and Blough, 1994). Decadal absorbance ($A(\lambda)$) is the unitless absorbance reading provided by most spectrophotometers. The decadal absorption coefficient is simply given as:

$$a(\lambda) = A(\lambda)/l \qquad (2.2)$$

From the absorption coefficients, spectral slope (S) can be calculated using a nonlinear fit of an exponential function to the absorption spectrum over the range of desired wavelengths using an equation such as:

$$\alpha_g(\lambda) = \alpha_g(\lambda_{\text{ref}})e^{-s(\lambda-\lambda_{ref})} \tag{2.3}$$

where, in this case, $\alpha_g(\lambda)$ is the Naperian absorption coefficient of CDOM at a specified wavelength, λ_{ref} is a reference wavelength and S is the slope fitting parameter (Twardowski et al., 2004). Specific UV absorbance (SUVA) was originally described as the decadal absorption coefficient in units of cm^{-1} at a given wavelength (commonly $\lambda = 254$ or 280 nm) divided by DOC concentration in units of mg C L^{-1} (Chin et al., 1994; Weishaar et al., 2003). Currently, it is common to express SUVA in units of L (mg C $m)^{-1}$ using the decadal absorption coefficient. For a discussion about ambiguity in the use of optical concepts, readers are referred to Hu et al. (2002) and the glossary of recommended terms provided by the IUPAC (Braslavsky, 2007). Helms et al. (2008) provide a discussion about the use of spectral slopes and slope ratios as indicators of DOM sources and reactivity. In all cases, it is useful to state whether Naperian or decadal absorption coefficients are being reported.

2.2.2 Fluorescence

Once excited to higher electronic and vibrational states, molecules return to the ground state by losing energy via a number of competing pathways. The most rapid relaxation pathway, occurring in 10^{-13} to 10^{-12} seconds, is by thermal deactivation resulting from the transfer of energy via collisions of the excited molecule with solvent molecules. Thermal deactivation includes the processes of vibrational relaxation, which is the loss of vibrational energy in the excited state, and internal conversion, the radiationless transition from a higher electronic state to a lower one. Many organic molecules that absorb light return completely to the ground state through thermal deactivation without emitting light (Schulman, 1985). In the case of some molecules, however, the transition from the lowest excited vibrational state to the ground state by thermal deactivation is slow enough to allow the emission of light by either fluorescence or phosphorescence. These phenomena differ in the lifetimes of the molecule in the excited state. Fluorescence occurs in a timeframe of 10^{-11} to 10^{-7} seconds, whereas phosphorescence requires more time (10^{-4} to 10 seconds). Excellent detailed descriptions of these phenomena can be found in a number of sources (e.g., Schulman, 1985). Fluorescence specifically refers to the phenomena wherein a molecule in the first excited singlet state (S_1) returns to the ground state by the direct emission of UV or visible light.

The ability of a molecule to fluoresce is, therefore, the result of competing pathways for relaxation. Molecular structure is an important factor controlling the pathways by which an excited molecule returns to the ground state, influencing both the intensity and the positions in the spectrum where molecules fluoresce (Schulman, 1985). Excited molecules that have greater degrees of rotational and vibrational freedom, such as alkenes and alicyclic molecules, relax efficiently by thermal deactivation pathways, and few of these molecules fluoresce in the UV-visible region. More rigid aromatic molecules have fewer rotational and vibrational degrees of freedom and are more likely to fluoresce than aliphatic and alicyclic molecules. Aromatic molecules that contain freely rotating substituents fluoresce less intensely than those without these substituents because energy can be lost from the molecule through these functional groups. Finally, highly conjugated molecules have smaller energy gaps between the excited and ground states and, therefore, fluoresce at longer, lower energy wavelengths than less conjugated systems.

Excitation and emission maxima and fluorescence intensity are sensitive to structural factors that can influence the energetics of the excited electronic state of the fluorophore. Among these factors, substituent effects are very important in controlling radiative relaxation rates. Electron-donating groups associated with conjugated aromatic molecules (such as –OH or $-NH_2$) can enhance fluorescence intensity by increasing rates of relaxation. The presence of these substituents extends fluorescence maxima to longer wavelengths compared to the parent molecule. Electron withdrawing groups (such as –COOH, –CHO, and $-NO_2$) tend to diminish fluorescence quantum yields in aromatic molecules. It is beyond the scope of this chapter to describe the reasons for these effects, and readers are referred to other sources for more detailed discussion of the chemistry of substituent effects (e.g., Schulman, 1985). It is important to realize, however, that, owing to aqueous solubility constraints, almost all DOM comprises polar molecules containing oxygen-, nitrogen- or sulfur-containing substituents (Thurman, 1985). The largest percentages of these molecules are organic acids with –COOH and –OH functional groups. Most of the fluorophores in DOM, therefore, are heavily substituted conjugated molecules. Senesi et al. (1991), in a thorough overview of the fluorescence behavior of soil-derived humic substances, describe the effects of conjugation and substituent effects on fluorescence data for 50 samples obtained from soils and related materials.

Because of the greater degree of specificity, fluorescence is less universally applicable in the study of DOM in comparison to UV-visible spectroscopy. UV-visible absorption is a general property of analytes that contain chromophoric groups, whereas most compounds are poor fluorophores. Therefore, a smaller set of compounds in DOM fluoresce relative to those that absorb light. Because fluorescence and absorption are related phenomena, however, it is possible to obtain more information on the chemistry controlling the optical properties of DOM when they are employed together, as recently demonstrated by Boyle et al. (2009).

2.3 DOM Fluorescence

Historically, organic matter in natural waters has been arbitrarily divided into dissolved and particulate organic carbon based on filtration, generally through 0.2- to 1.2-μm filters. No natural cut-off exists between these two fractions and the distinction is operational. Dissolved organic matter itself is a complex, heterogeneous continuum of high to low molecular weight species exhibiting different solubilities, reactivities, and optical properties depending on molecular structure. Overlapping the dissolved and particulate fractions is the colloidal fraction, which consists of suspended solids operationally considered as solutes (Morel and Gschwend, 1987). Colloidal organic matter in natural waters is composed of living and senescent organisms, cellular exudates, and partially to extensively degraded detrital material, all of which may be associated with mineral phases (Lead and Wilkinson, 2007). Distinctions between dissolved (filtered), colloidal, and particulate organic matters are important when measuring the optical properties of a sample. In some instances, samples can be analyzed without filtration. For example, the determination of chlorophyll associated with algae by fluorescence has long been used to study algal distributions and ecosystem dynamics in lakes and marine waters (Berman, 1972). Other examples include the application of fluorescence to the study of colloids and nanoparticles (Fatin-Rouge and Buffle, 2007). In general, however, spectroscopic methods, such as fluorescence and UV-visible absorption, are sensitive to the presence of particulate material in a sample, and, to obtain usable spectroscopic data, samples need to be appropriately filtered or optical data need to be corrected for the presence of particles (Karanfil et al., 2005; Saraceno et al., 2009).

DOM fluorescence data can be collected and presented in a variety of ways. The simplest approach is to measure fluorescence intensity of a single excitation–emission wavelength pair. This approach is the one used by modern in situ instruments designed to measure signals associated with DOM, chlorophyll, or specific fluorescent compounds, such as rhodamine. The commonly reported parameter fluorescence index (FI; Cory and McKnight, 2005) is simply the ratio of the emissions intensities determined at wavelengths of 470 and 520 with an excitation wavelength of 370 nm. Data can also be collected and presented as emission spectra measured at a single-excitation wavelength, or as the absorption or excitation spectrum at a single-emission wavelength. Another useful approach is the presentation of synchronous spectra, wherein emission data are offset by a constant amount from excitation wavelength (i.e., $\lambda + \Delta\lambda$). The offset in wavelength is determined by the fluorescence properties of the fluorophores of interest (Cabaniss and Shuman, 1987; Liu et al., 2006; Ziegmann et al., 2010). In theory, this approach allows measurement of fluorophores of interest and may provide less ambiguous data when properly constrained.

With the advent of modern spectrofluorometers, data are commonly collected and presented as excitation–emission matrices (EEMs). EEMs spectra contain a large amount of information and are usually displayed graphically (Figure 2.2). The large amount of data associated with EEMs spectra are often presented as the intensities of individual excitation–emission pairs or peaks associated with different compound classes, an approach commonly referred to as "peak picking." The most common peak assignments are given in Table 2.1 and graphically illustrated in Figure 2.2c. Statistically based approaches, such as hierarchal clustering (Jiang et al., 2008), partial least squares regression (Persson and Wedborg, 2001; Hall et al., 2005), principal component analysis (Persson and Wedborg, 2001; Hall and Kenny, 2007), and parallel factor analysis (PARAFAC; Stedmon et al., 2003), use all of the data contained in an EEM spectrum to identify spectral features and determine contributions to the spectrum of different areas or ex/em regions of fluorescence. These approaches employ curve fitting techniques and assume linear behavior of the different components of

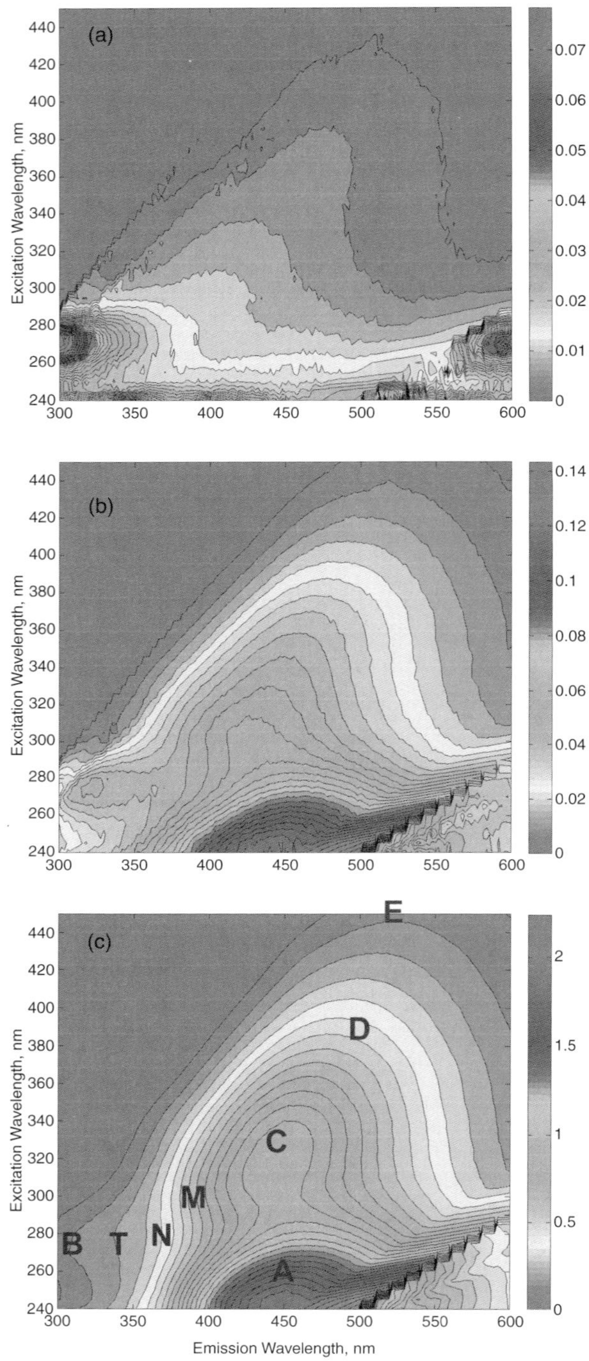

Figure 2.2. Excitation emission spectra for samples from (a) the Pacific Ocean, (b) the Gulf of Maine, and (c) the Penobscot River. Commonly reported peaks and region locations described in Table 2.1 are also graphically presented in (c). (See Plate 1.)

Table 2.1. *Commonly referenced peak and region locations for components of excitation–emission matrices (EEMs) obtained for aquatic humic substances and dissolved organic matter*

Peak label	Excitation maximum (nm)	Emission maximum (nm)	Description of fluorophores
B	275	305	Tyrosine-like, protein-like[a]
T	275	340	Tryptophan-like, protein-like[a]
A	260	400–460	Humic-like[a]
M	290–310	370–410	Marine humic-like[a]
C	320–360	420–460	Humic-like[a]
D	390	509	Soil fulvic acid[b]
E	455	521	Soil fulvic acid[b]
N	280	370	Plankton derived[b]

[a]Coble, 2007; [b]Stedmon et al., 2003.

a spectrum. The components used in these analyses are not actual fluorophores, and care is warranted in interpreting results. Further discussion of these approaches can be found in this volume and in recent papers (e.g., Fellman et al., 2010; Larsen et al., 2010).

EEMs spectra can vary widely depending on the nature of DOM fluorophores and the net effects of factors that influence fluorescence efficiency. The range of spectra for waters containing vastly different DOM compositions is apparent in comparing spectra presented in Figure 2.2 for two extreme environments types – the open ocean sample dominated by autochthonous DOM of microbial origin and a river sample with DOM predominantly derived from higher plants are presented. The Pacific Ocean sample, which was collected far from terrestrial sources, contains primarily proteinaceous fluorophores (Yamashita and Tanoue, 2003) and exhibits a relatively simple fluorescence spectrum. At the other extreme is the Penobscot River sample, which is heavily influenced by higher plant and soil sources of organic matter. This sample contains terrestrially derived DOM dominated by more complicated aromatic molecules that are excited and fluoresce at much longer wavelengths than the open ocean sample (e.g., Boyle et al., 2009). The sample from Maine coastal waters (Figure 2.2b) exhibits aspects of both of the end-member samples. For most waters, little is actually known about the chemistry of the different fluorophores in a sample, and, in general, interpreting spectra from all types of waters is complicated by the lack of information related to their concentrations. A complementary approach to studying whole water samples is to isolate functionally distinct DOM fractions from whole water samples to determine the fundamental chemical properties of each fraction, ultimately relating structural and chemical information to the biogenesis and environmental roles of these materials. Fractionation is often accomplished using solid-phase extraction on hydrophobic sorbents such as C_{18} (Green and Blough, 1994), styrene divinyl benzene polymers such as PPL resin (Dittmar et al., 2008), and XAD resins (Aiken et al., 1992). Using these approaches, the majority of chromophores and fluorophores present in the

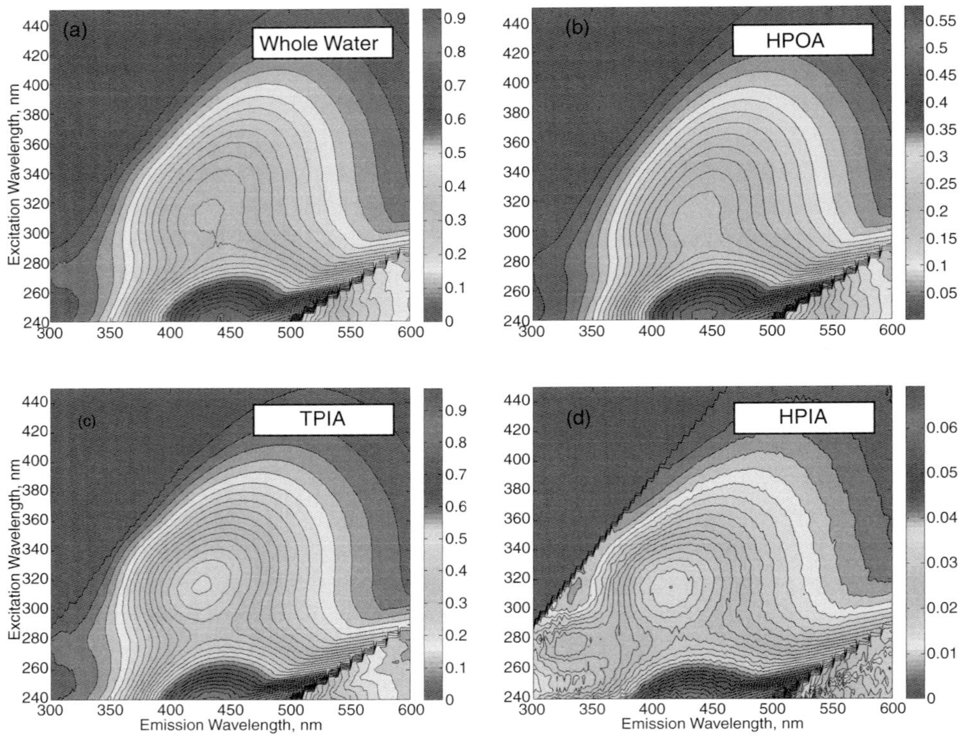

Figure 2.3. Excitation–emission spectra for the (a) whole water, and mass normalized EEMs for the (b) hydrophobic organic acid, (c) transphilic organic acid, and (d) hydrophilic fractions of a water sample from the Yukon River at Pilot Station, Alaska. (See Plate 2.)

whole water sample are often found in the hydrophobic fractions containing aquatic humic substances (Aiken et al., 1992). In most cases, however, the hydrophobic fractions of DOM do not contain all of the fluorpohores in a sample, and it cannot be assumed that this fraction is representative of whole water fluorescence (Green and Blough, 1994). To illustrate this point, fluorescence data for the whole water, hydrophobic acid (HPOA; humic fraction), transphilic acid (TPIA; intermediate polarity compounds), and hydrophilic fractions (HPIA) of the DOM for a sample from the Yukon River obtained by the XAD approach are presented (Figure 2.3). In this example, despite the strongest fluorescence intensity, the HPOA fraction does not account for all signals in the whole water fluorescence. In each of the more hydrophilic fractions, spectral maxima are blue shifted (to shorter wavelengths) relative to the HPOA fraction.

2.4 Fluorophores of Interest

With the exception of anthropogenic compounds found in wastewaters, organic fluorophores in soils and aqueous systems are largely limited to conjugated aromatic molecules substituted with polar functional groups (–COOH, –OH, $-NH_2$). The fluorophores themselves can be categorized by the central conjugated system (e.g., phenols, indoles, etc.). Understanding the positions of excitation/emission maxima for the common fluorophores associated with DOM or soil humic substances can assist with data interpretation and assignment of spectral regions in EEMs spectra. In considering the potential influence of model compounds on DOM optical properties, it is important that spectral properties, including absorptivity and quantum yields, are consistent with those exhibited by samples of interest. It is important to recognize two factors when considering the influences of potential compounds or compound classes on DOM optical properties. First, the mere similarity in fluorescence or absorbance behavior between model compounds or natural products and DOM is not, in itself, strong evidence for the presence of these materials in the pool of compounds comprising the DOM. In most studies, direct confirmation of the presence of a compound or compound class is rare. Second, although all chromophores and fluorophores in a sample contribute to its overall optical behavior, DOM absorbance and fluorescence can rarely be explained as simply the sum of chromophores or fluorophores (del Vecchio and Blough, 2004).

Fluorescence of naturally occurring organic molecules has been studied for many years. Wolfbeis (1985) published a thorough compendium of fluorescence data for organic natural products with more than 1400 references to original papers and experimental conditions for the data he summarized. Many of the compounds described by Wolfbeis are found throughout the microbial and plant world and can be considered as contributing to the pool of organic compounds comprising DOM and soil humic substances. To learn more about the chemistry, analyses, and biological functions of the compound classes described in Sections 2.4.1–2.4.9, readers are referred to Robinson (1991). EEMs spectra for select molecules described below are presented in Figures 2.4 and 2.5.

2.4.1 Amino Acids and Proteins

Among the common amino acids, phenylalanine, tyrosine, and tryptophan are fluorescent and have been heavily studied (Wolfbeis, 1985). Phenylalanine is weakly fluorescent and of little importance with regard to the fluorescence of DOM. The fluorophore associated with tyrosine is a simple phenol and tyrosine exhibits behavior similar to that of phenols, fluorescing strongly at ex 275 nm/em 303 nm (Lackowicz, 2006). However, tyrosine fluorescence is almost completely quenched when it occurs in proteins, being only about 10–50% as intense in the protein form compared to free tyrosine (Wolfbeis, 1985). Tryptophan, which contains an indole group, also fluoresces strongly (ex 287 nm/em 348 nm) in a region common for indole-containing compounds (discussed further in Section 2.4.3).

Free amino acids are present in low concentrations in natural waters. Reynolds (2003) reported low concentrations of free tryptophan ($0.82–3.44 \times 10^{-8}$ mol L^{-1}) in lake water samples using a method employing high-performance liquid chromatography (HPLC) analyses combined with synchronous fluorescence spectroscopy. In most cases, amino

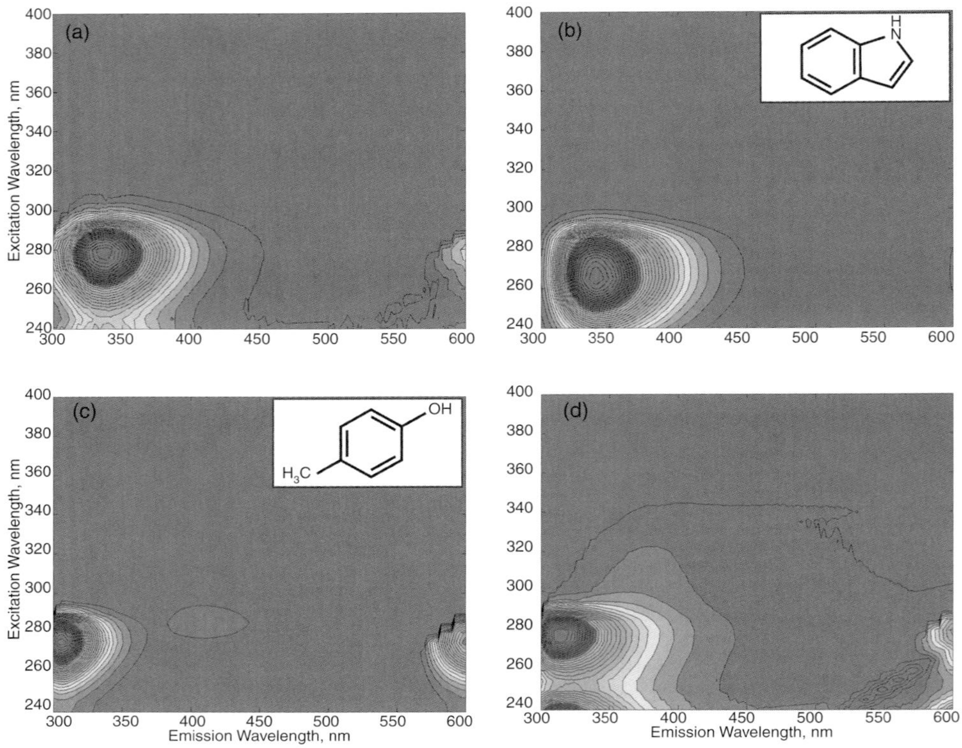

Figure 2.4. Excitation–emission spectra for (a) bovine serum albumin, (b) indole, (c) cresol, and (d) rhodendron tannin. (See Plate 3.)

acids are present in more complex, proteinaceous forms requiring hydrolysis (Cowie and Hedges, 1992). In proteins such as bovine serum albumin (Figure 2.4a), almost all fluorescence is due to the presence of tryptophan, the fluorescence of which can vary over a wide range of emission (308–350 nm) maxima (Schulman, 1985). The reason for this effect is that energy is transferred from tyrosine to tryptophan in protein molecules, even when excited at short wavelengths that excite both tyrosine and tryptophan (Lackowicz, 2006). Mayer et al. (1999) published spectra for tyrosine, tryptophan, and the protein bovine serum albumin, showing little difference between tryptophan and bovine serum albumin. Mayer et al. (1999) present a good review of the difficulties of accounting for protein and amino acid fluorescence in estuarine samples influenced by the presence of terrestrially derived organic matter. In addition to the complications presented by the sensitivity of tryptophan fluorescence within proteinaceous environments, there is spectral overlap between the fluorescence signals generally assigned to humic substances and background fluorescence associated with lower molecular weight materials and the signals for proteins. For instance, there is significant overlap of these peak centers with phenols (simple phenols, tannins, lignin phenols), and indoles (Figure 2.4b), compound classes that are common in natural

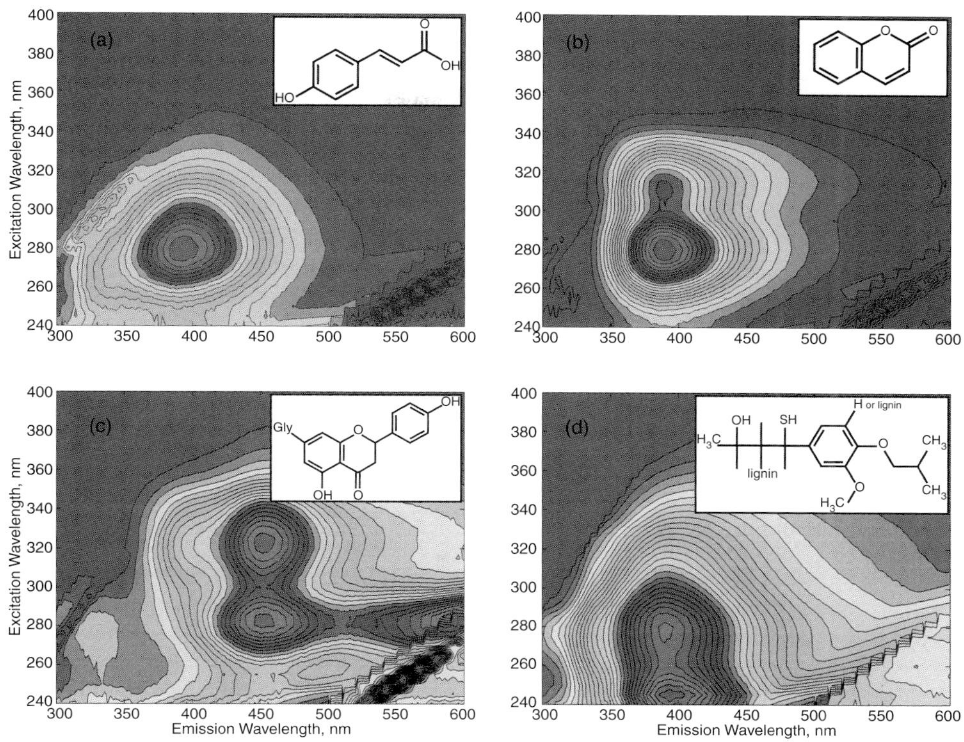

Figure 2.5. Excitation–emission spectra for (a) *p*-coumaric acid, (b) coumarin, (c) naringin hydrate, and (d) lignin, alkali, 2-hydroxypropyl ether. (See Plate 4.)

waters and wastewaters (see Section 2.4.3). Mayer et al. (1999) conclude that quantitative measurement of tyrosine, tryptophan, and protein by fluorescence alone is not possible.

First reported by Coble et al. (1990) for samples from the Black Sea, reports of protein-like, tyrosine-like, and tryptophan-like fluorescence are common in data reported for freshwater (Cory and McKnight, 2005; Fellman et al., 2009), marine (Coble et al., 1990), estuarine (Mayer et al., 1999; Maie et al., 2007), and wastewater systems (Baker, 2001). A complication of assigning peaks in this way is that the names imply composition and, as noted previously, there are a number of fluorophores present in water samples with spectral overlap in the regions with proteins and amino acids. Few studies provide supporting data based on compound specific analyses for the presence of amino acids or the array of compounds that fluoresce in these regions. Yamashita and Tanoue (2003) related the protein-like fluorophores for marine samples to tyrosine and tryptophan content obtained from analyses of hydrolyzable amino acids. In their system, amino acid components likely dominate their fluorescence signals; however, no other analyses were provided to rule out the possibility that other compound classes may have contributed to these regions of fluorescence. The influence of phenolic compounds (tannins and simple phenols) on protein-like fluorescence signals was demonstrated by Maie et al. (2007) for estuarine and Florida coastal waters. In this study, weak correlations were observed between the T-peak (associated commonly with tryptophan) and the high molecular weight fraction of ultra-filtered DOM. A large portion of the T-peak was found to be associated with nonproteinaceous compounds by using size-exclusion chromatography combined with a fluorometer tuned to measure T-peak fluorescence. Maie et al. (2007) proposed that phenols associated with tannins were the likely fluorophores contributing to T-peak fluorescence in their samples.

Because the fluorescence properties of tryptophan within a protein molecule are sensitive to local conditions, tryptophan fluorescence is important in many biochemical studies related to the properties of proteins (Lackowicz, 2006). Of particular significance in the study of DOM are the effects of reactions of proteins with other organic compound classes on protein fluorescence. It is well known, for instance, that tannins bind proteins (Robinson, 1991; Kraus et al., 2003). Labieniec and Gabryelak (2006) demonstrated that interaction of the protein bovine serum albumin with tannic, ellagic, and gallic acids (all polyphenols) resulted in a marked decrease in protein fluorescence. A representative, but not inclusive, list of DOM relevant compounds known to influence protein fluorescence includes caffeine (Kriško et al., 2005); cinnamic acid and its hydroxyl derivatives, coumaric acid and caffeic acid (Min et al., 2004; Bian et al., 2007); chlorogenic and ferulic acids (Kang et al., 2004); and flavanols, such as quercitin (Mishra et al., 2005). These studies commonly involve the measurement of the quenching of protein fluorescence resulting from binding of the compound of interest to the proteins. The results of these measurements provide information about the nature (e.g., number of binding sites, binding constants, binding distances, and conformational changes) of the interactions. Min et al. (2004) demonstrated that interactions of coumaric and caffeic acids with human serum albumin resulted in both quenching and a red-shift of the tryptophan fluorescence signal. In the case of the flavanol quercitin, the absorption spectrum of quercitin itself was red-shifted and its fluorescence intensity increased, while the fluorescence intensity of albumin was quenched. These results indicate an excitation energy transfer between tryptophan and the bound quercitin. Reactions such as these are possible in soils and natural waters and may strongly influence fluorescence of the protein-like fluorophores, as well as contributing to other portions of the fluorescence spectra of natural samples. Despite the sensitivity of different fluorophores to reactions such as those discussed, these reactions have received little attention in the study and interpretation of DOM fluorescence.

2.4.2 Simple Phenols

Simple phenols are important compound groups produced by higher plants (Robinson, 1991) and algae (Geiselman and McConnell, 1981; Connan et al., 2004) that comprise DOM. In addition, compounds such as cresol (methylphenol; Figure 2.4c) and salicylic acid (2-hydroxybenzoic acid) are often present in wastewater effluents. They are found as individual molecules (Thurman, 1985), as the fundamental structures in tannins

(Kraus et al., 2003), and as major constituents in the more complex molecules that comprise both soil and aquatic humic substances (Aiken et al., 1985). In molecules of this group, phenolic structures are the fluorophores of interest, and many have similar excitation and emission spectra (Figures 2.4c, d). As a point of interest, especially with regard to DOM studies, phenols are the subject of much interest in the wine industry, and fluorescence is an important analytical tool for detecting their presence (Bonerz et al., 2008).

Most of the aromatic carboxylic acids produced by plants also have phenolic groups and share properties with phenols (Robinson, 1991), including fluorescence. The presence of hydroxyl groups is important in controlling the fluorescent behavior of this class of compounds. Unsubstituted aromatic carboxylic acids, such as benzoic acid, have been shown to be nonfluorescent in aqueous solutions except under acidic conditions wherein the undissociated form of the acid is predominant (Martin and Clarke, 1978; Wolfbeis, 1985). Addition of a phenolic hydroxyl group results in stronger fluorescence. For example, salicylic acid (2-hydroxybenzoic acid), a compound commonly associated with humic substances and present in wastewaters (Flaherty et al., 2002), fluoresces under a range of pH conditions observed in natural waters (ex 296 nm/em 408 nm; Wolfbeis, 1985). Other simple hydroxylated benzoic acids of significance in the study of DOM include gentistic acid (2,5-dihydroxybenzoic acid; ex 318 nm/em 442 nm; Wolfbeis, 1985) and gallic acid (3,4,5-dihydroxybenzoic acid; ex 260 nm;/em 346 nm; Maie et al., 2007).

Tannins are the fourth most abundant class of molecules in terrestrial biomass (Hernes and Hedges, 2000) and have been identified as components of terrestrially derived DOM (Sleighter et al., 2010). Polyphenolic compounds similar to tannins are also produced by brown algae (Geiselman and McConnell, 1981). Tannins, which are classified as either condensed or hydrolyzable, are large molecules containing numerous phenolic moieties (Kraus et al., 2003). Condensed tannins are polymers of three-ring flavanols, whereas hydrolyzable tannins contain various phenolic acids linked to sugar moieties, the most common of which is gallic acid (3,4,5-dihydroxybenzoic acid; Robinson, 1991). Some tannins are a mixture of both types (Kraus et al., 2003). The fluorescence spectral properties for condensed and mixed tannins (Figure 2.4d) are similar to those of gallic acid and other lower molecular weight phenolic compounds (e.g., cresol; Figure 2.4c). Maie et al. (2007) presented compelling evidence that the usefulness of the T-peak intensity as a proxy for dissolved organic nitrogen (DON) was limited in coastal waters of Florida because of the overlap of this peak with gallic acid and tannin from red mangrove (*Rhizophora mangle*) leaves.

One of the most important phenolic compounds potentially contributing to the fluorescence of DOM is tyrosine. Although it is categorized as an amino acid, tyrosine could be considered a simple phenol from the perspective of fluorescence because the phenol group is the fluorescing portion of the molecule. Overlap of the fluorescence properties of tyrosine with other phenols complicates the interpretation of fluorescence data with regard to peak assignments. For example, Goldberg and Weiner (1993) employed specific compound analyses to demonstrate that cresol, a common wastewater component (e.g.,

Tertuliani et al., 2008), was responsible, in part, for the fluorescence signal commonly defined as the T-peak for samples from the South Platte River downstream from a water treatment plant near Denver, Colorado. In other work, Hernes et al. (2009) noted that the regions of EEMs spectra most useful for the prediction of lignin phenols were those commonly assigned to both tyrosine and tryptophan, but most notably with tyrosine. Lignin (discussed in Section 2.4.6) contains no amino acids and samples thought to be derived from lignaceous sources have been found to contain low amounts of nitrogen (McKnight and Aiken, 1998). Hernes et al. (2009) ascribed the strong relationship between protein-like regions and lignin phenols to the chemical similarities of fluorophores associated with the amino acids and the propylphenol monomers associated with lignin. Many fluorescence studies of natural waters and aquatic systems influenced by wastewaters refer to signals in tyrosine region as "tyrosine-like" (e.g., Hudson et al., 2007), a term that suggests chemical composition, without the measurement of either tyrosine or other compounds that also fluoresce in this region.

2.4.3 Indoles

Indole (2,3-benzopyrrole; Figure 2.4b) is a precursor compound for the biosynthesis of tryptophan, some alkaloids, and other biomolecules (Lehninger, 1970; Robinson, 1991). It is found in coal tar, oil shale, and feces (The Merck Index, 1996), and is a major product of the degradation of tryptophan. Indole and its simple derivatives (e.g., 3-methyl-1H-indole) are also constituents in personal care products and pharmaceuticals. Indoles are commonly found in wastewaters associated with agricultural feed lots (Harden, 2009), municipal wastewater treatment effluents (Goldberg and Weiner, 1993; Tertuliani at al., 2008) and oil shale and coal mining operation wastewaters (Gu and Berry, 1991). As for the case of phenols and flavanols, indole is also of interest in the production of wine, where its presence can be determined using fluorescence (Bonerz et al., 2008). Indole and 3-methyl-1H-indole fluoresce strongly and overlap significantly with the fluorescence of tryptophan, which is an indole-containing compound.

The indole fluorophore peak has been noted in wastewaters and is commonly referred to as the "T-peak" or as "tryptophan-like" (Baker, 2001; Hudson et al., 2007). Elliot et al. (2006) demonstrated that this peak can be microbially generated in simple systems containing bacteria; however, their samples were unfiltered and, as noted by the authors, the resulting signals were the combination of bacterial biomass and exudates. Whereas many reports of wastewater DOM fluorescence often describe this peak as "tryptophan-like," thereby inferring the presence of proteins and tryptophan, few actually measure for the presence of tryptophan or hydrolyzable amino acids. Indole and 3-methyl-1H-indole, on the other hand, are both routinely determined by direct methods in groundwater and surface water influenced by wastewaters. In one of the few studies of its kind, Goldberg and Weiner (1993) showed a good relationship between the presence of indole, measured directly, and fluorescence response in the indole region.

2.4.4 Phenylpropanes

Phenylpropanes are among the most widespread natural aromatic compounds (Robinson, 1991). They are found as precursors, structural components (Higuchi, 1980, Boerjan et al., 2003), and degradation products (Larsen and Rockwell, 1980) of lignin. In addition to its association with lignin, *p*-coumaric acid (Figure 2.5a) has also been found to play a role in the photoreactivity of bacteria (Putschögl et al., 2008). Many phenylpropanes can be described as hydroxylated cinnamic acids, but there are some whose stereochemistries are unrelated to cinnamic acid (Robinson, 1991). Cinnamic acids are also the aromatic centers associated with the chlorogenic acids, a related group of compounds (Robinson, 1991). Caffeic (3,4-dihydroxycinnamic), ferulic (4-hydroxy-3-methoxycinnamic), and *p*-coumaric acids (also known as *p*-hydroxycinnamic acid; spectrum presented in Figure 2.5a) are three common phenylpropanes.

The phenylpropanes have both carboxyl and phenolic functional groups, and fluorescence properties are pH dependent (Wolfbeis, 1985; Putschögl et al., 2008). Wolfbeis et al. (1986) identify the chemistries of the fluorophores active at different pH values for *o*-hydroxycinnamic acid. The fluorescence intensities of many compounds in this group are weak, being strongest under mildly acidic or basic conditions (Wolfbeis, 1985). Data presented by Wolfbeis (1985) and Larsen and Rockwell (1980) indicate that simple hydroxyl- and methoxycinnamic acids, including ferulic and caffeic acids, have similar fluorescence properties (approximate excitation maxima between 295–350 nm and emission maxima between 390 and 445 nm). Larson and Rockwell (1980) noted the similarities in fluorescence behavior between partially degraded lignin and ferulic and caffeic acid and suggested that the fluorescence characteristics of partially degraded lignin results from oxidative changes in phenylpropanoid side chains of lignin. In addition to potentially contributing directly to DOM fluorescence, phenylpropanes are reactive with other fluorophores, such as proteins, in ways that influence their fluorescence behavior (Kang et al., 2004; Min et al., 2004; Bian et al., 2007). Fluorescence methods have been developed for the analyses of *p*-coumaric acid in beer (Garcia-Sanchez et al., 1988) and a suite of hydroxycinnamic acids in wine (Bonerz et al., 2008).

2.4.5 Oxygen Ring Compounds

Molecules included in this group represent classes of biomolecules whose fluorescence properties are influenced by an oxygen atom that is part of the ring structure of the molecule. Included here are the coumarins and flavonoid groups of natural products. The optical properties of compounds in this group have long been of interest because they are common natural products, play important roles in the functioning of plants, and are often colored (Wolfbeis, 1985; Robinson, 1991; Quina et al., 2009).

Coumarins (not to be confused with coumaric acid) are lactones of *o*-hydroxycinnamic acids (Figure 2.5b; Robinson, 1991). These aromatic compounds, many of which contain hydroxyl functional groups, are common in plants (Robinson, 1991), are potential degradation products of lignin (Larson and Rockwell, 1980), are reasonably water soluble, and are highly fluorescent (Fink and Koehler, 1970; Larsen and Rockwell, 1980; Wolfbeis, 1985). As these compounds are lactones, opening of the ring structures (return to *o*-hydroxycinnamic acids) is favored at high pH whereas the lactone form is favored at low pH. Not surprisingly, the fluorescence properties of these molecules are strongly pH dependent (Fink and Koehler, 1970; Wolfbeis, 1985). Larsen and Rockwell (1980) demonstrated that the intensely fluorescent coumarin compound, esculetin, could be formed from caffeic acid (3,4-dihydroxycinnamic acid) by a photooxidation reaction, whereas Fink and Koehler (1970) suggest that cinnamic acid is a photolysis product of 7-hydroxycoumarin. Based on similarities in the fluorescence spectra of coumarins and water soluble humic substances, Larsen and Rockwell (1980) hypothesized that coumarins are important contributors to the fluorescence properties of both DOM and humic substances.

Among natural products, the flavonoids represent the largest group of oxygen ring compounds (Wolfbeis, 1985). They are benzopyran derivatives having a carbon skeleton that consists of two substituted benzene rings connected by a three-carbon aliphatic chain and multiple phenolic hydroxyl groups (Wolfbeis, 1985; Robinson, 1991). The flavonoid category includes flavanones, flavones, flavanols (e.g., quercitin, morin), and anthocyanins. The distinctions between compound groups are determined by oxidation states and variations of the three-carbon chain (Robinson, 1991). The flavonoids are common throughout the plant kingdom and include most of the common plant pigments (Robinson, 1991). Many flavonoids fluoresce strongly, especially in polar solvents, and exhibit large Stokes shifts (Wolfbeis, 1985). The fluorescence characteristics of individual flavonoids are influenced by the number and positions of hydroxyl groups (Wigand et al., 1992; Ale et al., 2002). These compounds and their fluorescence properties have long been of interest to natural products chemists. Wolfbeis (1985) presents a good summary of earlier studies of flavonoid fluorescence. Of particular interest, the anthocyanins, a major class of pigments in leaves, flowers, and fruits, are an important compound class associated with red wines, a connection that has resulted in the development of fluorescence approaches in their analyses (Figueiredo et al., 1990; Bonerz et al., 2008). Whereas the flavonoid group of compounds has not received the same attention as lignin or tannins as sources of DOM, it is possible that they contribute to DOM fluorescence given their prevalence in the plant world and aqueous solubility, although direct evidence for their presence is currently lacking.

The flavonoid group of compounds can exhibit complicated absorbance and fluorescence spectra (e.g., Figueiredo et al., 1990; Drabent et al., 2007). For example, multiple absorbance and emission bands are evident in the aqueous spectra for naringin hydrate, a flavanone (Figure 2.5c). Of the natural products reported on in this chapter, the flavonoids are the only ones with fluorescence behavior in the "C" and "A" peak regions commonly ascribed to humic materials (Table 2.1). Figueiredo et al. (1990) determined that fluorescence emission bands associated with malvidin 3,5-diglucoside, an anthocyanin, could be assigned to six forms of the molecule – the flavylium cation (λ_{max} = 620 nm), hemiacetal (λ_{max} = 370 nm), chalcone (λ_{max} = 435 nm), ionized chalcone (λ_{max} = 495 nm), quinonoidal base (λ_{max} = 660 nm), and ionized quinonoidal base (λ_{max} = 665 nm). The presence of any one of these forms is pH dependent; however, multiple forms exist at a given pH. Further complicating interpretation of spectra for these compounds is their capacity for interacting via H-bonding and charge transfer interactions with compounds such as gallic acid, caffeine, chlorogenic acid (Wigand et al., 1992), and other flavonoids (Santhanam et al., 1983). These interactions can result in increased absorption of light by anthocyanin chromophores and either fluorescence enhancement or quenching depending on the anthocynanin (Santhanam et al., 1983; Wigand et al., 1992).

2.4.6 Lignin

According to Kirk et al. (1980), lignin is a generic term used to describe the complex aromatic biopolymers that are major components of plant tissues. Lignin represents the second most abundant terrestrial biopolymer after cellulose (Kirk et al., 1980; Boerjan et al., 2003), and, given the absence of aromatic moieties in cellulose, the most abundant class of aromatic biopolymers. Its significant annual production is roughly balanced by its degradation by microorganisms (Kirk et al., 1980). It is a compound class that has long been thought to contribute to the composition of terrestrially derived humic substances (Stevenson, 1985). A large body of research has addressed lignin as a source of aromatic compounds to DOM, and there is substantial direct evidence that lignin-derived compounds are a major component of terrestrially derived DOM (e.g., Kujawinski et al., 2009; Sleighter et al., 2010). For example, Ertel et al. (1986) determined that 3–8% of the humic carbon in the Amazon River was present as lignin components. The fluorescence properties of lignin (Figure 2.5d) have long been the subject of study (e.g., Hartley, 1893), and assignment of structures responsible for lignin fluorescence has been elusive (Radotić et al., 2006). As is the case for DOM, identification of lignin fluorophores and understanding the dynamics responsible for lignin fluorescence are complicated by inherent structural complexity and variability between different samples. Olmstead and Gray (1997) provide a thorough review of the fundamentals of lignin fluorescence, applications of fluorescence to the analyses of industrial lignins and wood pulp, and novel fluorescence approaches for the study of nonfluorescent components in wood products. Phenylcoumarin, stilbene structures, coniferol alcohol, *p*-oxybenzaldehyde, biphenyl, and benzoquinone structures have been proposed to be the primary fluorophores in lignin (Lundquist et al., 1978; Olmstead and Gray, 1997; Albinsson et al., 1999; Machado et al., 2001). The fluorescence properties of these compound groups within lignin are thought to result from direct excitation of individual fluorphores, energy transfer interactions, and the presence of charge-transfer complexes. Lundquist et al. (1978) originally proposed that lignin fluorescence was the result of the transfer of energy from excited structural elements in lignin to other structures acting as "energy sinks" that subsequently emitted the energy as light. Phenylcoumarone-type and stilbene-type components in lignin, neither of which exhibit efficient fluorescence of themselves, were proposed as potential acceptor groups. In another example of energy transfer, to explain the observation that spruce lignin exhibited fluorescence at approximately 360 nm for excitation wavelengths ranging from 240–320 nm, Albinsson et al. (1999) proposed that energy absorbed by a range of chromophores (hence multiple excitation wavelengths) is transferred to small quantities of phenylcoumarone type structures and subsequently emitted by these fluorophores. Quinones and quinine methides have also been proposed to act as acceptor functional groups (Olmstead and Gray, 1997). Alternatively, Barsberg et al. (2003) proposed that lignin optical properties, including emission quenching, are influenced by charge-transfer interactions with quinones as receptors.

Fluorescence analyses of lignin and lignin sulfonates (byproducts of pulp mills) have been employed to identify the presence of these compounds in water samples (Christman and Minear, 1967; Thruston, 1970; Santos et al., 2000) and pulp mill liquors (Bublitz and Meng, 1978). Christman and Minear (1967) assessed fluorescence as a tool to detect lignin sulfonates in paper mill waste waters. They noted that excitation at three wavelengths (λ_{ex}=253, 293, 340 nm) all resulted in emission at λ_{em} =400 nm. This observation is similar to those noted for lignin by Albinsson et al. (1999) and others. Christman and Minear (1967) also noted that emission maxima for these samples were not a function of molecular weight as determined by gel permeation chromatography, although intensity did vary between the fractions. The fluorescence behavior of lignin compounds appears to be largely independent of wood source, with many samples having similar emission maxima, although different woods exhibit different fluorescence intensities (Bublitz and Meng, 1978). Lignin sulfonate fluorescence can be detected in waters influenced by lignin waste products. However, interferences from aquatic humic substances can be an issue (Wilander et al., 1974). The presence of lignin sulfonates from pulp mills has been traced in marine (Almgren et al., 1975) and freshwaters (Josefsson and Nyquist, 1976).

A number of studies have reported strong positive correlations between DOM absorbance coefficients at various wavelengths and lignin phenol contents for a variety of surface waters (e.g., Boyle et al., 2009; Spencer et al., 2009). Recent studies suggest that partially oxidized products of lignin exert a strong influence on the optical properties of DOM, especially at longer wavelengths. Del Vecchio and Blough (2004) provided evidence that the long-wavelength absorption and emission properties of aquatic humic substances largely resulted from the presence of intramolecular charge transfer interactions involving hydroxyl-aromatic donors and quinoid acceptors formed from the partial degradation of lignin precursors. These intramolecular interactions are proposed to involve both energy transfer and charge transfer mechanisms (Boyle et al., 2009).

2.4.7 Quinones

Quinones are an important class of redox reactive molecules produced by microbes, fungi, and higher plants whose presence in aquatic humic substances has been confirmed by electron spin resonance (ESR) (e.g., Scott et al., 1998) and nuclear magnetic resonance (NMR) studies (Thorn et al., 1992). The group consists of molecules ranging from simple to polynuclear quinones, with the anthroquinones being the largest group of naturally occurring quinones (Robinson, 1991). Quinones can exist in oxidized (quinone) and reduced (semiquinone, hydroquinone) forms, and are thought to play important roles in biogeochemical electron transfer reactions (abiotic and biotic) associated with DOM (Nurmi and Tratnyek, 2002; Fimmen et al., 2007). As noted earlier, quinones are thought to play important roles as acceptor molecules in charge transfer interactions that may influence lignin fluorescence (Barsberg et al., 2003) and DOM absorbance properties (Del Vecchio and Blough 2004).

Oxidized quinones are relatively poor fluorophores; however, the reduced forms, hydroquinones, are highly fluorescent, a property that has been utilized in developing photochemical-reaction fluorescence detectors for liquid chromatography (Poulson and Birks, 1989). Klapper et al. (2002) used the differences in fluorescence properties between oxidized and reduced samples of humic substances isolated from a range of environments to infer the presence of quinone-like moieties. These authors proposed that quinone moieties contribute significantly to the fluorescence of humic substances and that fluorescence analyses could be used to assess the redox status of humic substances.

Cory and McKnight (2005) analyzed fluorescence EEMs data for a large number of DOM samples and, using PARAFAC analyses, concluded that approximately 50% of the EEMs spectrum for a given sample could be accounted for by quinone-like fluorophores. However, the presence of quinones in these samples was not quantified and the nature of their influence on the EEMs spectra was not explored. Subsequently, Miller et al. (2006) proposed that a ratio of PARAFAC regions thought to be related to the presence of quinone moieties could be utilized as a proxy for DOM redox status (redox index).

Recently, a number of papers have provided data challenging the hypothesis that quinones directly influence the fluorescence properties of DOM. Ma et al. (2010) demonstrated that the fluorescence behavior of DOM on reduction with $NaBH_4$ was inconsistent with the behavior of a large number of model quinones and their corresponding hydroquinones. Further, based on the observation that subsequent reoxidation of the DOM samples did not completely return the fluorescence spectra to the original state (i.e., the reduction step was largely irreversible), Ma et al. (2010) suggested that aromatic ketones, known to be structural components of terrestrially derived aquatic humic substances (Leenheer et al., 1987), may play more important roles in DOM optical properties than quinones. MacCalady and Walton (2010) and Mauer et al. (2010) observed no changes in fluorescence EEMs in regions previously assigned to quinone activity on either chemical ($NaBH_4$) or electrochemical reduction and reoxidation of their DOM samples. In the case of the MacCalady paper, Miller et al. (2010) suggested that inner-filter effects may have obscured changes in the redox sensitive regions of their spectra. The potential influences of charge-transfer interactions involving quinone-like fluorophores on DOM fluorescence are still the subject of study and debate.

2.4.8 Alkaloids

The alkaloids are a widely distributed class of natural products that contain nitrogen, often in a heterocyclic ring (Robinson, 1991). Almost all alkaloids are fluorescent (Wolfbeis, 1985).

This group of molecules is common in the plant world, is not chemically well defined, and compound classification is often based on the nature of the nitrogen-containing ring structure (Robinson, 1991). Alkaloids have long been of interest to pharmacological chemists because they exhibit a wide range of physiological reactivities. The fluorescence properties of alkaloids are often determined by the central heterocyclic ring structure. Wolfbeis (1985) presents a thorough review of alkaloid fluorescence through 1985. Unlike most molecules comprising DOM, alkaloids are often basic in nature and can become more basic in the excited state (Wolfbeis, 1985).

Of major importance for DOM fluorescence are the indoles and quinolines. The indoles represent one of the largest groups of alkaloids. As is the case for tryptophan (see discussion in Section 2.4.3), simple indole derivatives exhibit fluorescence behavior similarly to that of indole itself (see Figure 2.4b). These compounds are efficiently excited in the wavelength range of 270–290 nm, with emission in the range of 330–350 nm (Wolfbeis, 1985). As noted previously, a number of indole-containing compounds are present in sewage-related wastewaters. Another wastewater compound of interest is caffeine, a compound that has found utility as a marker for the presence of sewage wastewater in surface water samples (Buerge et al., 2003). Caffeine, with a purine-type fluorophore, fluoresces in a region (ex 270 nm/em 320 nm) similar to that of "B" type fluorophores (Table 2.1) and simple phenols.

One of the best known and most studied of the fluorescent alkaloids is the isoquinoline alkaloid, quinine, which is responsible for the bluish hue of tonic water. Much of the early research defining the fluorescence phenomenon was carried out on solutions containing quinine (Lakowicz, 2006). Quinine sulfate (ex 331 nm/em 382 nm) is often used as a "quantum counter" to correct EEMs spectra (Lakowicz, 2006). Its use as a standard to quantify fluorescence response in DOM studies has been proposed to properly correct DOM spectra and improve comparability of results between different instruments and analysts (Hoge et al., 1993, Murphy et al., 2010).

2.5 Factors Influencing DOM Fluorescence

2.5.1 Quenching

Fluorescence quenching describes processes that reduce fluorescence intensity of a fluorophore. These processes generally involve interactions that influence one or more chemical aspects of the fluorophore, such as rates of decay, intermolecular energy transfer, or the population of molecules in the excited state that decrease the fluorescence intensity of the molecule (Lakowicz, 2006). The pathways leading to fluorescence quenching are described according to whether the quenching results from interactions of the quenching species with the ground state of the fluorescing molecule (static quenching), interactions with the excited state of the fluorophore (collisional quenching), or if the quenching results from nonmolecular mechanisms (Lakowicz, 2006). Each type of quenching can influence the fluorescence of DOM, and experiments designed to determine quenching effects have proven to be useful for studying the reactivity of both DOM and soil organic matter (SOM) with other fluorophores, such as anthropogenic compounds (Chen et al., 1994; Backhus et al., 2003). The effects of naturally occurring molecules comprising DOM or SOM on the fluorescence properties of either of these materials are less well defined.

Static quenching results from interactions, such as chemical binding or the formation of charge transfer complexes, of the ground state of a fluorescing molecule with another chemical species. In this case, the interaction results in the formation of a non-fluorescing "complex." Quenching efficiency is determined by the strength of the interaction and the concentration of the quencher. For studies of DOM and soil organic matter, interactions with metals, such as iron, resulting in the formation of a complex and reduced fluorescence of the DOM is an example of static quenching (e.g., Blaser and Sposito, 1987). Other examples involve the interactions of organic molecules with each other by weak electrostatic interactions resulting in decreased fluorescence of one or both molecules. For example, the quenching of protein fluorescence resulting from interactions with compounds such as cinnamic acids is an example of static quenching (Min et al., 2004). In environmental chemistry, "partitioning" interactions of nonpolar organic pollutants with DOM have been shown to result in static quenching of fluorescence probe molecules, such as fluorescent polycyclic aromatic hydrocarbons (Gauthier et al., 1986; Backhus and Gschwend, 1990; Backhus et al., 2003). In these studies, fluorescence quenching of the polycyclic aromatic hydrocarbon of interest was used to determine the equilibrium constant for the association of the compound with DOM. In another example, the reduction of the fluorescence intensities of Suwannee River fulvic acid (FA) and humic acid (HA) bound to cationic nitroxides was used to estimate surface potentials of the humic compounds (Green et al., 1992).

Collisional quenching, also known as "dynamic quenching" (Schulman, 1985), occurs when an excited fluorophore comes in contact (as via a collision) with the quenching species. As part of the interaction of the excited fluorophore and the quenching molecule, energy is transferred to the quenching molecule and the excited fluorophore is deactivated, returning to the ground state via a nonradiative pathway. In this case, the fluorophore returns to the ground state without a chemical reaction, and neither the quencher nor the fluorophore is chemically altered. Molecular oxygen, halogens, amines, and electron-deficient molecules are among the compounds that can act as quenchers (Lakowicz, 2006). For the case of DOM fluorescence, molecular oxygen could be an important collisional quencher; however, the effects of molecular oxygen on DOM fluorescence have not been well described, even in studies noting shifts in fluorophore intensities under different redox conditions (see, e.g., Klapper et al., 2002)

A third type of quenching, the "inner filter effect," does not involve direct interactions between the excited molecule and another chemical species that result in loss of energy or a change in the fluorophore energetics, but is a major problem associated with the collection of fluorescence data (Tucker et al., 1992; Lakowicz, 2006). It can result from molecules that absorb excitation light, reducing the intensity of light available to excite a fluorophore (primary effect), or by the absorption of light emitted by the fluorophore (secondary effect). Any light-absorbing chemical species can result in inner filtration of light; when these are the fluorophores themselves, this process is called solute self-absorption. In addition, particles or colloids in unfiltered samples can bring about the same effect through light scattering. Inner filter effects are an important issue with regard to the fluorescence of natural samples owing to the complexity of DOM composition. UV-visible absorption spectra of DOM samples (Figure 2.1) indicate that light is absorbed by most samples at the wavelengths of interest in fluorescence analyses (approximately 250–550 nm). A number of authors address this issue with regard to analyses of DOM (Mobed et al., 1996; McKnight et al., 2001). Care is needed to ensure that samples are optically dilute (A_{254} cm^{-1} < 0.2) before measuring fluorescence of samples that contain DOM (Miller et al., 2010). Even for optically dilute samples, fluorescence data must still be corrected for inner filter effects.

Finally, the relaxation energetics of an excited molecule returning to the ground state are temperature sensitive (Lakowicz, 2006). Thermal "quenching" of fluorophores results at higher temperatures because it is more likely that the excited molecules will return to the ground state via radiationless pathways; fluorescence intensities are generally enhanced at lower temperatures. Different fluorophores exhibit different temperature dependencies, and this phenomenon is useful for providing information about both chemical structure and reaction pathways in chemistry and biochemistry (Baker, 2005; Lackowicz, 2006). Measurement of thermal quenching behavior of DOM samples from a variety of environments has been explored as a method to provide structural information about DOM (Baker, 2005). In this approach, the thermal dependency of different fluorophores in DOM was used to provide structural information about the tryptophan-like groups in untreated and treated wastewaters. That paper demonstrated that different fluorophores in the samples exhibited different responses to increasing temperature. An important implication of this work was that the deployment of in situ fluorometers or the acquisition of laser-induced remote-sensing data may require different compensation equations for samples containing different DOM compositions.

2.5.2 pH Effects

Excitation emission spectra of most DOM fluorophores are pH sensitive owing to the influence of either deprotonation or protonation of acidic (–COOH, –OH) and basic ($–NH_2$) functional groups bound directly to aromatic fluorophores. Within the pool of molecules comprising DOM, the major acidic functional groups are –COOH and –OH, whereas $–NH_2$ is the major basic functional group. The presence or absence of lone pair electrons on these functional groups alters the lifetimes of the excited state by influencing the rates of relaxation. In addition, the properties of acids and bases differ between the ground and excited states. The equilibrium dissociation constants for –COOH groups are greater in the excited state (pK_a^*) than in the ground state (pK_a), indicating that –COOH groups are weaker acids in the excited state. Electron-donating groups (–OH, $–NH_2$) are stronger acids in the excited state relative to the ground state (Sharma and Schulman, 1999). For the case of phenols, the excited state may exhibit pKa^* values as low as 2–3, whereas in the ground state these compounds are generally weak acids ($pK_a \sim 10$). The effect exerted by pH depends on the nature of the acidic or basic substituent (Schulman, 1985). For electron withdrawing groups, such as –COOH, protonation shifts fluorescence to longer wavelengths (red shift) whereas deprotonation shifts the position of fluorescence to shorter wavelengths (blue shift). The opposite effect happens with electron-donating groups, such as –OH and $–NH_2$. Protonation of groups such as $–NH_2$ results in shifts to shorter wavelengths, whereas deprotonation or dissociation of phenolic –OH groups results in shifts to longer wavelengths. Changes in intensity and peak positions at different pHs are dependent on protonation/deprotonation of both the ground and excited states of the fluorophores. Kelly and Schulman (1988) and Sharma and Schulman (1999) provide more detailed discussion of the influence of pH on the fluorescence of aromatic acids and bases.

An excellent example of the complexities associated with pH effects on the ground and excited states of *o*-hydroxycinnamic acid is presented by Wolfbeis et al. (1986). This molecule contains one carboxyl and one phenolic group. Under basic conditions, the molecule is highly fluorescent due to deprotonation of the phenolic (pK_a 9.7) and carboxyl sites (pK_a 4.2) in the ground state. Fluorescence intensity is weaker at neutral pH where only the carboxyl group is deprotonated. Below pH 4, the main peak shifts and a second fluorophore appears resulting from deprotonation of the phenolic group in the *excited* state. In the excited state, the pK_a for the carboxyl group increased 2.2 units whereas the pK_a for the phenolic group decreased 8.3 units, making the phenolic group a much stronger acid (Wolfbeis et al., 1986). Similar results have been reported for other classes of phenolic molecules, such as anthocyanins (Moreira et al., 2003) and coumarins (Fink and Koehler, 1970).

The effects of pH on the fluorescence behavior of DOM and humic substances have been examined in a number of studies. Ghosh and Schnitzer (1980) reported that the fluorescence of soil humic materials was both ionic strength and pH sensitive. In this study, fluorescence intensity decreased with greater ionic strength and increased with higher pH for both humic and fulvic acid. The increasing intensity with higher pH was interpreted as evidence for the role played by phenolic groups in the fluorescence behavior of humic substances. Mobed et al. (1996) reported both changes in intensity and blue- and red-shifted portions of the EEMs spectra as a result of collecting spectra under different pH conditions. The red-shifted portions of the spectra were in the long-wavelength regions and attributed to the fluorescence characteristics of phenolic fluorophores. Westerhoff et al. (2001) also noted a slight blue shift for a wastewater sample and Suwannee River fulvic acid at pH 3 compared to pH 7. Lowering the pH from 7 to 3 in this study also resulted in 30–40% loss of intensity for most EEMs peaks, although the greatest change was noted for the peak near ex 250 nm/em 320 nm. The decreased intensities in this region were hypothesized to be due to protonation of phenolic fluorophores. The dependencies of different fluorophores on sample pH may pose a problem for comparing spectra obtained on different water samples at different pHs. Spencer et al. (2007b) examined the influence of pH over a range of 2–10 for 35 waters from different source areas. The effects of pH were found to be greater at high and low pH with different responses exhibited at different wavelengths. Given the unknown structures associated with DOM and humic substances, ascribing structural explanations to observed changes in fluorescence spectra is difficult (Spencer et al., 2007b), however, the observations reported for different compound classes (e.g., Fink and Koehler, 1970 and Wolfbeis et al., 1986) may be of relevance.

Natural water samples generally range in pH between 4 to 10 with most samples having pH between 6 and 8. For most samples, therefore, it is sufficient to measure fluorescence without pH adjustment.

2.5.3 Interactions with Metals

Metals bound to fluorescing ligands can influence the electronic state of the ligand in a manner analogous to protonation reactions. Metals are Lewis acids, and, as such, the coordination of the ligand with a metal ion is similar to protonation of the ligand (Sharma and Schulman, 1999). Fluorescence can be either quenched or enhanced by the coordination of metals with fluorescing ligands, such as aromatic compounds with electron-rich functional groups (–COOH, –OH, $-NH_2$), depending on the ligand and the effect the metal has on the nonradiative processes competing with fluorescence. Most interactions between main group transition metals and the fluorescing ligands in DOM or humic substances result in static quenching due to interactions of the π electrons of the ligand with the metal.

In some instances, the ability of an organic compound to form complexes with metals results in enhanced fluorescence. This is the case for a number of colorimetric indicator compounds used to indicate the presence of some metals and cations in water samples (Skoog and West, 1982). Flavones are one of the natural product classes that exhibit enhanced fluorescence when complexed by metals, and this reaction has been used to detect the presence of both flavanols and metals alike (Wolfbeis, 1985). A common example is the compound morin, which is nonfluorescent in the uncomplexed state but fluoresces in the presence of Al^{3+} and other metals (Brown et al., 1990). The reaction of morin with Al^{3+} has been used to study the speciation of Al^3+ in natural waters as well as in the study of the effects of Al^{3+} on plant materials (Eticha et al., 2005). In a final example, the fluorescence of salicylic acid is enhanced in the presence of As^{3+} and sodium dodecyl sulfate (Karim et al., 2006), whereas interactions with Fe^{3+} result in fluorescence quenching (Cha and Park, 1998).

Fluorescence quenching has been used to measure metal binding constants of a range of metals with humic substances (Saar and Weber, 1980) and plant extracts (e.g., Blaser and Sposito, 1987). Recent studies have used PARAFAC analyses to determine changes in different regions of organic matter fluorescence spectra resulting from interactions with metals (Ohno et al., 2008). However, while providing information about interactions between naturally occurring fluorophores and metals, fluorescence quenching suffers from limitations that restrict its application for the determination of environmentally relevant DOM-metal binding constants (Cabaniss and Shuman, 1988). The fluorescence quenching approach lacks sensitivity, requires larger concentrations of metals than are environmentally relevant, and provides information only about the fluorophores interacting with the metals, not the nonfluorescent ligands that comprise most of the DOM. Binding constants obtained using fluorescence quenching of soil extracts or DOM are usually many orders of magnitude less than those obtained using other, more sensitive methods. Difficulties encountered in the determination DOM-metal binding constants under environmentally relevant conditions are reviewed by Town and Fillela (2002) and Gasper et al. (2007).

The interactions of metals with DOM can pose a problem for the measurement of fluorescence data on natural samples. In particular, iron (Fe^{3+} and Fe^{2+}) is an issue because it is often present at concentrations sufficiently large enough to influence DOM optical properties. Ferric iron (Fe^{3+}) itself absorbs light at wavelengths important for both excitation and emission in fluorescence analyses of water samples and is more problematic when measuring absorbance than is Fe^{2+} (Doane and Horwáth, 2010). Weishaar et al. (2003) showed the influence of Fe^{3+} on the measurement of absorbance data at $\lambda = 254$ nm in the determination of specific UV absorbance of DOM. For fluorescence, the presence of Fe^{3+} could contribute to inner filter effects. Recently Doane and Horwáth (2010) described a method whereby Fe^{3+} is chemically reduced to Fe^{2+} to correct for its influence in absorbance measurements. However, both forms appear to influence the intensities and peak positions with humic fluorophores. Cory (2005) demonstrated that addition of both Fe^{2+} and Fe^{3+} in the range of 0.5–50 μM to a sample of aquatic fulvic acid reduced total fluorescence, resulted in peak position shifts, and influenced the distribution of components based on PARAFAC analyses. The presence of iron and its effects are often overlooked in reports describing the optical properties of DOM in natural systems. These effects may be substantial and need to be taken into account to improve data interpretation.

2.5.4 Charge Transfer Interactions

Charge transfer interactions are important in understanding both absorbance and fluorescence properties of organic molecules. Charge transfer interactions are weak electrostatic interactions between two molecules wherein electron density is partially transferred from one molecule (donor) to the other (acceptor), thereby forming a "charge transfer complex" and altering the energy levels of both species. These interactions influence the optical properties of the molecules that are involved in the complex, resulting in absorption and fluorescence spectra that cannot be expressed as the sum of the parts (March, 1968; Barsberg et al., 2003). For organic compounds, aromatic molecules with π-electrons can act as both donors and acceptors. In the classic case, the π system of one molecule overlaps that of the second molecule and yields an electron to the second molecule. This is the case for the formation of the quinhydrone complex between hydroquinone (donor) and quinone (acceptor) (Szent-Györgyi et al., 1961; D'Souza and Deviprasad, 2001). Other types of charge transfer interactions are possible involving both inorganic (e.g., some metals) and organic species (March, 1968). The optical properties of charge-transfer complexes will depend on chemical properties, such as oxidation and reduction potentials, of both the acceptor and donor compounds (Barsberg et al., 2003)

Charge transfer interactions have long been known to be important for a number of compound classes of interest for DOM fluorescence. For instance, indoles, including tryptophan, have been shown to be efficient electron donating molecules (Isenberg and Szent-

Györgyi, 1959; Green and Malrieu, 1965), as well as purines, pyrimidines, and flavins (Pullman and Pullman, 1958). Charge-transfer complexation is also thought to be important in color intensification reactions (copigmentation) involving anthocyanins, the major red, blue, and purple pigments in plants (Quina et al., 2009). da Silva et al. (2005) showed that flavylium ions associated with anthocyanins are excellent electron acceptors forming charge-transfer complexes with a number of naturally occurring polyphenols (ferulic acid, gallic acid, caffeic acid, protocatechuic acid, and quercitin) wherein the polyphenols were electron donors. As a final example, charge-transfer interactions are thought to be important in the absorption behavior of lignin, especially with regard to visible absorbance (Furman and Lonsky, 1988). The potential importance of charge-transfer interactions within the pool of molecules that comprise DOM is large. In an effort to elucidate the factors controlling the fluorescence behavior of Suwannee River humic substances, Del Vecchio and Blough (2004) performed experiments wherein the optical properties of the humic substances were altered by the selective photolytic destruction of chromophores using narrowly focused laser light. The observed changes were inconsistent with a model for humic substance fluorescence resulting from the simple, conservative addition of chromophores and flourophores. Rather, the results suggested the presence of charge-transfer interactions. Del Vecchio and Blough (2004) proposed that these interactions involve hydroxyl-aromatic donors and quinoid acceptors resulting from the partial oxidation of lignin precursors. In a subsequent paper, Boyle et al. (2009) present a strong argument for the importance of the partial oxidation products of lignin in this process, with aromatic ketones possibly playing important roles in absorption and emission properties of DOM (Ma et al, 2010). Other possibilities exist for donor (e.g., indoles, tryptophan, simple phenols, polyphenols) and acceptor compounds (e.g., flavonoids, anthocyanins,). It is, at this time, difficult to pinpoint the significance of these species owing to the complexity of DOM samples and analytical constraints. However, given the reactive natures of aromatic compounds likely to be precursors for DOM, it is possible that charge-transfer interactions associated with a myriad of compound types contribute to observed optical behavior of DOM and aquatic humic substances.

2.6 Conclusions

Fluorescence spectroscopy has been used as a method for studying organic matter in soils and aquatic systems for many years. In the last 20 years, however, technological advances in instrumentation and data management have resulted in greatly improved efficiency of fluorescence data collection. Associated with these advances have been developments of new approaches for analyzing the large amounts of data that can be generated for each sample. Together, these developments have resulted in a resurgence of fluorescence as an important tool for studying organic matter in a myriad of environments and there has been a large increase of fluorescence data reported throughout the water sciences. Ease of data collection and manipulation, however, do not, in themselves, result in improved data interpretation. The goals of this chapter have been twofold: (1) to survey the literature related to fluorescence of natural products to help increase understanding of DOM fluorescence and to provide the foundation to improve data interpretation; and (2) to address the chemical factors controlling fluorescence of organic molecules with an emphasis on their effects of DOM fluorescence.

From a chemical perspective, fluorescence analysis of DOM is more complicated than the ease of data collection and manipulation would suggest. Numerous types of compounds containing different fluorophores exhibit similar fluorescence properties. In the absence of additional, compound specific analyses, therefore, assignment of the fluorophores in a given sample to particular classes of compounds is difficult. In general, the power of fluorescence analyses in biogeochemical and ecological studies would be greater if accompanied by specific compound analyses, such as those for amino acids and lignin phenols. A further complication is that the fluorophores comprising DOM are well suited to engage in chemical reactions that influence the molecular energetics involved with the absorption of light and subsequent relaxation of excited molecules. In most studies reporting fluorescence data in the water sciences literature, these influences are largely ignored, even with regard to obvious conditions (such as temperature) and chemical interferences (such as iron and oxygen) influencing interpretation of data.

Finally, continued improvements in elucidating the roles of organic matter reactions on both the absorption and fluorescence properties of DOM are necessary to improve data interpretation, to develop new experiments designed to provide more organic matter structural information, and to develop new applications of fluorescence to address problems in the aquatic sciences. As described in this chapter, reactions such as charge transfer interactions and chemical bonding reactions can alter fluorescence spectra in non-conservative ways. The influences of these interactions are often overlooked in interpreting data obtained under various conditions or for samples with different distributions of reactive constituents. In these cases, the tacit assumption is that fluorescence spectra are conservative with regard to individual flourophores – an assumption that is not supported by the large amount of research on the fluorescence properties of biomolecules and natural products.

Acknowledgments

Thanks to Kenna Butler and Suzanne M. Bourret for providing fluorescence spectra, and to Tamara Kraus for providing well characterized tannin samples. Thanks also to Laurel Larsen, Robert Spencer, and two anonymous reviewers for providing critical feedback during the writing of this chapter. Support was provided by the U.S. Geological Survey, National Research Program.

References

Aiken, G.R. and Leenheer, L.A. (1993). Isolation and characterization of dissolved and colloidal organic matter. *Chem. Ecol.*, **8**, 135–151.

Aiken, G.R., McKnight, D.M., Wershaw, R.L., and MacCarthy, P., Eds. (1985). *Humic Substances in Soil, Sediment, and Water, Geochemistry, Isolation, and Characterization*. New York: John Wiley & Sons, 692 pp.

Aiken, G.R., McKnight, D.M., Thorn, K.A., and Thurman, E.M. (1992). Isolation of hydrophilic acids from water using macroporous resins. *Org. Geochem.*, **18**, 567–573.

Albinsson, B., Li, S., Lundquist, K., and Stomberg, R. (1999). The origin of lignin fluorescence. *J. Mol. Struct.*, **508**, 19–27.

Ale, R., Olives, E., Martin, L., Matin, M.A., del Castillo, B., Agnese, A.M., Ortega, M.G., Nuñez-Montoya, S., and Cabrera, J.L. (2002). Study of the solvatochromic effect on natural phenolic compounds. *Ars Pharmaceutica*, **43**, 57–71.

Almgren, T., Josefsson, B., and Nyquist, G. (1975). A fluorescence method for the studies of spent sulfite liquor and humic substances in seawater. *Anal. Chim. Acta.*, **78**, 411–421.

Backhus, D.A. and Gschwend, P.M. (1990). Fluorescent polycyclic aromatic hydrocarbons as probes for studying the impact of colloids on pollutant transport in groundwater. *Environ. Sci. Technol.*, **24**, 1214–1223.

Backhus, D.A., Golini, C., and Castellanos, E. (2003). Evaluation of fluorescence quenching for assessing the importance of interactions between nonpolar organic pollutants and dissolved organic matter. *Environ. Sci. Technol.*, **37**, 4717–4723.

Baker, A. (2001). Fluorescence excitation-emission matrix characterization of some sewage-impacted rivers. *Environ. Sci. Technol.*, **35**, 948–953.

Baker, A. (2005). Thermal fluorescence quenching properties of dissolved organic matter. *Water Res.*, **39**, 4405–4412.

Barsberg, S., Elder, T., and Felby, C. (2003). Lignin-quinone interactions: Implications for optical properties of lignin. *Chem. Mater.*, **15**, 649–655.

Berman, T. (1972). Profiles of chlorophyll concentrations by *in vivo* fluorescence: Some limnological applications. *Limnol. Oceanogr.*, **17**, 616–618.

Bian, H., Zhang, H., Yu, Qing, Chen, Z., and Liang, H. (2007). Studies of the interaction of cinnamic acid with bovine serum albumin. *Chem. Pharm. Bull.*, **55**, 871–875.

Blaser, P. and Sposito, G. (1987). Spectrofluorometric investigation of trace metal complexation by an aqueous chestnut leaf litter extract. *Soil Sci. Soc. Am. J.*, **51**, 612–619.

Blough, N. and Del Vecchio, R. (2002). Chromophoric DOM in the coastal environment. In D.A. Hansell and C.A. Carlson (Eds.), *Biogeochemistry of Marine Dissolved Organic Matter* (pp. 509–546). San Diego, CA: Academic Press.

Boerjan, W., Ralph, J., and Baucher, M. (2003). Lignin biosynthesis. *Annu. Rev. Plant Biol.*, **54**, 519–546.

Bonerz, D.P.M., Nikfardjam, M.S.P., and Creasy, G.L. (2008). A new RP-HPLC method for the analysis of polyphenols, anthcyanins, and indole-3-acetic acid in wine. *Am. J. Enol. Vitic.*, **59**, 106–109.

Boulton, R. (2001). The copigmentation of anthocyanins and its role in the color of red wine: A critical review. *Am. J. Enol. Vitic.*, **52**, 67–87.

Boyle, E.R., Guerriero, N., Thiallet, A., Del Vecchio, R., and Blough, N. (2009). Optical properties of humic substances and CDOM: Relation to structure. *Environ. Sci. Technol.*, **43**, 2262–2268.

Braslavsky, S.E. (2007). Glossary of terms used in photochemistry. *IUPAC Recommendations*, **3**rd ed., *Pure Appl. Chem.*, **79**, 293–465.

Brown, B.A., McColl, J.G., and Driscoll, C.T. (1990). Aluminum speciation using morin: I. Morin and its complexes with aluminum. *J. Environ. Qual.*, **19**, 65–72.

Bublitz, W.J. and Meng, T.Y. (1978). The fluorometric behavior of pulping waste liquors. *Tappi*, **61**, 27–30.

Buerge, I.J., Poiger, T., Müller, M.D., and Buser, H.R. (2003). Caffeine, an anthropogenic marker for wastewater contamination of surface waters. *Environ. Sci. Technol.*, **37**, 691–700.

Cabaniss, S.E. and Shuman, M.S. (1987). Synchronous spectra of natural waters: tracing sources of dissolved organic matter. *Mar. Chem.*, **21**, 37–50.

Cabaniss, S.E. and Shuman, M.S. (1988). Fluorescence quenching measurements of copper-fulvic acid binding. *Anal. Chem.*, **60**, 2418–2412.

Cha, K-W. and Park, K-W. (1988). Determination of iron (III) with salicylic acid by the fluorescence quenching method. *Talanta*, **46**, 1567–1571.

Chen, S., Inskeep, W.P., Williams, S.A., and Callis, P.R. (1994). Fluorescence lifetime measurements of fluyoranthene, 1-naphthol, and napropamide in the presence of dissolved humic acid. *Environ. Sci. Technol.*, **28**, 1582–1588.

Chin, Y.P. (2003). The speciation of hydrophoboic organic compounds by dissolved organic matter. In S.E.G. Findlay and R.L. Sinsabaugh (Eds.), *Aquatic Systems: Interactivity of Dissolved Organic Matter* (pp. 343–362). San Diego, CA: Academic Press.

Chin, Y., Aiken, G., and O'Loughlin, E. (1994). On the molecular weight, polydispersity and spectroscopic properties of aquatic humic substances. *Environ. Sci. Technol.*, **28**, 1853–1858.

Christman, R.F. and Minear, R.A. (1967). Fluorometric detection of lignin sulfonates. *Trends Eng.*, **19**, 3–7.

Coble, P.G. (1996). Characterization of marine and terrestrial DOM in seawater using excitation-emission matrix spectroscopy. *Mar. Chem.*, **51**, 325–346.

Coble, P.G. (2007). Marine optical biogeochemistry: The chemistry of ocean color. *Chem Rev*, **107**, 402–418.

Coble, P.G., Green, S., Blough, N.V., and Gagosian, R.B. (1990). Characterization of dissolved organic matter in the Black Sea by fluorescence spectroscopy. *Nature*, **348**, 432–435.

Connan, S., Goulard, F., Stiger, V., Deslandes, E., and Gall, E.A. (2004). Interspecific and temporal variation in phlorotannin levels in an assemblage of brown algae. *Botan. Mar.*, **47**, 410–416.

Cory, R.M. (2005). *Redox and Photochemical Reactivity of Dissolved Organic Matter in Surface Waters*. PhD thesis, University of Colorado, Boulder, CO, 252.

Cory, R.M. and McKnight, D.M. (2005). Fluorescence spectroscopy reveals ubiquitous presence of oxidized and reduced quinones in dissolved organic matter. *Environ. Sci. Technol.*, **39**, 8142–8149.

Cowie, G.L. and Hedges, J.I. (1992). Improved amino acid quantification in environmental samples: Charged matched recovery standards and reduced analysis time. *Mar. Chem.*, **37**, 223–238.

da Silva, P.F., Lima, J.C., Freitas, A.A., Shimizu, K., Maçanita, A.L., and Quina, F.H. (2005). Charge-transfer complexation as a general phenomenon in the copigmentation of anthocyanins. *J. Phys. Chem. A*, **109**, 7329–7338.

Del Vecchio, R. and Blough, N.V. (2004). On the origin of the optical properties of humic substances, *Environ. Sci. Technol.*, **38**, 3885–3891.

Dittmar, T., Koch, B., Hertkorn, N., and Kattner, G. (2008). A simple and efficient method for the solid-phase extraction of dissolved organic matter (SPE-DOM) from seawater. *Limnol. Oceanogr. Methods*, **6**, 230–235.

Doane, T.A. and Horwáth, W.R. (2010). Eliminating interference from iron(III) for ultraviolet absorbance measurements of dissolved organic matter. *Chemosphere*, **78**, 1409–1415.

Downing, B.D., Boss, E., Bergamaschi, B.A., Fleck, J.A., Lionberger, M.A., Ganju, N.K., Schoellhamer, D.H., and Fujii, R. (2009). Quantifying fluxes and characterizing compositional changes of dissolved organic matter in aquatic systems in situ using combined acoustic and optical measurements. *Limnol. Oceanogr. Methods*, **7**, 119–131.

Drabent, R., Pliszka, B., Huszcza-Ciolkowska, G., and Smyk, B. (2007). Ultraviolet fluorescence of cyaniding and malvidin glycosides in aqueous environment. *Spectrosc. Lett.*, **40**, 165–182.

D'Souza, F. and Deviprasad, C.R. (2001). Studies on porphyrin-quinhydronde complexes: Molecular recognition of quinine and hydroquinone in solution. *J. Org. Chem.*, **66**, 4601–4609.

Elliot, S., Lead, J.R., and Baker, A. (2006). Characterization of the fluorescence from freshwater, planktonic bacteria. *Water Res.*, **40**, 2075–2083.

Ertel, J.R., Hedges, J.I., Devol, A.H., Richey, J.E., and De Nazare Goes Ribeiro, M. (1986). Dissolved humic substances in the Amazon River system, *Limnol. Oceanogr.*, **31**, 739–754.

Eticha, D., Stass, A., and Horst, W.J. (2005). Localization of aluminum in the maize root apex: Can morin detect cell wall-bound aluminum? *J. Exp. Bot.*, **56**, 1351–1357.

Fatin-Rouge, N. and Buffle, J. (2007). Study of environmental systems by means of fluorescence correlation spectroscopy. In K.J. Wilkinson and J.R. Lead (Eds.), *Environmental Colloids and Particles, IUPAC Series on Analytical and Physical Chemistry of Environmental Systems*, Vol. 10 (pp. 507–553). West Sussex, UK: John Wiley & Sons.

Fellman, J.B., Hood, E., D'Amore, D.V., Edwards, R.T., and White, D. (2009). Seasonal changes in the chemical quality and biodegradability of dissolved organic matter exported from soils to streams in coastal temperate rainfall watersheds. *Biogeochemistry*, **95**, 277–293.

Fellman, J.B., Hood, E., and Spencer, R.G.M. (2010). Fluorescence spectroscopy opens new windows into dissolved organic matter dynamics of freshwater ecosystems: A review. *Limnol. Oceanogr.*, **55**, 2452–2462.

Figueiredo, P., Pina, F., Vilas-Boas, L., and Maçanita, A.L. (1990). Fluorescence spectra and decays of malvidin 3,5-diglucoside in aqueous solutions. *Photochem. Photobiol.*, **52**, 411–424.

Fimmen, R.L., Cory, R.M., Chin, Y-P, Trouts, T.D., and McKnight, D.M. (2007). Probing theoxidation-reduction properties of terrestrially and microbially derived dissolved organic matter. *Geochim. Cosmochim. Acta*, **71**, 3003–3015.

Findlay, S. (2003). Bacterial response to variation in dissolved organic matter. In S.E.G. Findlay and R.L. Sinsabaugh (Eds.), *Aquatic Systems: Interactivity of Dissolved Organic Matter (pp.* 363–379). San Diego, CA: Academic Press.

Fink, D.W. and Koehler, W.R. (1970). pH effects on fluorescence of umbelliferone. *Anal. Chem.*, **42**, 990–993.

Flaherty, S., Wark, S., Street, G., Farley, J.W., and Brumley, W.C. (2002). Investigation of CE/LIF as a tool in the characterization of sewage effluent for fluorescent acidics: Determination of salicylic acid. *Electrophoresis*, **23**, 2327–2332.

Furman, G.S. and Lonsky, W.F.W. (1988). Charge-transfer complexes in Kraft lignin part 2: Contribution to color. *J. Wood Chem. Technol.*, **8**, 191–208.

Garcia-Sanchez, F., Carnero, C., and Heredia, A. (1988). Fluorometric determination of p-coumaric acid in beer. *J. Food Agric. Chem.*, **36**, 80–82.

Gasper. J.D., Aiken, G.R., and Ryan, J.N. (2007). A critical review of three methods used for the measurement of mercury (Hg^{+2}) dissolved organic matter stability constants. *Appl. Geochem.*, **22**, 1583–1597.

Gauthier, T.D., Shane, E.C., Guerin, W.F., Seitz, W.R., and Grant, C.L. (1986). Fluorescence quenching method for determining equilibrium constants for polycyclic aromatic hydrocarbons binding to dissolved humic materials. *Environ. Sci. Technol.*, **20**, 1162–1166.

Geiselman, J.A. and McConnell, O.J. (1981). Polyphenols in brown algae *Fucus vesiculosus* and *Ascophyllum nodosum*: Chemical defenses against the marine herbivorous snail, *Littorina littorea*. *J. Chem. Ecol.*, **7**, 1115–1133.

Ghosh, K. and Schnitzer, M. (1980). Fluorescence excitation spectra of humic substances. *Can. J. Soil Sci.*, **60**, 373–379.

Goldberg, M.C. and Weiner, E.R. (1993). Fluorescence spectroscopy in environmental and hydrological sciences. In O.S. Wolfbeis (Ed.), *Fluorescence Spectroscopy – New Methods and Applications* (pp. 211–241). New York: Springer-Verlag.

Green, J.P. and Malrieu, J.P. (1965). Quantum chemical studies of charge-transfer complexes of indoles. *Proc. Natl. Acad. Sci. USA*, **54**, 659–664.

Green S.A., and Blough, N. V. (1994). Optical absorption and fluorescence properties of chromophoric dissolved organic matter in natural waters. *Limnol. Oceanogr.* **39**, 1903–1916.

Green, S.A., Morel, F.M.M., and Blough, N.V. (1992). Investigation of the electrostatic properties of humic substances by fluorescence quenching. *Environ. Sci. Technol.*, **26**, 294–302.

Gu, J. and Berry, D.F. (1991). Degradation of substituted indoles by an indole-degrading methanogenic consortium. *Appl. Environ. Microbiol.*, **57**, 2622–2627.

Haitzer, M., Aiken, G.R., and Ryan, J.N. (2002). Binding of mercury (II) to dissolved organic matter: The role of mercury to DOM concentration ratio. *Environ. Sci. Technol.*, **36**, 3564–3570.

Hall, G.J. and Kenny, J.E. (2007). Estuarine water classification using EEM spectroscopy and PARAFAC-SIMCA. *Anal. Chim. Acta*, **581**, 118–124.

Hall, G.J., Clow, K.E., and Kenny, J.E. (2005). Estuarial fingerprinting through multidimensional fluorescence and multivariate analysis. *Environ. Sci. Technol.*, **39**, 7560–7567.

Harden, S.L. (2009). Reconnaissance of organic wastewater compounds at a concentrated swine feeding operation in the North Carolina Coastal Plain, 2008, U.S. Geological Survey Open-File Report 2009–1128, 14 p.

Hartley, W.N. (1893). Observations on the origin of colour and of fluorescence. *J. Chem. Soc.*, **63**, 243–256.

Helms, J.R., Stubbins, A., Ritchie, J.D., Minor, E.C., Kieber, D.J., and Mopper, K. (2008). Absorption spectral slopes and slope ratios as indicators of molecular weight, source, and photobleaching of chromophoric organic matter. *Limnol. Oceanogr.*, **53**, 955–969.

Henderson, R.K., Baker, A., Murphy, K., Hambly, A., Stuetz, R.M., and Khan, S.J. (2009). Fluorescence as a potential monitoring tool for recycled water systems: A review. *Water Res.*, **43**, 863–881.

Hernes, P.J. and Hedges, J.I. (2000). Determination of condensed tannin monomers in environmental samples by capillary gas chromatography of acid depolymerization extracts. *Anal. Chem.*, **72**, 5115–5124.

Hernes, P.J., Bergamaschi, B.A., Eckard, R.S., and Spencer, R.G.M. (2009). Fluorescence-based proxies for lignin in freshwater dissolved organic matter. *J. Geophys. Res.*, **114**, doi:10.1029/2009JG000938.

Higuchi, T. (1980). Lignin structure and morphological distribution in plant cells walls. In Kirk, T.K., Higuchi, T., and Chang, H. (Eds.), *Lignin Biodegradation: Microbiology, Chemistry, and Potential Applications* (pp. 1–20). Boca Raton, FL: CRC Press.

Hoch, A.R., Reddy, M.M., and Aiken, G.R. (2000). Calcite crystal growth inhibition by humic substances with emphasis on hydrophobic acids from the Florida Everglades. *Geochim. Cosmochim. Acta*, **64**, 61–72.

Hoge, F.E., Vodacek, A., and Blough, N.V. (1993). Inherent optical properties of the ocean: Retrieval of the absorption coefficient of chromophoric dissolved organic matter from fluorescence measurements. *Limnol. Oceanogr.*, **38**, 1394–1402.

Hu, C., Muller-Karger, F.E., and Zepp, R.G. (2002). Absorbance, absorption coefficient, and apparent quantume yield: A comment on common ambiguity in the use of these concepts. *Limnol. Oceanogr.*, **47**, 1261–1266.

Hudson, N., Baker, A., and Reynolds, D. (2007). Fluorescence analysis of dissolve organic matter in natural, waste and polluted waters – a review. *River Res. Applic.*, **23**, 631–649.

Isenberg, I. and Szent-Györgyi, A. (1959). On charge transfer complexes between substances of biochemical interest. *Proc. Natl. Acad. Sci. USA*, **45**, 1229–1231.

Jiang, F., Lee, F.S., Wang, X., and Dai, D. (2008). The application of excitation/emission matrix spectroscopy combined with multivariate analysis for the characterization and source identification of dissolved organic matter in seawater of Bohai Sea, China. *Mar. Chem.*, **110**, 109–119.

Josefsson, B. and Nyquist, G. (1976). Fluorescence tracing of the flow and dispersion of sulfite wastes in a fjord system. *Ambio*, **5**, 183–187.

Kang, J., Liu, Y., Xie, M-X., Li, S., Jiang, M., and Wang, Y-D. (2004). Interactions of human serum albumin with chlorogenic acid and ferulic acid. *Biochim. Biophys. Acta*, **1674**, 205–214.

Karanfil, T., Erdogan, I., and Schlautman, M. (2005). The impact of filtrate turbidity on UV_{254} and $SUVA_{254}$ determinations. *J. Am. Water Works Assoc.*, **97**, 125–136.

Karim, M.M., Lee, H.S., Kim, Y.S., Bae, H.S., and Lee, S.H. (2006). Analysis of salicylic acid based on the fluorescence enhancement of the As(III)-salicylic acid system. *Anal. Chim. Acta*, **576**, 136–139.

Kelly, R.N. and Schulman, S.G. (1988). Proton transfer kinetics of electronically excited acids and bases. In S.G. Schulman (Ed.), *Molecular Luminescence Spectroscopy – Methods and Applications: Part 2* (pp. 461–510). New York: John Wiley & Sons.

Kirk, T.K., Higuchi, T., and Chang, H., Eds. (1980). *Lignin Biodegradation: Microbiology, Chemistry, and Potential Applications*, Vol. 1. Boca Raton, FL: CRC Press.

Klapper, L., McKnight, D.M., Fulton, J.R., Blunt-Harris, E.L., Nevin, K.P., Lovely, D.R., and Hatcher, P.G. (2002). Fulvic acid oxidation state detection using fluorescence spectroscopy. *Environ. Sci. Technol.*, **36**, 3170–3175.

Kraus, T.E.C., Yu, Z., Preston, C.M., Dahlgren, R.A., and Zasoski, R.J. (2003). Linking chemical reactivity and protein precipitation to structural characteristics of foliar tannins, J. *Chem. Ecol.*, **29**, 703–729.

Kraus, T.E.C., Bergamaschi, B.A., Hernes, P.J., Spencer, R.G.M., Stepanauskas, R., Kendall, C., Losee, R.F., and Fujii, R. (2008). Assessing the contribution of wetlands and subsided islands to dissolved organic matter and disinfection byproduct precursors in the Sacremento-San Joaquin River Delta: A geochemical approach. *Org. Geochem.*, **39**, 1302–1318.

Kriško, A., Kveder, M., Peáar, S., and Pifat, G. (2005). A study of caffeine binding to human serum albumin, *Croat. Chem. Acta*, **78**, 71–77.

Kujawinski, E.B., Longnecker, K., Blough, N.V., Del Vecchio, R., Finlay, L., Kitner, J.B., and Giovannoni (2009). Identification of possible markers in marine dissolved organic matter using ultrahigh resolution mass spectrometry. *Geochim. Cosmochim Acta*, **73**, 4384–4399.

Labieniec, M. and Gabryelak, T. (2006). Interactions of tannic acid and its derivatives (ellagic and gallic acid) with calf thymus DNA and bovine serum albumin using spectroscopic method. *J. Photochem. Photobiol B*, **82**, 72–78.

Lakowicz, J.R. (2006). *Principles of Fluorescence Spectroscopy*, 3rd ed. New York: Springer Science+Business Media, 954 pp.

Larsen, L.G., Aiken, G.R., Harvey, J.W., Noe, G.B., and Crimaldi, J.P. (2010). Using fluorescence spectroscopy to trace seasonal DOM dynamics, disturbance effects, and hydrologic transport in the Florida Everglades. *J. Geophys. Res.*, **115**, doi:10.1029/2009JG001140.

Larson, R.A. and Rockwell, A.L., (1980). Fluorescence spectra of water-soluble humic materials and some potential precursors. *Arch. Hydrobiol.*, **89**, 416–425.

Lead, J.R. and Wilkinson, K.J. (2007). Environmental colloids and particles: Current knowledge and future developments. In K.J. Wilkinson and J.R. Lead (Eds.), *Environmental Colloids and Particles – Behaviour, Separation and Characterisation* (pp. 1–16). IUPAC Series on Analytical and Physical Chemistry of Environmental Systems, Vol. 10. Chichester, England: John Wiley & Sons.

Lead, J.R., Wilkinson, K.J., Balnois, E., Cutak, B.J., Larive, C.A., Assemi, S., and Beckett, R., (2000). Diffusion coefficients and polydispersities of the Suwannee River fulvic acid: Comparison of fluorescence correlation spectroscopy, pulsed field gradient nuclear magnetic resonance, and field-flow fractionation. *Environ. Sci. Technol.*, **34**, 3508–3513.

Leenheer, J.A., Wilson, M.A., and Malcolm, R.L. (1987). Presence and potential significance of aromatic-ketone groups in aquatic humic substances, *Org. Geochem.*, **11**, 273–280.

Lehninger, A.L., 1970, *Biogeochemistry*, New York: Worth, 833 pp.

Liu, X., Tao, S., Deng, N., Liu, Y., Meng, B., Xue, B., and Liu, G. (2006). Synchronous-scan fluorescence as a selective detection method for sodiumdodecybenzene-sulfonate and pyrene in environmental samples. *Anal. Chem. Acta*, **572**, 134–139.

Lundquist, K., Josefsson, B., and Nyquist, G. (1978). Analysis of lignin products by fluorescence spectroscopy. *Holzforschung*, **32**, 27–32.

Ma, J., Del Vecchio, R., Golanoski, K.S., Boyle, E.S., and Blough, N.V. (2010). Optical properties of humic substances and CDOM: Effects of borohydride reduction. *Environ. Sci. Technol.*, **44**, 5395–5402.

MacCalady, D.L. and Walton- Day, K. (2009). New light on a dark subject: On the use of fluorescence data to deduce redox states of natural organic matter (NOM). *Aquat. Sci.*, **71**, 135–143.

MacCarthy, P. and Rice, J. A. (1985). Spectroscopic methods (other than NMR) for determining functionality in humic substances. In G. Aiken, D. McKnight, and R. Wershaw (Eds.), *Humic Substances in Soil, Sediment, and Water: Geochemistry, Isolation and Characterization* (pp. 527–559). New York: John Wiley & Sons.

Machado, A.E.H., Nicodem, D.E., Ruggiero, R., Perez, D.S., and Castellan, A. (2001). The use of fluorescent probes in the characterization of lignin: The distribution, by energy, of fluorophores in *Eucalyptus grandis* lignin. *J. Photochem. Photobiol. A: Chem.*, **138**, 253–259.

Maie, N., Scully, N.M., Pisani, O., and Jaffe, R. (2007). Composition of protein-like fluorophore of dissolved organic matter in coastal wetland and estuarine systems, *Water Res.*, **41**, 563–570.

March, J. (1968). *Advanced Organic Chemistry: Reactions, Mechanisms and Structure*, New York: McGraw Hill, 1098 pp.

Martin, R. and Clarke, G.A. (1978). Fluorescence of benzoic acid in aqueous acidic media. *J. Phys. Chem.*, **82**, 81–86.

Mauer, F., Christl, I., and Kretzscmar, R. (2010). Reduction and reoxidation of humic acid: Influence on spectroscopic properties and proton binding. *Environ. Sci. Technol.*, **44**, 5787–5792.

Mayer, L.M, Schick, L.L., and Loder, T.C. (1999). Dissolved fluorescence in two Maine estuaries, *Mar. Chem.* **64**, 171–179.

McKnight, D. M. and Aiken, G. R. (1998). Sources and age of aquatic humic substances. In D. Hessen (Ed.), *Aquatic Humic Substances: Ecology and Biogeochemistry* (pp. 9–40). Ecological Studies, Vol. 133. New York: Springer-Verlag.

McKnight, D.M., Boyer, E.W., Westerhoff, P.K., Doran, P.T., Kulbe, T., and Anderson, D.T. (2001). Spectrofluorometric characterization of dissolved organic matter for indication of precursor material and aromaticity. *Limnol. Oceanogr.*, **46**, 38–48.

Miller, W.L. (1994). Recent advances in the photochemistry of natural dissolved organic matter. In G.R. Helz, R.G. Zep, and D.G. Crosby (Eds.), *Aquatic and Surface Photochemistry* (pp. 111–127). Ann Arbor, MI: Lewis Publishers.

Miller, M.P., McKnight, D.M., Cory, R.M., Williams, M.W., and Runkel, R.L. (2006). Hyporheic exchange and fulvic acid redox reactions in an alpine stream/ecosystem, Colorado Front Range, *Environ. Sci. Technol.*, **40**, 5943–5949.

Miller, M.P., Simone, B.E., McKnight, D.M., Cory, R.M., Williams, M.W., and Boyer, E.W. (2010). New light on a dark subject: Comment. *Aquat. Sci.*, **72**, 269–275.

Min, J., Meng-Xia, X., Dong, Z., Yuan, L., Xiao-Yu, L., and Xing, C. (2004). Spectroscopic studies on the interaction of cinamic acid and its derivatives with human serum albumin. *J. Mol. Struct.*, **692**, 71–80.

Mishra, B., Barik, A., Priyadarsini, I., and Mohan, H. (2005). Fluorescent spectroscopic studies on binding of a flavonoid antioxidant quercitin to serum albumins. *J. Chem. Soc.*, **117**, 641–647.

Mobed, J.J., Hemmingsen, S.L., Autry, J.L., and McGown, L.B. (1996). Fluorescence characterization of IHSS humic substances: Total luminescence spectra with absorbance correction, *Environ. Sci. Technol.*, **30**, 3061–3065.

Moran, M.A. and Covert, J.S. (2003). Photochemically mediated linkages between dissolved organic matter and bacterioplankton. In S.E.G. Findlay and R.L. Sinsabaugh (Eds.), *Aquatic Systems: Interactivity of Dissolved Organic Matter* (pp. 244–262). San Diego, CA: Academic Press.

Moreira, P.F., Giestas, L., Yihwa, C., Vautier-Giongo, C., Quina, F.H., Maçanita, A.L., and Lima, J.C. (2003). Ground- and excited state proton transfer in anthocyanins: From weak acids to superphotoacids. *J. Phys. Chem.*, **107**, 4203–4210.

Morel, F.M.M. and Gschwend, P.M. (1987). The role of colloids in the partitioning of solutes in natural waters. In W. Stumm (Ed.), *Aquatic Surface Chemistry* (pp. 405–422). New York: John Wiley & Sons.

Murphy, K. R., Ruiz, G.M., Dunsmuir, W.T.M., and Waite, T.D. (2006). Optimized parameters for rapid fluorescence-based verification of ballast water exchange by ships. *Environ. Sci. Technol.*, **40**, 2357–2362.

Murphy, K.R., Butler, K.D., Spencer, R.G.M., Stedmon, C.A., Boehme, J.R., and Aiken, G.R. (2010). Measurement of dissolved organic matter fluorescence in aquatic environments: An interlaboratory comparison. *Environ. Sci. Technol.*, **44**, 9405–9412.

Nurmi, J.T. and Tratnyek, P.G. (2002). Electrochemical properties of natural organic matter (NOM), fractions of NOM, and model biogeochemical electron shuttles. *Environ. Sci. Technol.*, **36**, 617–624.

Ohno, T., Amirbahman, A., and Bro, R. (2008). Parallel factor analysis of excitation-emission matrix fluorescence spectra of water soluble organic matter as basis for the determination of conditional metal binding parameters. *Environ. Sci. Technol.*, **42**, 186–192.

Olmstead, J.A. and Gray, D.G. (1997). Fluorescence spectroscopy of cellulose, lignin and mechanical pulps: A review. *J. Pulp Paper Sci.*, **23**, J571–J581.

Pellerin, B.A., Saracen, J.F., Shanley, J.B., Sebestyn, S.D., Aiken, G.R., Wollheim, W.M., and Bergamaschi, B.A. (2012). Taking the pulse of snowmelt: In-situ sensors reveal seasonal, event and diurnal patterns of nitrate and dissolved organic matter variability in an upland forest stream, *Biogeochemistry*, **108**, 183–198, doi:10.1007/s10533–011–9589–8..

Perdue, E.M. (1998). Chemical composition, structure and metal binding properties. In D. Hessen (Ed.), *Aquatic Humic Substances: Ecology and Biogeochemistry* (pp. 41–62). Ecological Studies, Vol. 133. New York: Springer-Verlag.

Persson, T. and Wedborg, M. (2001). Multivariate evaluation of the fluorescence of aquatic organic matter. *Anal. Chim. Acta*, **434**, 179–192.

Poulson, J.R. and Birks, J.W. (1989). Photoreduction fluorescence detection of quinones in high-performance liquid chromatography. *Anal. Chem.*, **61**, 2267–2276.

Pullman, B. and Pullman, A. (1958). Electron-donor and -acceptor properties of biologically important purines, pyrimidines, pteridines, flavins, and aromatic amino acids. *Proc. Natl. Acad. Sci.USA*, **44**, 1197–1202.

Putschögl, M., Zirak, P., and Penzkofer, A. (2008). Absorption and emission behavior of *trans*-p-coumaric acid in aqueous solutions and some organic solvents. *Chem. Phys.*, **343**, 107–120.

Qualls, R.G. and Richardson, C.J. (2003). Factors controlling concentration, export and decomposition of dissolved organic nutrients in the Everglades, Florida. *Biogeochemistry*, **62**, 197–229.

Quina, F.H., Moreira, P., Vautier-Giongo, C., Rettori, D., Rodigues, R.F., Freitas, A.A., Silva, P.F., and Maçanita, A.L. (2009). Photochemistry of anthocyanins and their biological role in plant tissues. *Pure Appl. Chem.*, **81**, 1687–1694.

Radotić, K., Kalauzi, A., Djikanović, D., Jeremić, M., Leblanc, R.M., and Cerović, Z.G. (2006). Component analysis of the fluorescence spectra of a lignin model compound. *J. Photochem. Photobiol. B. Biol.*, **83**, 1–10.

Reynolds, D.M. (2003). Rapid and direct determination of tryptophan in water using synchronous fluorescence spectroscopy, *Water Res.*, **37**, 3055–3060.

Robinson, T. (1991). *The Organic Constituents of Higher Plants*, 6th ed. North Amherst, MA: Cordus Press, 346 p.

Saar, R.A. and Weber, J.H. (1980). Comparison of spectrofluorometry and ion-selective electrode potentiometry for determination of complexes between fulvic acid and heavy metal ions. *Anal. Chem.* **52**, 2095–2100.

Santhanam, M., Hautala, R.R., Sweeny, J.G., and Iacobucci, G.A. (1983). The influence of flavanoid sulfonates on the fluorescence and photochemistry of flavylium cations. *Photochem. Photobiol.*, **38**, 477–480.

Santos, E.B.H., Filipe, O.M.S., Duarte, R.M.B.O., Pinto, H., and Duarte, A.C. (2000). Fluorescence as a tool for tracing the organic contamination from pulp mill effluents in surface waters. *Acta Hydrochim. Hydrobiol.*, **28**, 364–371.

Saraceno, J.F., Pellerin, B.A., Downing, B.D., Boss, E., Bachand, P.A.M., and Bergamaschi, B.A. (2009). High-frequency in situ optical measurements during a storm event: Assessing relationships between dissolved organic matter, sediment concentrations, and hydrologic processes. *J. Geophys. Res. Biogeosci.* **114**, doi: 10.1029/2009JG000989

Schulman, S.G. (1985). Luminescence spectroscopy: An overview. In S.G. Schulman (Ed.), *Molecular Luminescence Spectroscopy – Methods and Applications; Part 1* (pp. 1–28). New York: John Wiley & Sons.

Scott, D.T., McKnight, D.M., Blunt-Harris, E.L., Kolesar, S.E., and Lovley, D.R. (1998). Quinnone moieties act as electron acceptors in the reduction of humic substances by humic-reducing microorganisms. *Environ. Sci. Technol.*, **32**, 2984–2989.

Senesi, N., Miano, T.M., Provenzano, M.R., and Brunetti, G. (1991). Characterization, differentiation, and classification of humic substances by fluorescence spectroscopy. *Soil Sci.*, **152**, 259–271.

Sharma, A. and Schulman, S.G. (1999). *Introduction to Fluorescence Spectroscopy*, New York: John Wiley & Sons, 173 pp.

Siegel, D.A., Maritorena, S., Nelson, N.B., Behrenfield, M.J., and McClain, C.R. (2005). Colored dissolved organic matter and its influence on the satellite-based characterization of the ocean biosphere. *Geophys. Res. Lett*, **32**, L20605.

Silverstein, R.M., Bassler, G. C., and Morrill, T.C. (1974). *Spectrometric Identification of Organic Compounds*, 3rd ed. New York: John Wiley & Sons.

Singer, P.C. (1994). Control of disinfection by-products in drinking water. *J. Environ. Engng. ASCE*, **120**, 727–744.

Skoog, D.A. and West, D.M. (1982). *Fundamentals of Analytical Chemistry*, New York: Saunders College, 859 pp.

Sleighter, R.L., Liu, Z., Xue, J., and Hatcher, P.G. (2010). Multivariate statistical approaches for the characterization of dissolved organic matter analyzed by ultrahigh resolution mass spectrometry. *Environ. Sci. Technol.*, **44**, 7576–7582.

Spencer R.G.M., Baker A., Ahad J.M.E., Cowie G.L., Ganeshram R., Upstill-Goddard R.C., and Uher G. (2007a). Discriminatory classification of natural and anthropogenic waters in two U.K. estuaries. *Sci. Tot. Environ.*, **373**, 305–323.

Spencer, R.G.M., Bolton, L., and Baker, A. (2007b). Freeze/thaw and pH effects of freshwater dissolved organic matter fluorescence and absorbance properties from a number of UK locations. *Water Res.*, **41**, 2941–2950

Spencer, R.G.M., Aiken, G.R., Butler, K.D., Dornblaser, M.M., Striegl, R.G., and Hernes, P.J. (2009). Utilizing chromophoric dissolved organic matter measurements to derive export and reactivity of dissolved organic carbon exported to the Arctic Ocean: A case study of the Yukon River, Alaska. *Geophys. Res. Lett.* **36**, L06401.

Stedmon, C.A., Markager, S., and Bro, R. (2003). Tracing dissolved organic matter in aquatic environments using a new approach to fluorescence spectroscopy. *Mar. Chem.*, **82**, 239–254.

Stevenson, F.J. (1985). Geochemistry of soil humic substances. In G. Aiken, D. McKnight, and R. Wershaw (Eds.), *Humic Substances in Soil, Sediment, and Water: Geochemistry, Isolation and Characterization* (pp. 13–52). New York: John Wiley & Sons.

Stubbins, A., Hubbard, V., Uher, G., Law, C.S., Upstill-Goddard, R.C., Aiken, G.R., and Mopper, K. (2008). Relating carbon monoxide photoproduction to dissolved organic matter functionality. *Environ. Sci. Technol.*, **42**, 3271–3276.

Szent-Györgyi, A., Isenberg, I., and McLaughlin, J. (1961). Local and π-π interactions in charge transfer. *Proc. Natl. Acad. Sci. USA*, **47**, 1089–1093.

Tertuliani, J.S., Alvarez, D.A., Furlong, E.T., Meyer, M.T., Zaugg, S.D., and Koltun, G.F. (2008). Occurrence of organic wastewater compounds in the Tinkers Creek watershed and two other tributaries in the Cayahoga River, Northeast Ohio, U.S. Geological Survey Scientific Investigations Report 2008–5173, 60 pp.

The Merck Index: An Encyclopedia of Chemicals, Drugs, and Biologicals, 12th ed. (1996). Budavari, S., O'Neal, M.J., Smith, A., Heckelman, P.E., and Kinneary, J.F. (Eds.). Whitehouse Station, NJ: Merck & Co., 1983, entry 4993.

Thorn, K.A., Arterburn, J.B., and Mikita, M.A. (1992). ^{15}N and ^{13}C NMR investigation of hydroxylamine-derivatized humic substances. *Environ. Sci. Technol.*, **26**, 107–116.

Thruston, A.D. (1970). A fluorometric method for the determination of lignin sulfonates in natural waters. *J. Water Pollut. Control Fed.*, **42**, 1551–1555.

Thurman, E.M. (1985). *Organic Geochemistry of Natural Waters*. Dordrecht, The Netherlands: Martinus Nijhoff/ Dr W. Junk Publishers, 497 pp.

Town, R.M., and Fillela, M. (2002). Implications of natural organic matter binding heterogeneity on understanding lead(II) complexation in aquatic systems. *Sci. Tot. Environ.*, **300**, 143–154.

Tranvik, L.J. (1998). Degradation of dissolved organic matter in humic waters by bacteria. In D. Hessen (Ed.), *Aquatic Humic Substances: Ecology and Biogeochemistry* (pp. 259–283). Ecological Studies, Vol. 133. New York: Springer-Verlag..

Tucker, S.A., Amszi, V.L., and Acree, W.E. (1992). Primary and secondary inner filtering. *J. Chem. Educ.*, **69**, A8–A12.

Twardowski, M.S., Boss, E., Sullivan, J.M., and Donaghay, P.L. (2004). Modeling the spectral shape of absorption by chromophoric dissolved organic matter. *Mar. Chem..*, **89**, 69–88.

Vodacek, A. (1989). Synchronous fluorescence spectroscopy of dissolved organic matter in surface waters: Application to airborne remote sensing. *Remote Sens. Environ.*, **30**, 239–247.

Vodacek, A., Hoge, F., Swift, R.N., Yungel, J.K., Peltzer, E.T., and Blough, N.V. (1995). The use of in-situ and airborne fluorescence measurements to determine UV absorption coefficients and DOC concentrations in surface waters, *Limnol. Oceanogr.*, **40**, 411–415.

Waples, J.S., Nagy, K.L., Aiken, G.R., and Ryan, J.N. (2005). Dissolution of cinnabar (HgS) in the presence of natural organic matter. *Geochim. Cosmochim. Acta*, **69**, 1575–1588.

Weishaar, J.L., Aiken, G.R., Bergamaschi, B.A., Fram, M.S., and Fujii, R. (2003). Evaluation of specific ultra-violet absorbance as an indicator of the chemical composition and reactivity of dissolved organic carbon. *Environ. Sci. Technol.*, **37**, 4702–4708.

Westerhoff, P., Chem, W., and Esparza, M. (2001). Fluorescence analysis of a standard fulvic acid and tertiary treated wastewater. *J. Environ. Qual.*, **30**, 2037–2046.

Wetzel, R.G. (2001). *Limnology: Land and River Systems* 3rd ed. San Diego: Academic Press, 1006 pp.

Wigand, M.C., Dangles, O., and Brouillard, R. (1992). Complexation of a fluorescent anthocyanin with purines and polyphenols. *Phytochemistry*, **31**, 4317–4324.

Wilander, A., Kvarnäs, and Lindell, T. (1974). A modified fluorometric method for measurement of lignin sulfonates and its in situ application in natural waters. *Water Res.*, **8**, 1037–1045.

Wolfbeis, O.S. (1985). The fluorescence of organic natural products. In S.G. Schulman (Ed.), *Molecular Luminescence Spectroscopy – Methods and Applications; Part 1* (pp. 167–370). New York: John Wiley & Sons.

Wolfbeis, O.S., Begum, M., and Hochmuth, P. (1986). An unusual excited state species of *ortho*-hydroxy cinnamic acid. *Photochem. Photobiol.*, **44**, 551–554.

Yamashita, Y. and Tanoue, E. (2003). Chemical characterization of protein-like fluorophores in DOM in relation to aromatic amino acids. *Mar. Chem.*, **82**, 255–271.

Ziegmann, M., Abert, M., Müller, M., and Frimmel, F.H. (2010). Use of fluorescence fingerprints for the estimation of bloom formation and toxin production of *Microcystis aeruginosa*. *Water Res.*, **44**, 195–204.

3

Aquatic Organic Matter Fluorescence

PAULA G. COBLE, ROBERT G. M. SPENCER, ANDY BAKER, AND DARREN M. REYNOLDS

3.1 Introduction

The earliest reference to the fluorescence of natural waters is attributed to Dienert (1910), according to Smart et al. (1976). Almost three decades later, Kurt Kalle (Kalle 1938, 1939) made observations of seawater fluorescence using direct observation of samples under a microscope (Duursma, 1974). Kalle recognized that the source of this material, which he called "gelbstoff" ("yellow stuff"), was primarily from terrestrial inputs in coastal regions. He described it as being "humic-like" and having a yellow color and a blue fluorescence. An early list of substances found to have the properties of gelbstoff included rain, natural source water, Bordeaux wine, light Italian vermouth, light export beer, dark malt beer, melanin, bees' honey, house dust, air, rocks, filter paper, and brown lime-tree leaves (Duursma, 1974). Fluorescent organic matter has since been observed universally in natural waters, including in bottled drinking water and in the bluest open ocean regions.

Early studies lacked both the sensitivity and wavelength specificity required to fully characterize aquatic organic matter (AOM) fluorescence. Broad-band excitation and emission provided increased sensitivity to measure fluorescence intensity at low concentrations, but without wavelength resolution to determine peak positions. Smart et al. (1976) cited several literature values for excitation and emission maxima, however, results of early studies must be examined carefully for inaccuracies due to instrument biases, including errors in peak positions as well as in the number of peaks present in a sample. Many of these errors were not recognized in the 1970s and 1980s, but are the subject of detailed discussion in Chapter 5 of this volume (see also Holbrook et al., 2006).

The development of the modern spectrofluorometers, equipped with grating monochromators for wavelength resolution, xenon arc lamps for generating high-energy excitation into the ultraviolet C (UVC, 100–289 nm) region, and photomultiplier tubes to boost sensitivity, permitted high-resolution characterization of AOM fluorescence in the late 1980s. The first study to show fluorescence spectral differences between marine and terrestrial colored dissolved organic matter (CDOM) was that of Cabaniss and Shuman (1987) using a synchronous scanning technique. Donard et al. (1989) were able to make the same discrimination using high-resolution fluorescence spectroscopy of individual spectra. Coble et al. (1990) introduced the use of excitation-emission matrix (EEM) spectroscopy for the study of CDOM composition. In this technique, multiple emission scans are collected at different excitation wavelengths and combined to form a three-dimensional matrix of data. The ability of spectrophotometers to collect data from λ_{ex} = 220–455 nm and λ_{em} = 230–700 nm has also revealed the presence of additional fluorescent materials in many samples, both natural and anthropogenic in origin. Reported fluorescent components of AOM include proteins, tyrosine (Tyr), tryptophan (Trp), pigments, lignin phenols, humic substances, and hydrocarbons. This chapter provides a broad review of the current understanding of AOM fluorescent components, starting with a general description of peaks identified in samples from all environments. This is followed by brief summaries of the distribution and characteristics of fluorescence in discrete environments, including freshwater, seawater, groundwater, wastewater, and drinking water.

3.1.1 Peak Nomenclature

Traditional spectroscopic techniques use parameters such as position of excitation and emission wavelength maximum and quantum efficiency to characterize fluorescence properties of a compound. The fluorescence of AOM is complicated because it results from a mixture of compounds, some of which have overlapping excitation and emission spectra. The position of maximum fluorescence is not constant across environments, but rather shifts along both excitation and emission axes not only in response to variation in the relative amounts of fluorophores and thus CDOM chemistry, but also due to matrix effects, water content of the matrix, and changes due to alteration of the tertiary physicochemical structure of fluorophores during sampling and sample handling (Zsolnay, 2003). Fluorescence maxima also can shift with variation in solvent properties (e.g., ionic strength, pH; see Osburn et al., Chapter 7, this volume). The following discussion of peak nomenclature is applicable to all aquatic environments, as well as to soil-derived waters, but was developed from analysis of freshwater and marine samples.

The earliest peak nomenclature, and the one still most widely used is that of Coble et al. (1990), which denotes two peaks for humic-like fluorescence, peaks A and C, and one for tyrosine-like fluorescence, peak B. Coble (1996) introduced peaks T (tryptophan-like) and M (marine humic-like). A similar naming scheme was proposed by Parlanti et al. (2000). Since the introduction and expanding use of the multicomponent analysis technique parallel factor analysis (PARAFAC; Bro, 1997; Stedmon et al., 2003), peak nomenclature has evolved into a numbering scheme based on the output of the model. PARAFAC models have now been developed for diverse environments, both freshwater and marine, and the outputs have resulted in an ever-increasing number of peak designators.

In the following section, we discuss the fluorescence properties of AOM in terms of peaks observable in the spectrally corrected EEM data according to the scheme of Coble (1996), while at the same time attempting to reconcile the myriad of peak tables published in the past 20 years, with some speculation as to commonality among results. For the purpose of this chapter, we use the term "peak" in the context of spectroscopy practice as anything that exceeds the signal to noise of the background of the spectrum. The term "component" refers to inferred fluorophores or fluorophore groups that may have one or more peaks. Thus, the humic-like component has two peaks, C and A_C.

One of the difficulties in reconciling results is that early EEM studies did not collect data below 250 nm excitation, and most PARAFAC models require that data in this range be omitted owing to the presence of large molecular scattering peaks. Another difficulty is that previous tables of CDOM fluorescence components organized by peak name or number fail to represent adequately the double peak characteristics of both humic-like and amino acid-like components. In Table 3.1, we have therefore imposed a solution acknowledging that UVC peaks exist for peaks C, M, T, and B, and have designated them as A_C, A_M, A_T, and A_B, respectively. In so doing, PARAFAC and bulk EEM components are much more readily matched.

3.1.2 Humic-like EEM Components

Humic-like fluorescence is the dominant signal in most AOM and is due to the presence of humic substances that arise from remineralization of organic matter occurring in soils on land and in the water column and sediments in aquatic environments, both freshwater and marine. The chemical nature of humic substances varies across environments, and this changing composition is reflected in the EEM, making it extremely difficult to generalize about humic-like fluorescence in AOM. In general, two types of humic-like fluorescence have been described – type C and type M. Because type C is the one most commonly found in both freshwater and seawater, the discussion that follows begins with its description.

The EEM for a typical AOM sample is not characterized by the symmetrical peaks of pure fluorescent compounds (fluorophores), which appear round in the EEM contour view, but rather by what may be compared to an elephant under a blanket in EEM 3-D view. Figure 3.1 shows two views of EEMs for quinine sulfate dihydrate and water from the Columbia River in Oregon. On the left, quinine sulfate fluorescence in contour view shows a peak in the UVC region at λ_{ex} = 250 nm and a peak in the UVA (315–400 nm) region at λ_{ex} = 350, with a shoulder at 300 nm. The emission maximum occurs at 450 nm and, as for all pure fluorophores, is independent of excitation energy. The variability in emission is thus due solely to the width of the peak. For pure compounds, the emission intensity at any given wavelength pair has a constant ratio to the intensity of emission at the fluorescence maximum. In contrast, the contour plot of the UVA peak from natural organic matter in the Columbia River sample is not round, but rather oval, with the long axis of the oval at an angle to both excitation and emission axes. This indicates that emission wavelength is dependent on excitation wavelength, and therefore the fluorescence of the sample is not due to a pure compound. Peak width is not constant, but rather is dependent on the relative amounts of fluorophores comprising the mixture in the sample. Data for fluorescence intensity of natural organic matter samples thus require inclusion of the wavelength pair at which the measurement was recorded, and there is no a priori relationship between different points of the matrix.

Table 3.1. *Nomenclatures for fluorophores in natural waters, showing chemical component, position of peak fluorescence (excitation/emission maximum in nm), peak name from literature sources, dual peak component name, and environmental source of the material*

Component ex/em	Peak name	Dual peak component name	Source
Tyrosine-like protein-like			
230/305	A_B		Autochthonous, resembles tyrosine but may be free or combined amino acids
275/305	B[1],8[3], 6[4],Y[5],P2[6],P5[6]		
Tryptophan-like protein-like			Autochthonous
230/340	A_T	A[4]	
275/340	T[1], 7[3], S[5], P6[6], P7[6]		
Unknown			Autochthonous?
280/370	N[1]		
Humic–like M			
240/350–400	A_M	5[2], 6[3], 3[3], 2[4]	Autochthonous, microbial
290–310/370–420	M[1],3[4], P1[6], β[5]		
Humic–like C			
260/400–460	A_c, A[1], 1[2], 1[3], α[5]	Humic-like C, 4[2], 4[3], P8[6]	Humic, terrestrial, allochthonous
320–365/420–470	C[1],5[3],5[4],α[5]		
Humic–like C+			Humic, terrestrial, allochthonous
250/470–504		3[2], 2[3], 1[4], 7[4], P3[6]	
385–420/470–504			
Pigment–like			Phytoplankton
398/660	P[1]		
Photobleached			Autochthonous
230/275–350	H, P4[6]		

Note: Dual peak component names arise from PARAFAC models in which components have more than one peak

References: [1]Coble et al. (1998); [2]Stedmon et al. (2003); [3]Stedmon and Markager (2005); [4]Stedmon and Markager (2005a); [5]Parlanti et al. (2000); [6]Murphy et al. (2008).

In samples containing only humic substances, two peaks are observed, one from excitation around 260 nm (peak A_C, formerly peak A), and one from excitation around 340 nm (peak C). Peak C shows elongated contours that result from a shift toward longer wavelength emission as excitation wavelength increases. This fluorescence response is typical of aquatic humic substances and is attributed to several factors, perhaps the most important of which is that CDOM is composed of a mixture of individual fluorophores. It is likely

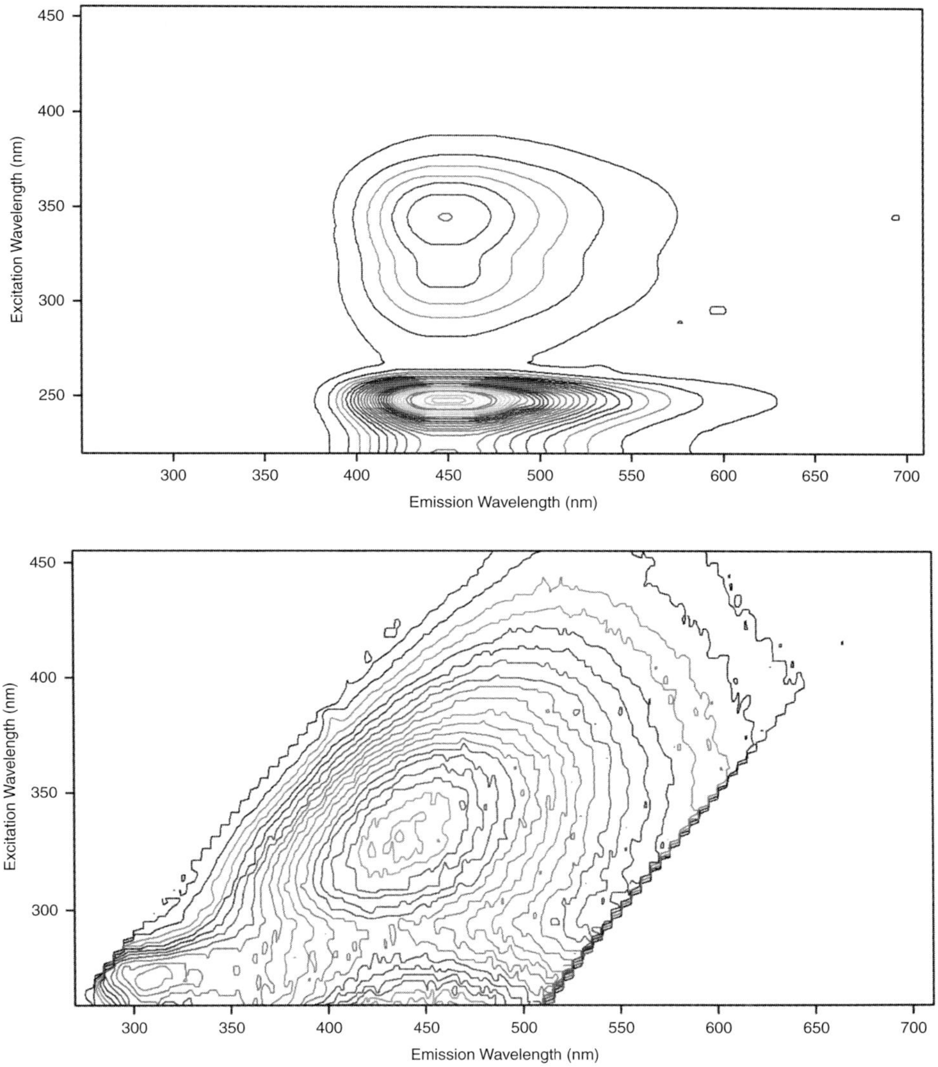

Figure 3.1a. EEMS for quinine bisulfate (top) and Columbia River water (bottom) shown in contour view.

that some of these are families of compounds that share a common fluorophore backbone but have different ring substitutions, giving them slightly different excitation and emission maxima. In addition, charge transfer reactions between compounds may explain the near continuum of excitation and emission in bulk water samples (Del Vecchio and Blough, 2004). Peak A_C has a wide range of emission but a narrow range of excitation. In the environment, peaks C and A_C are always observed together, although the ratio of the two peaks

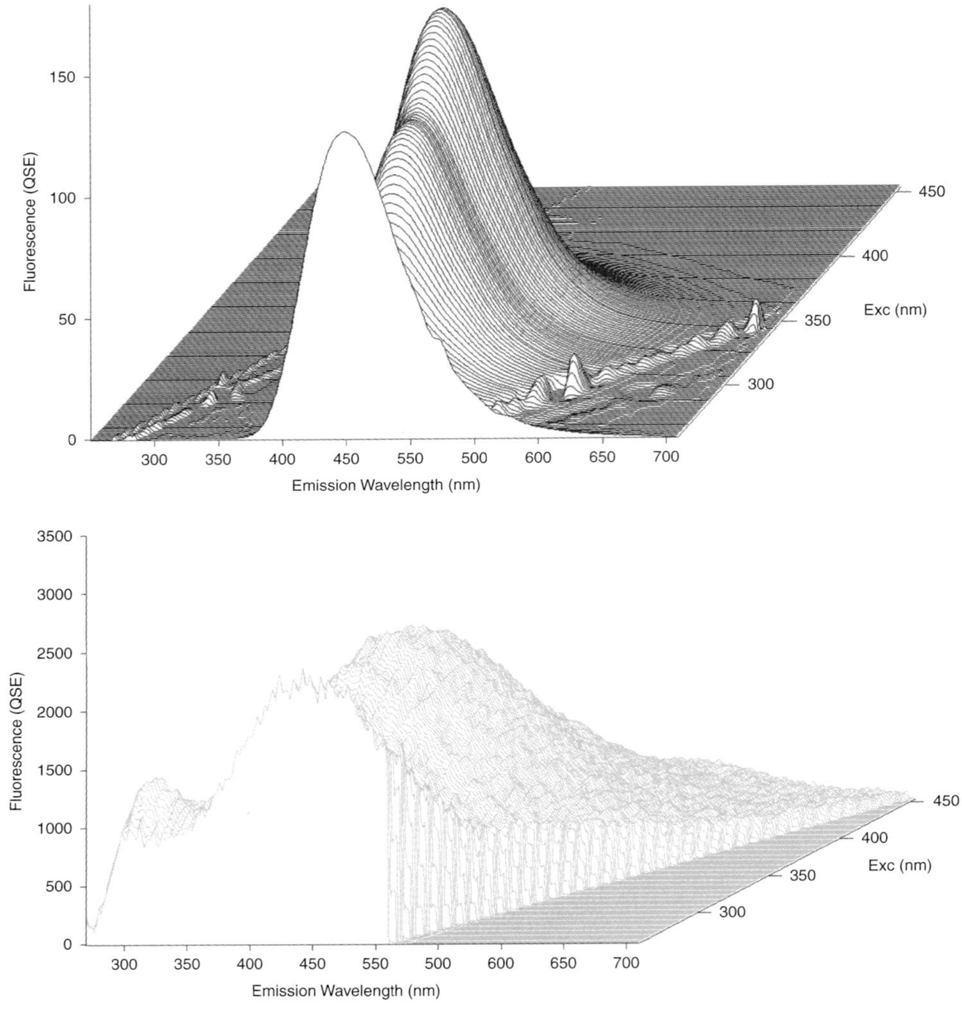

Figure 3.1b. EEMS for quinine bisulfate (top) and Columbia River water (bottom) shown in three-dimensional view.

exhibits some variability. For this reason, it is useful to specify a humic-like C component that includes both peaks.

Humic-like C component is the predominant type found in soils, most freshwater samples, rivers, and the deep ocean. The range in composition across these environments is reflected in the wide range of excitation and emission wavelengths given in Table 3.1 for the peak C maximum. Environments dominated by humic substances leached from soils tend to be the most complex pool observed. Across the range of environments, the position of the overall fluorescence maximum from the humic-like C component varies, with older or more terrestrial material showing the most red-shifted excitation and emission maxima (Figure 3.2). Some researchers working in peaty waters have designated subregions of peak

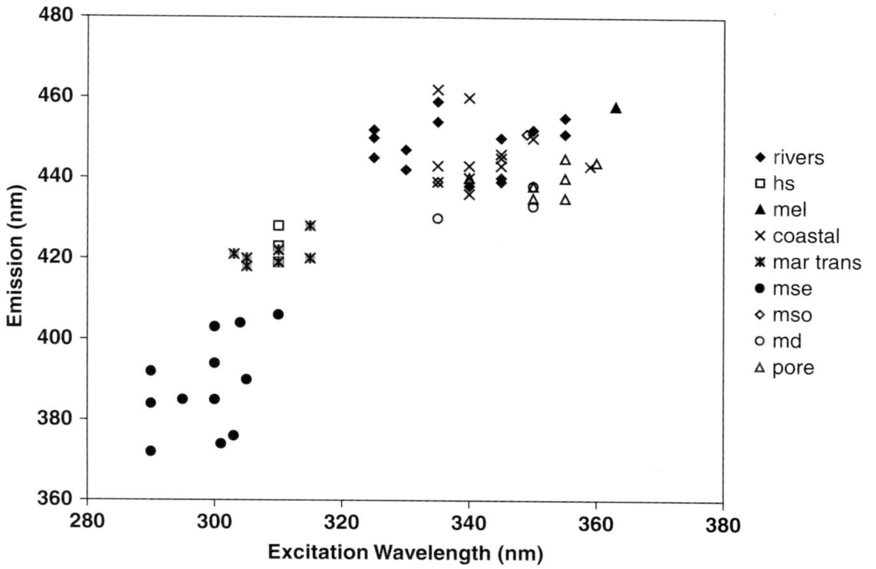

Figure 3.2. Position of overall fluorescence maximum showing dependence on sample origin. rivers = river waters, hs = humic substance extracts, mel = melanoidin, coastal = nearshore marine, mar trans = marine shallow transitional, mse = marine shallow eutrophic, mso = marine shallow oligotrophic, md = marine deep, pore = porewater. (After Coble 1996).

C as peaks C_1 and C_2 for the purpose of interpretation of their data (Henderson et al., 2009), however, evidence has not been published to establish these as "peaks" by the definition established herein. Rather, this interpretation is more similar to the use of fluorescence indices (see Chapter 9).

Marine surface waters as well as freshwater samples that have active microbial populations typically show the most blue-shifted maxima. Coastal seawater samples fall between these two extremes, reflecting the presence of both older, soil-derived AOM and newer, marine-derived AOM. Where AOM is additionally derived from wastewater discharges, one might observe a relative increase in fluorescence in the peak C region that is derived from optical brighteners and therefore not "humic-like." As an extreme example, Baker (2002) demonstrated that optical brighteners from recycled paper processed in a pulp mill increased the riverine AOM fluorescence in the peak C region downstream of the works (see Section 3.5).

Variability in the position of peak C is also due to transformation processes leading to its destruction. Once released from the soil matrix, humic substances are more readily transformed and degraded. Terrestrial humic substances in most cases persist long enough to be transported to the ocean, where transformation and degradation continues, further altering the fluorescence properties. Photochemical degradation processes cause a gradual shift in peak C position toward shorter wavelengths (blue-shift) with a concomitant decrease in fluorescence intensity. Biological activity during river transport or in the coastal ocean can result in production of new fluorescent materials, including a newer humic-like material. The fluorescence associated with new humic-like material is called peak M.

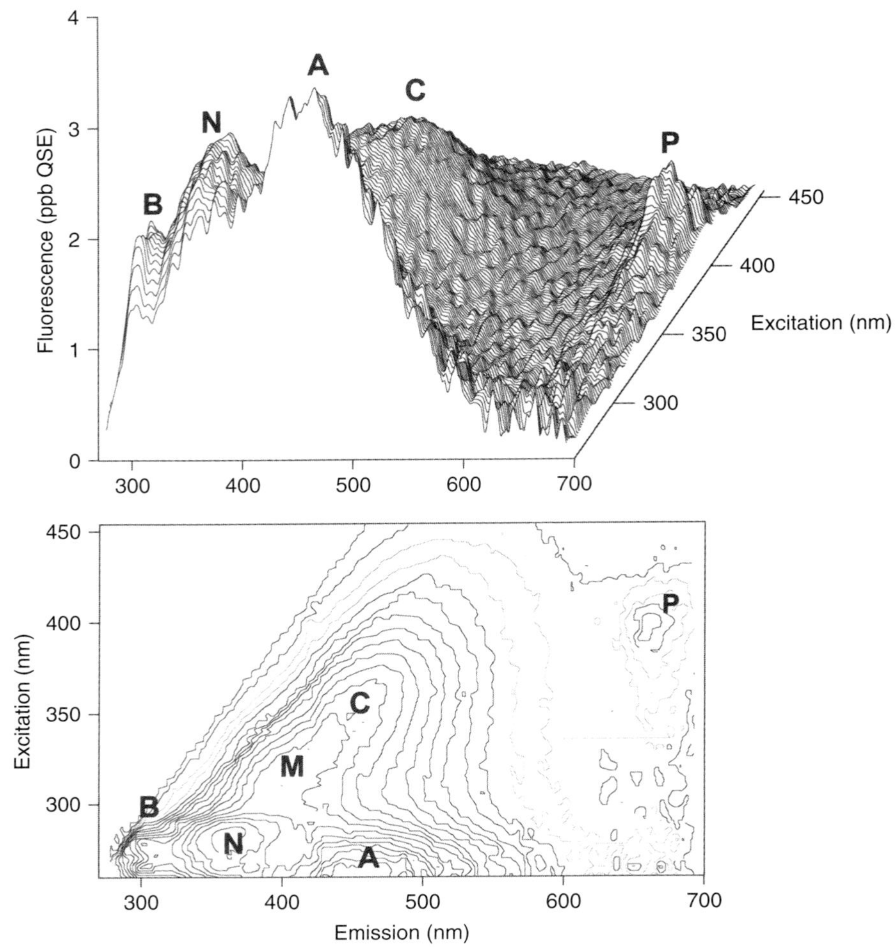

Figure 3.3. Fluorescence fingerprint representative of samples collected in the (Arabian Sea) upwelling. Water Mass Type 4. Although typical of most samples collected in the cool upwelling waters, not all had the chlorophyll-like peak at ex/em = 400/685. (Reprinted from Coble et al., 1998, with permission from Elsevier.) Note that the shortest excitation wavelength was 260 nm; therefore some of the peaks noted in Table 3.1 were not observed. The peak labeled A in the original figure is equivalent to peak A_C according to the revised nomenclature proposed in this work.

Peak M was originally described from bloom conditions in the Gulf of Maine (Coble, 1996) and was subsequently observed in the upwelling zone of the Arabian Sea (Coble et al., 1998). This humic-like component was considered to represent a pool of humic material newly formed in the marine environment. It meets the criteria of a peak distinct from peak C based on two observations. The first was its appearance in some samples as a nearly pure fluorophore, with round contours in map view, and the second was the appearance of distinct M and C peaks in some samples (Figure 3.3). Although the original reports of this component did not include data below 260 nm excitation, it is now clear that peak M

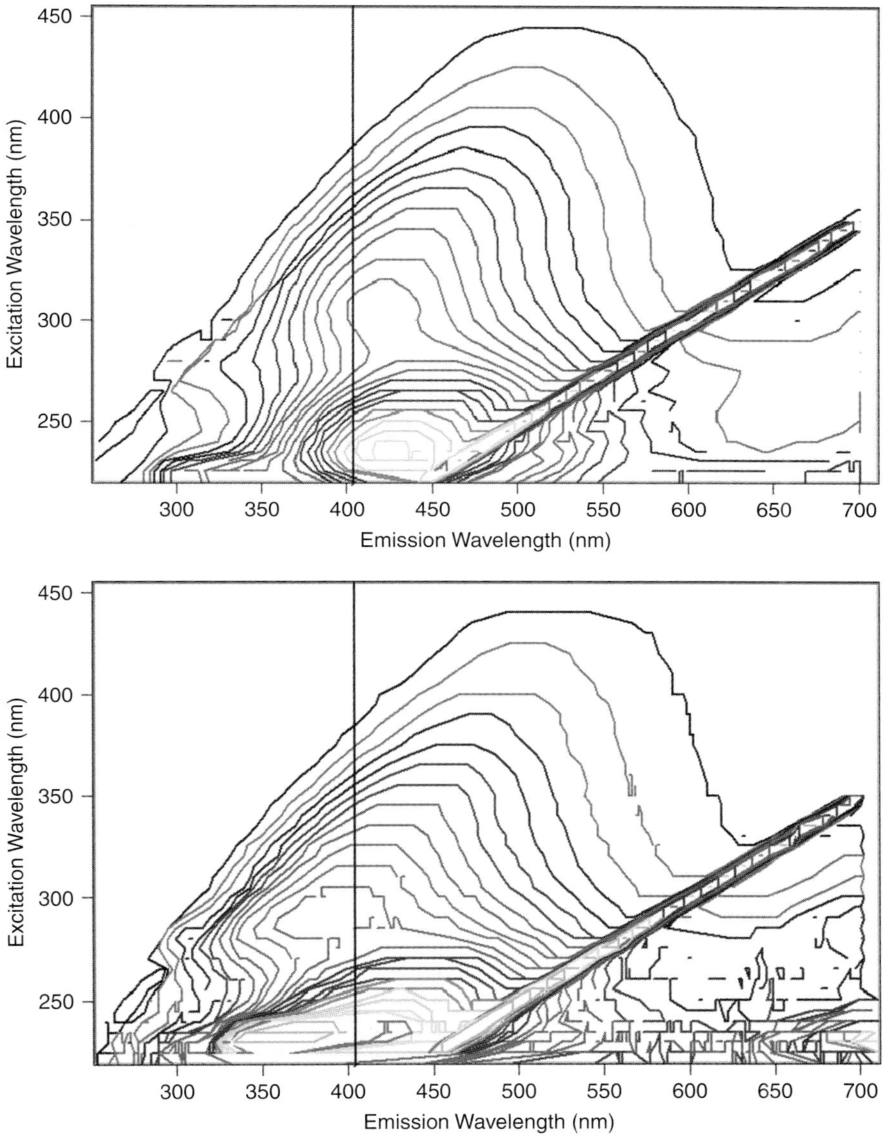

Figure 3.4. EEMs for sample showing humic-like C, with peaks C and A_C (above) and for a sample showing humic-like M component, with peaks M and A_M (bottom). Note that both peaks in the bottom panel are blue-shifted relative to the peaks in the top panel.

is also always associated with second peak from excitation in the UVC region called peak A_M. Taken together, these two peaks comprise humic-like component M. Peak A_M displays variability in peak position in parallel to peak M, such that excitation and emission maxima for both type M humic-like peaks are blue-shifted relative to the analogous maxima for type C humic-like fluorescence (Figure 3.4).

Component M has now been observed in nonmarine environments (Stedmon and Markager, 2005), and thus may be more indicative of recent microbial activity rather than arising exclusively from a marine source. In samples where there is mixing of old and new humic material, the resulting EEM most commonly exhibits a UVA fluorescence maximum at a position intermediate between peak M and peak C, rather than two separate peaks. Likewise, the UVC maximum appears as a mixture of A_C and A_M, with a much broader range along both excitation and emission axes than is the case when component M is absent. Coble (1996) reported the position of the wavelength-independent maximum fluorescence (ex_{max}/em_{max}) at which the overall fluorescence maximum (F_{max}) occurred and showed that it was a function of environment, with coastal samples transitional between river and marine samples. Thus, within any given sample, the position of the humic-like component peak provides information regarding source, CDOM composition, and environmental conditions.

3.1.3 Other EEM Components

In areas of active biological productivity, additional fluorescence peaks are observed that more readily resemble pure compounds (Figure 3.3). These include the amino acid–like peaks similar to tyrosine and tryptophan, a chlorophyll-like pigment peak, quinone-like peaks (Cory and McKnight, 2005), and several unidentified compounds, including fluorophore N (Coble 1996). These biological fluorophores can sometimes dominate the EEM signature, but appear not be environmentally persistent.

Peak positions for the amino acid–like components are much less variable than for humic-like components. Positions for tyrosine and tryptophan are $\lambda_{ex/em}$ = 230,275/305 nm and $\lambda_{ex/em}$ = 230,275/340 nm, respectively (Table 3.1). Both components have dual excitation maxima, although many studies do not present data for the 230 nm peaks (A_B, A_T). Reported ranges from PARAFAC models for environmental samples have a broader range of values for emission maximum, likely reflecting the fact that the signal in natural waters arises from some combination of free and combined amino acids. The fluorescence emission maximum for proteins is also dependent on the hydrophobicity of the site surrounding the amino acid moiety.

There is another fluorescence type that was not specified in early nomenclature schemes that has now been widely observed and also is a component in some PARAFAC models (Murphy et al. 2009; Component 9). This type of fluorescence has been observed in very clear open ocean surface waters (Coble 2006) and is most likely the result of extensive photobleaching. Figure 3.5 shows the EEM for a component from a Gulf of Mexico PARAFAC model that is representative of this fluorescence type.

3.1.4 Reconciling PARAFAC Model EEM Components

Comparison of peak positions between bulk sample EEMs and PARAFAC components is relatively straightforward for tyrosine-like and tryptophan-like components, but is more

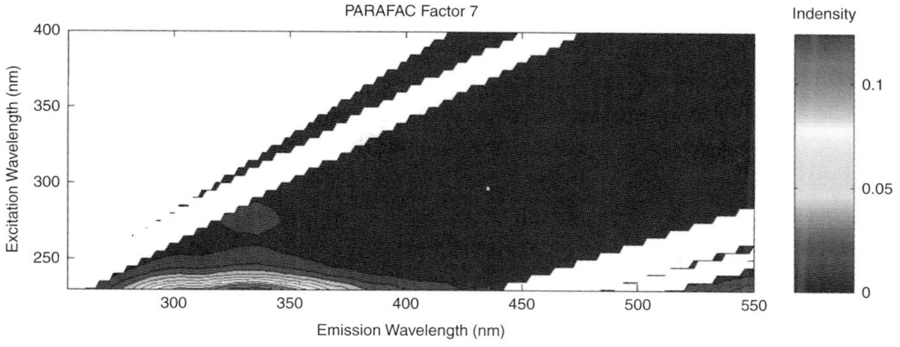

Figure 3.5. EEM for factor 7 from a PARAFAC model for samples from the eastern Gulf of Mexico. This component represents remaining fluorescence signal in highly photobleached samples.

challenging for humic-like components for several reasons. In PARAFAC modeling, EEMs are resolved into a number of components to maximize the level of fit to the data while minimizing the residual signal (Stedmon et al., 2003). The PARAFAC model can add humic-like components with single excitation maxima, either in the UVC or UVA regions, or add dual peak components that have maxima at wavelengths shorter and longer than components C or M in order to obtain a best fit to the data. These adjustments to the model account for the complexities of AOM fluorescence. In the first case, addition of single excitation peaks accounts for the fact that humic-like C fluorescence does not show a constant ratio between peak A_C and C intensities (Coble, 1996). This could be caused by variability in the A/C ratio for individual components, differential susceptibility to photobleaching, or complete uncoupling of the two peaks such that there are independent sources of C and A_C fluorophores, in which case the PARAFAC model would be reflecting the underlying chemistry of some samples. Prahl and Coble (1994) showed variability in the A/C ratio during a tidal cycle in the Columbia River estuary and attributed it to new organic matter produced on the tidal mud flats entering the main river flow on the ebb tide.

In the second case, the addition of dual maxima components with peaks at longer and shorter wavelengths than observed in bulk EEMs accounts for the fact that the bulk samples most likely do have multiple fluorophores, more than just a component M or C. The most common of these from the various models has an emission maximum at longer wavelengths than does peak C, and has been labeled humic-like C+ in Table 3.1. This component is often seen in soils and freshwaters close to terrestrial sources of CDOM. Although it is tempting to conclude that the PARAFAC components represent actual fluorophores in the environment, none have been demonstrated to exist in natural water samples. In lieu of additional chemical analyses, the PARAFAC components can provide valuable information regarding the relative contribution of components and how those contributions change with changing environmental parameters.

The discussion in this section has covered only one method of multivariate analysis of fluorescence data, that is, PARAFAC. Additional approaches are discussed in Chapter 10.

As analyses proliferate, care should be taken in use of terminology to describe both model and spectroscopic results. Designation of a new peak from spectroscopic data requires demonstration that it can be observed as a distinct peak in an EEM in the presence of previously identified peaks. Lacking that verification, our recommendation is to use the term "region" to refer to newly identified areas of interest. It is the ultimate goal of many researchers eventually to discover the chemical compound identities for all fluorophores in natural water samples. Only when this goal is attained will the underlying distinctions among peaks, components, and regions be clarified.

3.2 Fluorescence in Seawater

3.2.1 Introduction

The distribution and properties of CDOM in seawater has been the subject of several reviews. The original review by Blough and Green (1995) and its update by Blough and Del Vecchio in 2002 covers observations of CDOM optical properties in coastal regions collected worldwide since 1981. Nelson and Siegel (2002) reviewed CDOM in open ocean areas, including remote sensing applications, photochemistry, and photobiology. The review of Del Castillo (2005) emphasizes remote sensing applications for CDOM in the coastal ocean, including the chemical basis of the optical properties, the role of CDOM in ocean color and ocean optics, and development of empirical algorithms for separation of ocean color spectra into discrete components. The most recent review by Coble (2007) covers the topic of CDOM in both coastal and open ocean areas between 2004 and 2007, including the application of PARAFAC and other multiple-component analysis techniques for resolving EEMs into individual components.

The following section summarizes the key characteristics and distribution of CDOM in seawater and highlights a few case studies of recent developments and observations.

3.2.2 CDOM in Coastal Ocean and Estuaries

The distribution of CDOM in seawater is dominated by high concentrations along coastal margins contributed by inflowing freshwater from rivers. This CDOM has a terrestrial signature, consisting of humic-like fluorescence derived from soils. Open ocean regions have a CDOM concentration on the order of 1 ppb quinine sulfate equivalents (QSE), which is two or more orders of magnitude lower than in rivers; hence dilution of freshwater with seawater is the major process controlling nearshore distributions. The preponderance of studies of CDOM distribution in coastal regions have reported a robust negative correlation with salinity, and thus CDOM is largely considered to be a conservative tracer for mixing on the time scale of weeks for the surface ocean. This distribution is robust enough such that CDOM has been used as a tracer of coastal mixing in numerous past studies, and more recently enabled the development of ocean color algorithms for salinity and freshwater discharge (Del Castillo and Miller, 2008).

Conservative behavior has also been observed for the relationships among DOM absorbance, fluorescence, and dissolved organic carbon (DOC) concentration in many coastal regions, where freshwater runoff controls CDOM composition. Kowalczuk et al. (2010) used a six-component PARAFAC model of CDOM in the Cape Fear Estuary to determine separate relationships between DOC and optical properties for each component. They found strong relationships between DOC concentration and absorbance, total fluorescence intensity, and fluorescence intensity of all the individual PARAFAC components except the tryptophan-like component. Regression coefficients varied among components, with higher slopes and larger intercepts for the terrestrial humic-like components, although these were both smaller than slopes and intercepts for absorbance and total fluorescence intensity. Both carbon-specific absorbance and carbon-specific fluorescence were found to decrease rapidly with increasing salinity. The results of this study provide a significant improvement in the important efforts to estimate carbon export using optical sensors, both in water and on satellites.

Extreme events such as floods and hurricanes can drastically alter the expected coastal distributions of CDOM. Conmy et al. (2009) showed the influence of hurricanes to both greatly increase and greatly decrease CDOM distributions on the West Florida shelf, depending on storm track and thus wind direction. The composition of the CDOM over the shelf was also altered, as blue-shifted marine CDOM replaced the terrestrial CDOM following strong onshore winds. These events had some secondary consequences such as increased turbidity, increased nutrients from upwelling or river discharge, and decreased water clarity that interfere with satellite estimates of CDOM. This study highlights the fact that seasonal dynamics of CDOM distribution and composition have ecological consequences, as CDOM serves as a source of nutrients, provides UV protection, and absorbs critical photosynthetically available radiation (PAR; Zimmerman 2003, 2006). A recent study on health of nearshore corals found that CDOM absorption was higher and less variable along intact shorelines than along developed shorelines, providing significantly greater UV protection to corals within 5 km of the coast (Ayoub et al., 2008).

The overall observation of conservative behavior of CDOM does not preclude production and destruction in coastal areas, rather, these processes appear to occur on time and space scales that most studies fail to examine in detail. A few notable examples of production of CDOM during estuarine mixing and tidal cycles can be found, in which release of excess CDOM appears to be associated with drainage of tidal mud flats during ebb flow (Prahl and Coble, 1994; Gardner et al., 2005). A recent study in a shallow embayment in Florida showed production of a high salinity, high CDOM water mass in the dry season during spring tides (Milbrandt et al., 2010). EEM analysis confirmed that the material had blue-shifted fluorescence characteristic of newly formed CDOM and observed concentrations were eight times higher than previously reported seawater values. This new CDOM was flushed out of the bay at ebb tide, and was replaced by lower salinity, lower CDOM on flood tides.

Less is known of CDOM production by benthic communities such as mangroves, seagrasses, tidal marshes, and corals. Stabenau et al. (2004) reported CDOM production from seagrass detritus, but not from chambers placed over bare sand. The similarity of temperature dependence of CDOM production to temperature dependence of microbial decomposition led to the conclusion that production was microbially mediated. The new material was very readily photodegraded. Part of the refractory CDOM found in tropical estuaries can be attributed to production in mangrove porewaters (Trembly et al., 2007) based on molecular as well as optical methods. Shank et al. (2010b) found that production of CDOM from red mangrove leaf litter was highest from mid-senescent orange leaf litter during the wet season. They also found CDOM production for floating Sargassum mats at potentially significant rates for oligotrophic ocean waters. The CDOM exported from tidal marshes in Chesapeake Bay has a similar refractory nature, with red-shifted emission maxima and lower fluorescence per absorbance (Tzortziou et al., 2008). Release of CDOM has been observed from shallow coral reef environments in the Bahamas (Boss and Zaneveld, 2003; Otis et al., 2004), either from high productivity on the reef or from remineralization in the sediments (Burdige et al., 2004). Maie et al. (2006) found that the CDOM from Florida Bay was likely to be bacterial in origin, due to blue-shifted fluorescence maxima, whereas CDOM from nearby regions in the study area was more terrestrial in nature, reflecting a tide marsh and mangrove source.

Destruction of CDOM in the ocean is due primarily to photodegradation by sunlight. Several recent studies have shown differential photodegradation of pools of CDOM. In a study of photodegradation off Ria Vigo, Spain, it was found that peak T fluorescence was degraded in proportion to decrease in DOC concentration and resulted in the formation of peak M fluorescence. Humic-like peaks C and A were found to be tracers of refractory DOC (Lonborg et al., 2010). Another study comparing rates of photobleaching found that freshly produced CDOM from mangrove leaves and Sargassum bleached faster than did terrestrial or ambient marine CDOM (Shank et al., 2010a). Application of CDOM photochemistry has been extended to provide estimates of CDOM degradation rates (Belanger et al., 2006) CO production (Fichot et al., 2010) and surface seawater concentrations of the partial pressure of carbon dioxide (pCO_2) (Lohrenz and Cai, 2006) from satellite ocean color parameters.

3.2.3 CDOM in Open Ocean Waters

CDOM concentrations are at minimal levels in the surface of the ocean gyres due to long exposure to sunlight and low biological activity. Thus, both production of new fluorescent materials and the consequences of photobleaching are more readily observed in the open ocean away from the influence of rivers. In the Equatorial Pacific Ocean, diel variability in the composition and concentration of CDOM was observed in the surface waters (Coble, unpublished data). Samples collected at dawn showed the presence of both humic-like (peaks C and A_C) and protein-like fluorescence (peak B), which was greatly diminished or absent from the same station at noon (Figure 3.6). The mean concentration observed in these surfaces waters was 0.3 ppb QSE, but lower values of less than 0.1 ppb QSE were observed at the Hawaii Open Time Series (HOTS) station (Coble, 1996).

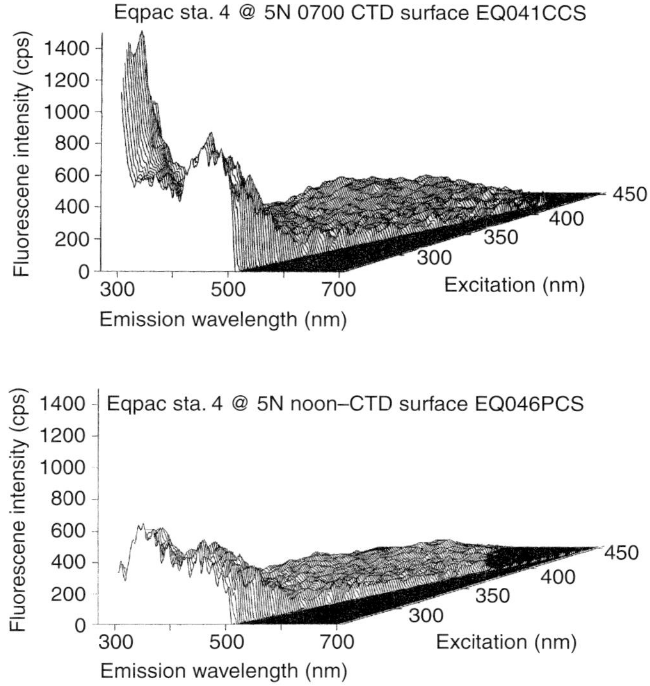

Figure 3.6. EEMs for surface waters collected at 0700 (top) and 1200 (bottom) at the same station in the Equatorial Pacific during the EQPAC expedition. Note the reduction in fluorescence in all peaks (B, A_C, and C).

Below the surface mixed layers, CDOM concentrations increase with depth in the open ocean. This distribution has allowed some investigators to use CDOM as a tracer for upwelling in open ocean gyres (Hoge and Lyon, 2005) as well as in coastal upwelling areas (Coble et al., 1998).

Several lines of evidence indicate that CDOM is very stable below the zone of photobleaching. Off the coast of Oman in the Arabian Sea upwelling, there is a strong negative correlation between CDOM and salinity in surface waters during most of the year due to photobleaching, but a strong positive correlation between CDOM and salinity in the subsurface salinity maximum (Coble et al., 1998). After periods of extreme mixing, the relationship in surface waters is reversed due to input of subsurface waters. More recently, analysis of CDOM and apparent oxygen utilization (AOU) data in the North Pacific Ocean has demonstrated a linear relationship between the two properties (Yamashita and Tanoue, 2008). This finding leads to the conclusion that the fluorescent material is produced from organic matter in the water column but is itself highly resistant (on the order of 100 to 1000 years) to degradation, as it accumulates in direct proportion to oxygen utilization. The authors estimated the rate of production in the interior of the ocean to be larger than input of fluorescent materials from global rivers. In a subsequent study, Yamashita and others (2010) performed EEM analysis on samples from two other Pacific Ocean sites and modeled fluorescence composition using PARAFAC. They found two humic-like components (M and C), one protein-like component (B), and an unknown component. The humic-like components reached maximum values in the mesopelagic zone and decreased below these depths. Both components likewise were correlated with AOU in the bathypelagic layer and the ratio of the two components showed a relatively narrow range of values in the deep waters. The implications of these findings are that both humic-like C and humic-like M are produced by bacterial respiration in the deep ocean in relatively constant proportions, and neither is significantly altered by biogeochemical processes subsequent to formation until returned to the surface ocean.

Increased interest in the Arctic Ocean as a site of rapid climate change has stimulated observations of CDOM in this region. Arctic rivers are high in DOC, with concentrations in the range of 1–10 mg L^{-1} (Retamel et al., 2007; Hessen et al., 2010). Warming due to climate change is already melting permafrost and is expected to increase freshwater discharge (ACIA, 2004). Several early studies showed that CDOM fluorescence was a tracer of DOC in the Arctic (Guay et al., 1999; Amon et al., 2003). Terrestrial CDOM dominates the surface waters and can be traced 50–400 km offshore in plumes (Retamel et al., 2007), but there is also evidence for autochthonous CDOM. A strong correlation exists between CDOM fluorescence and lignin phenol concentration (Amon et al., 2003; Walker, 2009). Walker et al. (2009) identified six PARAFAC components across the Arctic region. Four of these were related to lignin phenols, indicating a terrestrial source, while two others appeared to be of autochthonous origin. Three distinct water masses were identified using PARAFAC components, illustrating the potential usefulness of fluorescence in studies of Arctic Ocean circulation.

3.3 Fluorescence in Freshwater

The investigation of dissolved organic matter (DOM) fluorescence is now widely undertaken in the freshwater sciences. Such studies utilizing fluorescence techniques focusing on freshwater ecosystems have a range of aims such as examining the optical properties, composition, source, redox state, and biological and photochemical reactivity of DOM. Two discrete DOM fluorescent groups have been reported in freshwater ecosystems, with one group having properties similar to humic-like material and the other group with properties similar to protein-like material. Currently the exact chemical compounds responsible for DOM fluorescence are still unknown, however, a great deal of knowledge can be obtained about their environmental role in freshwater ecosystems by tracing their dynamics.

The incorporation of fluorescence measurements into DOM studies in freshwaters can be used to provide new insights into DOM biogeochemistry (Jaffe et al., 2008; Fellman et al., 2010). Here we highlight a number of examples in which studies have utilized fluorescence spectroscopy in freshwater ecosystems to examine the biogeochemical role of DOM.

3.3.1 Temporal Variation in DOM Source and Dynamics

The examination of DOM fluorescence in freshwater ecosystems with respect to temporal variability has been undertaken at a range of time scales from diurnal to seasonal. Seasonal variation in humic-like and protein-like fluorescence has been used to show increasing autochthonous inputs (i.e., peak in protein-like fluorescence), increasing allocthonous inputs (i.e., peak in humic-like fluorescence), and thus the dominance of biological or hydrological controls on the ecosystem (Jaffe et al., 2008; Miller and McKnight, 2010). Simple DOM measurements such as the fluorescence index (ratio of λ_{em} 470 to 520 nm at a λ_{ex} of 370 nm; McKnight et al., 2001; Cory et al., 2010) has been linked to aromaticity and indicates the relative contribution of low molecular weight (LMW) nonaromatic DOM versus high molecular weight (HMW) aromatic DOM (see Chapter 9). Therefore, fluorescence index (FI) has been used to track autochthonous versus allocthonous changes with respect to seasonal DOM inputs in freshwater ecosystems (Hood et al., 2005; Miller and McKnight, 2010). FI has also been shown to change due to increased flow path, residence time, and thus greater microbial mineralization of DOM, as well as due to extensive flushing of organic rich horizons (i.e., source materials) during wet periods in a tropical ecosystem (Figure 3.7; Spencer et al., 2010). The relatively lower FI during the flushing period shown in Figure 3.7a highlights that DOM during this period is enriched in HMW compounds and aromatic moieties compared to other times of the year due to the leaching of fresh organic rich layers (Spencer et al., 2010). This is further supported by the linear correlation observed between lignin carbon-normalized yields (Λ_8; which indicates the contribution of vascular plant–derived material to the DOM pool) and FI shown in Figure 3.7b (Spencer et al., 2010).

The impact of short-term events such as storms and salmon runs has also been examined via DOM fluorescence characterization. Hood et al. (2007) presented data from an Alaskan (USA) salmon spawning stream and showed that DOM derived from salmon carcasses during the spawning period is elevated in protein-like fluorescence. The DOM in the stream at the time of salmon spawning is therefore distinct from the humic-like material that dominates the stream DOM load during the rest of the year and that is derived predominantly from wetlands (Hood et al., 2007). Stormflows can represent a substantial portion of seasonal and annual catchment DOM export as concentrations of DOM typically increase with increasing discharge. Therefore, understanding the quality of DOM exported during stormflows is often critical for understanding DOM biogeochemical processes in a catchment. Utilizing DOM fluorescence measurements Fellman et al. (2009a) tracked the contribution of protein-like fluorescence and humic-like fluorescence during storm events and highlighted as hydrologic flowpaths changed to near surface soil horizons the DOM transported into streams was similar to that in soil solution. Recent developments with respect to in situ fluorescence techniques are now opening up the possibility of monitoring storm events at the temporal resolution required to capture these highly dynamic events with respect to DOM quantity and quality (Saraceno et al., 2009). The potential of high temporal resolution in situ monitoring of DOM fluorescence to capture diurnal DOM

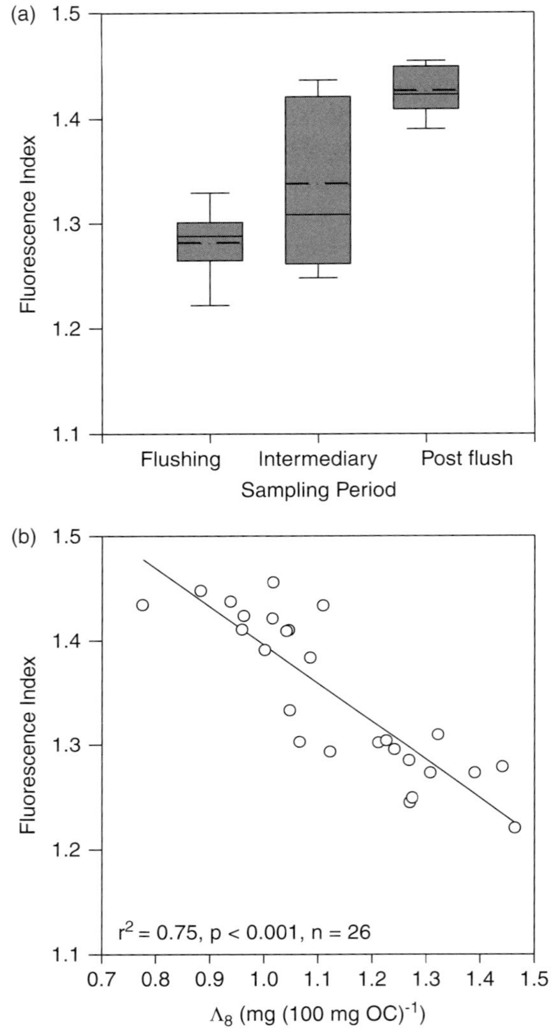

Figure 3.7. (a) Box plot of fluorescence index during the flushing, intermediary and post flush sampling periods. The black dash-dot line and the solid black line in the box represent the mean and the median respectively. The horizontal edges of the box represent the 25th and 75th percentiles and the error bars denote the 10th and 90th percentiles. (b) Relationship between lignin carbon-normalized yield (Λ_8) and fluorescence index. (Reproduced from Spencer et al., 2010, with permission of American Geophysical Union in the format Republish in book and "other" book via Copyright Clearance Center.)

variability has also been described and thus the ability to examine DOM processing in real-time (Spencer et al., 2007).

3.3.2 Anthropogenic and Land Use Impacts on DOM

Anthropogenic sources of pollution to freshwater ecosystems can be examined via fluorescence characterization of DOM. For example, Baker et al. (2002) highlighted the influence of wastewater from a large tissue paper mill through elevated protein-like fluorescence intensities and a fluorescence center attributed to fluorescent whitening agents. Similarly, Baker and Inverarity (2004) described an urban catchment with elevated wastewater inputs and showed protein-like fluorescence correlated to biological oxygen demand (BOD) and inorganic N and P concentrations. DOM fluorescence has also been used to investigate diffuse agricultural pollution resulting from slurry spreading (Naden et al., 2010). A strong linear relationship was found between fluorescence intensity and slurry concentration, and fluorescence ratios were described that allowed slurry impacted waters to be discriminated from uncontaminated waters (Naden et al., 2010). Such studies examining anthropogenic pollution in freshwater systems will benefit greatly from the development and utilization of in situ fluorescence sensors and portable monitoring devices that allow real-time monitoring of DOM fluorescent components (Baker et al., 2004; Spencer et al., 2007).

As well as examining point and diffuse anthropogenic sources of DOM, fluorescence characterization has also been utilized to investigate the current and future effects of land use and land management practices on freshwater biogeochemistry (Wilson and Xenopoulos, 2009; Williams et al., 2010). A study examining a land use gradient of increasing cropland coverage and decreasing wetland coverage showed that the structural complexity of DOM decreased as the ratio of continuous croplands to wetlands increased, as evidenced by the FI and humification index (Zsolnay et al., 1999; see Chapter 9; Figures 3.8a,b). Therefore, simple fluorescence ratios can be used to examine how land use changes have impacted on the biogeochemical role of DOM in freshwater ecosystems.

3.3.3 Transformations and Reactivity

The redox state and the rates of oxidation/reduction of DOM can be examined via the redox index (RI; derived from $Q_{red}/(Q_{red} + Q_{ox})$ where Q_{red} and Q_{ox} are the sum of the loadings of the reduced and oxidized quinone-like components respectively as defined by Cory and McKnight [2005]) in freshwater ecosystems. For example, Miller et al., (2006) investigated nutrient cycling in a wetland-stream ecosystem using RI to show that DOM in a wetland adjacent to the stream was reduced. In addition, dissolved iron concentrations were elevated and ammonium was the primary species of inorganic nitrogen in the wetland. Thus, as water from the wetland entered the stream, DOM was rapidly oxidized with a corresponding shift in RI, ammonium was converted to nitrate, and dissolved iron concentrations decreased (Miller et al., 2006). Therefore, fluorescence measurements can

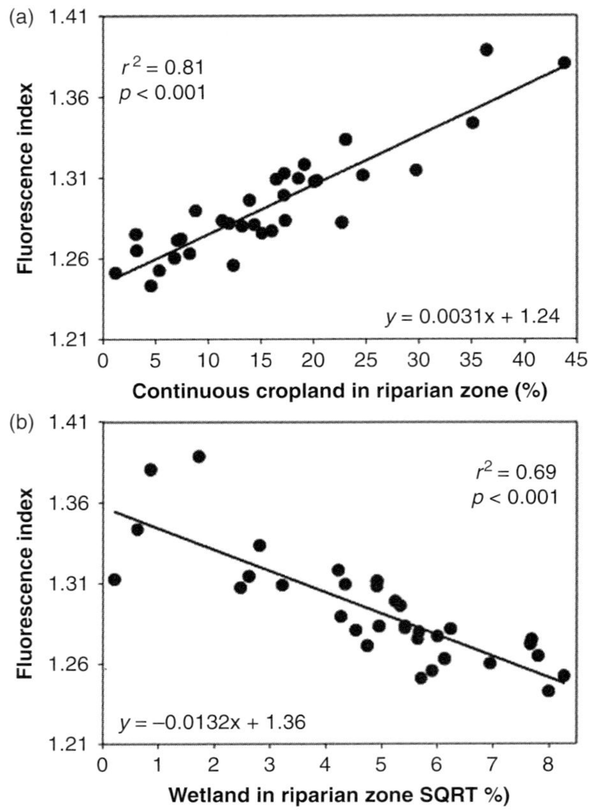

Figure 3.8. Relationships between indices of DOM character and landcover. The best predictors for fluorescence index (a,b). (Reprinted from Wilson and Xenopoulos, 2009, by permission from Macmillan Publishers Ltd: *Nature Geoscience*.)

be utilized to provide information about environmental conditions and the biogeochemical transformations that are taking place, which is very informative in studies examining nutrient cycling in hyporheic and riparian ecosystems (Fellman et al., 2010).

Protein-like fluorescence has been linked to biologically labile DOM in a range of freshwater ecosystems. For example, a study in *Lakes of Southern Quebec* by Cammack et al. (2004) correlated protein-like fluorescence with bacterial production, bacterial respiration, and community respiration. Laboratory incubation experiments from a diverse range of freshwater DOM sources have also found strong relationships between protein-like fluorescence and biodegradable DOM (Fellman et al., 2009b; Hood et al., 2009). Protein-like fluorescence can also be used with respect to examining in stream uptake of DOM as Fellman et al. (2009c) demonstrated that protein-like fluorescence decreased downstream during soil leachate addition experiments in forested headwater streams in Alaska (USA). In this leachate addition experiment, humic-like fluorescence did not change and protein-

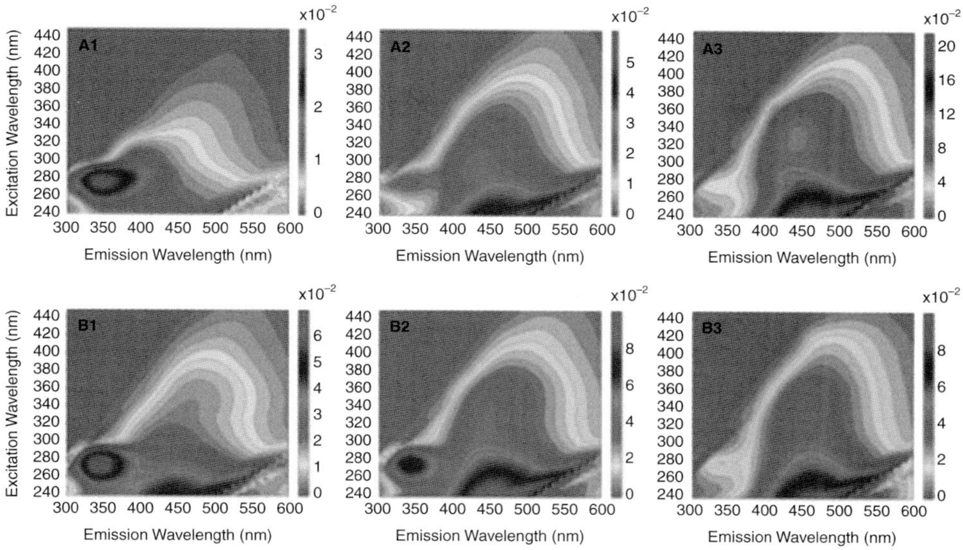

Figure 3.9. Change in fluorescence of moss leachate DOM over time: (A1–A3) *Sphagnum angustifolium* at 24 hours (A1), 194 hours (A2), and 3 months (A3); Feathermoss mix at 24 hours (B1), 194 hours (B2), and 3 months (B3). The color bar to the right indicates fluorescence intensity, which has been normalized to DOC concentration. (Reprinted from Wickland et al., 2007, With kind permission from Springer Science+Business Media.) (See Plate 5.)

like fluorescence was removed at a faster rate than bulk concentrations of DOC and DON, highlighting that certain DOM moieties were preferentially removed, which is consistent with the idea that different DOM pools turn over at different rates (Brookshire et al., 2005; Fellman et al., 2009c). Therefore, future studies seeking to examine DOM uptake in freshwater ecosystems could gain insight into the role of DOM pools with different reactivities by combining instream tracer releases with fluorescence characterization of DOM (Fellman et al., 2010). Changes in the composition of DOM during biodegradation experiments can also be evaluated via fluorescence spectroscopy. For example, Wickland et al. (2007) showed that in laboratory incubations of moss leachates, microbial degradation of DOM resulted in an increase in the humification and fluorescence indices as well as a loss of the protein-like fluorescence with time (Figure 3.9; Table 3.2).

The effect of photochemical degradation of DOM in freshwater systems can also be examined with fluorescence measurements. A typical response for freshwater DOM was observed in a study by Cory et al. (2007) with whole water samples from Alaskan (USA) stream and lake water, which showed a decrease in total fluorescence by the end of the short-term (12 hours) irradiation (Figure 3.10). On irradiation different fluorophores varied with respect to their percent change of fluorescence but humic-like fluorophores (SQ1 and SQ2; Figure 3.9) showed the greatest loss of fluorescence intensity. Overall the protein-like fluorophores (Tyr and Trp; Figure 3.10) showed little change with irradiation and for

Table 3.2. *Fluorescence properties and PARAFAC analyses of moss leachate DOC over time*

Time	*Sphagnum angustifolium*			*Feathermoss mix*		
	24 hours	194 hours	3 months	24 hours	194 hours	3 months
FI	1.14	1.26	1.38	1.17	1.30	1.38
HIX	0.61	0.83	0.89	0.68	0.79	0.90
Tryptophan-like fluorescence (%)	29	4	1	18	9	3
Tyrosine-like fluorescence (%)	15	9	5	12	11	3

Source: Modified from Wickland et al. (2007), Table 5, with kind permission from Springer Science+Business Media.

other samples even showed a small increase, resulting in post-irradiation waters having an increased contribution of protein-like relative to humic-like components in comparison to initial waters (Cory et al., 2007). Whole water samples in freshwater systems also typically show a decrease in FI with irradiation. The variation in FI has been shown to be related to relative amounts of SQ1 and SQ2 in a sample and thus the decrease in FI with irradiation time can be attributed to the greater loss of SQ2 in comparison to SQ1 in Figure 3.10 (Cory et al., 2007). Fluorescence characterization of DOM can help understand how photochemical processes influence DOM quantity and quality, and utilizing spectrophotometric techniques in comparison with other DOM characterization techniques shows promise for understanding how photochemical processes remove and modify DOM in aquatic systems (Spencer et al., 2009a; Stubbins et al., 2010).

3.3.4 Rainwater DOM Fluorescence

Fluorescence characterization of rainwater samples has shown the presence of both humic-like and protein-like materials in precipitation DOM (Kieber et al., 2006; Muller et al., 2008; Santos et al., 2009). Muller et al. (2008) highlighted that the highest humic-like fluorescence intensities are observed during convective events and events of continental origin reflecting the influence of terrestrial and anthropogenic sources. During well-mixed conditions Muller et al. (2008) found that humic-like fluorescence intensity decreases, whereas at low wind speeds stagnation of the atmosphere leads to higher fluorescence intensities that the authors attributed to an increase of localized sources, especially anthropogenically derived sources. No significant trends were observed by Muller et al. (2008) between protein-like fluorescence and meteorological variables and these fluorophores are likely to have been derived from a number of sources and processes. Rainwater DOM has been shown to be highly susceptible to photodegradation and variation in terms of photolability between humic-like and protein-like components may lead to a lack of a relationship with meteorological variables (Kieber et al., 2007). Finally, the integrated signal for entire

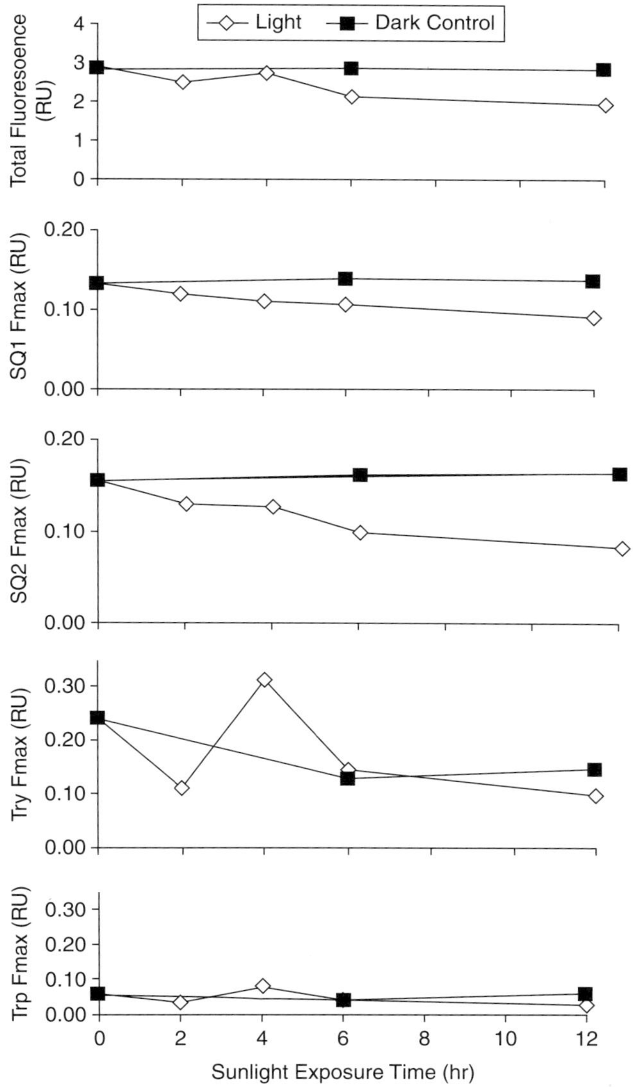

Figure 3.10. A typical response for freshwater DOM (Island Lake at Toolik Lake LTER) during short-term photochemical degradation highlighting a decrease in total fluorescence, a decrease in humic-like fluorophores (SQ1 and SQ2) and little change or a small increase in protein-like fluorophores (Tyr and Trp). (Reproduced from Cory et al., 2007, with permission of American Geophysical Union in the format Republish in book and "other" book via Copyright Clearance Center.)

fluorescence EEMs has been correlated with DOC in rainwater, with higher DOC concentrations resulting in an increased fluorescence signal of rainwater ($r = 0.699$, $p < 0.001$), for rainwater collected in southeastern North Carolina (USA), indicating that fluorescent DOM is a common component of the DOC pool at that location (Kieber et al., 2006).

3.3.5 Dissolved Organic Carbon vs. Fluorescence Relationships

A number of studies have examined relationships between fluorescence intensities of protein-like or humic-like fluorophores and DOC concentration in freshwater ecosystems (Baker, 2002; Cumberland and Baker, 2007; Hudson et al., 2007; Baker et al., 2008). The fluorescence intensity versus DOC relationship has been observed to differ in gradient (i.e., the fluorescence per gram of carbon) and in the strength of the correlation coefficient dependent on the source of the DOC and which fluorophore is being related to DOC concentration (Cumberland and Baker, 2007; Baker et al., 2008). Typically, the strongest correlations between fluorescence intensity and DOC have been observed at study sites where natural DOM is dominant (i.e., sites that have little anthropogenic impact) and greater fluorescence per gram of carbon is reported in samples dominated by HMW, aromatic DOM (e.g., peat catchments, wetlands) (Cumberland and Baker, 2007; Baker et al., 2008). There is clear potential for future studies to utilize DOC versus fluorescence intensity relationships, as has been shown for CDOM absorption coefficient versus DOC relationships, to derive fluxes of DOC from freshwater systems that have been appropriately ground-truthed (Spencer et al., 2009b). The prospect of deploying in situ fluorometers to derive real-time DOC fluxes as well as improve spatial and temporal resolution is an exciting prospect, particularly in watersheds that exhibit significant short-term export events (e.g., freshets, storms) (Saraceno et al., 2009; Spencer et al., 2009b).

3.4 Fluorescence in Groundwater

3.4.1 Introduction

Groundwater organic matter fluorescence has been investigated systematically since the work of Smart et al. (1976). They had previously observed "background" fluorescence variations during dye trace experiments in limestone aquifers, and the 1976 paper described this natural fluorescence signal in a combination of groundwater, rain, and surface water samples. Smart et al. observed a peak excitation at 340–350 nm, with emission spectra exhibiting broad peaks at 400–460 nm, equivalent to what is now considered peak C. In the experiment, samples were stored for up to 9 days in both light and dark conditions and filtered and unfiltered to assess sample stability. Temperature dependence, pH effects, and potential metal quenching effects were also considered. Smart et al. (1976) demonstrated a strong correlation between fluorescence intensity and dissolved organic carbon concentration. Although the relationship was specific to sample source, and their work was widely cited by the wider aquatic organic matter fluorescence community (e.g., Stewart and Wetzel, 1980, 1981; Zepp and Schlotzhauer, 1982; Laane, 1982) the use of fluorescence to characterize and quantify organic matter in groundwater was largely forgotten by that community.

A revitalization of interest in groundwater fluorescent organic matter came about after the observation that cave stalagmites contain annual fluorescent laminae (Baker et al., 1993;

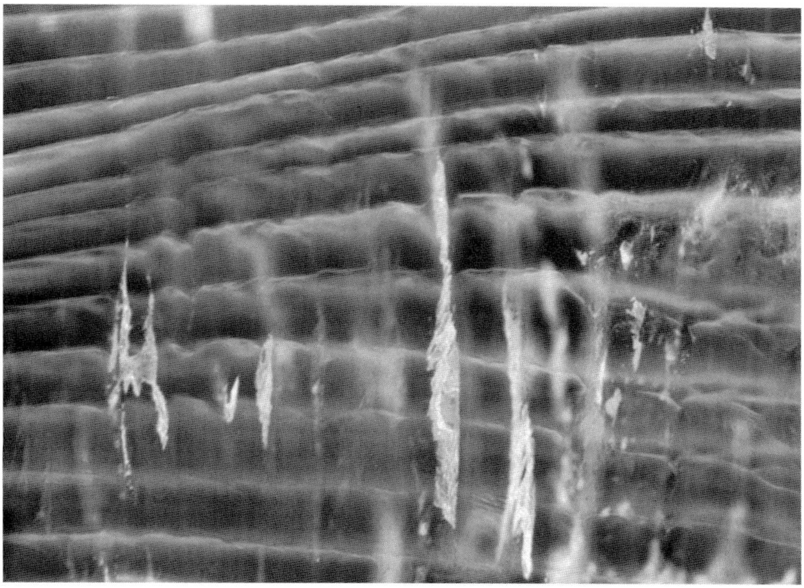

Figure 3.11. Fluorescent NOM preserved in a stalagmite to form annual laminae. This picture shows annual cycles of NOM fluorescence associated with seasonal groundwater recharge. Stalagmite is ER-77 from Ernesto Cave, Italy. (Photograph courtesy of Ian Fairchild, Birmingham.) (See Plate 6.)

Figure 3.11). Speleothems, cave calcium carbonate deposits that include stalagmites and stalactites, have the ability to preserve some or all of the contents of groundwater that supplies them. This includes fluorescent organic matter which is primarily derived from the overlying soil. The observation that cave speleothems are fluorescent was first made by cavers experimenting with underground flash photography, who noticed an "after-glow" effect (O'Brien, 1956). This was not theoretically explained until the experimental work of White and Brennan (1989), who examined speleothems under 365 and 253.7 nm ultraviolet light, and observed that all emitted fluorescence within the blue-green filter window region, which we now know to be the visible "tail" of peak C fluorescence. Using a mercury light source with a 320–420 nm excitation filter and a 420–500 nm emission filter, Baker et al. (1993) demonstrated annual variations in this peak C fluorescence in a cave stalagmite for the first time, opening the way for stalagmite organic matter fluorescence variations to be used as a chronological tool. This work also prompted new groundwater studies, in karstified limestone aquifers, of the fluorescence characteristics of groundwater organic matter, in particular as a tracer of surface water (e.g., soil)–groundwater connectivity (e.g., Baker et al., 1997, 1999). For reviews, with a specific focus on stalagmite forming groundwaters and the preserved records in speleothems, see Blyth et al. (2008) and McGarry and Baker (2000). Most recently, natural organic matter (NOM) fluorescence has been more widely applied in a wider variety of hydrogeological settings (Baker et al., 2001; Lapworth et al., 2008; 2009; Conmy 2008).

3.4.2 Groundwater NOM Fluorescence Characteristics

EEM analysis of groundwater natural organic matter fluorescence demonstrates that groundwaters have natural organic matter fluorescence characteristics very similar to those of freshwater and marine samples (Figure 3.3). In terms of specific peaks, the following general observations can be made:

1. Peak C and peak A_C fluorescence is the dominant fluorescence signature, typical of humic-like material. This material is presumed to be derived predominantly from the overlying soil, although systematic investigations of groundwaters have not been undertaken, and it is probable that some material is derived from organic matter production and degradation within the aquifer.
2. Although systematic data are sparse, peak C and peak A_C fluorescence is likely to be blue-shifted compared to the overlying soil water fluorescence (Baker and Genty, 1999), and definitely blue-shifted when compared with surface water samples in the same region (e.g., see Conmy 2008). The relatively hydrophobic properties of NOM would lead to the adsorption of truly dissolved material to the bedrock, leaving only the relatively hydrophilic fraction to be transported, which could explain some or all of this blue-shifted emission, as well as the processing of organic matter by groundwater microbial communities. Conmy (2008) compared surface, ground, and marine water samples, and demonstrated that peak C fluorescence emission wavelength is intermediate to marine and surface samples.
3. Peak C fluorescence intensity correlates with dissolved organic carbon concentration and decreases with depth in the aquifer. The latter reflects the combination of physical losses in the aquifer (e.g., adsorption) as well as potential biological degradation. A decrease in both peak C and peak T intensity with depth is well observed in Lapworth et al. (2009). The peak C fluorescence intensity–DOC relationship leads to the potential for the use of fluorescence as a surrogate measure of DOC, as first observed by Smart et al. (1976). However, the gradient and strength of this relationship varies from site to site (Mudarra et al., submitted), depending on the overall content of fluorescent material.
4. Peak T and peak B fluorescence peaks are rarely observed in groundwater organic matter fluorescence EEMs, unless a contaminant source is present (see Section 3.4.3). Peak M fluorescence has yet to be definitively observed in groundwater samples, but if derived from a microbial source it might also be expected to be observed in microbially contaminated groundwaters.

NOM fluorescence properties have therefore been mostly utilized as a natural tracer of surface water–groundwater connectivity, either from the soil directly to the groundwater (e.g., Baker et al., 1999), or through river–groundwater exchanges in the hyporheic zone (Lapworth et al., 2009). In the future, NOM fluorescence is likely to be used to understand better the chemical and biological processing of organic matter, or the character and function of groundwater organic matter, following research already undertaken in the marine and freshwater fields. For example, Lapworth et al. (2009) investigated hyporheic transformations of dissolved, colloidal, and particulate organic matter using fluorescence EEMs.

3.4.3 Groundwater Anthropogenic Organic Matter Characteristics

Contaminated groundwaters might be expected to comprise the same fluorescence EEM characteristics of the pollutant source that has undergone dilution and chemical and microbial processing in the aquifer. Almost certainly, a wider range of contaminant fluorescent signatures exist than those reported. The two reported contaminant fluorescence organic matter characteristics are:

1. Unusually intense peak T and/or peak B fluorescence, derived from organic effluents. For example, Lapworth et al. (2008) report high peak T intensities at a groundwater site that had historically undergone surface slurry applications.
2. A specific fluorescence peak at excitation 220–230 nm and emission between 340–370 nm, which is particularly intense and derived from landfill leachates. First reported in contaminated Missouri spring waters, Baker and Curry (2004) demonstrated a strong correlation between the intensity of this peak and geochemical parameters such as BOD and ammonia concentration, and postulated that it was derived from one or more polycyclic aromatic hydrocarbons present in the landfill leachate.

International peer review published applications of the use of fluorescence to detect groundwater organic matter pollution are not widespread. In the authors' experience, this reflects a greater amount of work undertaken in a consultancy setting. Given the increasing global interest in water recycling, including aquifer recharge schemes, future research might be expected to investigate the movement and transformation of recycled water within groundwater aquifers.

3.5 Fluorescence of Wastewater and Drinking Water

3.5.1 Wastewater Fluorescence

Despite the substantial body of literature that is concerned with the fluorescence and spectral properties of dissolved organic matter in marine, terrestrial, and fluvial environments, much less attention has been given to the investigation of these properties in relation to wastewaters. The generic term wastewater encompasses any water body that has been impacted by anthropogenic influence, so as to adversely affect the overall quality. Wastewater can comprise liquid waste discharged by residential, commercial, and industrial properties (including agriculture), and by definition incorporates a wide range of potential chemical contaminants at varying concentrations. Most commonly the term wastewater refers to municipal wastewater containing a broad mix of contaminants from a variety of wastewaters from different sources. Sewage is a subset of wastewater, referring to waste matter that is collected and passes through sewers. This includes both solid and liquid contaminants, originating from human wastes as well as surface water runoff.

To understand the nature of the fluorescence spectra of treated and untreated wastewaters, it is important first to consider the composition of sewage and wastewaters. Owing to the complexities associated with their origin, wastewater composition varies enormously and the exact composition of a particular wastewater sample is dependent upon a number of key factors, such as geography and the nature of the inputs entering the system. Inputs can exhibit both physical and aggregate properties (e.g., suspended particulates). All wastewaters contain some metal ions, such as aluminium, copper and iron at varying concentrations ($\mu g\ L^{-1}$ to $mg\ L^{-1}$) as well as inorganic nonmetallic constituents (e.g., chlorine, phosphate, and nitrate ions). Aggregate organic constituents including humic acid and fulvic acid, tannins, lignin-related material and a variety of surfactants are also present. Moreover disinfection by-products and pesticides are always present to some degree. Many wastewaters have a complex microbiota including bacteria and viruses, as well as plankton and algae.

Crude sewage is composed of a heterogeneous mixture of compounds including fulvic acids, proteins, carbohydrates, and lipids (with varying contributions from organic surfactants), nucleic acids, and volatile fatty acids (Ahmad and Reynolds, 1995). It is a mixture of domestic waste, industrial discharges, and the domestic elements from industrial premises such as kitchen and toilet wastes, in addition to surface runoff and storm flow. Composition of domestic sewage varies depending on the age and type of sewerage system in the catchment (separate or combined), time of day (Reynolds and Ahmad, 1997), prevailing and prior weather conditions, and type of incoming sewer (gravity or pumped). This complicated matrix is best summarized by Eaton et al. (2005) in the latest (21st) edition of *Standard Methods for the Examination of Water and Wastewaters*.

The complex composition of wastewaters implies that the observed fluorescence properties are a manifestation of innumerable types of fluorophores, including humic and fulvic acids and lignin-derived substances. In addition to these fluorophores there are variable amounts of steroids, phenols, nonvolatile acids, oils, and trace quantities of surface-active agents (Waggot and Butcher, 1976). Therefore, the spectral shape of the fluorescence spectra obtained from sewage and wastewater samples will be a composite of the overlapping contributions from a variety of these fluorophores and devoid of any sharp features. The interpretation of fluorescence spectra from wastewater samples is further complicated by variations in chemical and physical parameters that are known to affect fluorescence, namely pH, metal ion content, temperature, and suspended solids (Reynolds and Ahmad 1995a). This latter point is important as it is widely accepted that the fluorescence spectral properties observed at any one point in time, and from a particular site, are "quenched." Given the diverse nature of wastewater composition the direct comparison of fluorescence intensities of samples from different locations is problematic.

Up until the mid-1990s the fluorescence properties of wastewaters and sewage wastes were poorly understood. Bari and Farooq (1984) first reported the use of fluorescence in their investigation of the treatment efficiency of potassium ferrate and ozone, in various combinations, for the removal of organic matter from different wastewaters. In this study,

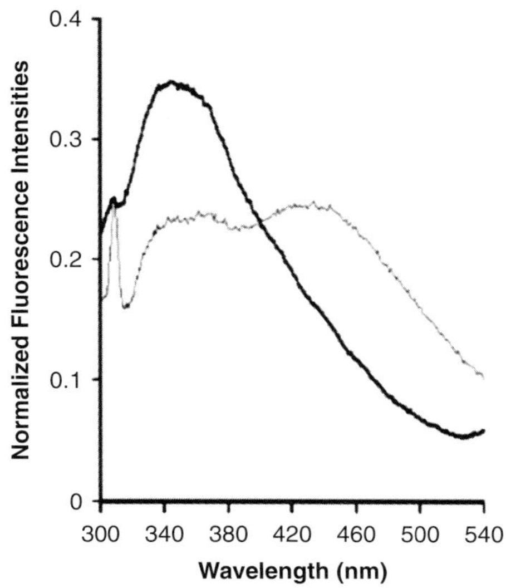

Figure 3.12. Typical fluorescence spectra of untreated (black line) and treated wastewater (gray line) using 280 nm excitation.

fluorescence was employed because of its specificity in measuring humic substances, aromatic compounds, and heterocyclic systems. Rather than obtaining fluorescence emission spectra, specific intensities at 490 nm were obtained using a fixed excitation wavelength of 365 nm. In addition to these fluorescence measurements the UV absorbance at 280 nm was also measured, and correlated with respective chemical oxygen demand (COD) values. The COD provides an estimation of the amount of oxidizable material present within the sample via oxidation with a strong acid (Eaton et al., 2005), and from this the amount of organic matter removal is estimated. The focus of this study was the removal efficiency of organic matter as opposed to understanding and interpreting the nature of the observed fluorescence spectra.

It is now accepted that all wastewaters exhibit characteristic fluorescing properties and this phenomenon was first reported in the mid-1990s. Research undertaken by Ahmad et al. (1994), Ahmad and Reynolds (1995), and Reynolds and Ahmad (1995) demonstrated the fluorescence emission spectra of wastewaters using a number of different excitation wavelengths. A typical fluorescence emission spectrum, using an excitation at 280 nm, is shown in Figure 3.12. From these early studies, research concerning the use of fluorescence as a tool for water treatment process optimization, water quality assessment, and pollution monitoring has emerged (Henderson et al., 2009). Further developments from this early work facilitated the use of synchronous fluorescence spectroscopy (SFS), which is the simultaneous scanning of both the excitation and emission wavelengths, and the recording of the distribution of intensities over the emission (or excitation) wavelengths.

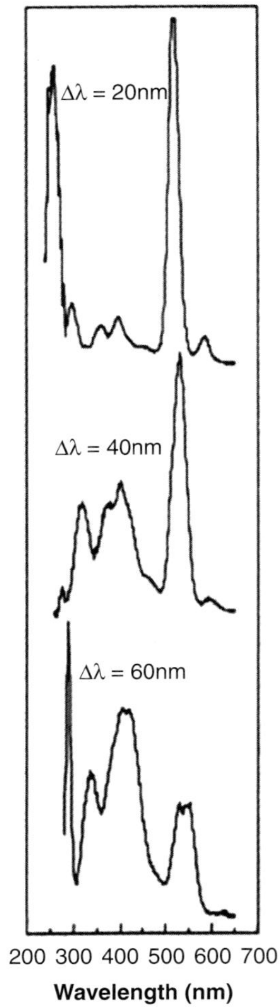

Figure 3.13. Synchronous fluorescence spectra of treated sewage effluent at 20 nm, 40 nm, and 60 nm offset values ($\Delta\lambda$). (Adapted from Galapate et al., 1998, with permission from Elsevier.)

For SFS, fluorescence intensities can be expressed as $I_f = \eta I_a$, where η is the quantum yield at λ_{ex} and I_a is the absorbed intensity (Reynolds and Ahmad, 1995b). From Beer's law, $I_a = I_0\{1 - \exp(-\alpha cl)\}$, where I_0 is the incident intensity, α is the absorption cross section and l is the effective sampling length. For $\alpha cl \ll l$, $I_f \approx I_0\, \eta cl$. In SFS the excitation wavelength (λ_{ex}) and the emission wavelength (λ_{em}) are scanned synchronously, with a fixed off-set such that $\Delta\lambda = (\lambda_{em} - \lambda_{ex})$. The normalized fluorescence intensity can be represented by: $I_f/I_0 = I_0\,\lambda'\alpha\,(\lambda')cl$, where the quantum yield and the absorption cross-section are functions of $\lambda' = (\lambda_{em} - \Delta\lambda)$. The intensity distribution will, therefore, show a pronounced peak when the absorption maximum and the quantum yield maximum overlap. For well defined

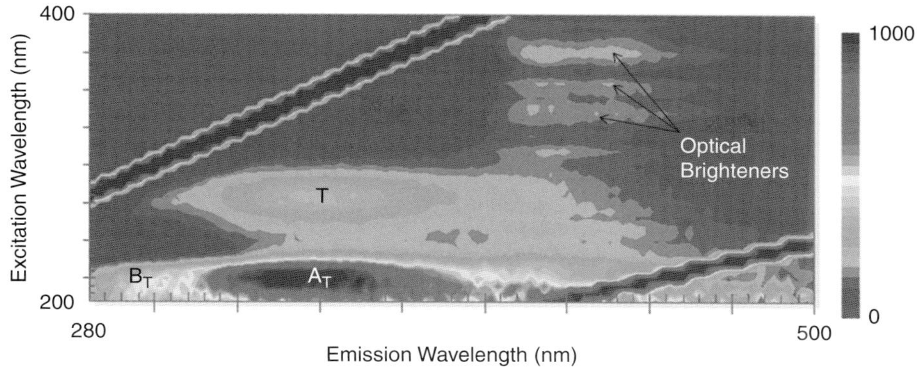

Figure 3.14. Fluorescence excitation–emission matrices (EEMs) for raw sewage where BT = tyrosine-like, T and A_T = tryptophan-like fluorescence. (Adapted from Henderson et al., 2009, with permission from Elsevier.) (See Plate 7.)

absorption and quantum yield maxima, the optimum value of $\Delta\lambda$ is set by the difference in wavelength of the emission and excitation maxima, which is known as Stoke's shift. This technique has the potential to resolve spectral components arising from different species, especially when typical offsets for $\Delta\lambda$ are between 20 and 60 nm (Ahmad and Reynolds, 1995; Wu et al., 2006). SFS has been used for tracing the detection of sewage impacted rivers (Galapate, 1998; Hur and Kong, 2008; Hur et al., 2008) and fingerprinting wastewaters (Wu et al., 2006). Typical synchronous fluorescence spectra are shown in Figure 3.13.

In addition to the T, A_C, B, and C peaks, EEMs acquired from sewage-derived DOM commonly exhibit additional peaks at λ_{ex} = 375 nm, 350 nm, and 330 nm, and corresponding λ_{em} = 410–450 nm (see Figure 3.14). These are now known to be characteristic of optical brighteners such as those used in detergents (Westerhoff et al., 2001; Hayashi et al., 2002; Takahashi and Kawamura, 2007; Hartel et al., 2008). As much as 80% of these optical brighteners are removed during wastewater treatment but they can still be detected at low concentrations of approximately 0.5 mg L^{-1} in river water (Poiger et al., 1998). Studies of EEMs of untreated wastewaters show that they commonly comprise a broad humic-type peak C with intense T peaks and B peaks which occur at the same position in optical space as standard solutions of tryptophan and tyrosine, respectively (Baker et al., 2004). Peak T generally contributes the highest intensity peaks in wastewaters (Reynolds and Ahmad, 1997), and previous researchers (Galapate et al., 1998; Baker et al., 2003, 2004; Reynolds, 2003) have demonstrated that it can be considered as a tracer and relic of anthropogenic material in natural waters owing to its intensity. Rivers that are impacted by sewage wastes do not typically show optical brightening agents in their EEMs because of the high level of background fluorescence (Baker, 2001). This background is often high because sewage-derived DOM is rich in proteinaceous material which is associated with intense tryptophan-like (T and A_T) fluorescence. Sewage-derived DOM is dominated by organic matter originating from microbial activity (Hudson et al., 2008). In contrast, the

DOM related to natural systems is derived predominantly from plant material and where microbial activity is slow and weak but nevertheless sustained. The unique origins of wastewater DOM are highlighted in their associated EEMs, which are most notably different from EEMs of DOM from marine and freshwater systems, where peak C predominates (Hudson et al., 2007). The unique spectral characteristics of wastewater DOM have facilitated the tracking of sewage contamination in aquatic systems (Galapate et al., 1998; Baker, 2001; Baker et al., 2003, 2005; Chen et al., 2003; Holbrook et al., 2005; Hudson et al., 2008; Carstea, 2010).

The intense fluorescence intensity associated with sewage-derived DOM has led to the investigation of fluorescence as a marker for existing biochemical and chemical parameters commonly used to determine wastewater quality and monitor wastewater treatment processes (Reynolds and Ahmad, 1997; Ahmad and Reynolds, 1999; Reynolds, 2002; Vasel and Praet, 2002; Lee and Ahn, 2004; Cumberland and Baker, 2007; Hudson et al., 2008; Hur et al., 2008). Relationships between the fluorescence intensity of various peaks (A, B, T, and C) and water quality parameters have been investigated. The most common wastewater quality parameters investigated include the 5-day BOD; COD of filtered and unfiltered samples; total organic carbon (TOC); dissolved organic carbon (DOC); nitrate (NO_3^-) and phosphate (PO_4^{3-}) ions; UV absorption at 254 nm, 340 nm, and 450 nm; dissolved oxygen levels (DO); and ammonia (NH_3).

Correlations of Pearson's coefficient (r or r^2) values of between 0.77 and 0.98 have been reported between BOD and peaks T and A_T (Reynolds and Ahmad, 1997; Ahmad and Reynolds, 1999; Baker, 2001; Ahmad et al., 2002; Hudson et al., 2008; Hur et al., 2008). The correlations of peaks A_C and C (which represent humic and fulvic-like fluorescence) with the 5-day BOD were lower (r^2 = 0.72–0.77) (Baker, 2001; Hudson et al., 2008). The strong correlation between the 5-day BOD and the tryptophan-like fluorescence at around 340–350 nm is expected, as BOD is an indirect measure of microbial growth. Furthermore, the tryptophan-like fluorescence has been associated with wastewater microbial activity (Reynolds, 2002; Elliott et al., 2006; Hudson et al., 2008), whereas fulvic and humic-like fluorescence has been reported to represent the nonreadily biodegradable organic fraction of wastewaters (Reynolds, 2002).

Reported correlations of Pearson's coefficient for peak T and the COD and DOC values of wastewaters typically range from values between 0.42 and 0.97 (Reynolds, 2002; Vasel and Praet, 2002; Lee and Ahn, 2004; Wu et al., 2006). This is unsurprising given the well established chemical relationships between TOC and DOC. The wide range of correlations that have been observed for fluorescence (especially peak T) and DOC, COD, and TOC are explained by the ratio of refractory DOM to labile DOM, of fluorescent and nonfluorescent character. Correlations can exhibit greater variation depending on the ratio of both humic/fulvic-like/tryptophan-like material and fluorescent/nonfluorescent DOM. For this reason, some investigators have found it useful to examine the T/C ratio in relation to wastewater and effluent samples. Baker (2001) found that the T/C ratios for rivers (1.0) were far lower than that of untreated sewage (2.7–31). A comprehensive data set representing the T/C ratios of waste, rivers, and drinking and deionized water is shown in the recent review by

Henderson et al. (2009). Limited data are available regarding the relationship with peak C, although a correlation of 0.87 was reported by Bari and Farooq in 1985. Overall, for both BOD and COD measurements, much stronger correlations have been found when sewage-derived fluorescence dominates the system under investigation. For example, Wu et al. (2006) noted that the strength of correlation between COD and peak T in river water increased when sewage-derived DOM was more dominant (correlation coefficients rising from 0.5 to 0.9). Strong correlations have also been found between peak T and PO_4^{3-}, NO_3^- (Baker and Inverarity, 2004), total Kjeldahl nitrogen (N_k), and NH_3 (Vasel and Praet, 2002). These relationships are generally considered to be indirect because wastewater-derived pollution is typically characterized by high phosphate and nitrate ions when advanced nutrient removal is not undertaken at the wastewater treatment plant. Relationships between ammonia and peak T are not always observed owing to the high removal rate of ammonia during sewage treatment. However, ammonia can be present in significant quantities during pollution events, as observed by Baker et al. (2003).

The monitoring of DOM levels through a treatment works allows process optimization, and it has been estimated that as much as 40% of energy costs could be saved through efficiency gains, particularly with respect to aeration (Ahmad and Reynolds, 1998). A number of articles by Ahmad and Reynolds (Ahmad and Reynolds, 1995, 1999; Reynolds and Ahmad, 1997; Reynolds, 2002) have determined that a decrease in normalized fluorescence intensities of peak T is observed from influent to effluent across a treatment process. Peak T at λ_{ex} = 280 and λ_{em} = 340 nm was identified as being most likely to relate to the biodegradable material. This phenomenon has been utilized by other researchers investigating wastewater treatment processes such as sludge dewatering (Yu et al., 2010), landfill leachates (Lu et al., 2009), membrane fouling (Moon et al., 2010), membrane bioreactors (Wang et al., 2009), organic matter removal via coagulation–flocculation processes (Gone et al., 2009) and the composting of municipal waste (He et al., 2011). Two of the key issues surrounding the analysis of wastewater samples using fluorescence spectroscopy are the correction for inner filter effects due to the highly absorbing nature of the samples and turbidity. Correction and normalization of fluorescence data are covered in Chapters 1 and 7, and more specifically the inner filter effects exhibited in wastewater samples is discussed by Reynolds and Ahmad (1997), Ahmad and Reynolds (1999), and Reynolds (2002).

A summary of the significant research that has established correlations between fluorescence and wastewater quality parameters over the last 25 years is shown in Table 3.3. What is evident from previous work is that strong correlations do exist between traditional water quality parameters and fluorescence, although there are issues with directly comparing fluorescence data between geographical locations and between sites. More recent literature (Hudson et al., 2007, 2008) indicates that future research should focus on utilizing and analyzing fluorescence measurements as a direct and independent parameter for water/wastewater quality, rather than as a surrogate for specific water quality parameters. It is widely accepted that further research is required to investigate fully the effects of advanced treatment process on peaks T and C, especially if fluorescence-based techniques are to be applied to wastewater treatment processes and the tracing of DOM within wastewater distribution systems.

Table 3.3. *Summary of correlations found between fluorescence peak intensities of sewage and wastewaters and common biochemical and chemical parameters*

Samples	Correlation (peak/parameter/Pearson's *r* unless stated)			References
Raw settled/treated sewage from three different treatment works ($n = 129$)	T_1	BOD_5	0.960	Reynolds and Ahmad (1997)
			0.970	
			0.960	
Raw settled/treated sewage ($n = 25$)	T_1–T_2	BOD_5	0.980	Ahmad and Reynolds (1999)
Synthetic sewage treated via a rotating bio-disc contactor ($n = 45$)	F_{Total}	BOD_5	0.890	Reynolds (2002)
	T_1	COD	0.920	
	F_{Total}–T_1	TOC	0.910	
F_{Total} = Total fluorescence intensity	F_{Total}	BOD_5	0.980	
	T_1	COD	0.980	
Settled and treated sewage samples over a 3 month period ($n = 56$)	F_{Total}–T_1	TOC	0.980	
		COD-BOD	0.840	
		BOD_5	0.790	
F_{Total} = Total fluorescence intensity		COD	0.820	
		TOC	0.800	
		BOD_5	0.930	
		COD	0.940	
		TOC	0.930	
		COD-BOD	0.710	
Filtered raw sewage	T_1	COD	0.420	Vasel and Praet (2002)
		TOC	0.410	
		N_k	0.690	
		NH_4-N	0.650	
		COD	0.560[a]	
		TOC	0.530[a]	
		N_k	0.760[a]	
		NH_4-N	0.840[a]	
Treated effluent samples (over a 3-month period)	T_1	COD	0.900	Lee and Ahn (2004)
Wastewater samples (96 in total) using $COD_{Dissolved}$ values	T_1	$COD_{Dissolved.}$	0.370	Wu et al. (2006)
		COD	0.510	
Sewage effluents ($n = 16$)	C_1	DOC	0.140	Cumberland and Baker (2007)
Wastewater effluents (223 samples – sewage, trade, and pollution incidents)	T_1	BOD_5	0.906[b]	Hudson et al. (2008)
	T_2	TOC	0.876[b]	
	C_2	BOD_5	0.848[b]	
	A	TOC	0.802[b]	
		BOD_5	0.771[b]	
		TOC	0.870[b]	
		BOD_5	0.720[b]	
		TOC	0.808[b]	

[a] Partial least squares where $n = 20$.

[b] Spearman's *p*.

3.5.2 Drinking Water Fluorescence

In comparison to wastewaters, the application of fluorescence spectroscopy for the detection and monitoring of organic matter (OM) in drinking water sources and water distribution and water systems is an emerging discipline. Organic matter, both dissolved (DOM) and natural (NOM), is ubiquitous in all waters that are used to supply drinking water systems (Matilainen et al., 2011). Even so, the recent advances of fluorescence-based techniques are sufficiently mature so as to be considered an alternative approach to more conventional methods for the characterization of DOM and/or NOM in drinking water systems. In a recent review by Matilainen et al. (2011), a comprehensive overview of current methods that are used to characterize NOM in relation to drinking water treatment, including fluorescence, is provided. Early work (Rosario-Ortiz et al., 2007) applied the use of EEMs as a tool for characterizing OM (inclusive of DOM or NOM) in drinking water sources. DOM characterization in drinking water sources is important, as it is known that DOM contributes to the formation of disinfection by-products (DBP), and therefore affects how water treatment facilities are optimized. The potential for DOM in drinking water sources to generate DBPs was investigated by Marhaba et al. (2009). Researchers were able to predict trihalomethanes formation potential by applying a principal component regression model to dimensionally reduced spectral fluorescent signatures. Beggs et al. (2009) studied the relationships between fluorescence intensities (total fluorescence intensities and fluorescence indexes), redox index, chlorine demand, and DBP formation during chlorination. This study used a PARAFAC model to extract 13 components (fluorophores or groups of fluorophores) from fluorescence EEMs. Quinone-like components were found to be strongly correlated to DBP formation. Johnstone and Miller (2009) investigated the correlation of water quality characteristics of Iowa River water and associated isolated fractions to the formation of DBP, specifically trihalomethanes and haloacetic acids, subsequent to chlorination using multifactor linear regression. In this work, defined regions within fluorescence EEMs were identified using fluorescence regional integration (Chen et al., 2003) and the changes in the fluorescence intensities of these identified regions, in conjunction with chlorine consumption, were reported to correlate to the formation of specific DBPs. This work was further developed in a study in which a three component PARAFAC model was used to assess drinking water DBP formation (Johnstone et al., 2009). This PARAFC model correlated with DOC, chlorine consumption, and individual DBP formation potential. Interestingly, the multifactor linear regression of selected component scores showed linear relationships to individual DBPs. According to the researchers, the specificity of this approach makes the prediction of DBP formation (DBP formation potential) possible.

A recent review by Henderson et al. (2009) concludes that the sensitive detection of contamination events in recycled water systems may be achieved by monitoring peak T and/or peak C fluorescence. Hambly et al. (2010) also examined the application of fluorescence spectroscopy as a monitoring tool in recycled water treatment plants and dual (recycled and drinking water) distribution systems. This work detected a 10-fold difference in

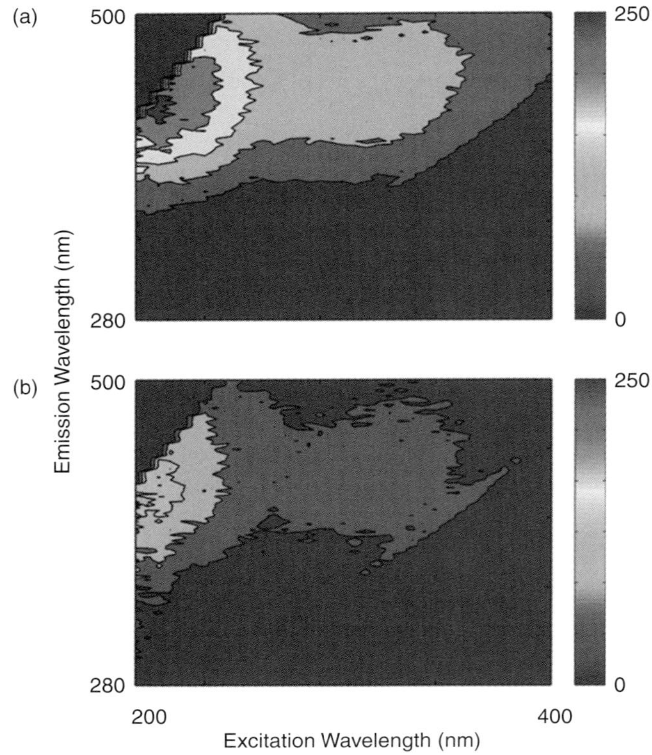

Figure 3.15. Excitation–emission matrices of waters at different water treatment stages. (a) Raw water (b) Clarified water. (Adapted from Bieroza et al., 2009a, with permission from Elsevier.) (See Plate 8.)

the mean fluorescence intensities observed for recycled water compared to drinking water, and concluded that fluorescence could be used for detecting cross connection. Bieroza et al. (2009a, 2009b, 2010) used an EEM technique for the assessment of TOC removal efficiency. Organic matter characterization of water samples was obtained for 16 UK surface water treatment works, and the fluorescence intensity of peak C was found to be a sensitive and reliable measure of OM content, providing both spatial and temporal variations (Bieroza et al., 2009a). Variations in EEMs were reported for samples from different sites, highlighting the importance of the nature of DOM present. Figure 3.15 shows EEMs for raw surface water and clarified surface water obtained from the same site. The same researchers (Bieroza et al., 2011a) also reported the use of fluorescence spectroscopy as a tool to assess the effect of changing coagulation pH on OM removal, character, and composition.

There is a great deal of interest in the application of fluorescence techniques for the monitoring of DOM in drinking water and drinking water treatment systems. Early research dealt with the influence of chlorination and oxidation of NOM and the prediction of disinfectant by-product (DBP) formation. Korshin et al. (1999) demonstrated that differential fluorescence–based indexes (namely the ratio of normalized fluorescence intensities at 500 and 450 nm using an excitation of 320 nm) used in conjunction with known treatment parameters (chlorine dose, chlorine reaction time, temperature, and NOM properties) can be used to determine DBP formation and speciation. Korshin et al. (1999) noted that the observed changes in fluorescence properties of the NOM after chlorination are consistent with the breakdown of NOM caused by both changes in the aromatic fluorophores and the conformations and molecular weights of NOM molecules. In conclusion, Korshin et al. (1999) hypothesized that the degradation of the aromatic chlorine attack sites, breakdown of the NOM, and release of DBPs occur simultaneously. Świetlik and Silorska (2004) used total luminescence studies, synchronous fluorescence techniques, and EEMs, to monitor the effects of chlorine dioxide and ozone on NOM. Their research showed that oxidation of NOM with chlorination dioxide resulted in a decrease in aromaticity and fragmentation of NOM fractions. Ozonation of NOM resulted in the formation of a significant amount of ozonation by-products. In both cases NOM fractions were shown to exhibit high reactivity. Yang et al. (2008) used EEMs to obtain fluorescence intensity data from 16 organic matter fractions isolated from a variety of sources including river water, wastewater effluents, water treatment works, lake water, and groundwater. As part of their work the researchers employed the fluorescence regional integration approach first developed by Chen et al. (2003) to analyze the fluorescence intensity data generated by the EEMs. Using this approach they were able to show relationships among fluorescence intensity data, organic matter properties, and DBP formation during chloramination. Specific UV absorbance (SUVA) values at 254 nm also correlated with DBPs formed during chloramination and these correlations were found to be significantly higher than those derived from the EEM data using fluorescence regional integration.

Research carried out in this field over the past 2 years has been concerned with data mining, data analysis, and use of multivariate analysis methods of the fluorescence spectra. Recently, Peiris et al. (2010) reported the use of principal component analysis (PCA) of EEMs for the performance monitoring of pretreatment stages (e.g., biological filtration), and to identify fouling events in membrane-based drinking water treatment processes. This work demonstrated that principal component score plots could be related to high fouling events resulting from elevated levels of particulates and/or colloidal material. The impact of this work could result in identification of key "foulants" and the provision of an early warning system that allows the implementation of suitable countermeasures.

Bieroza et al. (2009b) investigated the use of different multivariate analysis methods and artificial neural networks (ANNs) for the decomposition and calibration of EEMs obtained from DOM present in drinking waters. This research is the first to evaluate and compare the application of different data mining methods, including multiway analysis and artificial neural networks, for the analysis of EEMs. This research characterized the organic matter fluorescence properties and its removal in drinking water treatment. PARAFAC methodology and self-organizing maps were employed to analyze the EEM data, in order to obtain information about the organic matter present. From this it was possible to reduce the dimensionality of the data and therefore enhance the efficiency of calibration. Calibration of the fluorescence data with TOC values incorporated the use of partial least squares (PLS), multiple linear regression (MLR), and neural network with back-propagation. All models except PARAFAC-MLR produced consistent correlation coefficients for the validation data set. This research is the first to conduct such comparative analysis of fluorescence data modeling and addresses key issues regarding the suitability of different decomposition and calibration methods for the analysis of fluorescence intensity data. In a follow on from this work the same researchers (Bieroza et al., 2011b) have recently reported the application of robust data mining techniques for the assessment of water treatment performance. Again PCA and PARAFAC were used in addition to self-organizing maps. Bagoth et al. (2011) developed and validated a seven-component PARAFAC model using 147 EEMs of water samples obtained from two drinking water treatment plants. In this work NOM fractions (humics, building blocks, neutrals, biopolymers, and low molecular weight acids) correlated with maximum fluorescence intensities of seven PARAFAC components extracted from the EEM data. This work concluded that the fluorescent components derived from EEMs using PARAFAC can be related to defined NOM fractions, and therefore provide a tool for evaluating the removal of defined NOM fractions during water treatment.

The interest in this area is growing substantially and it is envisaged that this particular area of research will continue to emerge over the coming years such that it will impact considerably on the operational management of water treatment works.

References

Arctic Council and the International Arctic Science Committee (IASC). (2005). *ACIA (Arctic Climate Impact Assessment)*. New York: Cambridge University Press, 1042 pp.

Ahmad, S.R. and Reynolds, D.M. (1995). Synchronous fluorescence spectroscopy of wastewater and some potential constituents. *Water Res.*, **29**, 1599–1602.

Ahmad, S.R. and Reynolds, D.M. (1999). Monitoring of water quality using fluorescence techniques, prospect of on-line process control. *Water Res.*, **33**, 2069–2074.

Ahmad, S.R., Foster, V.G. and Reynolds D.M. (1993). Laser scattering technique for the non-invasive analysis of wastewater. SPIE Vol. 2092, *Substance Detection Systems* (pp. 353–359). Washington, DC: SPIE – International Society for Optical Engineering.

Ahmad, U.K., Ujang, Z., Yusop, Z., and Fong, T.L. (2002). Fluorescence technique for the characterization of natural organic matter in river water. *Water Sci. Technol.*, **46**(9), 117–125.

Amon, R.M.W. and Budeus, G. (2003). Dissolved organic carbon distribution and origin in the Nordic Seas: Exchanges with the Arctic Ocean and the North Atlantic. *J. Geophys. Res.*, **108**: (C7) 3221, doi:10.1029/2002JC001594.

Ayoub, L., Hallock, P., and Coble, P. (2009). Colored dissolved organic material increases resiliency of coral reefs by controlling exposure to UVR.. In *Proceedings of the 11th International Coral Reef Symposium*, Ft. Lauderdale, Florida, July 7–11, 2008, pp. 572–576.

Baghoth, S.A., Sharma, S.K., and Amy, G.L. (2009). Tracking natural organic matter (NOM) in a drinking water treatment plant using fluorescence excitation-emission matrices and PARAFAC. *Water Res.*, **45**, 797–809.

Baker, A. (2001). Fluorescence excitation–emission matrix characterization of some sewage-impacted rivers. *Environ. Sci. Technol.*, **35**(5), 948–953.

Baker, A. (2002). Fluorescence excitation-emission matrix characterization of river waters impacted by a tissue mill effluent. *Environ. Sci. Technol.*, **36**(7), 1377–1382.

Baker, A. and Genty, D. (1999). Fluorescence wavelength and intensity variations of cave waters. *J. Hydrol.*, **217**, 19–34.

Baker, A. and Lamont-Black, J. (2001). Fluorescence of dissolved organic matter as a natural tracer of groundwater. *Groundwater*, **39**, 745–750.

Baker, A. and Curry, M. (2004). Fluorescence of leachates from three contrasting landfills. *Water Res.*, **38**, 2605–2613.

Baker, A., and Inverarity, R. (2004). Protein-like fluorescence intensity as a possible tool for determining river water quality. *Hydrol. Process.*, **18**(15), 2927–2945.

Baker, A., Smart, P.L., Edwards, R.L., and Richards, D.A. (1993). Annual banding in a cave stalagmite. *Nature*, **364**, 518–520.

Baker, A., Barnes, W.L., and Smart, P.L. (1997). Stalagmite drip discharge and organic matter fluxes in Lower Cave, Bristol. *Hydrol. Process.*, **11**, 1541–1555.

Baker, A., Mockler, N.J., and Barnes, W.L. (1999). Fluorescence intensity variations of speleothem forming groundwaters: implications for palaeoclimate reconstruction. *Water Resourc. Res.*, **35**, 407–413.

Baker, A., Inverarity, R., Charlton, M., and Richmond, S. (2003). Detecting river pollution using fluorescence spectrophotometry: Case studies from the Ouseburn, NE England. *Environ. Pollut.*, **124**(1), 57–70.

Baker, A., Inverarity, R., and Ward, D. (2005). Catchment-scale fluorescence water quality determination. *Water Sci. Technol.*, **52**(9), 199–207.

Baker, A., Bolton, L., Newson, M., and Spencer, R.G.M. (2008). Spectrophotometric properties of surface water dissolved organic matter in an afforested upland peat catchment. *Hydrol. Process.*, **22**, 2325–2336.

Bari, A. and Farooq, S. (1985). Measurement of wastewater treatment efficiency by fluorescence and UV absorbance. *Environ. Monit. Assess.*, **5**(4), 423–434.

Beggs, K.M.H., Summers, R.S., and McKnight, D.M. (2009). Characterizing chlorine oxidation of dissolved organic matter and disinfection by-product formation with fluorescence spectroscopy and parallel factor analysis, *J. Geophys. Res.*, **114**, G04001.

Bieroza, M., Baker, A., and Bridgeman, J. (2009a). Relating freshwater organic matter fluorescence to organic carbon removal efficiency in drinking water treatment. *Sci.Tot. Environ.*, **407** (5), 1765–1774.

Bieroza, M., Baker, A., and Bridgeman, J. (2009b). Exploratory analysis of excitation-emission matrix fluorescence spectra with self organizing maps as a basis for determination of organic matter removal efficiency at water treatment works. *J. Geophys. Res. Biogeosci.*, **114**, G00F07.

Bieroza, M.Z., Bridgeman, J., and Baker, A. (2010). Fluorescence spectroscopy as a tool for determination of organic matter removal efficiency at water treatment works. *DWES*, **3**, 63–70.

Bieroza, M.Z., Baker, A., and Bridgeman, J. (2011a). Assessment of low pH coagulation performance using fluorescence spectroscopy. *J. Environ. Engng.*, **137**(7), 596.

Bieroza, M.Z., Baker, A., and Bridgeman, J. (2011b). New data mining and calibration approaches to the assessment of water treatment efficiency. *Adv. Eng. Software*, **44**, 126–135.

Black, A.P., and Christman, R.F. (1963). Characteristics of coloured surface waters. *J. Am. Water Works Assoc.*, **55**(6), 753–770.

Blough, N.V. and Green, S.A. (1995). Spectroscopic characterization and remote sensing of nonliving organic matter. In R.G. Zepp and C. Sonntag (Eds.), *The Role of Nonliving Organic Matter in the Earth's Carbon Cycle* (pp. 23–45). Dahlem Conference Report. Chichester: John Wiley & Sons.

Blough, N.V. and Del Vecchio, R. (2002). Chromophoric DOM in the coastal environment. In Hansell, D.S. and Carlson, C.A. (Eds.), *Biogeochemistry of Marine Dissolved Organic Matter* (pp. 509–546). Amsterdam: Academic Press.

Blyth, A.J., Baker, A., Collins, M.J., Penkman, K.E.H., Gilmour, M.A., Moss, J.S., Genty, D., and Drysdale, R.N. (2008). Molecular organic matter in speleothems and its potential as an environmental proxy. *Quat. Sci. Rev.*, **27**, 905–921

Boss, E. and Zaneveld, J.R.V. (2003). The effect of bottom substrate on inherent optical properties: Evidence of biogeochemical processes. *Limnol. Oceanogr.*, **48**, 346–354.

Bro, R. (1997). PARAFAC. Tutorial and applications. Chemom. *Intell. Lab. Syst.*, **38**, 149–171.

Brookshire, E.N., Valett, J.H.M., Thomas, S.A., and Webster., J.R. (2005). Coupled cycling of dissolved organic nitrogen and carbon in a forest stream. *Ecology*, **86**, 2487–2496.

Burdige, D.J., Kline, S.W., and Chen, W. (2004). Fluorescent dissolved organic matter in marine sediment pore waters. *Mar. Chem.*, **89**, 289–311.

Cabaniss, S.E. and Shuman, M.S. (1987). Synchronous fluorescence spectra of natural waters: Tracing sources of dissolved organic matter. *Mar. Chemi.*, **21**, 37–50.

Cammack, W.K.L., Kalff, J., Prairie, Y.T., and Smith, E.M. (2004). Fluorescent DOM in lakes: Relationships with heterotrophic metabolism. *Limnol. Oceanogr.*, **49**, 2034–2045.

Carstea, E., Baker, A., Johnson, R., and Reynolds, D.M. (2010). Real-time monitoring of river water quality using in-line continuous acquisition of fluorescence excitation and emission matrices. *Water Res.*, **44**(18), 5356–5366.

Chen, W., Westerhoff, P., Leenheer, J.A., and Booksh, K. (2003). Fluorescence excitation-emission matrix regional integration to quantify spectra for dissolved organic matter. *Environ. Sci. Technol.*, **37**(24), 5701–5710.

Coble, P.G. (1996). Characterization of marine and terrestrial DOM in seawater using excitation-emission matrix spectroscopy. *Mar. Chemi.*, **51**, 325–346.

Coble, P.G. (2007). Marine optical biogeochemistry: the chemistry of ocean color. *Chem. Rev.*, **107**, 402–418.

Coble, P.C., Green, S. Blough, N.V., and Gagosian, R.B. (1990). Characterization of dissolved organic matter in the Black Sea by fluorescence spectroscopy. *Nature*, **348**, 432–435.

Coble, P.G., Del Castillo, C.E., and Avril, B. (1998). Distribution and optical properties of CDOM in the Arabian Sea during the 1995 Southwest Monsoon. *Deep-Sea Res. Pt. II*, **45**, 2195–2223.

Conmy, R.N. (2008). *Temporal and Spatial Patterns in Optical Properties of Colored Dissolved Organic Matter on Florida's Gulf Coast: Shelf to Stream to Aquifer*. Ph.D. dissertation, College of Marine Science, University of South Florida, St. Petersburg, 134 pp.

Conmy, R.N., Coble, P.G. Cannizzaro, J.P., and Heil, C.A. (2009). Influence of extreme storm events on West Florida Shelf CDOM distributions. *J. Geophys. Res.*, **114**, G00F04, doi:10.1029/2009JG000981.

Cory, R.M. and McKnight, D.M. (2005). Fluorescence spectroscopy reveals ubiquitous presence of oxidized and reduced quinones in dissolved organic matter. *Environ. Sci. Technol.*, **39**, 8142–8149.

Cory, R.M., McKnight, D.M., Chin, Y.P., Miller, P., and Jaros, C.L. (2007). Chemical characteristics of fulvic acids from Arctic surface waters: Microbial contributions and photochemical transformations. *J. Geophys. Res. Biogeosci.*, **112**, G04S51, doi:10.1029/2006JG000343.

Cory, R.M., Miller, M.P., McKnight, D.M., Guerard, J.J., and Miller, P.L. (2010). Effect of instrument-specific response on the analysis of fulvic acid fluorescence spectra. *Limnol. Oceanogr. Methods*, **8**, 67–78.

Cumberland, S.A. and Baker, A. (2007). The freshwater dissolved organic matter fluorescence – total organic carbon relationship. *Hydrol. Process.*, **21**, 2093–2099.

Del Castillo, C. (2005). Remote sensing of organic matter in coastal waters. In R.L. Miller, C.E. Del Castillo, and B.A. McKee (Eds.), *Remote Sensing of Coastal Aquatic Environments: Technologies, Techniques, and Applications* (pp. 157–180). Dordrecht, The Netherlands: Springer Science+Business Media.

Del Castillo, C.E. and Miller, R.L. (2008). On the use of ocean color remote sensing to measure the transport of dissolved organic carbon by the Mississippi River Plume. *Remote Sens. Environ.*, **112**(3), 836–844.

Del Vecchio, R. and Blough, N.V. (2004). On the origin of the optical properties of humic substances. *Environ. Sci. Technol.*, **38**, 3885–3891.

Dienert, F. (1910). De la recherche des substances fluorescentes dans le controle de la sterilisation des eaux. *C. R. Hehd. Seanc. Acad. Sci. Paris*, **150**(8), 487–488.

Donard, O.F.X., Lamotte, M., Belin, C., and Ewald, M. (1989). High sensitivity fluorescence spectroscopy of Mediterranean waters using a conventional or a pulsed laser excitation source. *Mar. Chem.*, **27**, 117–136.

Duursma, E.K. (1974). The fluorescence of dissolved organic matter in the sea. In N.G. Jerlov and E. Steemann Nielsen (Eds.), *Optical Aspects of Oceanography* (pp. 237–256). New York: Academic Press.

Eaton, A.D., Clesceri, L.S., Rice, E.W., and Greenberg, A.E. (2005). *Standard Methods for the Examination of Water and Wastewaters* (21st ed.). Washington, DC: American Public Health Association, Water Environment Federation, and American Water Works Association.

Elliott, S., Lead, J.R., and Baker, A. (2006). Characterisation of the fluorescence from freshwater planktonic bacteria. *Water Res.*, **40**(10), 2075–2083.

Fellman, J.B., Hood, E., Edwards, R.T., and D'Amore, D.V. (2009a). Changes in the concentration, biodegradability, and fluorescent properties of dissolved organic matter during stormflows in coastal temperate watersheds. *J. Geophys. Res. Biogeosci.*, **114**, G01021, doi:10.1029/2008JG000790.

Fellman, J.B., Hood, E., D'Amore, D.V., Edwards, R.T., and White, D. (2009b). Seasonal changes in the chemical quality and biodegradability of dissolved organic matter exported from soils to streams in coastal temperature watersheds. *Biogeochemistry*, **95**, 277–293.

Fellman, J.B., Hood, E., Edwards, R.T., and Jones, J.B. (2009c). Uptake of allochthonous DOM from soil and salmon in coastal temperate rainforest streams. *Ecosystems*, **12**, 747–759.

Fellman, J.B., Hood, E., and Spencer, R.G.M. (2010). Fluorescence spectroscopy opens new windows into dissolved organic matter dynamics in freshwater ecosystems: A review. *Limnol. Oceanogr.*, **55**, 2452–2462.

Fichot, C.G., and Miller, W.L. (2010). An approach to quantify depth-resolved marine photochemical fluxes using remote sensing: Application to carbon monoxide (CO) photoproduction. *Remote Sens. Environ.*, **114**, 1363–1377.

Galapate, R.P., Baes, A.U., Ito, K., Mukai T., Shoto, E., and Okada, M. (1998). Detection of domestic wastes in Kurose River using synchronous spectroscopy. *Water Res.*, **32**, 2232–2239.

Gardner, G.B., Chen, R.F., and Berry, A. (2005). High-resolution measurements of chromophoric dissolved organic matter (CDOM) in the Neponset River Estuary, Boston Harbor, MA. *Mar. Chem.*, **96**, 137–154.

Ghassemi, M. and Christman, R.F. (1968). Properties of the yellow organic acids of natural waters. *Limnol. Oceanogr.*, **13**, 583–597.

Gone, D.L., Seidel, J.L., Batiot, C., Bamory, K., Ligban, R., and Biemi, J. (2009). Using fluorescence spectroscopy EEM to evaluate the efficiency of organic matter removal during coagulation–flocculation of a tropical surface water (Agbo reservoir). *J. Hazard. Mater.*, **172**(2–3), 693–699.

Guay, C.K., Klinkhammer, G.P., Falkner, K.K., Benner, R., Coble, P.G., Whitledge, T.E., Black, B., Bussell, F.J., and Wagner, T.A. (1999). High-resolution measurements of dissolved organic carbon in the Arctic Ocean by in situ fiber-optic spectrometry. *Geophys. Res. Lett.*, **26**(8), 1007.

Hambly, A., Henderson, R.K., Storey, M.V., Baker, A., Stuetz, R.M., and Khan, S.J. (2010). Fluorescence monitoring at a recycled water treatment plant and associated dual distribution system – Implications for cross-connection detection. *Water Res.*, **44**, 5323–5333.

Hartel, P.G., Rodgers, K., Moody, G.L., Hemmings, S.N.J., Fisher, J.A., and McDonald, J.L. (2008). Combining targeted sampling and fluorometry to identify human fecal contamination in a freshwater creek. *J. Water Health*, **6**(1), 105–116.

Hayashi, Y., Managaki, S., and Takada, H. (2002). Fluorescent whitening agents in Tokyo Bay and adjacent rivers: Their application as anthropogenic molecular markers in coastal environments. *Environ. Sci. Technol.*, **36**(16), 3556–3563.

He, X., Xi, B., Wei, Z., Guo, X., Li, M., An, D., and Liu, H. (2011). Spectroscopic characterization of water extractable organic matter during composting of municipal solid waste. *Chemosphere*, **82**(4), 541–548.

Henderson, R.K., Baker, A., Murphy, K.R., Hambly, A.C., Stuetz R.M., and Khan S.J., (2009). Fluorescence as a potential monitoring tool for recycled water systems: A review. *Water Res.*, **43**, 863–881.

Hessen, D.O., Carroll, J., Kjeldstad, B., Korosov, A.A., Pettersson, L.H., Pozdnyakov, D., and Sørensen, K. (2010). Input of organic carbon as determinant of nutrient fluxes, light climate and productivity in the Ob and Yenisey estuaries. *Estuar. Coast. Shelf Sci.*, **88**, 53–62.

Hoge, F.E., and Lyon, P.E. (2005). New tools for the study of oceanic eddies: Satellite derived inherent optical properties. *Remote Sens. Environ.*, **95**, 444–452.

Holbrook, R.D., Breidenich, J., and DeRose, P.C. (2005). Impact of reclaimed water on select organic matter properties of a receiving stream-fluorescence and perylene sorption behavior. *Environ. Sci. Technol.*, **39**(17), 6453–6460.

Holbrook, R.D., DeRose, P.C., Leigh, S.D., Rukhin, A.L., and Heckert, N.A. (2006). Excitation–emission matrix fluorescence spectroscopy for natural organic matter characterization: A quantitative evaluation of calibration and spectral correction procedures. *Appl. Spectrosc.*, **60**(7), 791–799.

Hood, E., Williams, M.W., and McKnight, D.M. (2005). Sources of dissolved organic matter (DOM) in a Rocky Mountain stream using chemical fractionation and stable isotopes. *Biogeochemistry*, **74**, 231–255.

Hood, E., Fellman, J.B., and Edwards, R.T. (2007). Salmon influences on dissolved organic matter in a coastal temperate brown-water stream. *Limnol. Oceanogr.*, **52**, 1580–1587.

Hood, E., Fellman, J.B., Spencer, R.G.M., Hernes, P.J., Edwards, R., David D'Amore, D., and Scott, D. (2009). Glaciers as a source of ancient and labile organic matter to the marine environment. *Nature*, **462**, 1044–1048.

Hudson, N., Baker, A., and Reynolds, D. (2007). Fluorescence analysis of dissolved organic matter in natural, waste and polluted waters – a review. *River Res. Appl.*, **23**, 631–649.

Hudson, N., Baker, A., Ward, D., Reynolds, D.M., Brunsdon, C., Carliell-Marquet, C., and Browning, S. (2008). Fluorescence spectrometry as a surrogate for the BOD5 test in water quality assessment: An example from South West England. *Sci. Tot. Environ.*, **391**(1), 149–158.

Hur, J. and Kong, D.S. (2008). Use of synchronous fluorescence spectra to estimate biochemical oxygen demand (BOD) of urban rivers affected by sewage treatment. *Environ. Technol.*, **29**(4), 435–444.

Hur, J., Hwang, S.J., and Shin, J.K. (2008). Using synchronous fluorescence technique as a water quality monitoring tool for an urban river. *Water, Air, Soil Pollut.*, **191** (1–4), 231–243.

Jaffe, R., McKnight, D., Maie, N., Cory, R., McDowell, W.H., and Campbell, J.L. (2008). Spatial and temporal variations in DOM composition in ecosystems: The importance of long-term monitoring of optical properties. *J. Geophys. Res. Biogeosci.*, **113**, G04 032,doi:10.1029/2008JG000683.

Johnstone, D.W. and Miller, C.M. (2009). Fluorescence excitation–emission matrix regional transformation and chlorine consumption to predict trihalomethane and haloacetic acid formation. *Environ. Engng. Sci.*, **26**(7), 1163–1170.

Johnstone, D.W., Sanchez, N.P., and Miller, C.M. (2009). Parallel factor analysis of excitation–emission matrices to assess drinking water disinfection byproduct formation during a peak formation period. *Environ. Eng. Sci.*, **26**(10), 1551–1559.

Kalle, K. (1937). Nahrstoff untersuchengen als hydrographisches Hilfsmittel zur unterscheidung von Wasserkorpern. *Ann. Hydrogr.*, **65**, 276–282.

Kalle, K. (1938). Zum Problem der Merreswasserfarbe. *Ann. Hydrogr.*, **66**, 1–13.

Kieber, R.J., Whitehead, R.F., Reid, S.N., Willey, J.D., and Seaton, P.J. (2006). Chromophoric dissolved organic matter (CDOM) in rainwater, southeastern North Carolina, USA. *J. Atmos. Chem.*, **54**, 21–41.

Kieber, R.J., Willey, J.D., Whitehead, R.F., and Reid, S.N. (2007). Photobleaching of chromophoric dissolved organic matter (CDOM) in rainwater. *J. Atmos.Chem.*, **58**, 219–235.

Korshin, G.V. Kumke, M.U., Li, C-W., and Frimmel, F.H. (1999). Influence of chlorination on chromophores and fluorophores in humic substances. *Environ. Sci. Technol.*, **33**(8), 1207–1212.

Kowalczuk, P., Cooper, W.J., Durako, M.J., Kahn, A.E., Gonsior, M., and Young, H. (2010). Characterization of dissolved organic matter fluorescence in the South Atlantic Bight with use of PARAFAC model: Relationships between fluorescence and its components, absorption coefficients and organic carbon concentrations. *Mar. Chem.*, **118**, 22–36.

Laane, R.W.P.M. (1982). Influence of pH on the fluorescence of dissolved organic-matter. *Mar. Chem.*, **11**, 395–401.

Lapworth, D.J., Gooddy, D.C., Butcher, A.S., and Morris, B.L. (2008). Tracing groundwater flow and sources of organic carbon in sandstone aquifers using fluorescence properties of dissolved organic matter (DOM). *Appl. Geochem.*, **23**, 3384–3390.

Lapworth, D.J., Gooddy, D.C., Allen, D., and Old, G.H. (2009). Understanding groundwater, surface water, and hyporheic zone biogeochemical processes in a Chalk catchment using fluorescence properties of dissolved and colloidal organic matter. *J. Geophys. Res.*, **114**, G00F02.

Lee, S. and Ahn, K.H. (2004). Monitoring of COD as an organic indicator in wastewater and treated effluent by fluorescence excitation–emission (FEEM) matrix characterization. *Water Sci. Technol.*, **50** (8), 57–63.

Lohrenz, S.E., Cai, W-J., Chen, F., Chen, X., and Tuel, M. (2009). Seasonal variability in air-sea fluxes of CO_2 in a river-influenced coastal margin. *J. Geophys. Res.*, 115, IC10, doi: 10.1029/2009JC005608

Lonborg, C., Alvarez-Salgado, X.A., Davidson, K., Martinex-Garcia, S., and Teira, E. (2010). Assessing the microbial bioavailability and degradation rate constants of dissolved organic matter by fluorescence spectroscopy in the coastal upwelling system of the Ria de Vigo. *Mar. Chem.*, **119**, 121–129.

Lu, F., Chang, C-H., Lee, D-J., He, P.J., Shao, L.M., and Su, A. (2009). Dissolved organic matter with multi-peak fluorophores in landfill leachate. *Chemosphere*, **74** (4), 575–582.

Maie, N., Boyer, J.N., Yang, C., and Jaffé, R. (2006). Spatial, geomorphological, and seasonal variability of CDOM in estuaries of the Florida Coastal Everglades. *Hydrobiologia*, **569**, 135–150.

Marhaba, T.F., Borgaonkar, A.D., and Punburananon, K. (2009). Principal component regression model applied to dimensionally reduced spectral fluorescent signature for the determination of organic character and THM formation potential of source water. *J. Hazard. Mater.*, **169**, 998–1004.

Matilainen, A., Gjessing, E.T., Lahtinen, T., Hed, L., Bhatnagar, A., and Sillanpää, M. (2011). An overview of the methods used in the characterisation of natural organic matter (NOM) in relation to drinking water treatment. *Chemosphere*, **83**, 1431–1442.

McGarry, S.F. and Baker, A., (2000). Organic acid fluorescence: Applications to speleothem palaeoclimate reconstruction. *Quat. Sci. Rev.*, **19**, 1087–1101

McKnight, D.M., Boyer, E.W., Westerhoff, P.K., Doran, P.T., Kulbe, T., and Andersen, D.T. (2001). Spectrofluorometric characterization of aquatic fulvic acids for determination of precursor organic material and general structural properties. *Limnol. Oceanogr.*, **46**, 38–48.

Milbrandt, E.C., Coble, P.G., Conmy, R.N., Martignette, A.J., and Siwicke, J.J. (2010). Evidence for the production of marine fluorescent dissolved organic matter in coastal environments and a possible mechanism for formation and dispersion. *Limnol. Oceanogr.*, **55**, 2037–2051.

Miller, M.P. and McKnight, D.M. (2010). Comparison of seasonal changes in fluorescent dissolved organic matter among aquatic lake and stream sites in the Green Lakes Valley. *J. Geophys. Res.*, **115**, G00F12.

Miller, M.P., McKnight, D.M., Cory, R.M., Williams, M.W., and Runkel, R.L. (2006). Hyporheic exchange and fulvic acid redox reactions in an alpine stream/wetland ecosystem, Colorado Front Range. *Environ. Sci. Technol.*, **40**, 5943–5949.

Moon, J. Lee, S., Song, J.H., and Cho, J. (2010). Membrane fouling indicator of effluent organic matter with nanofiltration for wastewater reclamation, as obtained from flow field-flow fractionation. *Sep. Purif. Technol.*, **73**(2) 164–172.

Muller, C.L., Baker, A., Hutchinson, R., Fairchild, I.J., and Kidd, C. (2008). Analysis of rainwater dissolved organic carbon compounds using fluorescence spectrophotometry. *Atmos. Environ.*, **42**, 8036–8045.

Naden, P.S., Old, G.H., Eliot-Laize, C., Granger, S.J., Hawkins, J.M.B., Bol, R., and Haygarth, P. (2010). Assessment of natural fluorescence as a tracer of diffuse agricultural pollution from slurry spreading on intensely-farmed grasslands. *Water Res.*, **44**, 1701–1712.

Nelson, N.B. and Siegel, D.A. (2002). Chromophoric DOM in the open ocean. In D.A. Hansell and C.A. Carlson (Eds.), *Biogeochemistry of Marine Dissolved Organic Matter* (pp. 547–578). Amsterdam: Academic Press.

O'Brien, B.J. (1956). "After-glow" of cave calcite. *Bull. Natl. Speleol. Soc.*, **18**, 50–51.

Otis, D.B., Carder, K.L., English, D.C., and Ivy, J.E. (2004). CDOM transport from the Bahamas Banks. *Coral Reefs*, **23**, 152–160.

Parlanti, E., Worz, K., Geoffroy, L., and Lamotte, M. (2000). Dissolved organic matter fluorescence spectroscopy as a tool to estimate biological activity in a coastal zone submitted to anthropogenic inputs. *Org. Geochem.*, **31**, 1765–1781.

Peiris, R.H., Halle´, C., Budman, H., Moresoli, C., Peldszus, S., Huck, P.M., and Legge, R.L. (2010). Identifying fouling events in a membrane-based drinking water treatment process using principal component analysis of fluorescence excitation-emission matrices. *Water Res.*, **44**, 185–194.

Poiger, T., Field, J.A., Field, T.M., Siegrist, H., and Giger, W. (1998). Behavior of fluorescent whitening agents during sewage treatment. *Water Res.*, **32** (6), 1939–1947.

Prahl, F.G. and Coble, P.G. (1994). Input and behavior of dissolved organic carbon in the Columbia River Estuary. In K.R. Dyer and R.J. Orth (Eds.), *Changes in Fluxes in Estuaries: Implications from Science and Management* (pp. 451–457). ECSA22/ERF Symposium, Plymouth, England. September 1992, Denmark: Olsen and Olsen. Fredensborg.

Retamal, L., Vincent, W.F., Martineau, C., and Osburn, C.L. (2007). Comparison of the optical properties of dissolved organic matter in two river-influenced coastal regions of the Canadian Arctic. *Estuar. Coast. Shelf Sci.*, **72**, 261–272.

Reynolds, D.M. (2002). The differentiation of biodegradable and non-biodegradable dissolved organic matter in wastewaters using fluorescence spectroscopy. *J. Chem. Technol. Biotechnol.*, **77**(8), 965–972.

Reynolds, D.M. (2003). Rapid and direct determination of tryptophan in water using synchronous fluorescence spectroscopy. *Water Res.*, **37**, 3055–3060.

Reynolds, D.M. and Ahmad, S.R. (1995a). The effect of metal ions on the fluorescence of sewage wastewater. *Water Res.*, **29**, 2214–2216.

Reynolds, D.M., and Ahmad, S.R. (1995b). Synchronous fluorescence spectroscopy of wastewater and some potential constituents. *Water Res.*, **29**, 1599–1602.

Reynolds, D.M. and Ahmad, S.R. (1997). Rapid and direct determination of wastewater BOD values using a fluorescence technique. *Water Res.*, **31**(8), 2012–2018.

Rosario-Ortiz, F.L., Snyder, S.A., and Suffet, I.H. (2007). Characterization of dissolved organic matter in drinking water sources impacted by multiple tributaries. *Water Res.*, **41**, 4115–4128.

Santos, P.S.M., Duarte, R.M.B.O., and Duarte, A.C. (2009). Absorption and fluorescence properties of rainwater during the cold season at a town in Western Portugal. *J. Atmos. Chem.*, **62**, 45–57.

Saraceno, J.F., Pellerin, B.A., Downing, B.D., et al. (2009). High frequency in situ optical measurements during a storm event: Assessing relationships between dissolved organic matter, sediment concentrations, and hydrologic processes. *J. Geophys. Res. Biogeosci.*, **114**, G00F09, doi:10.1029/2009JG000989.

Shank, G.C., Zepp, R.G., Vähätalo, A., Lee, R., and Bartels, E. (2010a). Photobleaching kinetics of chromophoric dissolved organic matter derived from mangrove leaf litter and floating *Sargassum* colonies. *Mar. Chem.*, **119**, 162–171.

Shank, G.C., Lee, R., Vähätalo, A., Zepp, R.G., and Bartels, E. (2010b). Production of chromophoric dissolved organic matter from mangrove leaf litter and floating *Sargassum* colonies. *Mar. Chem.*, **119**, 172–181.

Smart, P.L., Finlayson, B.L., Rylands, W.D., and Ball, C.M. (1976). The relation of fluorescence to dissolved organic carbon in surface waters. *Water Res.*, **10**, 805–811.

Spencer, R.G.M., Pellerin, B.A., Bergamaschi, B.A., Downing, B.D., Kraus, T.E.C., Smart, D.R., Dahlgren, R.A., and Hernes, P.J. (2007). Diurnal variability in riverine dissolved organic matter composition determined by in situ optical measurement in the San Joaquin River (California, USA). *Hydrol. Process.*, **21**, 3181–3189.

Spencer, R.G.M., Stubbins, A., Hernes, P.J., Baker, A., Mopper, K., Aufdenkapme, A.K., Dyda, R.Y., Mwamba, V.L., Mangangu, A.M., Wabakanghanzi, J.N., and Six, J. (2009a). Photochemical degradation of dissolved organic matter and dissolved lignin phenols from the Congo River. *J. Geophys. Res. Biogeosci.*, **114**, G03010, doi:10.1029/2009JG000968.

Spencer, R.G.M., Aiken, G.R., Butler, K.D., Dornblaser M.M., Striegl, R.G., and Hrnes, P.J. (2009b). Utilizing chromophoric dissolved organic matter measurements to derive export and reactivity of dissolved organic carbon to the Arctic Ocean: A case study of the Yukon River, Alaska. *Geophys. Res. Lett.*, **36**, L06401, doi:10.1029/2008GL036831.

Spencer, R.G.M., Hernes, P.J., Ruf, R., et al. (2010). Temporal controls on dissolved organic matter and lignin biogeochemistry in a pristine tropical river, Democratic Republic of Congo. *J. Geophys. Res. Biogeosci.*, **115**, G03013, doi:10.1029/2009JG001180.

Stabenau, E.R., Zepp, R.G., Bartels, E., and Zika, R.G. (2004). Role of the seagrass *Thalassia testudinum* as a source of chromophoric dissolved organic matter in coastal south Florida. *Mar. Ecol. Prog. Ser.*, **282**, 59–72.

Stedmon, C.A., Markager, S., and Bro, R. (2003). Tracing dissolved organic matter in aquatic environments using a new approach to fluorescence spectroscopy. *Mar. Chem.*, **82**, 239–254.

Stewart, A.J. and Wetzel, R.G. (1980). Fluorescence: Absorbance ratios – a molecular-weight tracer of dissolved organic matter. *Limnol. Oceanogr.*, **25**, 559–564.

Stewart, A.J. and Wetzel, R.G. (1981). Asymmetrical relationships between absorbance, fluorescence, and dissolved organic-carbon. *Limnol. Oceanogr.*, **26**, 590–590.

Stubbins, A., Spencer, R.G.M., Chen, H., Hatcher, P.G., Mopper, K., Hernes, P.J., Mwamba, V.L., Mangangu, A.M., Wabakanghanzi, J.N., and Six, J. (2010). Illuminated darkness: Molecular signatures of Congo River dissolved organic matter and its photochemical alteration as revealed by ultrahigh precision mass spectrometry. *Limnol. Oceanogr.*, **55**(4), 1467–1477.

Świetlik, J. and Silorska, E. (2004). Application of fluorescence spectroscopy in the studies of natural organic matter fractions reactivity with chlorine dioxide and ozone. *Water Res.*, **38**, 3791–3799.

Takahashi, M., and Kawamura, K. (2007). Simple measurement of 4,4-bis(2-sulfostyryl)-biphenyl in river water by fluorescence analysis and its application as an indicator of domestic wastewater contamination. *Water, Air, Soil Pollut.*, **180** (1–4), 39–49.

Tremblay, L.B., Dittmar, T., Marshall, A.G., Cooper, W.J., and Cooper, W.T. (2006). Molecular characterization of dissolved organic matter in a North Brazilian mangrove porewater and mangrove-fringed estuaries by ultrahigh resolution Fourier Transform-Ion Cyclotron Resonance mass spectrometry and excitation/emission spectroscopy. *Mar. Chem.*, **105**, 15–29.

Tzortziou, M., Neale, P.J., Osburn, C.L., Megonigal, J.P., Maie, N., and Jaffé, R. (2008). Tidal marshes as a source of optically and chemically distinctive coloured dissolved organic matter in the Chesapeake Bay. *Limnol. Oceanogr.*, **53**,148–159.

Vasel, J.L. and Praet, E. (2002). On the use of fluorescence measurements to characterize wastewater. *Water Sci. Technol.*, **45** (4–5), 109–116.

Waggot A. and Butcher H.V. (1976). Analysis of the organic carbon content of sewage effluent: general and specific group analysis. Technical Report TR29. Swindon, UK: Water Research Centre.

Walker, S.A., Amon, R.M.W., Stedmon, C., Duan, S., and Louchouarn, P. (2009). The use of PARAFAC modeling to trace terrestrial dissolved organic matter and fingerprint water masses in coastal Canadian Arctic surface waters. *J. Geophys. Res.*, **114**, G00F06.

Wang, Z., Wu, Z., and Tang, S. (2009). Characterization of dissolved organic matter in a submerged membrane bioreactor by using three-dimensional excitation and emission matrix fluorescence spectroscopy, *Water Res.*, **43**(6), 1533–1540.

Westerhoff, P., Chen, W., and Esparza, M. (2001). Fluorescence analysis of a standard fulvic acid and tertiary treated wastewater. *J. Environ. Qual.*, **30** (6), 2037–2046.

White, W.B. and Brennan, E.S. (1989). Luminescence of speleothems due to fulvic acid and other activators. In *Proceedings of the 10th International Conference of Speleology*, August 13–20, 1989, Budapest, pp. 212–214.

Wickland, K.., Neff, J.C., and Aiken, G.R. (2007). DOC in Alaskan boreal forests: sources, chemical characteristics, and biodegradability. *Ecosystems*, **10**, 1323–1340.

Williams, C.J., Yamashita, Y., Wilson, H.F., Jaffé, R., and Xenopoulos, M.A. (2010). Unraveling the role of land use and microbial activity in shaping dissolved organic matter characteristics in stream ecosystems. *Limnol. Oceanogr.*, **55**, 1159–1171.

Wilson, H.F. and Xenopoulos, M.A. (2009). Effects of agricultural land use on the composition of fluvial dissolved organic matter. *Nature Geosci*, **2**, 37–41.

Wu, J., Pons, M.N., and Potier, O. (2006). Wastewater fingerprinting by UV–visible and synchronous fluorescence spectroscopy. *Water Sci. Technol.*, **53** (4–5), 449–456.

Yamashita, Y. and Tanoue, E. (2008). Production of bio-refractory fluorescent dissolved organic matter in the ocean interior. *Nature Geosci.*, **1**, 579–582.

Yamashita, Y., Cory, R.M., Nishioka, J., Kuma, K., Tanoue, E., and Jaffé, R. (2010). Fluorescence characteristics of dissolved organic matter in the deep waters of the Okhotsk Sea and the northwestern North Pacific Ocean. *Deep-Sea Res. PtII*, doi:10.1016/j.dsr2.2010.02.016.

Yanga, X., Shanga, C., Lee, W., Westerhoff, P., and Fan, C. (2008). Correlations between organic matter properties and DBP formation during chloramination. *Water Res.*, **42**, 2329–2339.

Yu, G-H., He, P-J., and Shao, L-M. (2010). Novel insights into sludge dewaterability by fluorescence excitation–emission matrix combined with parallel factor analysis. *Water Res.*, **44**, 797–806.

Zepp, R.G. and Schlotzhauer, P.F. (1981). Comparison of photochemical behavior of various humic substances in water. 3. Spectroscopic properties of humic substances. *Chemosphere*, **10**, 479–486.

Zimmerman, R.C. (2003). A biooptical model of irradiance distribution and photosynthesis in seagrass canopies. *Limnol. Oceanogr.*, **48**, 568–585.

Zimmerman, R.C. (2006). Light and photosynthesis in seagrass meadows. In A.W.D Larkhum, R.J. Orth, and C.M. Duarte (Eds.), *Seagrasses: Biology, Ecology, and Conservation*. (pp. 303–321). Dordrecht, The Netherlands: Springer Science+Business Media.

Zsolnay, A. (2003). Dissolved organic matter: Artefacts, definitions, and functions. *Geoderma*, **113**, 187–209.

Zsolnay, A., Baigar, E., Jimenez, M., Steinweg, B., and Saccomandi, F. (1999). Differentiating with fluorescence spectroscopy the sources of dissolved organic matter in soils subjected to drying. *Chemosphere*, **38**, 45–50.

Part II

Instrumentation and Sampling

4

Sampling Design for Organic Matter Fluorescence Analysis

ROBERT G. M. SPENCER AND PAULA G. COBLE

4.1 Introduction

The development of a robust and appropriate sampling design is fundamental to the success of any study involving the collection of field samples, and many of the aspects are also applicable to laboratory studies. The importance of careful sample collection and processing is imperative to underscore, as these are the first steps undertaken in studies based in aquatic ecosystems. Clearly any issues introduced during these early stages result in a wasted effort as data quality is compromised from the outset of the study.

In the vast majority of studies in aquatic ecosystems that utilize absorbance and fluorescence analyses to examine dissolved organic matter (DOM), the goal of the study is to relate the measurements to the actual value in the natural environment. Therefore, such studies are interested in producing accurate data, and the primary aim of this chapter is to outline strategies that support this goal. For researchers aiming to relate measurements made in the laboratory back to aquatic ecosystems two points are fundamental for DOM sampling: first, collection of a representative sample and second, avoidance of any contamination (U.S. Geological Survey, 2006). Collection of a representative sample is determined by study objectives and the expected or known spatial and temporal variability of DOM at the study location. Contamination of a sample via the procedures of collection, processing, and storage and contamination avoidance are discussed in detail in this chapter. As with all measurements in analytical chemistry, quality assurance and control protocols should be implemented to ensure high-quality data and this chapter aims to outline useful information to aid with sample collection and processing.

4.2 Sample Collection

4.2.1 Contamination Sources

There are three main types of contamination to be aware of when collecting samples. The first are airborne and waterborne contaminants such as tobacco smoke, dust, exhaust fumes, cleaning and lubricating solutions as well as hydrocarbons from a research vessel or other vessels in the water. This category of contaminant can normally be avoided by paying careful attention to the sampling site and sampling upwind (or on the lee side of a research vessel) of atmospheric contaminants and avoiding any surface layers of the water body under study (see Section 4.2.4). Always record any potential sources of contamination when collecting samples in the field.

Second, careless handling is a major source of contamination including working in a dirty environment, careless sampling procedures, and not wearing disposable gloves. Any water that enters the sample bottle should not have been in contact with either a gloved or a bare hand, and latex gloves should be avoided as they can leach absorbing compounds. Powderless polypropylene or nitrile gloves are recommended. Careless sample handling is easily avoided by the investigator, and gloves should be worn at all times; however, working in a dirty environment (e.g., muddy wetlands) is common for many investigators. One simple solution is to employ the clean hands/dirty hands approach as outlined by the U.S. Geological Survey (U.S. Geological Survey, 2006) if two people are available. Briefly stated, one person is designated as clean hands (CH) and one person as dirty hands (DH) and both wear disposable gloves. CH is responsible for all operations involving equipment that comes into contact with the sample (e.g., changing filters), whereas DH is responsible for all operations involving contact with potential sources of contamination (e.g., preparing sampling equipment such as pumps; handling of equipment such as multiparameter instruments for ancillary field measurements). Another alternative if working in a dirty environment is to set up a clean workspace (e.g., in a clean laboratory on a research vessel or inside a vehicle) where sample processing can be carried out without further undue exposure to potential contaminants.

Finally, the sampling equipment itself may represent a source of contamination. Leaching of organic substances from new plastics as well as the sorption of organics to some plastics means that some fluorocarbon polymers, polypropylene, polyethylene (linear), polyvinyl chloride, silicone, and nylon sampling equipment are not recommended for DOM work (Lane et al., 2003). Nylon tubing in particular has been shown to leach absorbing compounds (Figure 4.1) and so is to be avoided. Glass, Teflon, and some "aged" plastic containers are recommended when cleaned appropriately (see Section 4.2.3). Metals such as stainless steel are appropriate (e.g., pressure filtration unit) if not corroded and cleaned appropriately. It is strongly recommended that all sampling equipment be blank checked with laboratory grade water (preferably type I ultrapure water, 18.2 MΩ cm^{-1} at 25°C with DOC < 10 ppb) at each step of possible contamination.

4.2.2 Blanks and Replicate Samples

The easiest way to catch any potential contamination issues is through the use of appropriate blanks utilizing laboratory-grade water. Blanks can address a range of issues as highlighted by the U.S. Geological Survey (2006). For example, field blanks should be collected and processed at the field site in exactly the same way as the samples of interest, replacing the sample water with laboratory grade water. This allows an overall contamination effect of all the collection and processing to be observed and represents a basic quality control on the whole sampling design. This can then be broken down into subcomponents of the

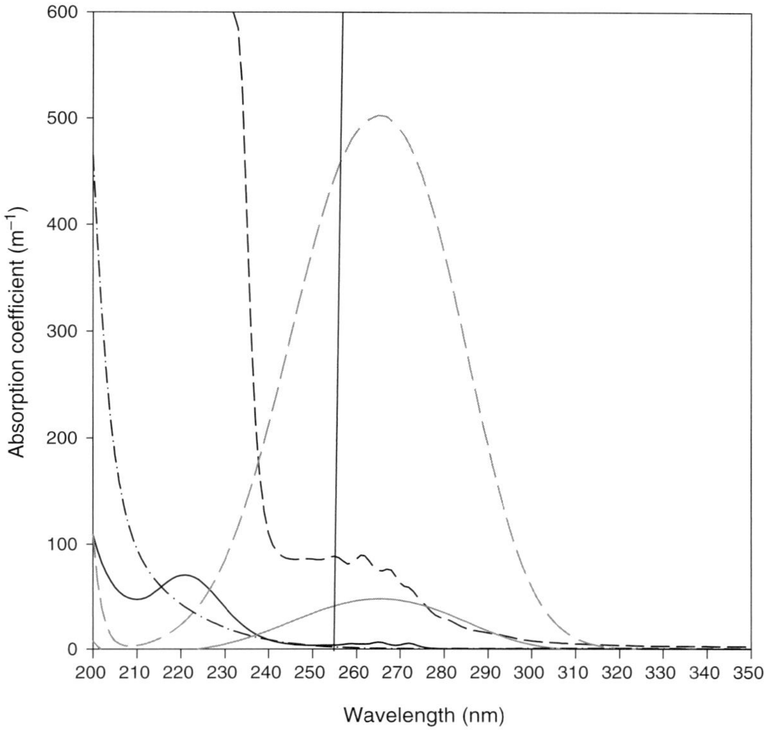

Figure 4.1. CDOM absorption properties of nylon tubing leachate (3.2 L of laboratory grade water flushed nylon through tubing at 200 ml per minute and then sampled; solid black line), soap (10% diluted antibacterial soap; dashed black line), acetone (12 drops in 200 mL of laboratory-grade water and cuvette rinsed with acetone and subsequently filled with laboratory grade water; solid gray line and dashed gray lines respectively) and methanol (dash-dot black line). Note the solid black line at 254 nm highlighting the potential for interference with $SUVA_{254}$ measurements.

sampling design to trace the specific source of any contamination observed in the field blank if required. Individual equipment, filter, and sampler blanks can be used to test for contamination from equipment, filters, or sampling devices (e.g., pumps, Niskin bottles) and to assess the equipment, filter and sampler cleaning protocols. Ambient blanks that are exposed to atmospheric outfall or other relevant conditions (e.g., inside a fume hood) can be used to examine if the sample exposure was the source of any contamination. As such, blanks are both important in observing any potential contamination and also discovering any contamination so that it may be eliminated and are fundamental to successful sampling design.

Replicate samples are collected at the same time or as close to one another as possible via identical equipment and procedures. Replicate samples are also important for quality control as they allow for identification and also quantification of any variability in the overall sampling, or parts thereof if broken down into individual sections (e.g., filtration replicates). It is recommended that replicates be collected in a minimum of triplicates (if only duplicates are collected and the data vary the replicates are of little aid). Different types of replicates can be used dependent on the quality assurance needs and the type of study (U.S. Geological Survey, 2006). For example, concurrent replication refers to a number of samples of water collected at the same time or as close to one another as possible and is the most common type of replication utilized. These replicates provide quality control data and can be used to examine the variability introduced from sample collection, handling, and processing and also the variability in laboratory handling and analysis of samples. Sequential replication refers to a number of samples that are collected consecutively and thus it varies from concurrent replication as it allows for examination of any variation in the water sampled between sequential replicates. Thus sequential replication examines the temporal variability of the water under study as well as variability introduced from sample collection, handling, and processing and also the variability in laboratory handling and analysis of samples. Finally, split replication refers to samples derived from splitting one bulk sample that is designated for a specific analysis (e.g., fluorescence) into a number of subsamples. These subsamples are then submitted to different laboratories for identical analyses (e.g., Sharp et al., 1995, 2004; Jaffe et al., 2008; Murphy et al., 2010).

4.2.3 Equipment Cleaning

The aim of equipment cleaning is to remove any existing organic matter and any other substances that may interfere with DOM fluorescence or absorbance analyses. After cleaning, sampling equipment should be stored dry and covered (e.g., wrapped in aluminum foil, storage in freezer bags) to limit microbial growth and to avoid contamination. Typical equipment cleaning protocols involve an initial cleaning with a detergent (e.g., soapy water), followed by extensive rinsing with laboratory-grade water, then a cleaning step with a solvent (typically methanol or acetone), again followed by extensive rinsing with laboratory grade water. The main issue here is that soap, acetone, and methanol all interfere with the optical properties of chromophoric DOM (CDOM) (Figure 4.1) and thus it is imperative that if they are used in the cleaning process, that all trace of them is removed. Sampling equipment may also be effectively cleaned by soaking in a 10% hydrochloric acid (HCl) solution and then extensively rinsed with laboratory grade water. Where possible glass is recommended for sampling equipment (e.g., bottles), as it can be guaranteed to have no organic matter contamination if it is combusted at 450°C for greater than four hours in a muffle furnace. Before combustion it is recommended to cover parts of the glassware that may come into contact with the sample (e.g., the opening on bottles) with aluminum foil so that they can be stored covered with foil (e.g., screw bottle caps on over the foil, thus not exposing the inside to air until the sample is taken). Glassware that cannot be combusted (e.g., volumetrics, pipettes) and plastics (including bottle caps) can be cleaned as described earlier and then after the final rinsing with laboratory grade water dried in a drying oven (60°C). Teflon tape can be very useful for sealing sampling equipment after cleaning to avoid any contaminants getting inside cleaned materials. While conducting cleaning procedures, inspect sampling equipment for signs of wear and tear that may lead to sources of contamination and replace as required. All cleaning procedures should be carried out in a designated area that does not have any airborne or other sources of contamination. Cleaning procedures preferably should be performed in a clean laboratory environment while wearing appropriate safety wear and powderless polypropylene or nitrile disposable gloves. Finally, all equipment cleaning procedures should be verified by filling or passing laboratory-grade water through the equipment post the cleaning process and analyzing for both fluorescence and absorbance.

4.2.4 Water Samplers

A description of the appropriateness for CDOM sampling of the numerous commercially available water samplers is beyond the scope of this chapter; however, a number of general points and a few specific examples are worth mentioning as they apply to a broad range of sampling equipment. The initial sampling of water via any water sampler should be undertaken with clean equipment that has been extensively rinsed/flushed with sample water from that study site. This may mean pumping water through a submersible pump head and tubing for a few minutes to flush the system or the collection of water on the upcast after a CTD profile. Whatever sampler is used for water collection it should not be made of materials as highlighted in Section 4.2.1, as these are likely to result in contamination issues (e.g., use Nalgene but not nylon tubing). Water samplers specifically designed for trace element work such as Niskin bottles (Figure 4.2) or similar with silicone seals and Teflon-coated springs are ideal. One attractive option if thoroughly cleaned before deployment in the field is the use of a GO-FLO water sampler (General Oceanics) (Figure 4.2), as these samplers operate on the close–open–close principle and thus pass the organic-rich surface microlayer closed. This avoids contamination from the organic-rich surface microlayer and also any potential contamination from hydrocarbons from the research vessel. If Niskin bottles or similar are used to collect water near the surface, care should also be taken to avoid the surface microlayer. Discrete water samplers are commonly used in a number of studies as they allow for water samples to be collected via an auto-sampler at a number of time points without the investigator being present. With respect to DOM absorbance and fluorescence analyses one drawback with the utilization of discrete water samplers is modification of the optical characteristics of DOM may occur in the hold time of the water between collection and subsequent analysis (see Section 4.4).

4.3 Sample Preservation

4.3.1 Filtration Techniques

Filtration directly after sample collection is recommended, as ongoing biological processes may either cause an increase in CDOM via release from organisms (e.g., lysis, sloppy feeding) or a decrease in CDOM due to metabolization by bacterioplankton. Filtration is also

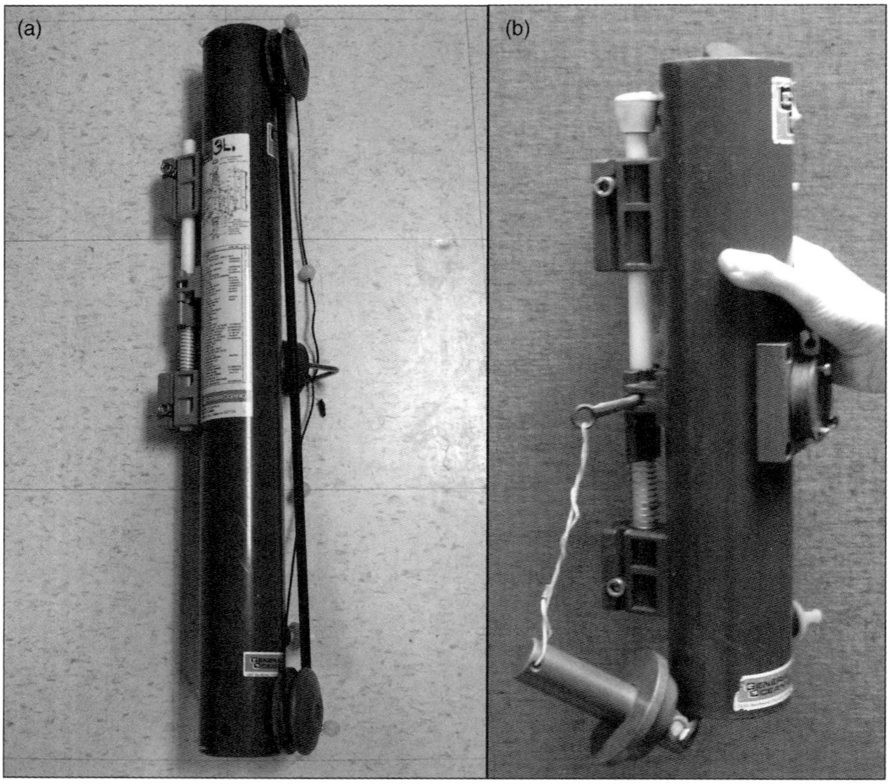

Figure 4.2. (a) GO-FLO water sampler. (b) Niskin bottle.

recommended as particles can interfere with CDOM measurements, for example, scattering of light (Blough et al., 1993; Chen and Gardner, 2004). There is no consensus on the filter cutoff size to define DOM operationally, and studies have ranged in the filter cutoff size used in the aquatic sciences from 0.1 to 1.2 μm, dependent largely on the field of research and the goals of the study. The vast majority of studies though have used filters in the 0.2–0.7 μm cutoff range. The requirements of an ideal filter have been previously described by Kremling and Brugmann (1999); modified for DOM analysis an ideal filter should have a uniform and reproducible pore size, a high filtration rate and not clog quickly, not adsorb any of the DOM to be determined or contain any constituents to be determined, have reasonable mechanical strength, not shed fibers, and in addition it should be easy to clean. Filters fall into two main categories; depth filters or sieve filters (Kremling and Brugmann, 1999). With respect to depth filters the pore sizes are poorly defined and separation of particles from the solution is dependent on physical trapping and the amount of surface contact. Such filters have a nominal pore size (i.e., in reality there is a range of pore sizes around the reported cutoff), although this can often be very close to the nominal cut-off (e.g., 0.6 μm–0.8 μm for Whatman GF/F 0.7 μm filters) and are typically made of cellulose, metal oxides, or glass fibers. Sieve filters have much more consistent pore sizes and are typically made from plastic films (e.g., polycarbonate or polysulfone capsule filters).

Filters can be a significant source of organic matter contamination and must be cleaned appropriately. Glass fiber filters such as Whatman GF/F are one of the most popular choices of filters owing to their rapid flow rate, high loading capacity and ease of cleaning (combust at 450°C > 4 hours). All polycarbonate or polysulfone capsule filters should be cleaned as recommended by the manufacturer and flushed with copious amounts of laboratory-grade water. Silver filters have also been employed successfully with respect to DOM absorbance and fluorescence analyses (Lapworth et al., 2009). It is recommended for all filter types that a suitable volume of sample water (dependent on the loading capacity of the filter) is rinsed through before collection of any sample water and that all surfaces that come into contact with the sample should be triple rinsed with filtrate (e.g., sample bottles). Filter holders must be precleaned as described in Section 4.2.3 and the filtration system thoroughly tested for any contamination (see Section 4.2.2). Vacuum, pressure, and gravity filtration are all routinely applied by researchers in their filtration units. The filtration pressure should be maintained at the lowest necessary to pass the filter in a reasonable period of time with respect to any potential microbial action and filtration should be conducted out of direct light. High pressures have been shown to result in lysis of cells during filtration (Rosenstock and Simon, 1993) and subsequent "contamination" of the filtrate. Clogging of filters reduces the flow rate and also may reduce the nominal pore size of the filter and over time lead to the lysis of cells into the filtrate and so should be avoided. In-line filtration typically of a large prescreen pore size filter (e.g., 10, 1.2 μm) before a small pore size filter (e.g., 0.2 μm) is recommended in many freshwater systems to avoid clogging and associated issues if desiring to filter to small pore sizes (e.g., Ahad et al., 2006; Saraceno et al., 2009).

4.3.2 Effects of Filtration on Fluorescence

The impact of filtration has been examined with respect to DOM fluorescence both in laboratory and in situ–based studies. Fluorescence of organic matter was examined in six contrasting U.K. freshwaters in unfiltered, 1.2 μm filtered, and 0.2 μm filtered waters to compare the effects of filtered to unfiltered water on organic matter fluorescence and to examine the effects of filtration at different pore sizes on DOM fluorescence in a study by Baker et al. (2007). The six chosen sites ranged from those with very good through those with bad water quality and ranged from free-flowing rivers to slow moving and regulated canal waters and also lake water. The study by Baker et al. (2007) focused on two fluorophores in detail: tryptophan-like fluorescence at an excitation of 225–230 nm and an emission range of 335–350 nm and humic-like fluorescence at an excitation of 230–245 nm and an emission range of 395–430 nm. All of the samples in the study showed for the tryptophan-like fluorescence a significant decrease between the unfiltered water and the 1.2 μm filtered water (5–71% decrease, mean = 35 %; Table 4.1) and the unfiltered water and the 0.2 μm filtered water (32–86% decrease, mean = 58%; Table 4.1). The results also show

Table 4.1. *Changes in the fluorescence intensity of tryptophan-like fluorescence (excitation of 225–230 nm and an emission range of 335–350 nm) and humic-like fluorescence (excitation of 230–245 nm and an emission range of 395–430 nm) with filtration in six freshwater samples*

Percent decrease from initial tryptophan-like fluorescence

Sample	Treatment		
	Raw water	1.2 μm filtered	1.2 and 0.2 μm filtered
1	0	–5	–32
2	0	–32	–68
3	0	–71	–79
4	0	–7	–32
5	0	–52	–86
6	0	–43	–50

Percent decrease from initial humic-like fluorescence

Sample	Treatment		
	Raw water	1.2 μm filtered	1.2 and 0.2 μm filtered
1	0	–2	–8
2	0	2	–4
3	0	–11	–14
4	0	–22	–30
5	0	–9	–13
6	0	–12	–9

Source: Modified from Baker et al. (2007).

a removal of tryptophan-like fluorescence between the 1.2 μm filtered water and the 0.2 μm filtered water. However, in the majority of samples a greater decrease in tryptophan-like fluorescence was observed between the unfiltered and the 1.2 μm filtered rather than between the 1.2 and 0.2 μm filtered waters. The authors suggest this means a significant portion of tryptophan-like fluorescence derives from particulate and the larger colloidal material as well as that a fraction of tryptophan-like fluorescence is found in the <0.2 μm fraction. The majority of the samples in the study also showed for the humic-like fluorescence a decrease between the unfiltered water and the 1.2 μm filtered water (2–22% decrease, mean = 10%; Table 4.1) and the unfiltered water and the 0.2 μm filtered water (4–30 % decrease, mean = 13%; Table 4.1). These results suggest that although some humic-like fluorescence is in the particulate and colloidal fractions the majority is truly dissolved in the <0.2 μm fraction (Lead et al., 2006; Baker et al., 2007; Seredynska-Sobecka et al., 2007). One clear outcome from this study is that filter pore size impacts unevenly on fluorophores and thus emphasizes the need to standardize filter size within individual studies (i.e., do not change filter pore size within a study as it likely means results from different filters will not be comparable). It also raises significant concerns about the comparability of data across studies, especially if comparing studies with samples collected with extremely different filter cutoffs (e.g., 0.2 μm vs. 1.2 μm) and comparison of the tryptophan-like fluorescence or any ratios involving this fluorophore.

The influence of suspended particulates on in situ fluorometers is of great interest as one of the great attractions of these instruments is the ability to provide high-frequency data with the need for little if any sample processing, thus avoiding any potential contamination from filtration or sample storage (Del Castillo et al., 2001; Spencer et al., 2007a; Conmy et al., 2009). Utilizing a WETStar fluorometer (WET Labs), Belzile et al. (2006) showed a strong correlation between WETStar fluorometer measurements on unfiltered samples from a range of aquatic environments and filtered samples measured using a spectrofluorometer with suspended sediment concentrations of up to 35 mg L^{-1}. However, a recent study by Saraceno et al. (2009) shows evidence that at high concentrations of suspended sediments that may occur in turbid rivers or during flushing events (e.g., freshets) a reduction in DOM fluorescence as measured by a WETStar fluorometer was reported. Saraceno et al. (2009) state that the combined effects of scattering and absorption, due to high suspended particle concentrations, resulted in an underestimate in their DOM fluorescence on unfiltered water. Correction for particle interference, though, is possible, but as with most optical measurements, relationships will need to be "ground-truthed" to individual sites and water/sediment types as appropriate.

4.4 Storage

4.4.1 General Comments

The primary causes of sample instability are due to microbial and photochemical degradation, and the effects of these processes on DOM absorbance and fluorescence in aquatic samples are well documented (Moran et al., 2000; Osburn et al., 2001, 2009; Del Vecchio and Blough, 2002; Stedmon and Markager, 2005; Cory et al., 2007; Tzortziou et al., 2007; Wickland et al., 2007). Therefore, samples should be filtered (see Section 4.3.1) and stored in the cold and dark (e.g., refrigeration in the dark at approximately 4°C) for short-term storage and analyzed as soon as possible after collection. It is not always realistic, however, to analyze samples immediately after collection for a range of reasons (e.g., remote field sites, large number of samples collected in a short time period) and so researchers routinely have to store samples for DOM analyses. Aside from analyzing samples as soon as possible there is little general consensus on the length of time for which it is appropriate to store samples. The application of different storage methods and appropriate storage time-frames should be tested by the researcher on the range of DOM samples they are examining, as results are likely to vary with the source and history of different samples (e.g., highly colored allochthonous dominated DOM samples vs. optically clear autochthonous dominated DOM samples) and also the filtration technique and cutoff utilized. It is strongly recommended that investigators then document their storage procedures and any tests they conducted on storage effects in the methods sections of reports and publications. As there is no general consensus on a preferred storage method, Section 4.4.2 highlights a number of current common practices and describes the effects of the storage method used for each study on specific DOM types.

4.4.2 Refrigeration and Freezing

Refrigerators and freezers used to store DOM samples should not be used to store other biological samples, as some volatile organic compounds can pose a contamination risk. Caps on bottles should be screwed on tightly to prevent leakage, evaporation and contamination and bottles are recommended to be stored in an upright position. Bottle caps can be wrapped with Teflon tape after filling to provide a more secure seal. To avoid breaking bottles on freezing, do not overfill (i.e., allow room for expansion on freezing) and if possible cool the samples first via refrigeration to avoid breakage due to thermal shock. Storage blanks should be carried out to examine any potential artefacts. Although glass is a preferred storage container due to its ease of cleaning (see Section 4.2.3) Teflon and some "aged" plastic containers when cleaned appropriately (see Section 4.2.3) are more suitable for frozen storage owing to the greatly reduced chance of breakage in comparison to glass.

Refrigeration in the dark at approximately 4°C is commonly used for short-term storage of filtered DOM samples (e.g., Coble et al., 1998; Baker, 2002; Stedmon et al., 2003; Wickland et al., 2007; Fellman et al., 2009; Hood et al., 2009; Lapworth et al., 2009) and a NASA study showed no changes in CDOM absorption for samples refrigerated for less than 24 hours (Mitchell et al., 2000). Figure 4.3 shows that in a range of freshwater samples the protein-like fluorophore (tryptophan-like fluorophore; typical excitation maxima 270–280 nm and emission maxima 335–360 nm) and the prominent fulvic-like fluorophore (typical excitation maxima 310–370 nm and emission maxima 410–460 nm) (Coble et al., 1998; McKnight et al., 2001; Baker, 2002; Stedmon et al., 2003; Spencer et al., 2007b) showed no change within analytical reproducibility in the first 7 days of storage (Figure 4.3). After 7 days of storage all of the freshwater samples exhibited a decrease in both fluorophores outside analytical reproducibility and at the 2-month stage show a decrease of 10–35% in relation to initial fluorescence intensity dependent on sample type and fluorophore under investigation (Figure 4.3) with typically a greater decrease in the tryptophan-like fluorophore relative to the fulvic-like fluorophore. This is similar to results reported by Hudson et al. (2009), who also observed a decline in fluorescence intensity over time in refrigerated samples and found greater removal of protein-like fluorescence compared to humic and fulvic-like fluorescence. It is therefore suggested that refrigerated samples be analyzed within 1 week. This guideline though will be very dependent on water type, and relevant hold times should be determined for each study specific to the range of DOM types under investigation. For example, open-ocean 0.2 μm filtered seawater samples from the Equatorial Pacific have been stored at 4°C with no measurable change in

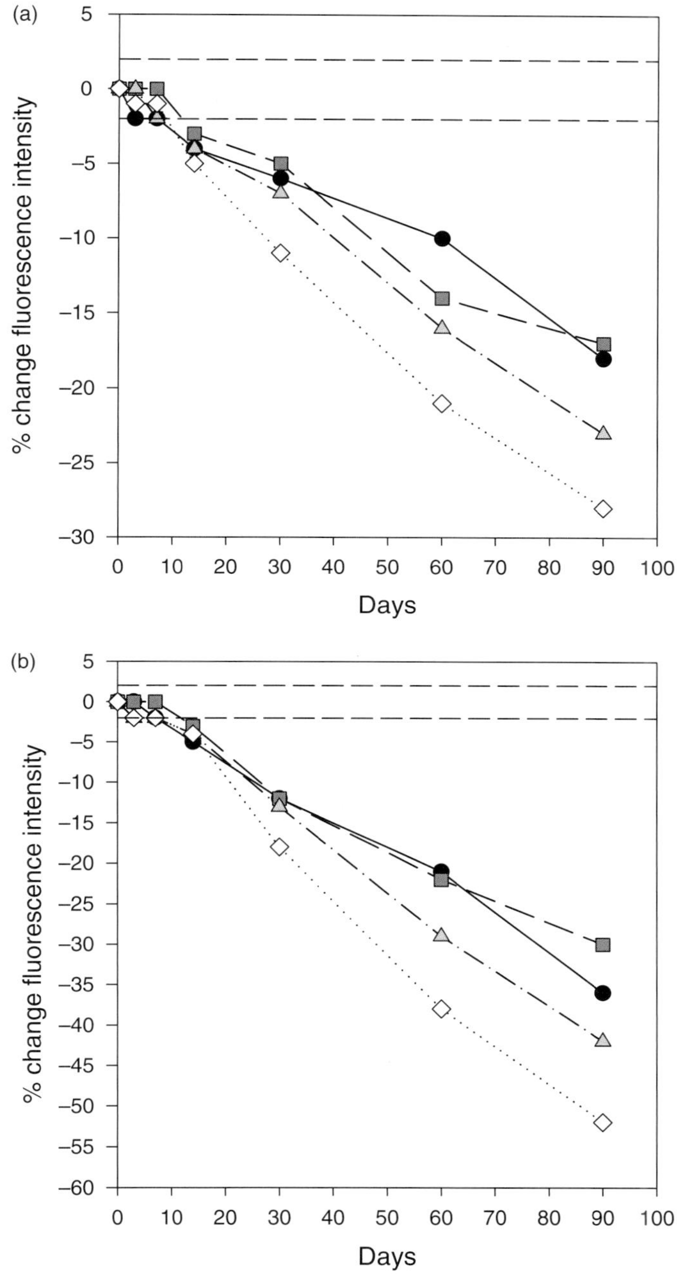

Figure 4.3. The percent change in fluorescence intensity of four freshwater samples over 90 days of storage via refrigeration (4°C) in the dark; horizontal dashed black lines represent analytical reproducibility. (a) Fulvic-like fluorophore (typical excitation maxima 310–370 nm and emission maxima 410–460 nm). (b) Protein-like fluorophore (tryptophan-like fluorophore; typical excitation maxima 270–280 nm and emission maxima 335–360 nm). Black circles, solid black lines = peatland river (DOC = 14.6 mg/L); dark gray squares, dashed black lines = organic-rich river (DOC = 9.8 mg L^{-1}); light gray triangles, dash-dot black lines = agricultural river (DOC = 4.8 mg L^{-1}); white diamonds, dot black lines = urban river (DOC = 3.2 mg L^{-1}).

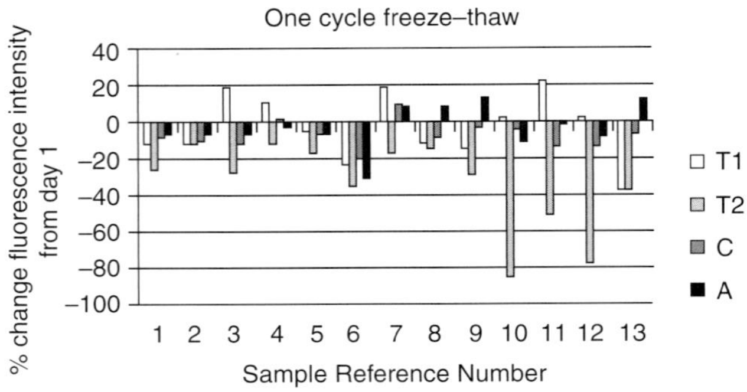

Figure 4.4. The percent change in fluorescence intensity after one cycle of freezing/thawing in 13 freshwater samples. T1 and T2 = tryptophan-like fluorescence in the regions of λex/em 280/350 nm and 215–220/340 nm respectively; C and A = humic-like fluorescence in the regions λex/em 380/420–480 nm and 260/380–460 nm respectively. (Reproduced from Hudson et al. 2009.)

absorbance properties for over 1 year (C. Swan, University of California, Santa Barbara, unpublished data). Likewise long-term (up to 1 year) frozen storage has been routinely used for CDOM fluorescence, absorbance and DOC analyses on open ocean samples with no observable changes outside of analytical error (P. Coble, unpublished data).

Freezing of filtered water samples to –20°C is also a commonly used storage method for DOM analyses (e.g., Coble et al., 1998; Murphy et al., 2008; Conmy et al., 2009; Walker et al., 2009; Gao et al., 2010; Spencer et al., 2010; Yamashita et al., 2010a). Clearly the quantity and quality of DOM in a sample will impact on its response to freezing and as a "general rule of thumb" it seems that highly colored allochthonous dominated DOM samples see much greater effects as a result of the freeze/thaw process than optically clear autochthonous dominated DOM samples (e.g., marine waters), and a number of marine DOM studies have shown minimal effects of freeze/thaw on DOM optical properties (Conmy et al., 2009; Yamashita et al., 2010b; P. Coble, unpublished data). Spencer et al. (2007c) and Hudson et al. (2009) found for a range of freshwater samples on freeze/thaw that fluorescence intensities and absorption coefficients exhibited both increases and decreases although on the whole a loss of CDOM was observed (e.g., for a_{340} the majority of samples in Spencer et al. [2007c] showed a decrease and 77% of the samples exhibited a change outside of the analytical reproducibility). As Figure 4.4 (Hudson et al., 2009) highlights, typically both protein-like fluorophores and humic and fulvic-like fluorophores intensities decline after one freeze/thaw cycle. Spencer et al. (2007c) also examined the influence of freeze/thaw on the excitation/emission properties of a number of fluorophores and for both the fulvic and humic-like fluorophores examined (excitation maxima 320–350 and 340–390 nm and emission maxima 400–450 and 440–500 nm respectively) mean changes across the samples were within the analytical error but individual samples exhibited up to ± 20 nm shifts. For the humic and fulvic-like fluorophores the greatest proportion of change was a blue shift potentially indicating a breakdown in aromatic moieties or loss due to precipitation during freeze/thaw. Both Spencer et al. (2007c) and Hudson et al. (2009) concluded that from their knowledge of original sample properties no simple relationship could be found between initial sample characteristics and the amount of change that will occur with freezing and subsequent thawing, and as such it was not possible to apply a correction factor to their data. Guidelines have been suggested though with respect to when it might be appropriate to freeze samples and DOC loss on freezing has been shown to be as high as 30% in organic rich freshwater samples and to be related to initial DOC concentration with higher initial DOC concentrations exhibiting greater loss during freeze/thaw (Fellman et al., 2008). Fellman et al. (2008) suggest that in freshwater samples with low DOC concentrations (<5 mg L^{-1}) and/or low $SUVA_{254}$ values (<3.5–4 L mg C^{-1} m^{-1}) freezing is potentially a practical choice for sample preservation. However, samples with higher DOC and/or $SUVA_{254}$ values should not be frozen if possible, as freezing these samples not only causes a loss of concentration due to precipitation but also changes the chemical composition and thus spectrophotometric characteristics of the sample. This is further supported by the observed lack of change in DOC concentrations after freeze/thaw in marine samples (Tupas et al., 1994).

Although the impact of freeze/thaw appears to be greater in organic-rich waters a study by Otero et al. (2007) investigating the effect of freezing and thawing on fluorescence and DOC properties from a number of organic-rich sediment pore waters observed no change in pore water fluorescence characteristics on freezing and thawing. Thus, further highlighting the variable response of DOM to freeze/thaw as observed in the Spencer et al. (2007c) and Hudson et al. (2009) studies. This highlights the need for researchers to evaluate the impacts of freezing if it is to be employed as a storage method on a study- and site-specific basis. Yamashita et al. (2010a) in a study examining optical properties of DOM in Venezuelan tropical rivers examined the differences before and after freezing for the CDOM absorption coefficient at 350 nm (a_{350}), the spectral slope ratio (S_R) parameter (Helms et al., 2008), and fluorescence index (FI; McKnight et al., 2001; Jaffe et al., 2008; Cory et al., 2010) and found minor changes of 2.5 ± 6.9%, –0.4 ± 1.5%, and 2.8 ± 2.5% respectively. On the contrary, Yamashita et al. (2010a) observed significant increases and decreases in fluorescence intensity of different PARAFAC components after freeze/thaw in their study and so chose not to use the frozen samples in their PARAFAC data set but showed the validity of including the CDOM absorption, S_R, and FI data from the samples that were frozen. Similarly, Gao et al. (2010) examined the effect of freeze/thaw on optical properties of DOM as samples were stored frozen in their study of Zhejiang coastal waters (southeast China) and found that for their measurements of interest the percent change was less than 15%. Gao et al. (2010) therefore concluded that the main conclusions of their study were not due to measurement biases introduced from the freeze/thaw process. Finally, Spencer et al. (2010) in an examination of DOM dynamics in a pristine organic-rich Congolese tropical river investigated the impact of freeze/thaw as samples were stored frozen and found that after freezing and subsequent thawing DOC, $a350$, $S_{275-295}$, $S_{350-400}$, SR, $SUVA_{254}$, and FI values exhibited changes typically within analytical error and always less than ± 2%. Therefore, although extra care is warranted in organic-rich samples if freezing is to be employed, it seems it can be a suitable storage method and should be examined with suitable testing of samples relative to those encountered in the study.

4.4.3 Poisoning – Acidification

The addition of chemicals to poison organisms that may alter analytes of interest (e.g., dissolved organic matter optical properties) has been applied in a range of studies. Poisoning of samples can be very useful in preserving sample integrity in situations where no refrigeration or frozen storage is possible (e.g., remote field locations without access to power). Four poisoning chemicals are commonly used to preserve natural water samples; acidification to pH ~ 2–3 (typically HCl or H_3PO_4), chloroform ($CHCl_3$), sodium azide (NaN_3), and mercuric chloride ($HgCl_2$) (Kaplan, 1992; Kirkwood, 1992; Benner and Hedges, 1993; Ferrari et al., 1996; Wiebinga and de Baar, 1998; Kattner, 1999; Gardolinski et al., 2001; Aufdenkampe et al., 2007; Hur et al., 2007; Bouillon et al., 2009; Stubbins et al., 2010). When adding any chemicals as potential preservatives conduct necessary blanks and utilize high purity grade chemicals (e.g., reagent or A.C.S. grade). Of the four commonly described poisoning chemicals only the effects of acidification have been examined extensively on DOM optical properties. Chloroform is viewed as problematic owing to issues with volatility that can result in losses from poorly fitting seals and even directly through some types of plastic bottles (Kremling and Brugmann, 1999). Mercuric chloride has lost favor recently owing to issues with contamination of concurrent measurements of mercury when any form of mercury is used in restricted areas (e.g., on board a research vessel) (Kremling and Brugmann, 1999). Furthermore, both mercuric chloride and sodium azide are very toxic to aquatic organisms and may cause long-term adverse effects in the aquatic environment, and as such any water containing them should be treated as hazardous waste.

A number of studies have shown addition of mercuric chloride to inhibit microbial growth in samples without any effect on CDOM absorption spectra (Kratzer et al., 2000; Helms et al., 2008; Spencer et al., 2009), whereas Hg(II) has been shown to quench DOM fluorescence, particularly protein-like fluorescence (Fu et al., 2007; Yamashita and Jaffe, 2008; Osburn et al., Chapter 7, this volume). Sodium azide has been utilized to prevent degradation of CDOM absorption during storage and has been shown to have no effect in some studies (Ferrari et al., 1996; Astoreca et al., 2009) but also to cause up to a 10% increase in a_{442} (Tilstone et al., 2002). Patel-Sorrentino et al. (2002) examined the effect of sodium azide addition to fluorescence excitation-emission matrices (EEMs) from a range of rivers (black and white waters) in the Amazon Basin and reported no effect on the fluorescence intensity of the two humic-like fluorophores defined in their study (excitation maxima 220–260 and 320–350 nm and emission maxima 420–450 and 420–500 nm). Acidification of CDOM samples is often carried out to avoid microbial degradation during storage and because low pH reduces the potential complexation between DOM and metals (Westerhoff et al., 2001; Chen et al., 2003; Hur et al., 2007; Hiriart-Baer et al., 2008). A classic response in freshwater DOM fluorescence to pH was observed by Patel-Sorrentino et al. (2002), who reported an increase in fluorescence intensity with increasing pH over the

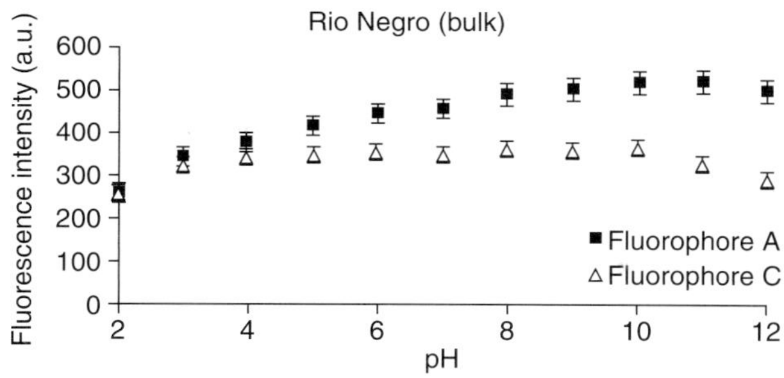

Figure 4.5. Variation of fluorescence intensities with pH for the characteristic peaks C and A; points are represented with their absolute error bar. Peaks C and A both have an emission maximum between 430 and 460 nm and an excitation at 325–340 nm and 245–260 nm respectively. (Reproduced from Patel-Sorrentino et al., 2002.)

range 1 to 10–11, with a decrease at pH 12 (Figure 4.5). Spectral shifts are also observed in fluorophores in response to changing pH. Mobed et al. (1996) observed a red shift in fluorescence intensity maxima with increasing pH at long wavelengths (excitation ~ 390 nm) and a similar red shift at shorter wavelengths (excitation ~ 320 nm) in soil derived humic substances. Conversely, in aquatic derived DOM, shorter wavelength fluorescence peaks have been reported to blue shift with increasing pH (Mobed et al., 1996). Other studies have found no wavelength change with pH (Tam and Sposito, 1993; Patel-Sorrentino et al., 2002). CDOM absorption has previously been observed to increase with increasing pH (Andersen et al., 2000).

Spencer et al. (2007c) examined pH effects on DOM absorption and fluorescence properties from a range of U.K. freshwater sites. The study looked at three fluorophores defined as: peak A (excitation maxima 320–350 nm and emission maxima 400–450 nm), peak B (excitation maxima 340–390 nm and emission maxima 440–500 nm respectively), and peak C (excitation maxima 270–275 nm and emission maxima 340–360 nm). Peak A and peak B previously have been described as related to fulvic and humic-like substances respectively and peak C has been attributed to protein-like (tryptophan-like fluorescence (Baker, 2001, 2002; Newson et al., 2001; Baker and Inverarity, 2004). The results from Spencer et al. (2007c) for spectrophotometric properties observed on modification of solution pH over the pH range 2–10 are summarized in Table 4.2. These results and Baker et al. (2007) both emphasize the greater impact of pH modification on humic and fulvic-like fluorescence in comparison to protein-like fluorescence. A decrease in pH to ~ 2–3 would generally result in a decrease in CDOM absorption and peak A and B fluorescence intensity, and this change was found to be greater if the original sample had higher values of these parameters. In the 35 freshwater samples examined by Spencer et al. (2007c) different responses to pH were apparent; for example, the increase in peak B fluorescence intensity with increasing pH (range 2–10) ranged from 32.1% to 74.8%. Therefore, changing the solution pH may result in varying responses between different DOM samples, and spectrophotometric properties of DOM samples are particularly sensitive to extremes of pH (Reynolds and Ahmad,

Table 4.2. *Summary of the response in spectrophotometric properties observed on modification of solution pH (range of pH: 2–10).*

Spectrophotometric Properties	Response to Increase in pH (2–10)
Peak C Variables	No response.
Peak $A_{EX}\lambda$ and peak $B_{EX}\lambda$	No response.
Peak $A_{EM}\lambda$	No consistent response or variation outside the reproducibility of the method.
Peak $B_{EM}\lambda$	A significant (95% confidence level) red shift was observed in all samples, over a different pH range and magnitude for each sample.
Peak A_{Fint}	Increase, to a maximum at variable pH, decrease at higher pH, mean difference between minimum and maximum 15.75% (SD 5.38).
Peak B_{Fint}	Increase, mean difference between minimum and maximum 41.82% (SD 7.43).
Peak B_{Fint}/Peak A_{Fint}	Increase, some samples exhibited a constant level below pH ~ 8.
a_{340} cm^{-1}	Increase, mean difference between minimum and maximum 17.79% (SD 3.45).

Fluorophores are defined as: peak A (excitation maxima 320–350 nm and emission maxima 400–450 nm), peak B (excitation maxima 340–390 nm and emission maxima 440–500 nm respectively), and peak C (excitation maxima 270–275 nm and emission maxima 340–360 nm).
Source: Reproduced from Spencer et al. (2007c).

1995; Patel-Sorrentino et al., 2002; Spencer et al., 2007c). At the typical pH levels found in most aquatic systems (4.5–8.5) little change in DOM spectrophotometric parameters have been reported indicating that changes in spectrophotometric parameters in aquatic systems are typically due to other processes and are not merely due to pH change (Reynolds and Ahmad, 1995; Spencer et al., 2007b,c; Hudson et al., 2009; Peiris et al., 2010). It is therefore recommended that for studies aiming to examine the role of DOM in ecosystem biogeochemistry that spectrophotometric analyses be undertaken at natural sample pH. It is further recommended that if possible the range of pH values of the samples should be reported to show they are not at an extreme of pH where significant variations in DOM optical properties may be observed.

4.5 Summary and Future Needs

The application of standard organic geochemical protocols utilizing rigorous cleaning, blanks and replicates are fundamental to ensuring high-quality CDOM absorbance and fluorescence data. Filtration of samples immediately after collection is recommended to stop biological processes and to remove particles that can interfere with CDOM absorbance and fluorescence measurements. If possible samples should be analyzed immediately or within 24 hours and stored refrigerated (5°C) in the dark in the interim. The optical properties of DOM samples change over time, in particular fluorescence intensities and after 7 days of refrigerated storage the optical properties no longer reflect the natural signal from the limited data available. The hold time, though, is clearly dependent on DOM type and specific hold time should be determined for each study relevant to the range of DOM under examination. If samples have to be preserved for longer periods of time, freezing and poisoning represent options but as with refrigeration they should be tested as to their suitability for the range of DOM under investigation. Acidification is not recommended as a preservation method. Within natural pH levels found in the majority of aquatic systems the response of CDOM absorbance and fluorescence measurements has been shown to be limited. There is a general paucity of data on poisons such as $HgCl_2$ and NaN_3 and whether they may be appropriate as preservatives for utilizing with the optical investigation of certain DOM types. Similarly, the possibility of utilizing silver filters with antibacterial properties and the difference in DOM optical results due to variation in filter pore sizes used needs further research. Systematic studies utilizing a broad range of DOM samples undergoing a range of storage protocols would provide highly useful data, particularly if the original DOM samples were well characterized and thus observed changes could be related back to the original composition/sample matrix, allowing informed choices on sample storage protocols to be made for specific DOM sample types.

Acknowledgments

We thank Gareth Old, Dan Lapworth, Kenna Butler, and George Aiken for sharing data with us and everyone who attended the AGU Chapman Conference on Organic Matter Fluorescence in Birmingham, U.K. in 2008 for fruitful discussion. We also thank four anonymous reviewers whose comments and suggestions improved this chapter. R.G.M.S acknowledges support from NSF grants DEB-1145932, OPP-1107774, OCE-1333157 and ANT-1203885. Finally, we thank the publishers who gave permission for figures and tables to be used.

References

Ahad, J.M.E., Ganeshram, R.S., Spencer, R.G.M., Uher, G., Gulliver, P., and Bryant, C.L. (2006). Evidence for anthropogenic 14C-enrichment in estuarine waters adjacent to the North Sea. *Geophys. Res. Lett.*, **33**, L08608, doi:10.1029/2006GL025991.

Andersen, D.O., Alberts, J.J., and Takacs, M. (2000). Nature of natural organic matter (NOM) in acidified and limed surface waters. *Water Res.*, **34**(1), 266–278.

Astoreca, R., Rousseau, V., and Lancelot, C. (2009). Coloured dissolved organic matter (CDOM) in Southern North Sea waters: Optical characterization and possible origin. *Estuar., Coast. Shelf Sci.*, **85**, 633–640.

Aufdenkampe, A.K., Mayorga, E., Hedges, J.I., Llerena, C., Quay, P.D., Gudeman, J., Krusche, A.V., and Richey, J.E. (2007). Organic matter in the Peruvian headwaters of the Amazon: Compositional evolution from the Andes to the lowland Amazon mainstem. *Org. Geochem.*, **38**, 337–364.

Baker, A. (2001). Fluorescence excitation–emission matrix characterization of some sewage impacted rivers. *Environ. Sci. Technol.*, **35**(5), 948–953.
Baker, A. (2002). Fluorescence excitation-emission matrix characterization of river waters impacted by a tissue mill effluent. *Environ., Sci. Technol.*, **36**(7), 1377–1382.
Baker, A. and Inverarity, R. (2004). Protein-like fluorescence intensity as a possible tool for determining river water quality. *Hydrol. Process.*, **18**(15), 2927–2945.
Baker, A., Elliott, S., and Lead, J.R. (2007). Effects of filtration and pH perturbation on freshwater organic matter fluorescence. *Chemosphere*, **67**, 2035–2043.
Belzile, C., Roesler, C.S., Christensen, J.P., Shakhova, N., and Semiletov, I. (2006). Fluorescence measured using the WETStar DOM fluorometer as a proxy for dissolved matter absorption. *Estuar., Coast. Shelf Sci.*, **67**, 441–449.
Benner, R. and Hedges, J.I. (1993). A test of the accuracy of freshwater DOC measurements by high-temperature catalytic oxidation and UV-promoted persulfate oxidation. *Mar. Chem.*, **41**, 161–165.
Blough, N.V., Zafiriou, O.C., and Bonilla, J. (1993). Optical absorption spectra of waters from the Orinoco River outflow: Terrestrial input of colored organic matter to the Caribbean. *J. Geophys. Res.*, **98**(C2), 2271–2278.
Bouillon, S., Abril, G., Borges, A.V., Dehairs, F., Govers, G., Hughes, H.J., Merckx, R., Meysman, F.J.R., Nyunja, J., Osburn, C., and Middelburg, J.J. (2009). Distribution, origin and cycling of carbon in the Tana River (Kenya): A dry season basin-scale survey from headwaters to the delta. *Biogeosciences*, **6**(11), 2475–2493.
Chen, R.F. and Gardner, G.B. (2004). High-resolution measurements of chromophoric dissolved organic matter in the Mississippi and Atchafalaya River plume regions. *Mar. Chem.*, **89**(1–4), 103–125.
Chen, W., Westerhoff, P., Leenheer, J.A., and Booksh, K. (2003). Fluorescence excitation-emission matrix regional integration to quantify spectra for dissolved organic matter. *Environ. Sci. Technol.*, **37**, 5701–5710.
Coble, P.G., Del Castillo, C.E., and Avril, B. (1998). Distribution and optical properties of CDOM in the Arabian Sea during the 1995 Southwest Monsoon. *Deep-Sea Res. Pt. II*, **45**(10–11), 2195–2223.
Conmy, R.N., Coble, P.G., Cannizzaro, J.P., and Heil, C.A. (2009). Influence of extreme storm events on West Florida Shelf CDOM distributions. *J. Geophys. Res-Biogeosci.*, 114, G00F04, doi:10.1029/2009JG000981.
Cory, R.M., McKnight, D.M., Chin, Y.P., Miller, P., and Jaros, C.L. (2007). Chemical characteristics of fulvic acids from Arctic surface waters: Microbial contributions and photochemical transformations. *J. Geophys. Res.-Biogeosci.*, **112**, doi:10.1029/2006JG000343.
Cory, R.M., Miller, M.P., McKnight, D.M., Guerard, J.J., and Miller, P.L. (2010). Effect of instrument-specific response on the analysis of fulvic acid fluorescence spectra. *Limnol. Oceanogr. Methods*, **8**, 67–78.
Del Castillo, C.E., Coble, P.G., Conmy, R.N., Muller-Karger, F.E., Vanderbloemen, L., and Vargo, G.A. (2001). Multispectral in situ measurements of organic matter and chlorophyll fluorescence in seawater: Documenting the intrusion of the Mississippi River in the West Florida Shelf. *Limnol. Oceanogr.*, **46**(7), 1836–1843.
Del Vecchio, R. and Blough, N.V. (2002). Photobleaching of chromophoric dissolved organic matter in natural waters: Kinetics and modeling. *Mar. Chem.*, **78**(4): 231–253.
Fellman, J.B., D'Amore, D.V., and Hood, E. (2008). An evaluation of freezing as a preservation technique for analyzing dissolved organic C, N, and P in surface water samples. *Sci. Tot. Environ.*, **392**, 305–312.

Fellman, J.B., Hood, E., Edwards, R.T., and D'Amore, D.V. (2009). Changes in the concentration, biodegradability, and fluorescent properties of dissolved organic matter during stormflows in coastal temperate watersheds. *J. Geophys. Res. Biogeosci.*, **114**, G01021, doi:10.1029/2008JG000790.

Ferrari, G.M., Dowell, M.D., Grossi, S., and Targa, C. (1996). Relationship between the optical properties of chromophoric dissolved organic matter and total concentration of dissolved organic carbon in the southern Baltic Sea region. *Mar. Chem.*, **55**, 299–316.

Fu, P., Wu, F., Liu, C., Wang, F., Li, W., Yue, L., and Guo, Q. (2007). Fluorescence characterization of dissolved organic matter in an urban river and its complexation with Hg(II). *Appl. Geochemis.*, **22**, 1668–1679.

Gao, L., Fan, D., Li, D., and Cai, J. (2010). Fluorescence characteristics of chromophoric dissolved organic matter in shallow water along the Zhejiang coasts, southeast China. *Mar. Environ. Res.*, **69**, 187–197.

Gardolinski, P.C.F.C., Hanrahan, G., Achterberg, E.P., Gledhill, M., Tappin, A.D., House, W.A., and Worsfold, P.J. (2001). Comparison of sample storage protocols for the determination of nutrients in natural waters. *Water Res.*, **35**(15), 3670–3678.

Helms, J.R., Stubbins, A., Ritchie, J.D., Minor, E.C., Kieber, D.J., and Mopper, K. (2008). Absorption spectral slopes and slope ratios as indicators of molecular weight, source, and photobleaching of chromophoric dissolved organic matter. *Limnol. Oceanogr.*, **53**(3), 955–969.

Hiriart-Baer, V.P., Diep, N., and Smith, R.E.H. (2008). Dissolved organic matter in the Great Lakes: Role and nature of allochthonous material. *J. Great Lakes Res.*, **34**, 383–394.

Hood, E., Fellman, J.B., Spencer, R.G.M., Hernes, P.J., Edwards, R., D'Amore, D.V., and Scott, D. (2009). Glaciers as a source of ancient and labile organic matter to the marine environment. *Nature*, **462**, 1044–1048.

Hudson, N., Baker, A., Reynolds, D.M., Carliell-Marquet, C., and Ward, D. (2009). Changes in freshwater organic matter fluorescence intensity with freezing/thawing and dehydration/ rehydration. *J. Geophys. Res. Biogeosci.*, **114**, G00F08, doi:10.1029/2008JG000915.

Hur, J., Jung, N.C., and Shin, J.K. (2007). Spectroscopic distribution of dissolved organic matter in a dam reservoir impacted by turbid storm runoff. *Environ. Monit. Assess.*, **133**, 53–67.

Jaffe, R., McKnight, D., Maie, N., Cory, R., McDowell, W.H., and Campbell, J.R. (2008). Spatial and temporal variations in DOM composition in ecosystems: The importance of long-term monitoring of optical properties. *J. Geophys. Res. Biogeosci.*, **113**, G04032, doi:10.1029/2008JG000683.

Kaplan, L.A. (1992). Comparison of high-temperature and persulfate oxidation methods for determination of dissolved organic carbon in freshwaters. *Limnol. Oceanogr.*, **37**(5), 1119–1125.

Kattner, G. (1999). Storage of dissolved inorganic nutrients in seawater: Poisoning with mercuric chloride. *Mar. Chem.*, **67**, 61–66.

Kirkwood, D.S. (1992). Stability of solutions of nutrient salts during storage. *Mar. Chem.*, **38**, 151–164.

Kratzer, S., Bowers, D., and Tett, P.B. (2000). Seasonal changes in colour ratios and optically active constituents in the optical Case-2 waters of the Menai Strait, North Wales. *Int. J. Remote Sens.*, **21**(11), 2225–2246.

Kremling, K. and Brugmann, L. (1999). 2. Filtration and storage. In K. Grashoff, K. Kremling, and M. Ehrhardt, (Eds.), *Methods of Seawater Analysis*, 3rd ed. Berlin, Germany: Wiley-TCH.

Lane, S.L., Flanagan, S., and Wilde, F.D. (2003). Selection of equipment for water sampling (ver. 2.0): U.S. Geological Survey Techniques of Water-Resources Investigations, book 9, chapter A2, March. Retrieved from http://pubs.water.usgs.gov/twri9A2/ (Accessed December 1, 2009).

Lapworth, D.J., Gooddy, D.C., Allen, D., and Old, G.H. (2009). Understanding groundwater, surface water, and hyporheic zone biogeochemical processes in a Chalk catchment using fluorescence properties of dissolved and colloidal organic matter. *J. Geophys. Res. Biogeosci.*, **114**, G00F02, doi:10.1029/2009JG000921.

Lead, J.R. and Wilkinson, K.J. (2006). Natural aquatic colloids: current knowledge and future trends. *Environ. Chem.*, **3**, 159–171.

McKnight, D.M., Boyer, E.W., Westerhoff, P.K., Doran, P.T., Kulbe, T., and Andersen, D.T. (2001). Spectrofluorometric characterization of aquatic fulvic acids for determination of precursor organic material and general structural properties. *Limnol. Oceanogr.*, **46**, 38–48.

Mitchell, B.G., Bricaud, A., Carder, K., Cleveland, J., Feraari, G.M., Gould, R., Kahru, M., Kishino, M., Maske, H., Moisan, T., Moore, L., Nelson, N., Phinney, D., Reynolds, R.A., Sosik, H., Stramski, D., Tassan, S., Trees, C., Weidemann, A., Wieland, J.D., and Vodacek, A. (2000). Determination of spectral absorption coefficients of particles, dissolved material and phytoplankton for discrete water samples. In G.S. Fargion and J.L. Mueller (Eds.), *Ocean Optics Protocols for Satellite Ocean Color Sensor Validation*, Revision 2 (pp. 125–153). NASA/TM-2000–209966. Greenbelt, MD: NASA Goddard Space Flight Center.

Mobed, J.J., Hemmingsen, S.L., Autry, J.L., and McGown, L.B. (1996). Fluorescence characterisation of IHSS humic substances: Total luminescence spectra with absorbance correction. *Environ. Sci. Technol.*, **30**(10), 3061–3066

Moran, M.A., Sheldon, W.M., and Zepp, R.G. (2000). Carbon loss and optical property changes during long-term photochemical and biological degradation of estuarine dissolved organic matter. *Limnol. Oceanogr.*, **45**, 1254–1264.

Murphy, K.R., Stedmon, C.A., Waite, T.D., and Ruiz, G.M. (2008). Distinguishing between terrestrial and autochthonous organic matter sources in marine environments using fluorescence spectroscopy. *Mar. Chem.*, **108**, 40–58.

Murphy, K.R., Butler, K.D., Spencer, R.G.M., Stedmon, C.A., Boheme, J.R., and Aiken, G.R. (2010). Measurement of dissolved organic matter fluorescence in aquatic environments: An intercalibration study. *Environ. Sci. Technol.*, **44**, 9405–9412.

Newson, M., Baker, A., and Mounsey, S. (2001). The potential role of freshwater luminescence measurements in exploring runoff pathways in upland catchments. *Hydrol. Process.*, **15**(6), 989–1002.

Osburn, C.L., Morris, D.P., Thorn, K.A., and Moeller, R.E. (2001). Chemical and optical changes in freshwater dissolved organic matter exposed to solar radiation. *Biogeochemistry*, **54**(3), 251–278.

Osburn, C.L., Retamal, L., and Vincent, W.F. (2009). Photoreactivity of chromophoric dissolved organic matter transported by the Mackenzie River to the Beaufort Sea. *Mar. Chem.*, **115**(1–2), 10–20.

Otero, M., Mendonca, A., Valega, M., Santos, E.B.H., Pereira, E., Esteves, V.I., and Duarte, A. (2007). Fluorescence and DOC contents of estuarine pore waters from colonized and non-colonized sediments: Effects of sampling preservation. *Chemosphere*, **67**, 211–220.

Patel- Sorrentino, N., Mounier, S., and Benaim, J.Y. (2002). Excitation-emission fluorescence matrix to study pH influence on organic matter fluorescence in the Amazon basin rivers. *Water Res.*, **36**, 2571–2581.

Peiris, R.H., Halle, C., Budman, H., Moresoli, C., Peldszus, S., Huck, P.M., and Legge, R.L. (2010). Identifying fouling events in a membrane-based drinking water treatment process using principal component analysis of fluorescence excitation-emission matrices. *Water Res.*, **44**, 185–194.

Reynolds, D.M. and Ahmad, S.R. (1995). The effect of metal ions on the fluorescence of sewage water. *Water Res.*, **29**(9), 2214–2216.

Rosenstock, B. and Simon, M. (1993). Use of dissolved combined and free amino acids by planktonic bacteria in Lake Constance. *Limnol. Oceanogr.*, **38**(7), 1521–1531.

Saraceno, J.F., Pellerin, B.A., Downing, B.D., Boss, E., Bachand, P.A.M., and Bergamaschi, B.A. (2009). High frequency in situ optical measurements during a storm event: Assessing relationships between dissolved organic matter, sediment concentrations, and hydrologic processes. *J. Geophys. Res. Biogeosci.*, **114**, G00F09, doi:10.1029/2009JG000989.

Seredynska-Sobecka, B., Baker, A., and Lead, J.R. (2007). Characterisation of colloidal and particulate organic carbon in freshwaters by thermal fluorescence quenching. *Water Res.*, **41**(14), 3069–3076.

Sharp, J.H., Benner, R., Bennett, L., Carlson, C.A., Fitzwater, S.E., Peltzer, E.T., and Tupas, L.M. (1995). Analyses of dissolved organic carbon in seawater – The JGOFS EQPAC methods comparison. *Mar. Chem.*, **48**(2), 91–108.

Sharp, J.H., Beuregard, A.Y., Burdige, D., Cauwet, G., Curless, S.E., Lauck, R., Nagel, K., Ogawa, H., Parker, A.E., Primm, O., Pujo-Pay, A., Savidge, W.B., Seitzinger, S., Spyres, G., and Styles, R. (2004). A direct instrument comparison for measurement of total dissolved nitrogen in seawater. *Mar. Chem.*, **84**(3–4), 181–193.

Spencer, R.G.M., Pellerin, B.A., Bergamaschi, B.A., Downing, B.D., Kraus, T.E.C., Smart, D.R., Dahlgren, R.A., and Hernes, P.J. (2007a). Diurnal variability in riverine dissolved organic matter composition determined by in situ optical measurement in the San Joaquin River (California, USA). *Hydrol. Process.*, **21**, 3181–3189.

Spencer, R.G.M., Baker, A., Ahad, J.M.E., Cowie, G.L., Ganeshram, R., Upstill-Goddard, R.C., and Uher, G. (2007b). Discriminatory classification of natural and anthropogenic waters in two U.K. estuaries. *Sci. Tot. Environ.*, **373**, 305–323.

Spencer, R.G.M., Bolton, L., and Baker, A. (2007c). Freeze/thaw and pH effects on freshwater dissolved organic matter fluorescence and absorbance properties from a number of UK locations. *Water Res.*, **41**, 2941–2950.

Spencer, R.G.M., Stubbins, A., Hernes, P.J., Baker, A., Mopper, K., Aufdenkampe, A.K., Dyda, R.Y., Mwamba, V.L., Mangangu, A.M., Wabakanghanzi, J.N., and Six, J. (2009). Photochemical degradation of dissolved organic matter and dissolved lignin phenols from the Congo River. *J. Geophys. Res. Biogeosci.*, **114**, G03010, doi:10.1029/2009JG000968.

Spencer, R.G.M., Hernes, P.J., Ruf, R., Baker, A., Dyda, R.Y., Stubbins, A., and Six, J. (2010). Temporal controls on dissolved organic matter and lignin biogeochemistry in a pristine tropical river, Democratic Republic of Congo. *J. Geophys. Res. Biogeosciences*, doi:10.1029/2009JG001180.

Stedmon, C.A., Markager, S., and Bro, R. (2003). Tracing dissolved organic matter in aquatic environments using a new approach to fluorescence spectroscopy. *Mar. Chem.*, **82**, 239–254.

Stedmon, C.A. and Markager, S. (2005). Tracing the production and degradation of autochthonous fractions of dissolved organic matter by fluorescence analysis. *Limnol. Oceanogr.*, **50**(5), 1415–1426.

Stubbins, A., Spencer, R.G.M., Chen, H., Hatcher, P.G., Mopper, K., Hernes, P.J., Mwamba, V.L., Mangangu, A.M., Wabakanghanzi, J.N., and Six, J. (2010). Illuminated darkness:

Molecular signatures of Congo River dissolved organic matter and its photochemical alteration as revealed by ultrahigh precision mass spectrometry. *Limnol. Oceanogr.*, **55**(4), 1467–1477.

Tam, S.-C. and Sposito, G. (1993). Fluorescence spectroscopy of aqueous pine litter extracts: effects of humification and aluminium complexation. *Eur. J. Soil Sci.*, **44**(3), 513–524.

Tiltstone, G.H., Moore, G.F., Sorensen, K., Rottgers, R., Jorgensen, P.V., Vicente, V.M., and Ruddick, K.G. (2002). Regional validation of MERIS chlorophyll products in North Sea coastal waters. REVAMP Inter-calibration report. Retrieved from https://earth.esa.int/workshops/mavt_2003/MAVT-2003_802_REVAMPprotocols3.pdf

Tupas, L.M., Popp, B.N., and Karl, D.M. (1994). Dissolved organic carbon in oligotrophic waters: experiments on sample preservation, storage and analysis. *Mar. Chem.*, **45**, 207–216.

Tzortziou, M., Osburn, C.L., and Neale, P.J. (2007). Photobleaching of dissolved organic material from a tidal marsh-estuarine system of the Chesapeake Bay. *Photochem. Photobiol.*, **83**(4), 782–792.

U.S. Geological Survey (2006). Collection of water samples (ver. 2.0): U.S. Geological Survey Techniques of Water-Resources Investigations, book 9, chapter A4. September. Retrieved from http://pubs.water.usgs.gov/twri9A4/ (Accessed December 1, 2009).

Walker, S.A., Amon, R.M.W., Stedmon, C., Duan, S.W., and Louchouarn, P. (2009). The use of PARAFAC modeling to trace terrestrial dissolved organic matter and fingerprint water masses in coastal Canadian Arctic surface waters. *J. Geophys. Res.-Biogeosci.*, **114**, G00F06, doi:10.1029/2009JG000990.

Westerhoff, P., Chen, W., and Esparza, M. (2001). Fluorescence analysis of a standard fulvic acid and tertiary treated wastewater. *J. Environ. Qual.*, **30**, 2037–2046.

Wickland K.P., Neff J.C., and Aiken G.R. (2007). DOC in Alaskan boreal forests: sources, chemical characteristics, and biodegradability. *Ecosystems*, **10**, 1323–1340.

Wiebinga, C.J. and de Baar, H.J.W. (1998). Determination of the distribution of dissolved organic carbon in the Indian sector of the Southern Ocean. *Mar. Chem.*, **61**, 185–201.

Yamashita, Y. and Jaffe, R. (2008). Characterizing the interactions between trace metals and dissolved organic matter using excitation-emission matrix and parallel factor analysis. *Environ. Sci. Technol.*, **42**, 7374–7379.

Yamashita, Y., Maie, N., Briceno, H., and Jaffe, R. (2010a). Optical characterization of dissolved organic matter in tropical rivers of the Guayana Shield, Venezuela. *Journal of Geophysical Research-Biogeosciences*, **115**, G00F10, doi:10.1029/2009JG000987.

Yamashita, Y., Cory, R.M., Nishioka, J., Kuma, K., Tanoue, E., and Jaffe, R. (2010b). Fluorescence characteristics of dissolved organic matter in the deep waters of the Okhotsk Sea and the northwestern North Pacific Ocea. *Deep-Sea Res. Pt. II*, **57**, 1478–1485.

5

Optical Spectroscopy Instrumentation Design, Quality Assurance, and Control: Bench-Top Fluorimetry

JOHN R. GILCHRIST AND DARREN M. REYNOLDS

5.1 Introduction

The scope of optical spectroscopy methods is very wide and encompasses many aspects of life- and material sciences. Using optical spectroscopy one can analyze spectrochemical events to monitor, for example, human health issues, food and water quality, environment quality, materials for whitening agents, lighting, and light-emitting diodes. The instrumentation used comprises optical, mechanical, electrical as well as signal processing and data analysis components. The data recorded are a convolution of the effects of each of these components in addition to the sample and its behavior.

Fluorescence spectroscopy, although not a new technique, is still, compared to other analytical methods, relatively immature in terms of standardization of measurement. As mentioned in Chapter 1, the birth of fluorescence spectroscopy was marked by the work of Sir George Gabriel Stokes, who in 1852 reported his studies on quinine bisulfate using what today would be considered a filter fluorimeter arrangement, as shown in Figure 5.1.

More than 100 years later, commercial fluorimeter systems emerged owing to the advancement in light sources, scanning monochromators, detector technology, and analog signal recording mechanisms. In the last 50 years there have been considerable improvements in each electro-optical component used in fluorimeter systems but, equally, there has also been a rapid advancement to computer-controlled, often termed "black-box," technology. As a result, in recent decades less attention has been given to the measurement of instrument performance and in particular to rigorous calibration and testing. The use of quinine sulfate as a reference standard is widespread and therefore it is often not clear if the measurements demonstrated by instruments are the true fluorescence spectra of the sample (in terms of its behavior) or that of the measuring instrument (e.g., light source, intensity of fluorescence, spectral features, etc.). More specifically, the observed measured spectrum is the convolution of both the experimental system and the true fluorescence spectrum. Some experimentalists who are not fully acquainted with measuring fluorescence spectra may, understandably, take a black box approach to these important aspects of measuring fluorescence spectra. Even so, it is important that we appreciate this convoluted system if we are to gain a quantitative insight from the fluorescence measurements that is related to the samples analyzed as opposed to the measurement system.

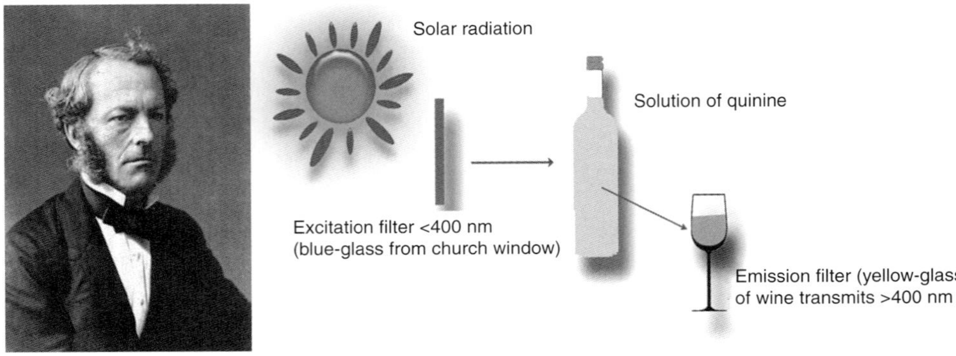

Figure 5.1. Sir George Gabriel Stokes (1852) and his filter fluorimeter.

For example, how does the instrument change in time, how does it compare with other instruments, and how do measurements compare from laboratory to laboratory? Furthermore, what is the meaning of fluorescence units? These questions and others are examined in this chapter with the aim of providing a simple and clear understanding of what is inside the "black box" and an appreciation of its relevance to achieving fluorescence measurements that are truly representative of the sample.

5.2 Methods of Optical Spectroscopy

Optical spectroscopy is the science of the interaction of optical radiation with matter. In many cases this involves specific transitions between the energy levels (states) of a sample, which are monitored experimentally by the absorption or emission of electromagnetic radiation. In these types of interactions the radiation is considered to be composed of packets of energy called photons, and they have a dual character in that they have both particle- and wave-like properties. As discussed in Chapter 1, the photon energy is related to its wavelength and frequency by:

$$E = h\nu = \frac{hc}{\lambda} \tag{5.1}$$

where E is the photon energy, h is Planck's constant, ν is the frequency, c is the speed of light, and λ is the wavelength.

There are many types of radiation–matter interactions, such as diffraction, refraction, reflection, and scattering that do not involve transitions between energy levels. However, these interactions may cause changes in the measured optical radiation due to direction or polarization and are often the result of bulk material properties rather than specific chemical species.

In the ultraviolet to infrared spectral regions the instrumental needs are similar in terms of the optical materials required for focusing, steering, and dispersing light. "Spectrometry"

Table 5.1. *Methods of optical spectroscopy and the measured quantity*

Method	Measured quantity	Examples
Absorption	Absorbance or the ratio of transmitted to incident radiant power, $A = -\log(\phi/\phi_O)$	Atomic absorption, UV-visible molecular absorption, IR absorption
Optical Emission	Radiant power of emission, ϕ_E	ICP and DCP emission, spark emission, laser-induced breakdown emission, flame emission, DC arc emission
Photoluminescence	Radiant power of luminescence,	Molecular fluorescence and phosphorescence, chemi- and bioluminescence, atomic fluorescence
Scattering	Radiant power of scattering, ϕ_S	Raman scattering, Mie scattering, turbidity

can be defined as the quantitative measurement of the intensity of an optical signal at one or more wavelengths using a photodetector.

Spectroscopic information about a species and its microenvironment is usually obtained by stimulating the sample with some energy source such as electrical energy, radiation, particles, or heat. Before stimulation the sample is normally residing in its lowest energy level, or ground state. After the stimulation the sample is momentarily induced to a higher energy level or excited state. Spectroscopic measurements are made from either the radiative emission from the excited species as it returns to the ground state or by the amount of absorbed radiation. The magnitude of the optical signal as a function of wavelength describes the sample in terms of its electronic, rotational, and vibrational energy levels and any associated transitions that have taken place. These, in turn, provide information on the molecular structure of the molecules that are within the sample that has been analyzed, such as characteristic absorption frequencies of known specific bonds. The signal intensity can be described in radiometric terms, now exclusively used in optical spectroscopy, or in photometric terms that have been based on the human eye response. In the radiometric system, the basic quantity measured is the actual amount of radiative energy in joules (J) either being emitted by a source or incident upon a detector. In fluorescence terms, the radiative energy emitted by a fluorescing sample will be a function (per unit) of exciting radiation. The four main methods associated with optical spectroscopy are absorption, emission, luminescence, and scattering (see Table 5.1). The emission process can be stimulated in a variety of ways, including the use of light (photoluminescence) or by collision with energetic electrons in a plasma (inductively coupled plasma). Scattering of electromagnetic radiation by matter (molecules and atoms) can be either *elastic* or *inelastic*. Elastic scattering occurs when the energy (and thus wavelength and frequency) of the light is not substantially changed, such as Mie and Rayleigh scattering. Inelastic scattering occurs when the kinetic energy of an incident particle, such as a photon, is not conserved and, more specifically, where some of the energy of the incident particle is lost or gained.

Although this chapter specifically concentrates on photoluminescence, specifically fluorescence spectroscopy, it is helpful, for completeness, to consider the other three main methods in optical spectroscopy in the context of the instrumentation and measurement requirements: absorption, emission and scattering (see also Chapter 1).

5.2.1 Absorption Spectroscopy

The absorption method involves measuring the ratio of two radiant powers (of the incident and transmitted light), calculating the absorbance, and relating the absorbance to concentration. For absorption to occur, the frequency of the incident light must correspond to the energy difference between two states, which allows the species to be excited from the ground state to some higher energy state. The energy absorbed is dissipated as luminescence (radiant energy), photochemical reaction (chemical energy), or thermal energy. For many experiments, the absorption of light follows the Beer–Lambert law (Eq. 5.2) previously outlined in Chapter 1.

$$= -\log(T) = -\log\left[\frac{\phi}{\phi_0}\right] = \varepsilon c l \qquad (5.2)$$

where A is the absorbance, T is the sample transmittance; I_0 and I represent the intensity (or radiant power) of the incident light and the transmitted light, ε is the extinction coefficient of the sample, c is the concentration of the absorbing species, and l is the optical path length through the sample (see Figure 5.2).

The absorption method is based on the following assumptions:

- The absorbers act independently of each other.
- The incident light intensity is not so high as to cause saturation or bleaching effects.
- The incident light beam is perpendicular to the absorbing surface.
- The path length is uniform, and the sample is homogeneous and does not scatter the light.

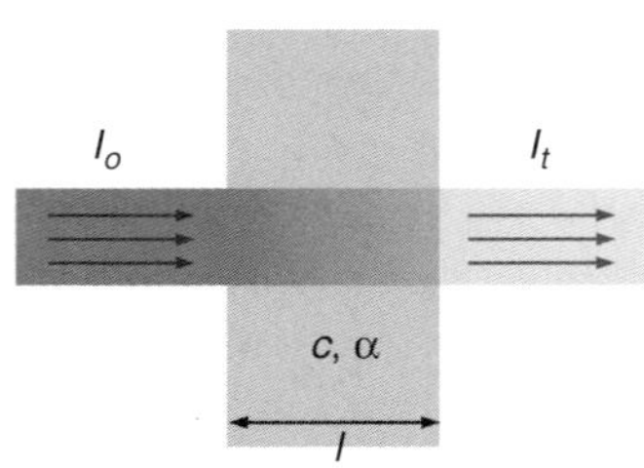

Figure 5.2. The absorption (Beer–Lambert) of a beam of light through a sample cuvette.

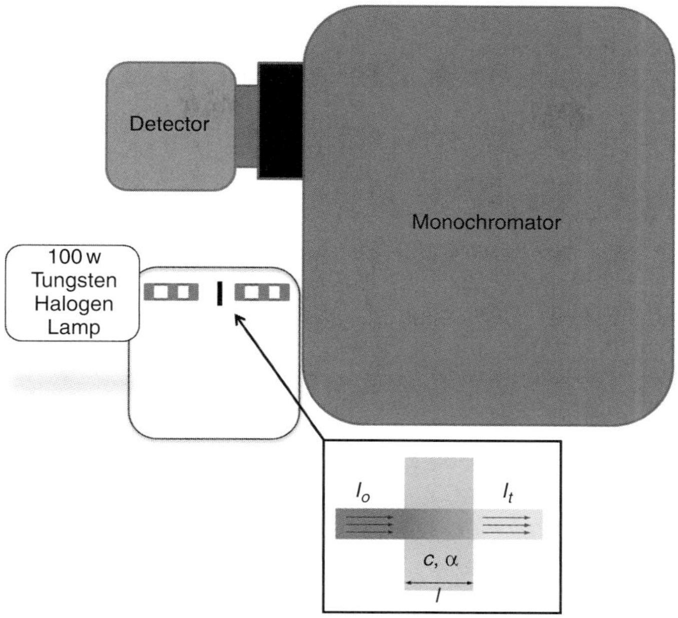

Figure 5.3. Single-channel absorption spectrophotometer.

For the measurement of absorption of a sample a typical arrangement of a single-channel absorption spectrophotometer in relation to the sample is shown in Figure 5.3.

5.2.2 Optical Emission Spectroscopy

Atoms of molecules can be brought to a so-called excited state via the interaction with a radiation source (electromagnetic radiation) as in photoluminescence or a nonradiational source such as a flame, DC current, high-voltage spark, or pulsed laser heating. Photoluminescence relies on a radiative emission from the sample after excitation by excited atoms or molecules. Figure 5.4 demonstrates a typical instrument arrangement for measuring optical emission spectroscopy.

Ideally, sodium atoms emit a characteristic radiation frequency when excited in a flame by a collisional process. When in a thermal equilibrium (an ideal scenario), it generates a statistical distribution of excited states. Once generated, the number of atoms, n, at a particular energy level, *i*, is given by the Boltzmann distribution:

$$n_i = \frac{n_t g_i e^{-E_i / kT}}{\sum_{i=0}^{\infty} g_i e^{-E_i / kT}} \tag{5.3}$$

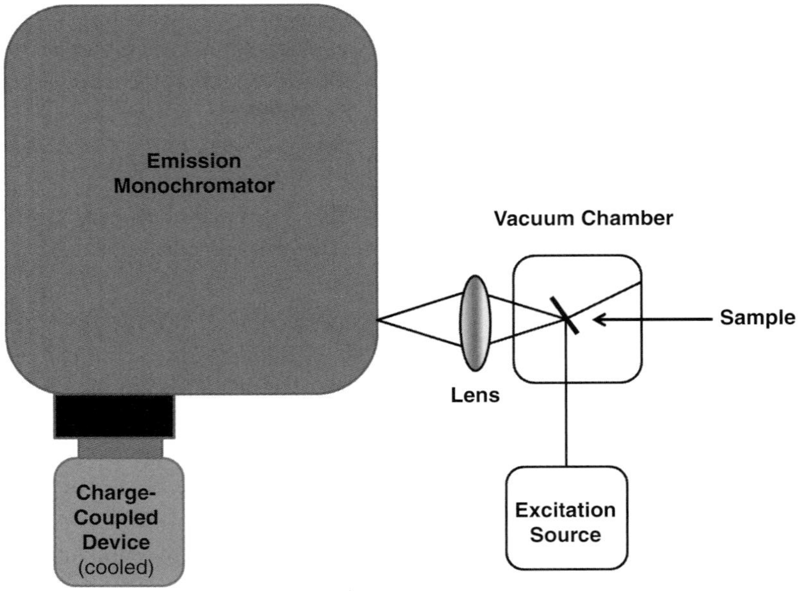

Figure 5.4. Typical arrangement of an optical emission spectrometer.

where n_i is the total atom density, k is Boltzmann's constant, T is the absolute temperature, and g_i is the statistical weight for state i. The frequency of the radiative emission corresponds to the difference in energy levels of the excited analyte.

Alkali metals such as sodium and potassium have excited levels close to their ground states and, as such, are relatively easy to excite using flames. Such atoms have resonance lines generally in the visible and near-infrared (IR) regions of the spectrum. The product of the excitation cross section as a function of both energy and the energy distribution of the exciting source is important, as transitions may involve energy levels other than just the ground and one excited level. As such, many other elements have energy levels significantly farther from their ground states and require considerably more intense excitation from sources such as plasmas to generate a suitable emission signal. Owing to the high energy difference between higher and lower energy states, such elements generally exhibit emission spectra in the ultraviolet (UV).

The radiant emissive power, ϕ_E, from state j to i is given by:

$$\phi_E = A_{ji} h \nu_{ji} n_{ji} \tag{5.4}$$

where A_{ji} is the transitions probability, $h\nu_{ji}$ is the energy of the emitted photon, and n_j is the number density of the higher energy state j.

For a system that is in thermal equilibrium, the radiant power of the emission is directly related to the population densities of the excited states and hence to the analyte concentration

Table 5.2. *Classes of scattering phenomena*

Class	Scatter Method
Kayleigh	Elastic scattering from small particles such as atoms or molecules, resulting in scattered radiation that occurs in all directions uniformly
Debye or Mie	Elastic scattering from large particle with dimensions comparable to the incident radiation wavelength (Debye) or much larger than the incident wavelength, where the result from scattering is nonuniform
Brilliouin	Inelastic scattering where the frequency of the reflected radiation is changed by thermal sound waves
Raman	Inelastic scattering where the frequency of the reflected radiation is changed by the gain or loss of vibrational/rotational quantum of energy by the analyte molecule

through the Boltzmann distribution. Thus by measuring the emission spectrum from a sample, one can apply known spectra from a library to identify and quantify the concentration of species in the sample. It is important to note here that many systems do not exhibit thermal equilibrium and therefore the measured radiant power does not always directly relate to the population densities of the excited states and or to the analyte concentration.

5.2.3 Scattering

Radiation incident on an analyte may be scattered as well as absorbed by the sample. The intensity, angular distribution, and radiation frequency of the scattered light can be used as a means of analysis. Several classes of scattering are possible and are simplistically described in Table 5.2.

The most widely used scattering phenomenon in spectroscopy is Raman scattering. Incident photons scattered inelastically from an analyte may either gain or lose energy (Raman and Krishnan, 1929; Long, 2002). Observed energy differences typically correspond to one quantum of vibrational energy of the species and the wavelength of the scattered light shifts accordingly. In simple terms, the most common form of Raman, called Stokes scattering, involves a loss of energy causing a red shift in the scattered photon while a gain or increase in energy is called anti-Stokes scattering. In certain circumstances (anharmonicity) overtone and combination bands can be observed in a Raman spectrum. For example, in the gas phase there are rovibrational structures whereas in the liquid phase these signals are far too weak and instead the full width at half maximum (FWHM) is related to the molecular reorientation time.

Raman spectroscopy, much like IR absorption spectroscopy, provides a method of determining the unique "fingerprint" of a species. Unlike in IR absorption spectroscopy, however, there is no need to detect the incident light. In most cases, light scatters in all directions, making Raman spectroscopy especially useful for the analysis of opaque solids;

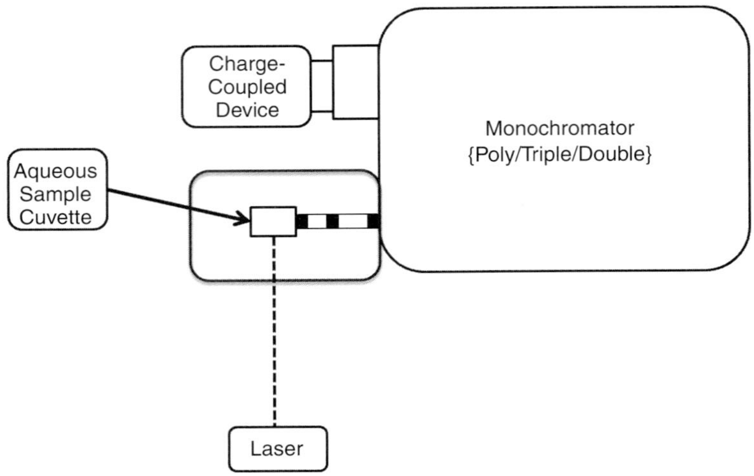

Figure 5.5. Basic configuration of a Raman spectrometer.

one can look at backscattered Raman, for example, in the Raman microscope. In Raman spectroscopy, there are several important factors to consider when designing an experimental system. Because the analyte scatters light with a frequency shifted with respect to the input light, the excitation source is almost always a monochromatic laser. Also, because Raman scattering is a relatively weak phenomenon, it is necessary to avoid simultaneous detection of the scattered light and of the input laser (incident light), which would dwarf the desired scattered signal. Figure 5.5 demonstrates a typical configuration of a Raman spectrometer for liquid samples. It is important to understand that polarization of the input and output is important, such as vertical polarization (VV) or where the input and output exhibit vertical and horizontal polarization (VH) respectively.

Also competing with Raman scattering is the much more intense Rayleigh scattering. Because Raman and Rayleigh scattering are of different wavelengths, Rayleigh scattered light can be filtered at relatively little expense to the Raman signal using a notch filter or a double or triple monochromator. For some materials, their highly fluorescent properties may dominate the Raman scattering. In such cases the use of longer wavelength lasers (to a certain extent), time-resolved techniques, and fast Fourier transform Raman eliminate this problem. Techniques such as surface-enhanced Raman scattering, resonance Raman scattering, and coherent anti-Stokes Raman scattering greatly enhance the signal with respect to spontaneous Raman scattering. One area where Raman spectroscopy is particularly useful is in the determination of species in aqueous media. In these solutions, water is the dominant species with the molecules of interest existing at lower concentrations. In this scenario infrared absorption may be ineffective because water absorption may dominate the spectrum and obscure the desired results. Water molecules exhibit intrinsically weak Raman signals; even so, Raman spectroscopy provides a useful tool for solute determination.

5.2.4 Photoluminescence (Fluorescence and Phosphorescence)

Molecular and atomic fluorescence are examples of widely accepted photoluminescent phenomena that are used to characterize analytes that encompass an extremely wide and diverse range of applications. Fluorescence and phosphorescence spectrometry are examples of photoluminescence where the quantity of interest in the measurement is the radiant power luminesced from the sample after absorption of a monochromatic incident light. Fluorescence involves the emission from singlet to singlet states; that is, of the same multiplicity. Phosphorescence, however, involves a radiative transition from triplet to singlet states; that is, of different multiplicity. The probability that a fluorophore will emit a photon (i.e., the quantum yield) is the number of times that a defined event occurs per photon absorbed by the system. Therefore the quantum yield is a combination of the absorption probability, in the form of a cross section, and the probability that the excited state will decay by radiative emission. Such processes are, in turn, related to the wavelength or energy of the incident radiation and the particular energy levels inherent within the sample, and therefore quantum (photon) yields are a function of several parameters.

The probability of absorption at a given wavelength is expressed by the molar extinction coefficient $\varepsilon(\lambda)$, $m^2\ mol^{-1}$. From this one can deduce that there will be a wavelength or wavelengths of maximum absorption and spectral shape. The microenvironment of the fluorophore, such as the solvent, presence of other ions and molecules, fluorophore concentration, and the surrounding temperature can all affect either the absorption or fluorescing properties of the bulk sample. For many situations, the emitted radiant power, ϕ_l, is proportional to the absorbed radiant power. Thus,

$$\phi_l = 2.303k\phi_0\varepsilon cl = k'\phi_0 c \tag{5.5}$$

where k is dependent on the species, its environment, and the efficiency with which the excited molecule or atom returns to its ground state (possibly via intermediate energy levels) by the emission of a photon. For low absorbances ($\varepsilon cl < 0.01$), the luminescence radiant power is directly proportional to the absorbed power and the sample concentration.

In the experimental diagram of a fluorimeter (Figure 5.6), the lamp provides broadband light to the excitation monochromator, which selects the excitation wavelength. Subsequent fluorescence is resolved and detected through the emission monochromator and photomultiplier tube detector. The emission intensity of luminescent materials can strongly depend on a range of a given material's parameters, and some examples of these parameters are shown in Table 5.3.

5.3 The Fluorescence Spectrometer

It is the measurement of fluorescence emission that is the fundamental quantity in all fluorescence studies. This quantity is affected not only by the dependence on other sample-related parameters but also by a series of instrument specific effects that can have significant

Table 5.3. *Luminescence dependence*

Parameters	Φ, Quantum yield I, Fluorescence Intensity P, fluorescence polarization r, fluorescence anisotropy
Dependence on one parameter	I(λ), spectra: excitation, emission, synchronous I(c), concentration or titration $I(T)$, temperature $r(\lambda)$, spectrally resolved anisotropy I(t), time-resolved
Dependence on two parameters	$I(\lambda,t)$, time-resolved emission spectra $I(\lambda_{exc}, \lambda_{em})$, excitation–emission spectral maps (EEM) $I(\lambda, T)$, temperature resolved spectra $R(\lambda, t)$, time and spectrally resolved anisotropy

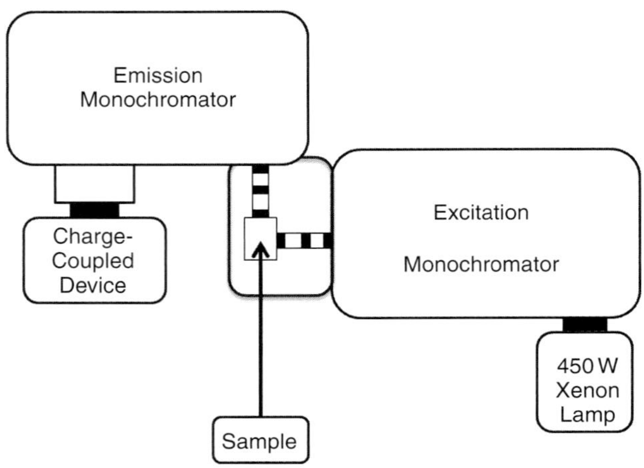

Figure 5.6. Basic fluorimeter arrangement.

bearings on the results, not only from day to day, but also instrument-to-instrument and laboratory-to-laboratory. It is these drawbacks, such as intensity, wavelength, polarization and time-resolved behaviors, of all fluorescence- based detection methods and their proper removal that make accurate measurement of the absolute fluorescence intensities extremely difficult. At the same time there are still relatively few fluorescence standards readily available that can be used to demonstrate instrument performance validation through rigorous characterization of the instrument. The basic measurements of excitation and emission spectral measurements, once mastered, are the foundation of all other measurements such as those referred to in Table 5.4.

Table 5.4. *Most common fluorescence scanning techniques*

Excitation scan	Select emission wavelength and bandpass, λ_{em}, $\Delta\lambda_{em}$, and hold fixed. Select excitation bandpass, $\Delta\lambda_{exc}$. Scan excitation monochromator across excitation spectral region of interest.
Emission scan	Select excitation wavelength and band, λ_{exc}, $\Delta\lambda_{exc}$, and hold fixed. Select emission bandpass, $\Delta\lambda_{em}$. Scan emission monochromator across emission spectral region of interest.
Synchronous scan	Select excitation and emission bandpasses, $\Delta\lambda_{exc}$ and $\Delta\lambda_{em}$. Select wavelength offset between excitation and emission wavelengths, $\Delta\lambda$. Scan excitation and emission monochromators in synchronization with the offset $\Delta\lambda$.
Excitation–emission matrices	Select excitation and emission bandpasses, $\Delta\lambda_{exc}$ and $\Delta\lambda_{em}$. Scan emission spectra (over a spectral range) as a function of excitation wavelength.

5.3.1 The Ideal Fluorescence Spectrometer System

Obtaining meaningful and quantitative fluorescence spectral intensity data from a given sample requires an understanding of instrumentation, design, operation, and calibration requirements. The ideal fluorescence spectrometer is one that allows the true spectra of the sample to be measured; in other words, it should possess the following qualities:

- Have high sensitivity and ideally no noise signals
- Measure excitation and emission spectra, that is, the photon flux emitted at each wavelength
- Be nonresponsive to interfering signals such as Raman and Rayleigh scattering, stray light, fluorescence from solvents, and so forth
- Measure the "true sample spectra" that are fully corrected for the nonuniform spectral output of light sources and the wavelength-dependent efficiencies of monochromators and detectors.

Thus, from an instrument design perspective, the ideal fluorimeter should possess the following attributes:

- A light source that yields constant photon output at all wavelengths
- Monochromators that pass photons at all wavelengths with equal efficiency
- Monochromators that are not sensitive to polarization effects, and
- Detectors that must detect photons of all wavelengths with equal efficiency

Unfortunately, there are so many variables in the optical path of any fluorimeter, or any other spectrometer system, that such systems simply *do not exist* and it is for this reason that careful consideration to instrumental system design, calibration, and correction methodologies is essential.

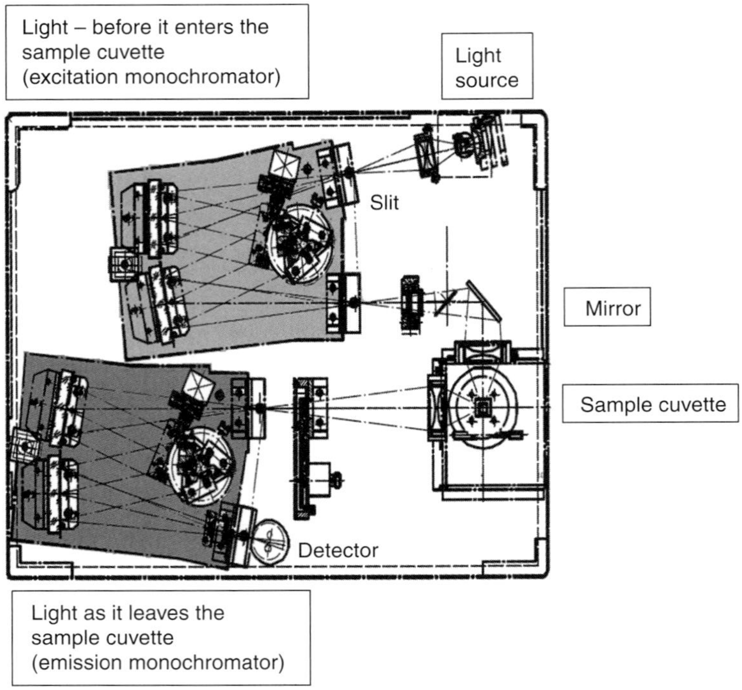

Figure 5.7. A typical spectrofluorimeter configuration.

5.3.2 Basic Spectrofluorimeter Design

Many spectrofluorimeter systems have a typical configuration similar to that presented in Figure 5.7. Illumination sources are typically steady-state xenon arc discharge lamps, coupled to scanning monochromators to make a tunable light source. The sample is illuminated at a certain wavelength, λ_{exc}, and the resulting luminescence measured using a second monochromator to select an emission wavelength, λ_{em}, and detected usually with a photomultiplier detector.

5.4 Measuring Fluorescence

5.4.1 Defining the Sensing Volume and Inner Filter Effects

All fluorescence measurements require that the instrument should provide a source of photons. These photons are generated by a light source, selected for an excitation wavelength, λ_{exc}, by the excitation monochromator, which also selects a specific pass-band of wavelengths $\Delta\lambda_{exc}$. For aquatic samples a geometrically defined volume of sample, usually in a cuvette, will be illuminated via an excitation source. During illumination light from a

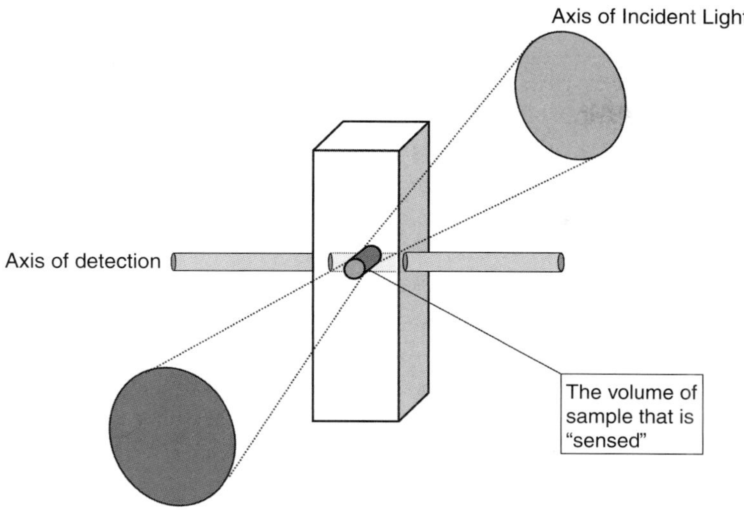

Figure 5.8. The sensing volume.

defined sampled volume is collected via optical components and then detected and analyzed. The overlap of the excitation and detection volumes *defines the sensing volume* of the system (Figure 5.8). The chemical, physical, and molecular nature of the sample within this volume will dictate the contribution to the fluorescence signal. At the same time the measured fluorescence intensity is also dependent on the fluorophore concentration, the absorption coefficient, and its quantum yield.

In general, fluorescence measurements are made at rather dilute concentrations, for example, with $\varepsilon cl < 0.01$. An increasing concentration of fluorophore usually increases the measured fluorescence signal levels; however, there comes a point where the increased concentration results in a decrease in the fluorescence signal. This is termed *the inner filter effect*, which manifests itself either by reabsorption of the fluorescence signal by the sample or its matrix (for example, where the emission from one fluorophore is reabsorbed by another fluorophore molecule) or at the same time the sensing volume changes because the light intensity in the excitation is not constant throughout the defined sensing volume. This can result in little or no excitation light reaching the detection volume. This phenomenon is a direct result of the Beer–Lambert law. A high concentration of fluorophore means high attenuation of light at the cuvette wall and hence less light through the defined sensing volume or pathlength to the cuvette center, as demonstrated by Figure 5.9. These inner filter effects can change not only the intensity of the emitted light but also the emission spectrum and can be a significant source of error. Elimination of these effects by suitable sample geometry or correction but other means (Parker, 1968) must be seriously considered if high concentration measurements are to be performed.

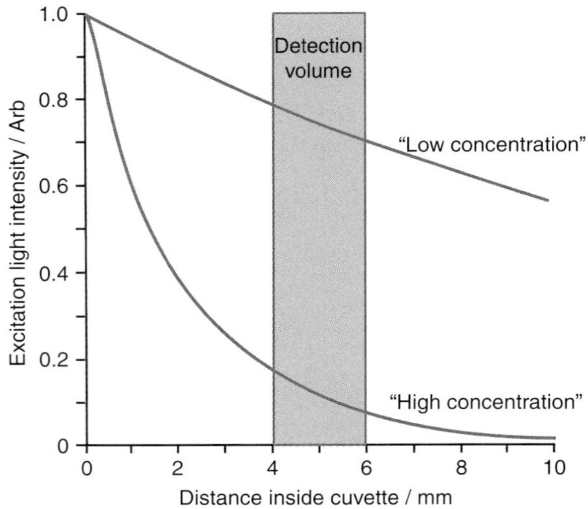

Figure 5.9. How the inner filter effect impacts on the illumination of the sensing volume.

5.4.2 Continuum Light Sources

Most photoluminescence instruments employ the use of high-intensity continuum sources with UV and visible output. Continuum light sources have wavelength-dependent intensity profiles and may also change in both intensity and spectral output as a function of time. For this reason many fluorimeter systems are now equipped with a reference silicon photodiode detector that monitors a portion of the excitation light that is incident upon the sample. The most common light source that is used in bench-top fluorimeter systems is the xenon short arc lamp. This has an intense, almost continuous spectral output from the UV (220 nm) to the near-IR (800 nm) with some fine structure superimposed. Line spectra are prominent in the 800- to 1000-nm range, and optical output can be observed well into the IR (2.6 μm). Xenon lamps have a small arc length (0.5–2.7 mm for 75- to 450-W lamps) and a high flux density. They are available in a variety of envelope materials allowing output from 185 nm and have relatively long operating lifetimes of up to 2000 hours.

Typically, xenon lamps are operated with DC power supplies, but pulsed versions are available with 1- to 10-μs pulse widths for use in time-resolved measurements such as phosphorescence studies. Xenon lamps offer the highest spectral flux density short of a laser. Their spectral output is close to that of the sun at 6000°C, and they make ideal solar simulators in many experiments. Special considerations when using an arc lamp include envelope materials (materials that allow the transmission of light within a given or defined spectral range); ozone dissipation; and convection current effects on noise, cooling, arc gap, and lamp life. Although they are the lamp of choice for most fluorescence instruments,

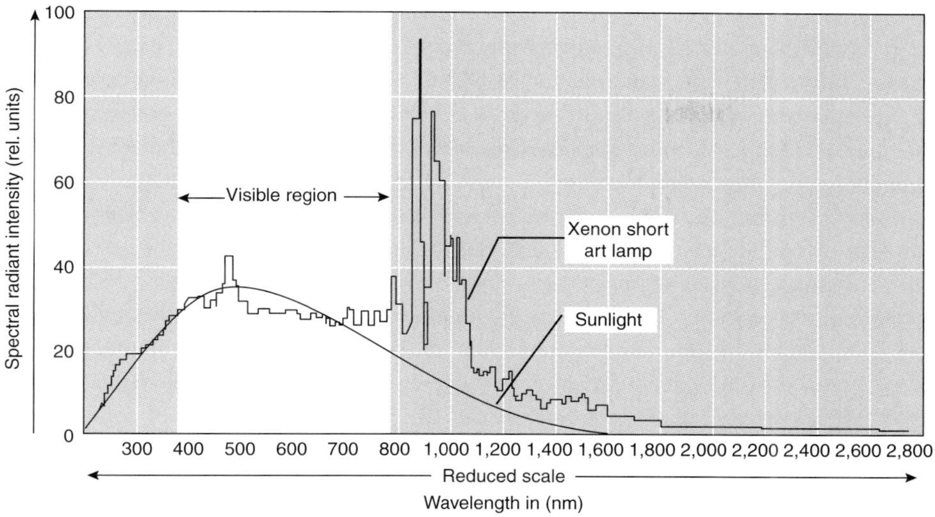

Figure 5.10. Spectral distribution of radiant intensity of a typical Osram XBO lamp and sunlight.

these lamps require a complex power supply with an ignition circuit to strike the arc and a well-designed lamp housing to provide proper lamp cooling. In addition lamps that exhibit intense UV characteristics require additional attention given their possible effects on eyes and skin in particular.

The spectral profile of the lamp is dependent on the operating conditions and the age of the lamp (number of hours used). At the same time the design of the power supply should be such as to ensure stability in the light output. Typically, the short- and long-term drifts in xenon lamp outputs are on the order 0.2–5% depending on manufacturers and the exact use of the lamp. For fluorimeter measurements 5% light stability is useless, as the fluorescence signal will shift in synchronization with the lamp fluctuations by at least this value in routine measurements, making quantitative spectroscopy difficult. A high-quality fluorimeter should demonstrate at least <0.5% lamp fluctuations and even at this level the reference photodiode is still a very useful and necessary monitor of performance. Typical spectral output of a xenon lamp is demonstrated in Figure 5.10. It is therefore self-evident that if the intensity of the lamp varies then the fluorescence signal level will also vary. Lamp drift, noise fluctuations, and changes in the lamp's spectral output as a function of time need to be considered.

5.4.3 Monochromators and Filters

All of the optical spectroscopy techniques require a means of delivering a particular wavelength of light to the sample and a means to analyze light emanating from a sample during illumination. Of particular importance is the separation of the "true" analyte signal from all of the potentially interfering ones. This process is, in most cases, based on wavelength discrimination. There are numerous methods that can be used to increase the selectivity of the measurement, including time-resolved techniques, polarization, and position sensitivity; however, the most common approach is based on wavelength discrimination by either dispersive or nondispersive means.

Monochromators and spectrographs are the most widely used dispersive instruments. They consist of a dispersive element, such as a prism or diffraction grating (www.newport.com), and image transfer optics, which separate a small wavelength from a polychromatic source (e.g., xenon lamp). Detectors used with monochromators are usually single-channel large-area devices. Single spectrographs use a fixed grating geometry to monitor a spectral range dispersed over a linear array that is made up of multiple detector elements. Grating-based monochromators and spectrographs are available in a wide range of configurations for applications in the 10-nm to 20-μm range. The manufacture of commercially available prism-based monochromators is on the decline and as a result the availability of such systems is scarce or specialized.

A diffraction grating is a plane or concave element with closely spaced grooves. The grating acts as a multislit source when illuminated by collimated radiation (www.newport.com). Different wavelengths are diffracted and constructively interfere at different angles. Most modern spectrometers use reflection gratings with groove densities from 75 to 3600 grooves per millimeter, depending on the spectral range and resolution required. Grating types are separated by their method of production, "ruled" or "holographic." Replicating a master grating prepared by a high-precision ruling engine makes ruled gratings. Projecting an interference pattern onto a photoresist plate and developing this to produce the pattern makes holographic gratings. Holographic gratings have essentially perfect groove patterns, almost perfect elimination of false lines or ghosts, and significantly improved stray light rejection compared with ruled gratings (www.horiba.com).

One of the most common configurations of a monochromator is the Czerny–Turner configuration (Figure 5.11). Although many other designs are also available most adopt the same operating principle (Figure 5.11). A Czerny–Turner monochromator. Light (a) is focused onto an entrance slit (b) and is collimated by a curved mirror (c). The collimated beam is diffracted from a rotatable grating (d) and the dispersed beam refocused by a second mirror (e) at the exit slit (f). Each wavelength of light is focused to a different position at the slit, and the wavelength that is transmitted through the slit (g) depends on the rotation angle of the grating.

The incident light is passed through the entrance slit and hits a collimating mirror that produces a parallel polychromatic light beam onto the diffraction grating. The grating rotates around a plane through the centre of its face and spatially separates the spectrum in the incident light. The resulting diffracted light is focused onto the exit slit using a focusing mirror. Each wavelength is incident upon the exit plane at a specific angle and, by rotating the grating position, one can scan these wavelengths across the exit slit and discriminate

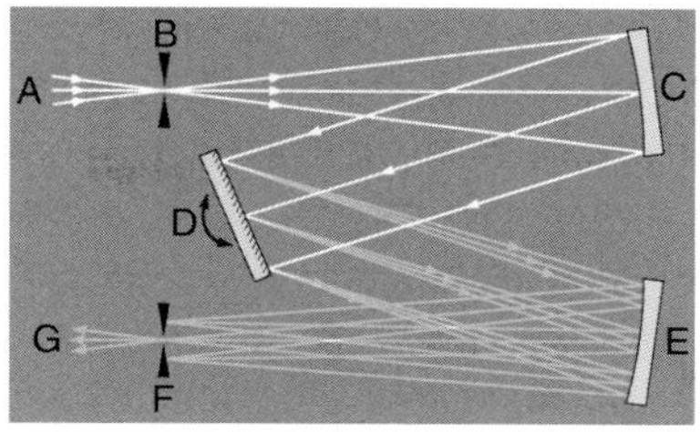

Figure 5.11. A Czerny–Turner monochromator. Light (a) is focused onto an entrance slit (b) and is collimated by a curved mirror (c). The collimated beam is diffracted from a rotatable grating (d) and the dispersed beam refocused by a second mirror (e) at the exit slit (f). Each wavelength of light is focused to a different position at the slit, and the wavelength that is transmitted through the slit (g) depends on the rotation angle of the grating. (http://upload.wikimedia.org/wikipedia/commons/e/e8/Czerny-turner.png)

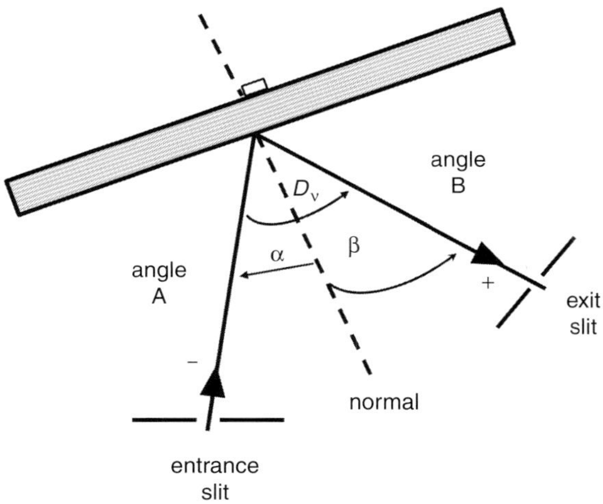

Figure 5.12. Graphical representation of the grating equation (Fortin, 2008).

between each wavelength. The grating equation specifies the angle required to bring each wavelength through the exit slit (Figure 5.12):

$$\sin\alpha + \sin\beta = 10^{-6} kn\lambda \tag{5.6}$$

where k is the diffraction order, n the grating groove density (grooves per millimeter), and λ the vacuum wavelength in nanometers.

Equation (5.6) illustrates the presence of overlapping spectral orders at a grating angle but with higher diffraction orders. For example, if a signal is measured at 750 nm in the first order then it is possible to also measure the light from 375 nm and 250 nm in their second and third orders respectively. This is an important phenomenon that can be eliminated by using either an additional dispersive element or a filter to separate or remove these orders from the acquired signal. In addition, the diffraction order can be either positive or negative in direction (www.newport.com). In the case of a negative order the monochromator can exhibit retro-diffraction effects that can also superimpose signals of the wrong spectral output at the exit slit (www.horiba.com). For illustration purposes the grating equation can be illustrated graphically (Fortin, 2008) as shown in Figure 5.12.

The slits themselves play an important role in determining the spectral resolution and throughput of the monochromator. In most cases, the positions of the entrance and exit slits are fixed but the width is adjustable. Typical slit widths can vary from a few microns to several millimeters but it is usual for the exit and entrance slits to be the same width. Several characteristics are important in a monochromator, such as the linear dispersion, *f*-number of solid angle, resolution, stray light rejection, and throughput factors. These factors are described in more detail below.

- The linear dispersion (D_L) is how far apart spatially two wavelengths are in the focal plane, $D_L = dx/d\lambda$, that is, at the exit slit in Figure 5.11. The more commonly quoted figure is the reciprocal linear dispersion (R_L), as this represents the wavelength range within a unit distance in the focal plane:

$$R_L = \frac{1}{D_L} = \frac{d\lambda}{dx} \tag{5.7}$$

- The limiting aperture in the actual instrument determines the *f*-number and solid angle (Arecchi et al., 2007). The *f*-number of an optical system can be simply defined as the focal length divided by the effective aperture diameter. Often this is the diffraction grating itself, as this is usually the most expensive element in the instrument. Lower *f*-numbers are usually associated with higher light gathering power or throughput because the light collected (or flux) is inversely proportional to the square of the *f*-number. The light collection efficiency is the solid angle that an optic makes with an object. The *f*-number describes this angle: *f*-number: $f/\# = l/d$, where l is distance and d is the diameter of the lens.
- With a limiting aperture diameter of L, a projected area of A, and the focal length of the collimating mirror of, the *f*-number ($f/\#$) is approximately $f/\# = f/L$, the solid angle (Ω) is then:

$$\Omega = \frac{A}{f^2} = \frac{\pi}{4(f/\#)^2} \tag{5.8}$$

- The spectral bandpass ($S\lambda$) is the full-width half-maximum of the wavelengths passed across the exit slit. The bandpass is controlled by the dispersion of the monochromator (R_D) except at very small slit widths, where both diffraction effects and aberrations need

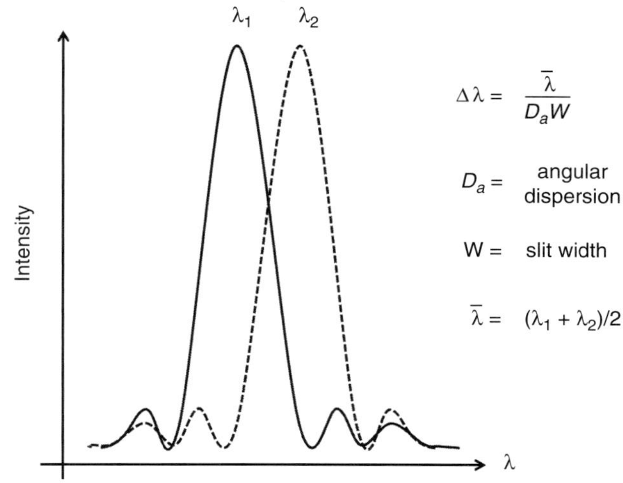

Figure 5.13. Rayleigh criterion for the resolution of two peaks.

to be considered. The spectral bandpass ultimately defines the resolution of the instrument and for a given slit width, *W*, this is given by: (5.9)

$$R_{\lambda} = R_{\mathrm{D}} W \tag{5.9}$$

- The resolution of the monochromator is closely related to the spectral dispersion. The dispersion governs how far apart two wavelengths are, while the resolution specifies whether the separation can be distinguished. The Rayleigh criterion states that two wavelengths, $\lambda 1$ and $\lambda 2$, are resolved if the central maximum of one line falls on a diffraction minimum of the other (see Figure 5.13). Thus, the spectral resolution can be defined by:

$$\Delta\lambda = \frac{\bar{\lambda}}{D_{\mathrm{a}} W} \tag{5.10}$$

where $\bar{\lambda}$ is the average wavelength between the two lines and D_a is the angular dispersion of the system, and *W* is the slit width.
- Stray light can be defined as any radiation passed by the monochromator that is outside the selected spectral position and bandpass. In many cases, the specification of stray light is made by reference to the relative amount of radiation being passed at a spectral position defined as an integer number of bandpass values from the test source – often a laser line. For example, a typical measurement involves filling the grating of a monochromator. Then, an intensity measurement is made both at the wavelength of the laser and at another wavelength eight bandpasses away from the laser wavelength. The ratio of the latter to the former is considered the stray light of the system under this criterion.

However, this type of approach is not realistic to the normal use of monochromators as broad-band tunable light sources or broad-band detection system. Typically, stray light is very difficult to measure, as it strongly depends on the wavelength, bandpass used, and the type of source. A further discussion of this important point is made in Section 5.4.8 on instrument performance validation.

- The optical throughput of a monochromator depends on the source, the slit height, the collected solid angle, the transmission factor of the optics, and the convolution of the entrance and exit slit widths (dispersion). The light gathering capacity (LGC) is defined by:

$$\mathrm{LGC} = \frac{\mathrm{height}_{\mathrm{slit}}\,(\mathrm{mm})}{(f\,/\,\#)^2 * \mathrm{dispersion}\left(\dfrac{\mathrm{mm}}{\mathrm{mm}}\right)} \tag{5.11}$$

5.4.4 Polarization Effects

Polarization effects in monochromators and other optical components can present significant difficulties in the overall operation and calibration of fluorimeter systems. At the same time, the introduction of polarizing elements such as polarizing filters or Glan–Thompson or Glan–Taylor polarizing optics are essential to the measurement of fluorescence polarization or anisotropy.

Reflection diffraction gratings are well described (www.horiba.com, www.newport.com) and can exhibit both strong and complex polarization effects. At certain wavelengths the grating may exhibit diffraction efficiencies in the S- or P-polarization planes. On average, this has little effect on the overall power transmitted by the monochromator for generally unpolarized light inputs but the monochromatic beam that exits the monochromator will, to some extent, be partly polarized, the extent of which can be strongly wavelength dependent. These effects lead to spectral shifts, signal loss, and a whole range of other *misleading* artefacts in the spectral signal. Figure 5.14 shows two sets of typical diffraction grating efficiency curves, from two master gratings, for 1200 g mm^{-1} and 500 nm blazed gratings with polarization angles for 45 degrees (upper boxes) and also S- (perpendicular) and P- (parallel) planes (lower boxes).

Nondispersive elements such as filters are also widely used in all of the optical spectroscopy methods, especially fluorescence measurements, and are based on either absorption or interference. Filters are commercially available for wavelengths above 200 nm and come in many forms, some of which are bandpass, cutoff, heat-absorbing, heat-reflecting, etc. The most common types are:

- Bandpass filters of the interference type are defined by the bandpass wavelength and the width of the bandpass. It is common to use such filters to select a wavelength; that is,

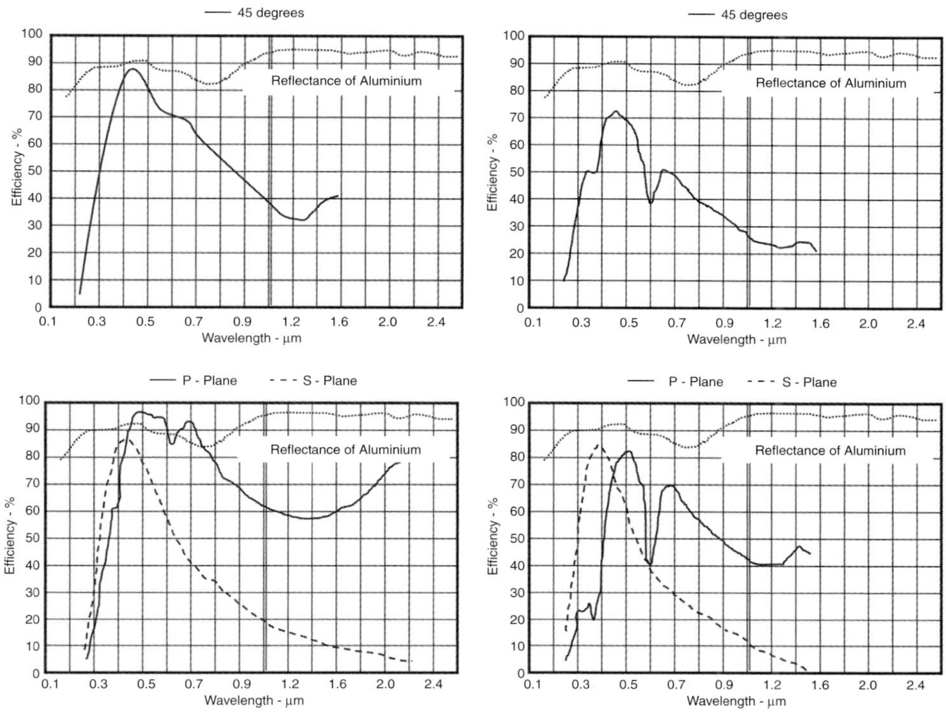

Figure 5.14. Examples of diffraction grating efficiency curves for 1200 g mm^{-1}, 500 nm blazed gratings.

excitation wavelength for fluorescence and a different filter for detection of an emission wavelength. Typical values for bandpass are 10–50 nm. Filters can offer excellent out-of-bandpass rejection, but do not offer the flexibility of a spectrometer.

- Cut-on and cutoff filters absorb all radiation at wavelengths either shorter or longer than the transition wavelength. The cutoff wavelength is defined as the spectral position where 50% of the maximum transmission of the filter is observed.
- Neutral density (ND) filters have relatively small wavelength dependence in the 180 nm to 2.5 µm spectral range. They transmit a specific percentage of the incident light; the exact transmission percentage can be very precise to allow strong signals to be measured on detectors that would otherwise be saturated. Use of ND filters enables greater dynamic range in a measurement system.

All filters require calibration for their transmission properties if used in the optical channels of a fluorimeter, as they also have spectral responses that are not always constant with wavelength, even the neutral density types.

5.4.5 Detectors

A radiation detector converts the radiant power in the measurement to an electrical signal that can be processed, recorded, and displayed. Detectors have the widest range of performance and cost of all the subcomponents of an optical spectroscopy system.

There are three main detector categories: photon, thermal, and multichannel. Photon detectors respond to the arrival rates of the photons and have a spectral response that changes with wavelength rather than responding to photon energies such as thermal detectors, which exhibit a near-uniform wavelength response. The performance criteria for all detectors include the spectral range, arrangements as single elements or arrays, the required detector area, acceptable signal-to-noise ratio (SNR), and dynamic range.

The largest item driving cost is the SNR requirement. Wide selections of detectors are available from a host of companies. There are literally thousands of photomultiplier tube types, array detectors (line and 2-D array), and semiconductor light sensors and thermal infrared detectors covering a spectral range from 150 nm to more than 40 μm. The wavelength range often determines the choice of detector type.

The responsivity of a detector is a function of the ratio of the output signal level to the incident radiant power (e.g., amps/watts), while the sensitivity of a detector can be characterized from the rate of change of signal output with respect to changes in the incident radiant power. This subtle difference is important when the response exhibits nonlinear properties. For all photon detectors, and many thermal detectors, both the responsivity and sensitivity of the detector are wavelength dependent properties and, hence, define the spectral response of the detector. The sensitivity of the detector may also change with other variables such as temperature, applied bias voltage, and other components in the signal processing circuit. Some detectors, for example, photomultiplier tubes, can have different output modes – as a charge or current (analog mode) or as a pulse rate (photon counting). Thus, care must be taken not only in the detector choice, but also in how it is going to be used in the data acquisition scheme.

Optical detectors can vary widely in their speed of response and in their ability to record quickly changing signals. The rise time of the detector can be important for many applications and can vary from <1 ns to seconds, depending on detector type. In the absence of any input optical signal, a detector will still produce an output, which is related to the inherent dark signal and dark noise of the detector or system. The dark signal is caused by a relatively small electric current that flows through all photosensitive devices even when no photons are entering the device. The dark noise relates to the fluctuations in the number of photons contributing to the current inherent within the detector, especially because these photons are all independent to each other (random). As a result the characteristics of these signals vary greatly with detector type and mode of operation. An understanding of the noise and possible drifts associated with the dark signal is important before this inherent "system noise" can be subtracted from the optical signal of interest.

The main photon detectors used in commercial fluorimeter systems are based on either photomultiplier or photodiode devices. A photomultiplier is a vacuum device with a

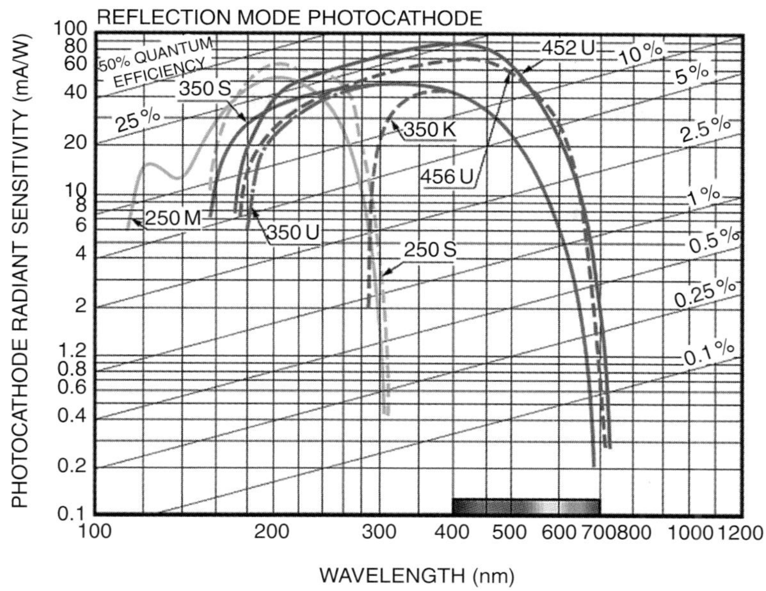

Figure 5.15. Radiant sensitivity of a selection of photomultiplier photocathode types; the dotted lines refer to a specific gain in the dynode chain.

photocathode that converts an absorbed photon to an emitted electron. This electron is drawn toward an electron multiplier stage. The multiplier gain is dependent on the number of stages in the photomultiplier (dynode stages) where secondary electrons are released. The signal is collected at the anode and output as a current pulse lasting a few nanoseconds. Typical photomultiplier gains are in the range 10^4–10^7, where single-photon sensitivity is possible. For photon counting the dynode chain is designed to give isolated short pulses for each photon.

The absorbing surface of the photomultiplier, its photocathode material, defines the spectral range and quantum efficiency of the detector. Photocathodes that respond to longer wavelengths have lower work functions and are therefore more susceptible to noise signals from thermally generated electrons. Thus, if working above 650 nm in photon-counting mode, it may be necessary to cool the detector to –20°C or less in a suitable housing.

Photomultiplier tube detectors are still the detector of choice for fluorimeter application owing to their excellent sensitivity and operational gain. Figure 5.15 shows typical spectral responsivity of a selection of photomultiplier detectors.

Photodiode detectors are based upon the absorption of a photon in a P–N junction diode (Skoog, 1998) and the creation of an electron-hole pair in the depletion layer by promotion of electrons across the valence band to the conduction band. The spectral response of silicon photodiodes extends from the UV to the near-IR (1.1 μm), and the responsivity of such

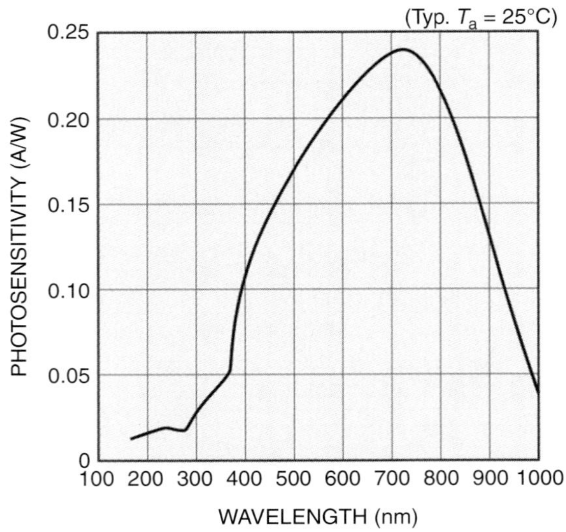

Figure 5.16. Radiant sensitivity of a large area silicon photodiode detector.

photodiodes is often much less than that of photomultipliers because they have no internal gain. However, single-photon avalanche diode (SPAD) identifies a class of solid-state photodetectors based on a reverse biased P–N junction in which a photogenerated carrier can trigger an avalanche current owing to the impact ionization mechanism. This device is able to detect low-intensity signals (down to the single photon) and to signal the arrival times of the photons with a jitter of a few tens of picoseconds (Niclass et al., 2005). Even so, silicon photodiodes are rugged, exhibit excellent linearity over seven decades of incident radiant power, and can be small in size with excellent time responses. As a result they are ideal as the detector used as the excitation channel monitor. Figure 5.16 demonstrates typical spectral responsivity of such detectors.

5.4.6 Measurement Systems: Data Acquisition Electronics and Software

The output signal from the detector must be monitored, stored, and analyzed to make a spectroscopic measurement. As such, the signal-processing and readout system is extremely important to the overall performance of the system. The particular type of signal processing depends on the form of the output signal, the noise sources expected, and the signal level itself. The signal-processing step can perform many conversions such as current-to-voltage, analog-to-digital conversion, amplification, or some mathematical operation designed to improve the measurement of SNR.

The outputs of most detectors are used in the analog mode. The exception to this is the photomultiplier, which may be also be used in the photon-counting mode. Nearly all measurements are made by using an analog-to-digital converter (ADC) to convert the analog output of a detector to the digital domain for future processing, analysis, and display. Even in digital measurements some conditioning, such as filtering or amplifying the analog signal, is usually necessary to make it suitable for recording by the ADC. In photon counting systems, the recorded signal is the number of pulses observed or counted. Photon counting is particularly useful in applications where the signal irradiance is relatively low, such as photoluminescence measurements.

5.4.7 Data Collection, Display, and Analysis Software

At the most basic level, spectroscopic software must allow for reliable, efficient collection and storage of data with minimal sacrifice of hardware performance. As spectroscopy increases in complexity and importance in everyday activities, software is becoming more important. Today, software often needs to be sufficiently flexible to adapt to varied applications, yet simple enough for the increasing number of users with limited spectroscopic backgrounds. More sophisticated software packages provide users with several options in data display and analysis. This must include the possibility to record, display, calculate, and then utilize spectral response curves on the excitation and emission channels. In addition, the software must be suitably flexible to provide these correction curves under a variety of different operating conditions such as changes in spectral bandpass, scan steps, signal levels, and so on. The task is not straightforward and attention to detail is critical. Too often the instrument software from manufacturers will assume "black box" operation and present a "result" without the rigor of knowing whether it is the "true" result or whether it is influenced by the optical spectral variations in instrument to instrument caused by spectral properties of the light source, the illumination and collection optics, and the wavelength performance of the analyzing spectrometer and detector.

It is the successful combination of the light sources, light discriminators, coupling optics, and sampling and detection methods that provides an optimal system. This hardware package, coupled with a complete robust software analysis package and the correct instrument calibration and correction functions, is the path to a successful instrument and successful fluorescence measurements.

5.4.8 Instrument Performance Validation

In many laboratories, the performance validation of a fluorimeter instrument is still in its infancy, and the influence of the many parameters involved in the instrument operation is often not considered principally because it is thought that the instrument manufacturers have taken care of this problem in their software. In reality instrument performance validation must also be the responsibility of the users to ensure the instrument meets their needs and has not drifted with time. With this in mind, users might undertake a series of standard measurements using one or more of the following:

- Instrument calibration standards – physical sources such as line and tungsten lamp emissions, known instrument independent luminescence standards, and so forth

Table 5.5. *Measurement and analysis of signal-to-noise using the water Raman test*

Measurement conditions	Excitation:	λ_{exc} = 350 nm, with $\Delta\lambda_{exc}$ = 5 nm
	Emission scan:	370 < λ_{em} < 460 nm, with $\Delta\lambda_{em}$=5nm, and scan in $\delta\lambda$=1 nm steps
	Integration time:	1 s per step
Analysis method	Signal = (peak signal at 397 nm) – (Average back ground signal)	
	SNR: Signal/background noise in range 450–460 nm	

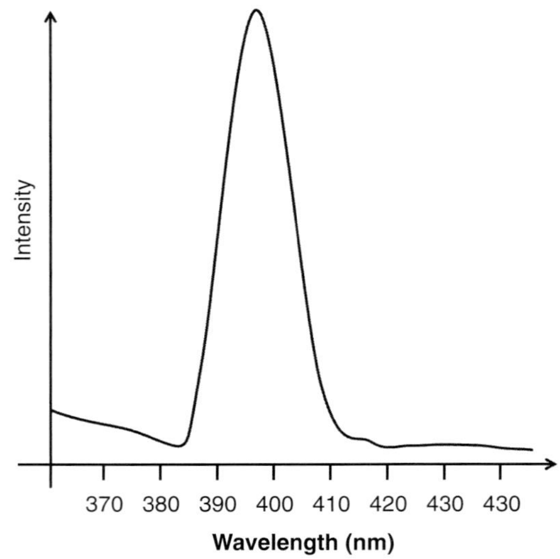

Figure 5.17. Typical spectra showing the water Raman measurement using an excitation wavelength of 350 nm.

- Application standards – such as spectrally matched and known fluorophores whose properties are close to those of the samples being measured.

In many cases, quantitative fluorescence assessment often consists of undertaking some intensity calibration of the instrument response with respect to the concentration of some "standard" fluorophore in a certain solvent. These titration type experiments are a common way to "calibrate" an instrument but they rely critically on very precise determinations of the fluorophore purity, solvent purity, initial concentration determinations, dilutions methods, pH and temperature control, and so forth. The whole process is fraught with possible sources of both random and systematic errors and of course is applicable only to that particular instrument at that time.

In reality, the long-term stability of an instrument is one of the key concerns of both the manufacturer and the user. Thus, day-to-day and even instrument-to-instrument validation is needed. One of the best known methods to achieve this is to undertake the so-called "water

Raman test." Assuming the water sample is of very high purity and is contained in a fully sealed cuvette to prevent ingress of contaminants, this test can be reliable at wavelengths below 400 nm. The reason for this is that the intensity of the Raman signal changes by $1/\lambda^4$, thus at longer wavelengths of excitation the signal levels are considerably reduced. For many manufacturers the water Raman test and the resulting SNR become a measure of their instrument performance guarantee. Not all manufacturers record this signal nor calculate the resulting SNR in the same manner, so considerable care should be applied in the exact definitions and comparison. Presented in Table 5.5 are some suggested measurement conditions and analysis methods for routine assessment of the instrument performance in terms of total signal and water Raman SNR. A water Raman spectra is shown in Figure 5.17.

5.4.9 Linearity, Signal to Noise, and Dynamic Range

The dynamic range of a spectroscopic instrument is of considerable importance because if allows the measurement of both weak and strong signals that might be closely spaced in wavelengths. There are several ways to consider the dynamic range and it is often quoted in a simple way that might not always reflect reality. For any instrument, the key figures of merit are the maximum signal that can be measured without distortion and what the noise level is. This gives a SNR maximum level, or the dynamic range of the instrument.

For a fluorimeter instrument the noise level is determined by several factors:

- The noise level from the photomultiplier detector
- The stability of the xenon lamp, that is, light fluctuation or noise in the excitation light level
- The stray light performance of the instrument at the wavelengths of interest

Single-photon counting fluorimeters exhibit exceptional dynamic ranges compared to their analog counterparts. Typical photon counters in modern fluorimeters are capable of up to 100 Mega counts per second (Mcps) if the signal is repetitive. Ideally, two photon events are identified by appropriate discriminators and counted as two events. However, in a practical situation two problems exist. First, the incoming photon rate is random and as such the available counting rate is reduced, as we need to be able to distinguish between photon pulses; and second, the pulse width of each photon pulse is of finite value because of the detector time responses. Therefore, it is necessary to distinguish between two pulses, that is, the pulse-pair resolution. If the time between photon pulses is less than or equal to the time to resolve the two photon pulses then they look like a "single" event and the signal is lost. This is the time after a first pulse within which the system cannot distinguish the next photon pulse, that is, a pile up phenomenon and is referred to as the dead-time of the system. In this case, the available count rate for a random signal is given by:

$$I_p = \frac{I_m}{2.718(I_m \, . \, D_t - 1)} \tag{5.12}$$

where I_m is the maximum possible count rate and D_i is the effective dead-time of the system.

Table 5.6. *Typical signal, noise and dynamic range characteristics of photon counting devices*

Method	Peak signal	Noise	Dynamic range
Typical photon counting systems	4–22 Mcps	100 cps	40,000–200,000: 1

Therefore if the maximum count rate is 100 Mcps, and the dead time is about 26–27 ns, this would give a maximum practical signal count rate of approximately 22 Mcps.

For photon counting, the photomultiplier dark noise level defines the noise level of the instrument, and this is determined by the operating conditions of the detector itself: type of detector photocathode, applied high voltage, voltage divider arrangement, and discriminator threshold value. Typically, for blue-sensitive photomultipliers the noise level is under 100 counts per second (cps), and for red-sensitive detectors it may be 10 times higher at room temperature owing to the lower work function photocathode and higher intrinsic dark noise level from such a detector. However, routine cooling can reduce dark count levels to a photon per second. Therefore the performance of any fluorescence detection system is usually defined in terms of its signal-to-noise performance over a dynamic range (Table 5.6).

5.4.10 Speed and Sensitivity

The speed of wavelength scanning is often considered a prime specification in a fluorescence spectrometer system. Many users equate this to the ability to perform a large range of measurements in a short time. Thus high scan speed is considered advantageous. However, high scan speed is of no importance if the instrument is not able to achieve a reasonable SNR level during sample measurement at that scan speed. This is the well known speed versus accuracy or speed versus sensitivity issue. The terms "scan rate" and "slew rate" are often interchanged although they have distinctly different meanings (see Table 5.7). Manufacturers may quote slew rates to demonstrate high rates of change in wavelength, but from a practical point of view it is the scan rate that is the important parameter.

The scan rate is determined by the

- Mechanical rotation speed of the monochromator grating
- Focal length of the monochromator
- Groove density of the grating

The focal length and groove density determine the reciprocal linear dispersion of the monochromator, or how much of the spectrum is spread across a specific distance in the output plane of the monochromator. Thus, short focal length monochromators with coarsely ruled diffraction gratings can produce relatively high scan rates.

Table 5.7. *Definitions of slew rate and scan rate*

Scan rate	The **scan rate** represents the continuous speed of scanning a spectrum between two wavelengths.
Slew rate	The **slew rate** represents the maximum rate of change of wavelength.

Short focal length monochromators can have poor resolution compared to longer focal length units. Also, coarsely ruled gratings produce low spectral resolution outputs. So to determine the equivalent scan rate of instruments an understanding and comparison of focal lengths and grating parameters is necessary. Ultimately, there must be a trade-off between scan speed, spectral resolution, and system sensitivity. Many fluorimeters specify their scan rates as very slow, slow, medium, fast, and very fast and at the same time they claim 30,000 nm min^{-1} or higher slew rates. As a result, the scan and/or the slew rates have little practical meaning and it is almost impossible to determine the real instrument performance in this regard. In practice, most users will select a scan speed and slit width commensurate with their measurement requirements from an SNR point of view, although the choices they make are often qualitative assessments of spectral quality rather than considered quantitative requirements.

The single-photon counting fluorimeter has many performance advantages over conventional fluorimeter instruments. One of the most important is that it uses single-photon counting as the measurement technique. This delivers unparalleled sensitivity. It means you can collect data faster, measure more samples, or work at lower concentrations with more accuracy. At the same time, the scanning capability of the instrument means more data can be measured and exposure time to sample is limited, which reduces the possibility of photo-bleaching or sample degrading with time and hence damaging the integrity of the sample and results. Single-photon sensitivity allows the user to analyze samples at low concentrations that is simply not possible with non-photon-counting instruments. If the emission signal from the samples is strong, then short integration times permit fast scanning, while maintaining the same level of accuracy and saving measurement time. The stronger the measured signal, the lower the statistical noise, the better the SNR and hence the greater the accuracy. Many fluorescence detection systems employ photodiode array detectors (e.g., high-performance liquid chromatography). These offer single-photon sensitivity, albeit at lower resolution, as the exit slit width is effectively a single element in the actual array, e.g., 25 μm. Unfortunately, an in-depth discussion concerning photodiode arrays detection systems is beyond the scope of this chapter.

5.4.11 Wavelength Accuracy

Wavelength accuracy is a fundamental requirement in any spectrometer system and one that should be checked on a periodic basis. The normal method to check wavelength accuracy

Table 5.8. *Emission lines from a low-pressure mercury pen lamp with the common calibration wavelengths highlighted in bold*

Wavelength/nm	Wavenumber cm^{-1}	Wavelength/nm	Wavenumber cm^{-1}
253.652	39,424.14	434.749	23,001.76
265.204	37,706.84	**435.833**	22,944.58
265.368	37,683.53	**546.074**	18,312.55
296.728	33,700.90	**576.960**	17,332.23
302.150	33,096.17	**579.066**	17,269.18
312.567	31,993.16	708.190	14,120.50
313.155	31,933.09	1013.976	9862.17
313.184	31,930.12	1357.021	7369.08
365.015	27,396.11	1367.351	7313.41
365.484	27,361.01	1529.582	6537.73
366.328	27,297.95	1707.279	5857.27
404.656	24,712.33	2325.307	4300.51
433.922	23,045.60		

is to undertake measurements using a suitable line source lamp such as a low-pressure mercury lamp. This lamp provides a range of discrete lines that can be used to demonstrate the wavelength calibration of the system as well as wavelength linearity. At the same time it can be used to demonstrate the correct and repeatable operation of the grating turret.

The emission lines for a low-pressure mercury lamp are demonstrated in Table 5.8, with the common wavelengths for calibration highlighted in bold.

Most monochromator systems utilize drive mechanisms that use stepping motors. These stepping motors operate in discrete angular increments called steps. Thus a wavelength calibration can be made by recording the position of the calibration lines from the lamp as a function of the number of steps driven from the zero order of the grating and hence the angular position of the grating. Often this mapping of spectral position with step position is linear or close to linear in form. A suitable fitting with either a straight line or some form of polynomial will yield a calibration curve of wavelength versus step (angle) position With this approach care must be taken to ensure that the correct wavelength lines have been identified and used for calibration. A common error is the use of emission lines that occur in the second order of the grating, for example, thinking there is calibration line at 507.3 nm, which is not a calibration wavelength but results from the emission line occurring at 253.65 nm.

The elemental emission lines from mercury also act as a very useful means to check the instrument spectral resolution and reproducibility of scanning. For example, Figure 5.18 shows the optical resolution obtainable from a research quality fluorimeter.

Scanning reproducibility is intrinsically a function of the quality of the mechanical design, build, and testing. For research instruments, accuracy of wavelength position and

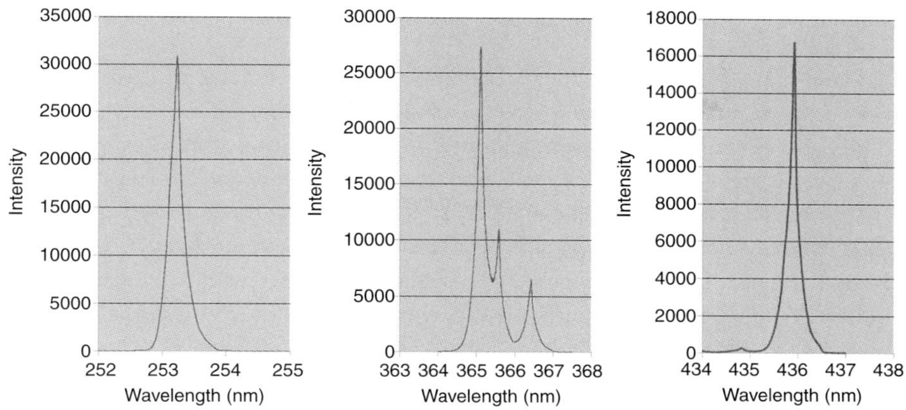

Figure 5.18. Typical resolution plots at 253.6, 365.0, and 435.8 nm showing optical resolutions of better than 0.1 nm FWHM. 435.8 nm is a common wavelength for manufacturers to specify spectral resolution, usually using a 1200 g/mm grating.

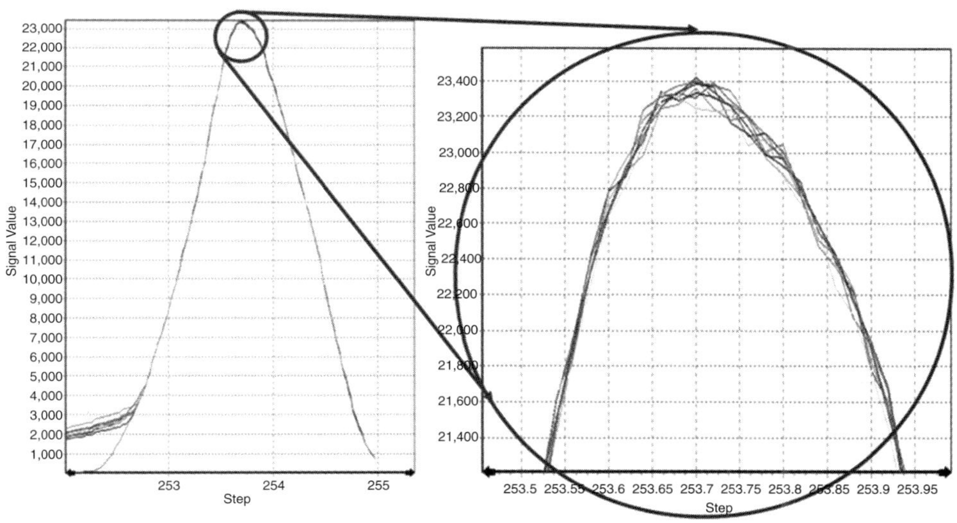

Figure 5.19. Determination of scanning reproducibility by repetitive scanning of the 253.6 nm Hg line. Typical examination of the peak feature indicates excellent reproducibility (>0.2 nm) in wavelength scanning.

repeatability in scanning should be equal to or exceed 0.2 nm. Figure 5.19 highlights the importance of scanning reproducibility.

5.4.12 Bandpass Selection

The spectral bandpass applied to the monochromators will determine the capability of the fluorimeter to accurately, or not, record the excitation or emission spectrum shape of a luminescent sample. The bandpass is determined by the slit width used and also by the diffraction grating parameters in the monochromator. A larger (or wider slits) bandpass means lower spectral resolution whereas a smaller bandpass (i.e., narrower slits) will result in a higher spectral resolution. At the same time, the light throughput will change with different bandpass. As a rule of thumb, doubling the bandpass will increase the signal level by a factor of four. For this reason, many have used the spectral bandpass selection of an instrument as a means to control the light level presented to the analyzing detector, rather than select an appropriate neutral density filter. Simplistically, it is quite easy to derive this empirically by adjusting the slit width and observing the resultant spectra in terms of its intensity and spectral shape. However, great care must be taken when using this approach such that important spectral information is not lost. For example, if the spectra show structure, such as multiple peaks or small shoulder effects, then the bandpass should be adjusted to an appropriate size to accurately record and maintain the integrity of these spectral features

The bandpass appropriate for the features of the spectrum that one wishes to resolve should be used. This is determined by the photo-physics, chemistry, and biology of the samples, as well as the closeness (in terms of spectral proximity) of the excitation and emission peak wavelengths and any scattering caused by the sample. Ideally, the widest bandpass that does not distort the spectral features should be used to ensure the best possible SNR in the measurement. Finally, it is important to note that bandpass and resolution are not the same quantity. If the bandpass is equal to the slit width multiplied by the reciprocal linear dispersion and the slit becomes so narrow that no improvement in the bandpass can be observed, then, in effect, the bandpass is equal to the resolution.

5.4.13 Stray Light

Stray light in a monochromator or spectrometer system is all light that reaches the image plane of the monochromator from anywhere other than from diffraction by the grating (according to the grating equation), that is, any light that is passed by the monochromator that is outside of the interval $\lambda_0 \pm \Delta\lambda$, where λ_0 is the wavelength setting and $\Delta\lambda$ is the spectral bandpass. Usually, stray light is expressed as a ratio of the total light passing through the exit slit at a specified wavelength compared to another wavelength. In some cases, the stray light is specified by the relative amount of light that is passed x nm from a specified laser line. Other methods of demonstrating stray light performance can be made using near-end or far-end methods using either narrow line or broadband spectral sources. What is clear is that manufacturers of optical systems measure and specify the stray light performance of their products differently. Thus it is often difficult for users to directly compare the performance of one device against another. In the end, the practical demonstration of the monochromator stray light performance in the specific illumination and spectral

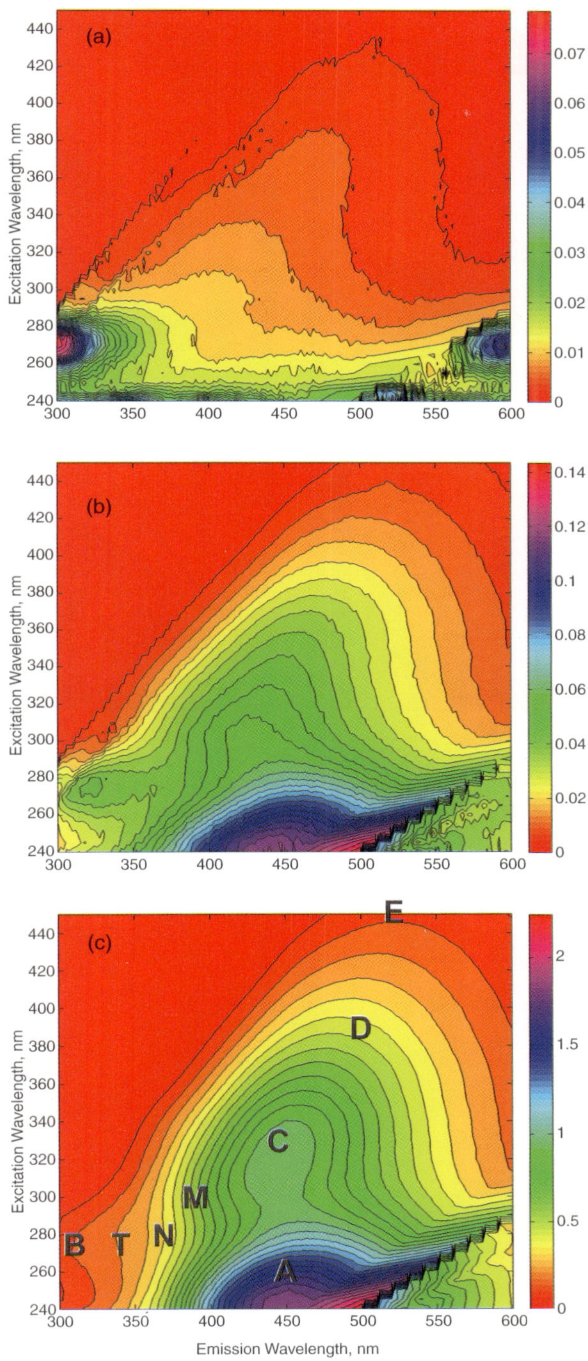

Plate 1. Excitation emission spectra for samples from (a) the Pacific Ocean, (b) the Gulf of Maine, and (c) the Penobscot River. Commonly reported peaks and region locations described in Table 2.1 are also graphically presented in (c).

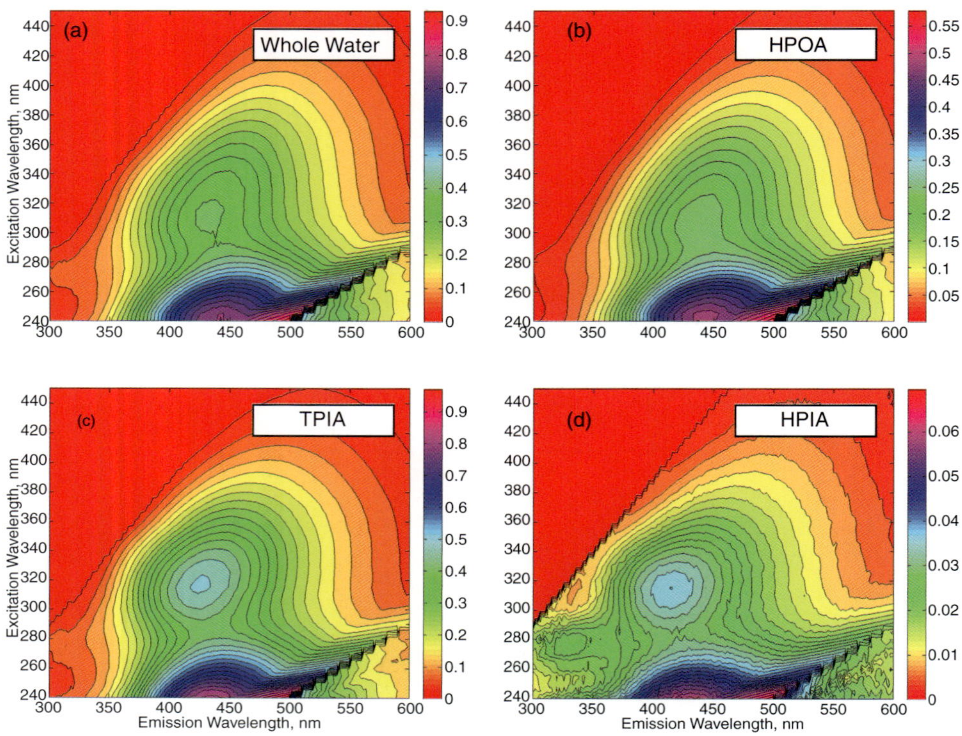

Plate 2. Excitation–emission spectra for the (a) whole water, and mass normalized EEMs for the (b) hydrophobic organic acid, (c) transphilic organic acid, and (d) hydrophilic fractions of a water sample from the Yukon River at Pilot Station, Alaska.

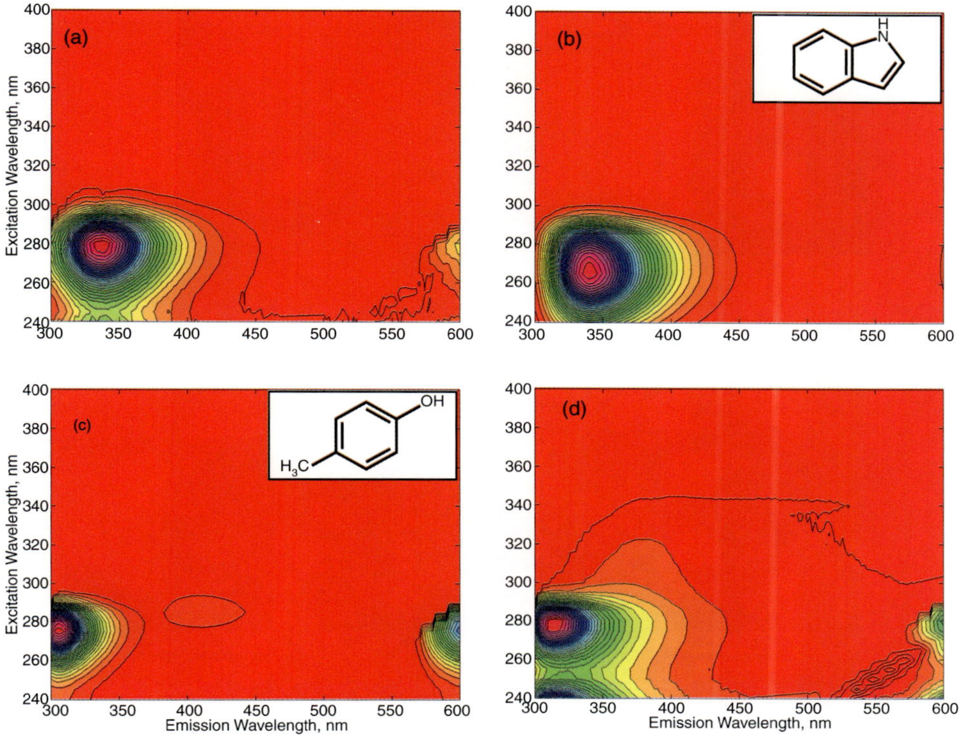

Plate 3. Excitation–emission spectra for (a) bovine serum albumin, (b) indole, (c) cresol, and (d) rhodendron tannin.

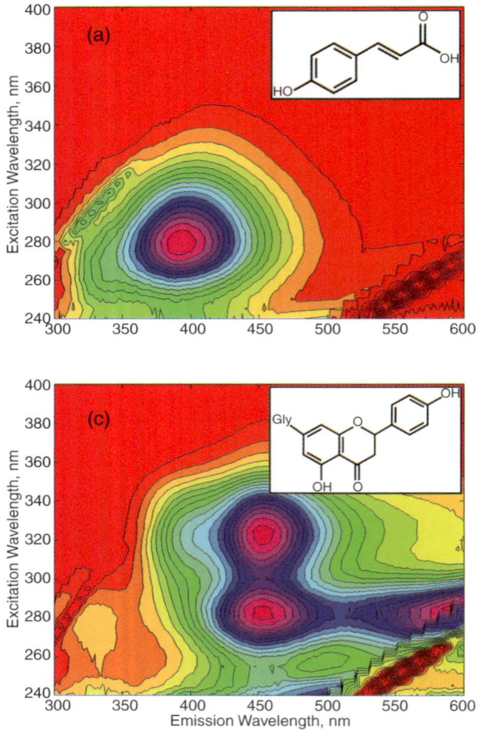

Plate 4. Excitation–emission spectra for (a) *p*-coumaric acid, (b) coumarin, (c) naringin hydrate, and (d) lignin, alkali, 2-hydroxypropyl ether.

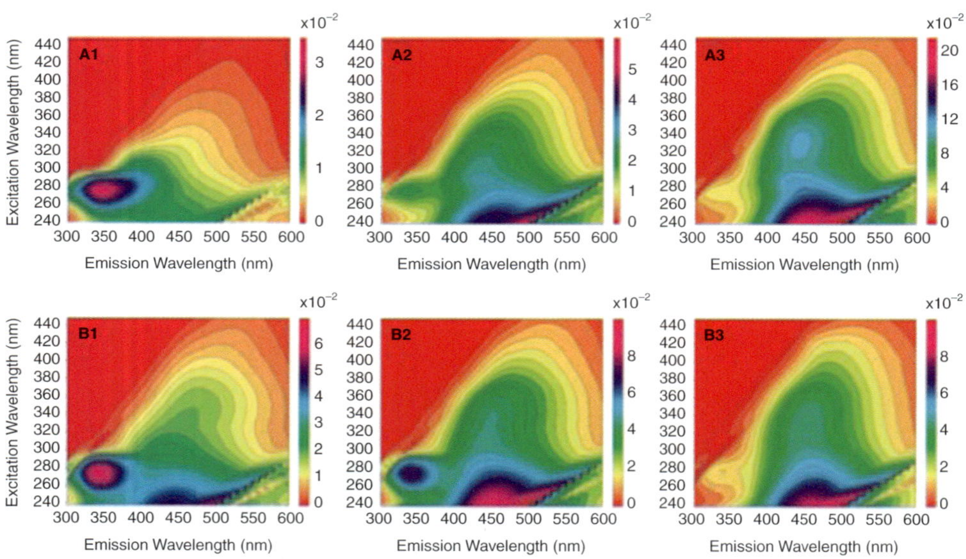

Plate 5. Change in fluorescence of moss leachate DOM over time: (A1–A3) *Sphagnum angustifolium* at 24 hours (A1), 194 hours (A2), and 3 months (A3); Feathermoss mix at 24 hours (B1), 194 hours (B2), and 3 months (B3). The color bar to the right indicates fluorescence intensity, which has been normalized to DOC concentration. (Reprinted from Wickland et al., 2007, With kind permission from Springer Science+Business Media.)

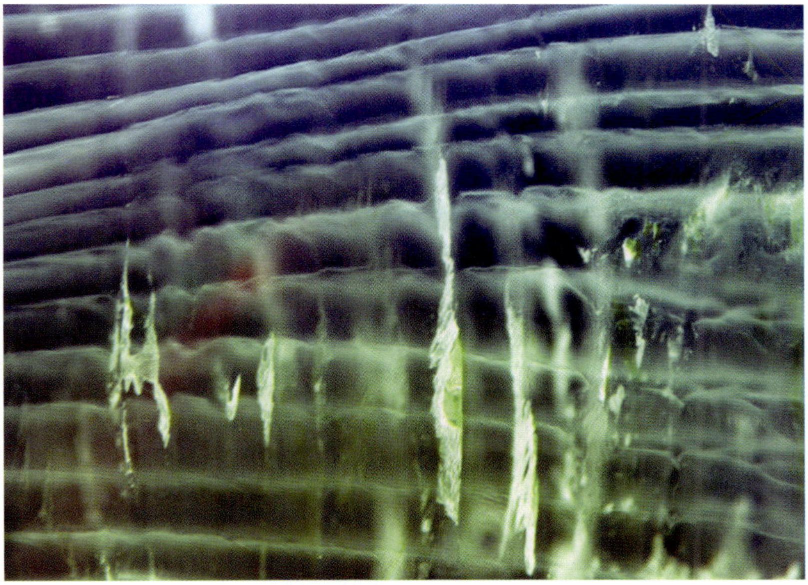

Plate 6. Fluorescent NOM preserved in a stalagmite to form annual laminae. This picture shows annual cycles of NOM fluorescence associated with seasonal groundwater recharge. Stalagmite is ER-77 from Ernesto Cave, Italy. (Photograph courtesy of Ian Fairchild, Birmingham.)

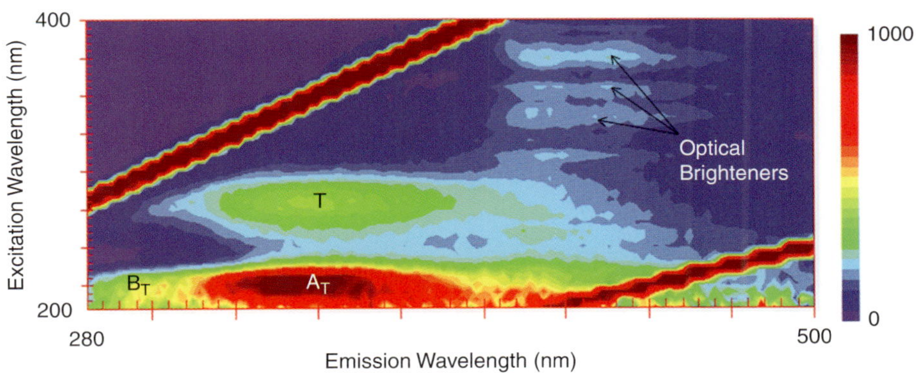

Plate 7. Fluorescence excitation–emission matrices (EEMs) for raw sewage where BT = tyrosine-like, T and A_T = tryptophan-like fluorescence. (Adapted from Henderson et al., 2009, with permission from Elsevier.)

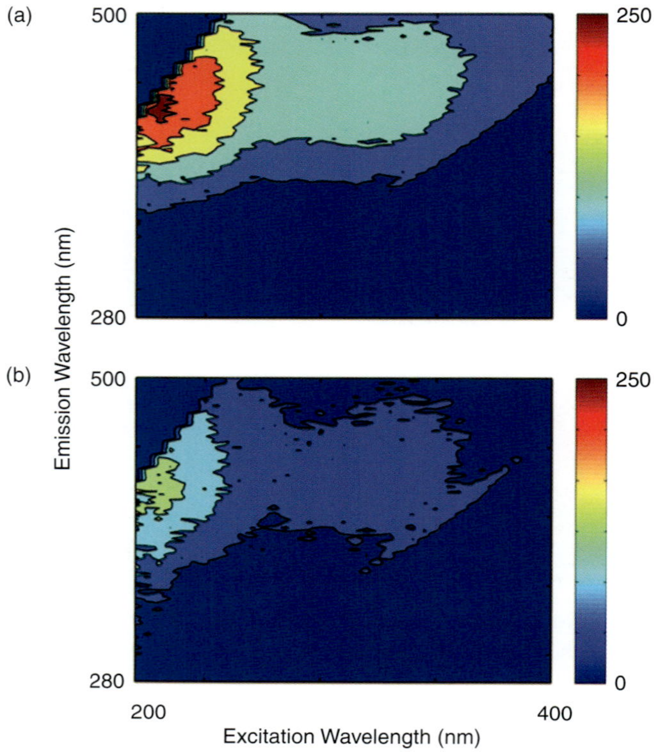

Plate 8. Excitation–emission matrices of waters at different water treatment stages. (a) Raw water (b) Clarified water. (Adapted from Bieroza et al., 2009a, with permission from Elsevier.)

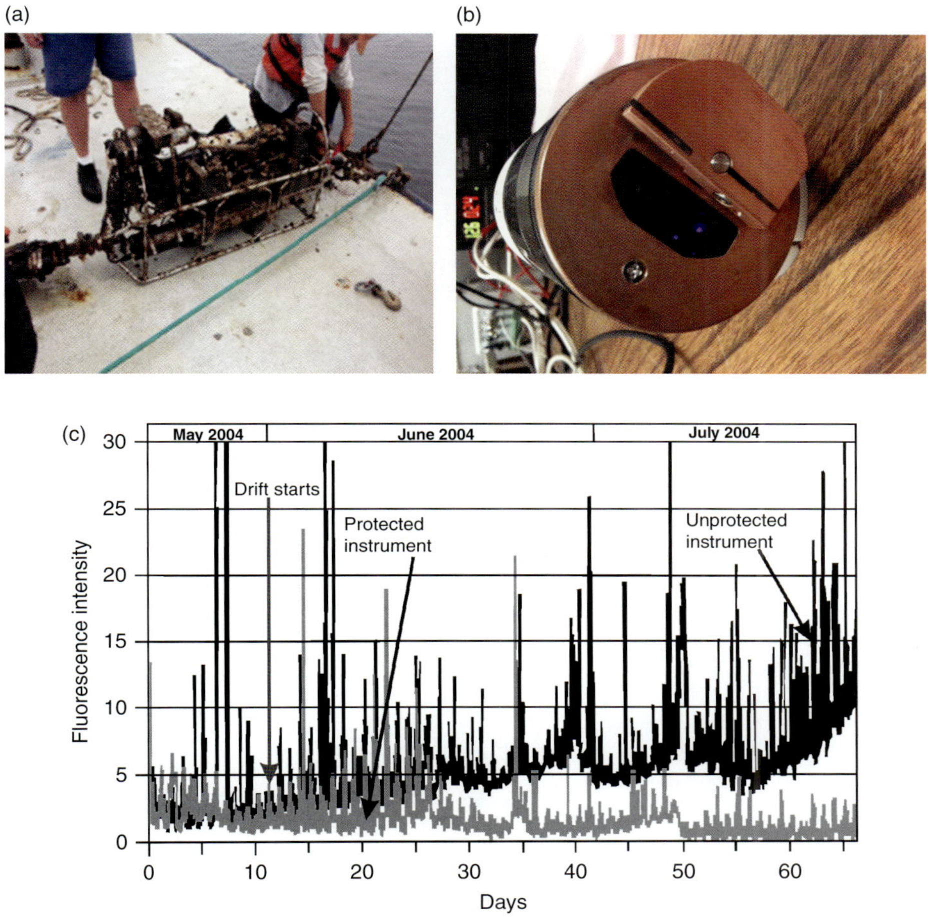

Plate 9. (a) Retrieval of an optical package after a deployment in the Penobscot River, Maine. (Photo courtesy of C. Roesler and A. Barnard.) (b) Example of a copper biofouling shutter on a sensor. (Photo courtesy of B. Downing.) (c) Drift of unprotected fluorometer due to biofouling. (Redrawn from Delauney et al., 2010.)

(a)

(b)

Plate 10. Fluorescence sensor deployment platforms. (a) CTD Rosette vertical profiler. Water was pumped in series through a fluorometer and other environmental sensors, where instruments were mounted at the bottom of the Rosette and CDOM fluorometer. (Courtesy of R. Conmy.) (b) Towed vehicle with various optical and chemical sensors. (Courtesy of R.F. Chen.) (c) Minishuttle tow-yo vehicle deployed with chlorophyll and NOM fluorometers, dissolved oxygen sensor, and CTD. (Courtesy of R.F. Chen.) (d, e) Buoys, moorings, and gliders are also platforms for optical and environmental sensors. (Courtesy of Cefas.)

(c)

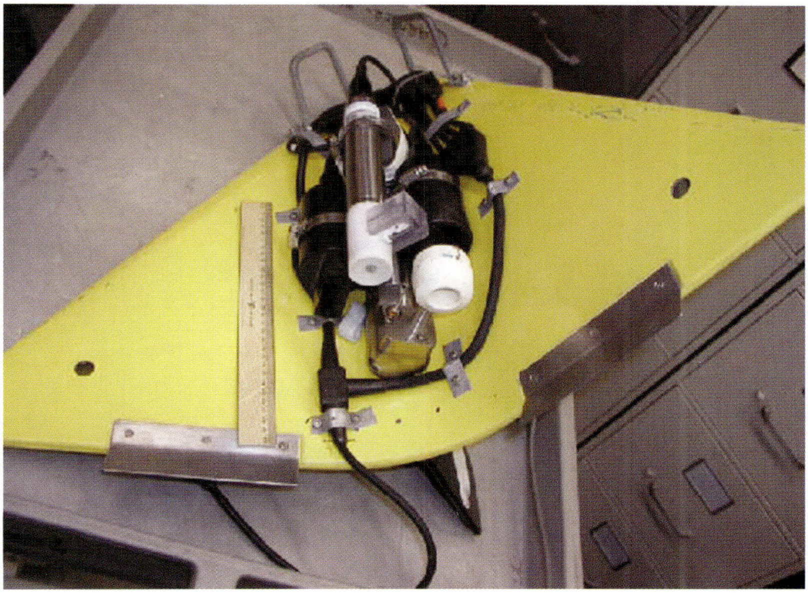

(d)

Plate 10. (*cont.*)

(e)

Plate 10. (*cont.*)

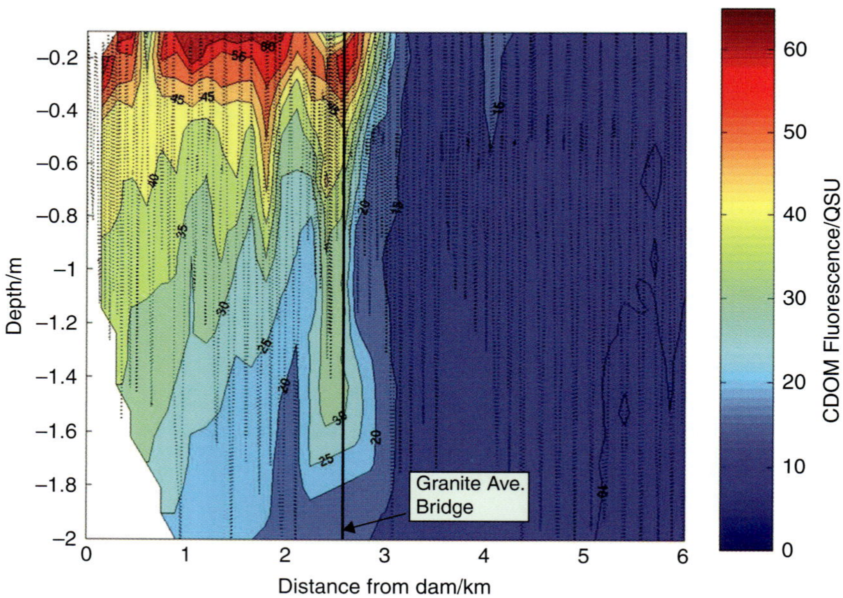

Plate 11. CDOM fluorescence in the Neponset Estuary at high tide. Dotted lines represent path of towed vehicle Bridge is located 2.6 km from dam.

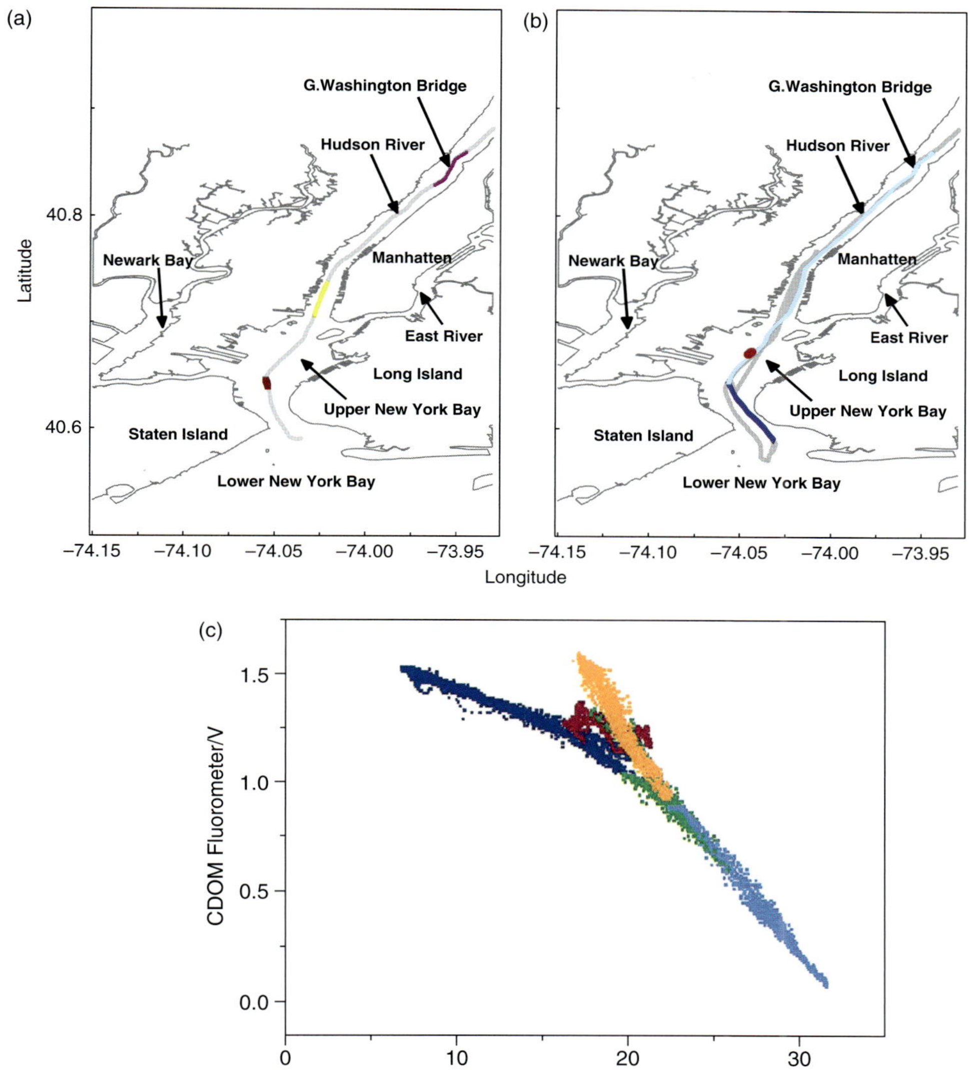

Plate 12. (Top) CDOM as a function of salinity in the Hudson Estuary during ebb and flood tides, respectively. (Bottom) Colors of data symbols correspond to colored portions of cruise tract on the map. Note the yellow CDOM contribution (presumably sewage-derived CDOM) during the ebb tide (top left) that is not observed in the shipping channel during the flood tide (top right).

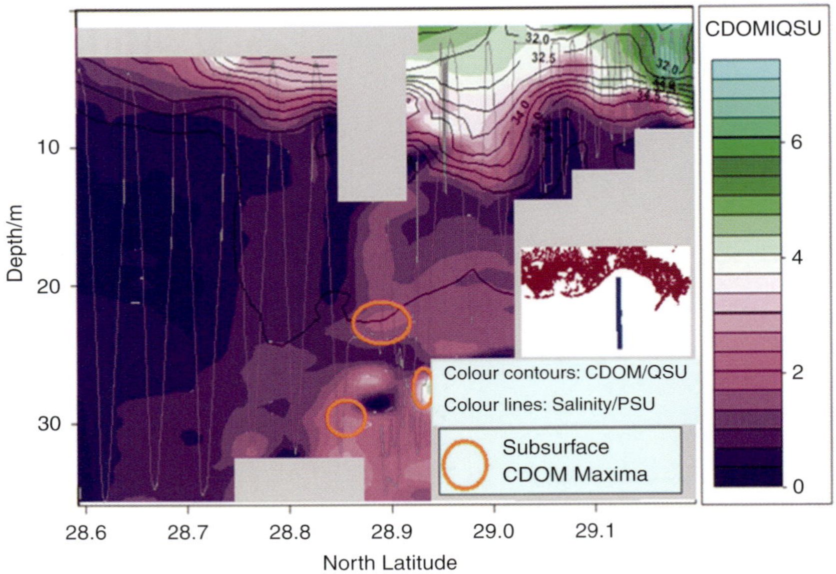

Plate 13. CDOM (grayscale colors) and salinity (black lines) contours across the Mississippi plume in June, 2000. Salinity is represented by the contour lines. Inset map shows the location of the survey transect in relation to the Mississippi Delta. Gray lines represent path of towed vehicle and gray boxes represent missing data. Bottom depths were approximately 5 m below the maximum depths of the fluorescence and salinity measurements. Subsurface CDOM maxima are highlighted with circles. (Redrawn from Chen and Gardner, 2004.)

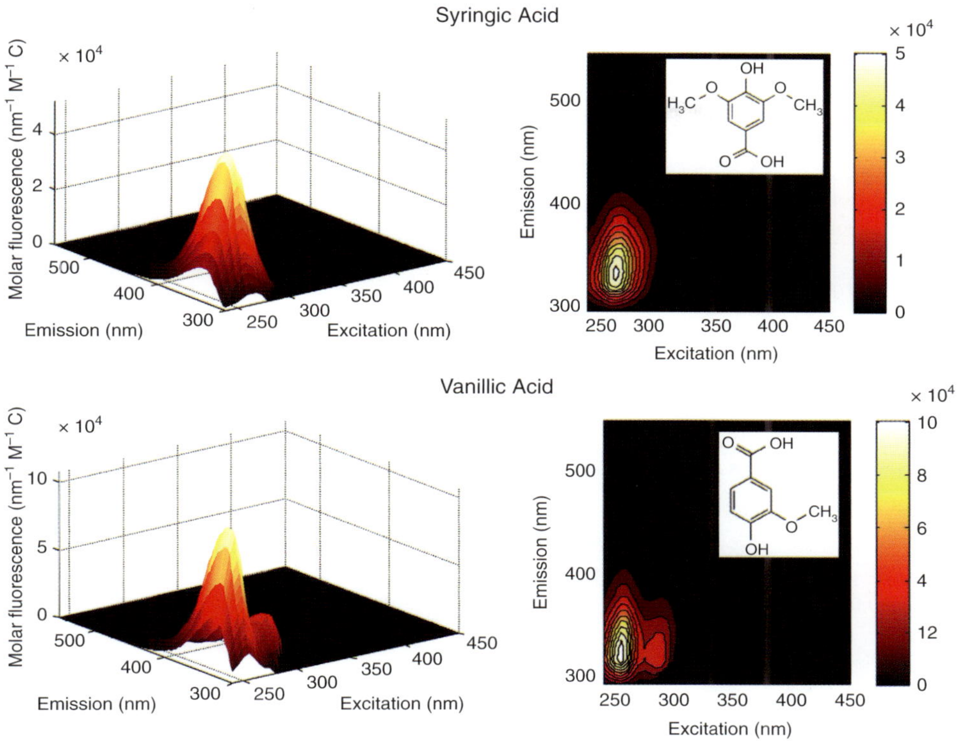

Plate 14. Molar fluorescence properties of Syringic (Sigma S 6881) and vanillic acid (Aldrich H3,600–1) in MilliQ water at room temperature. (Stedmon, unpublished data.)

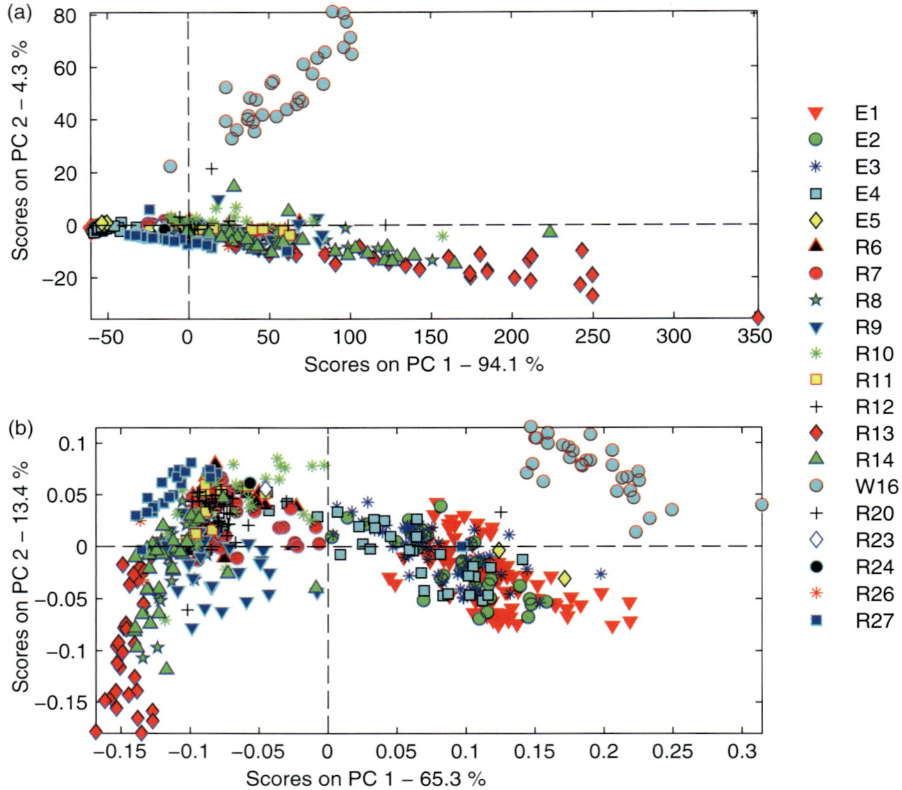

Plate 15. Principal components analysis of the Horsens catchment data set showing the effect of different preprocessing methods on model scores. (a) Auto-scaling. (b) Row normalization followed by mean centering.

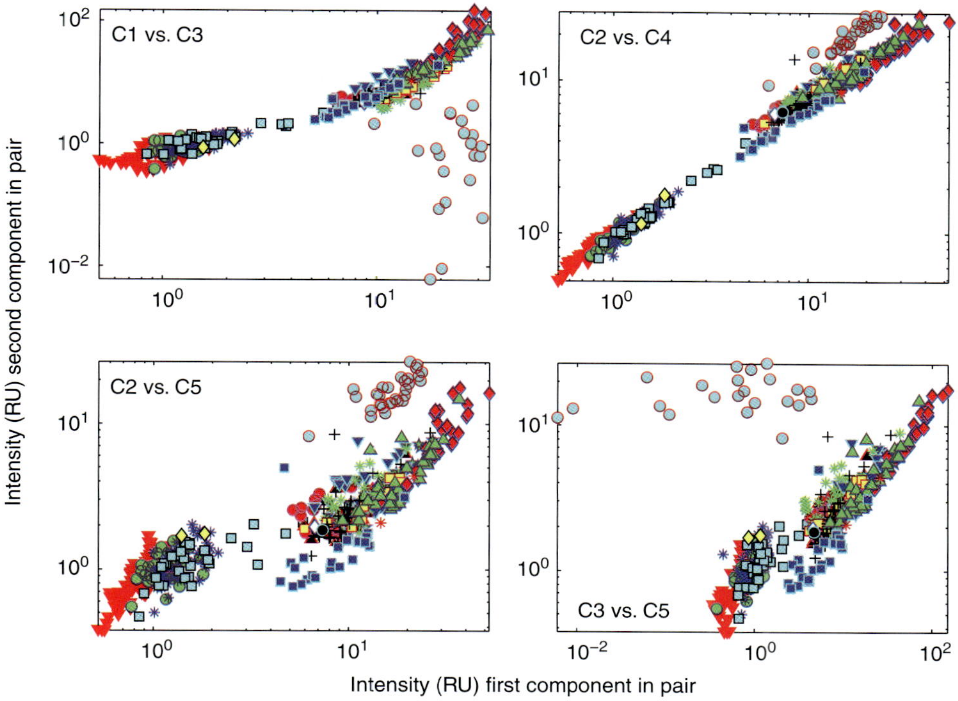

Plate 16. Paired comparisons between log-transformed fluorescence intensities of PARAFAC components in the Horsens catchment data set.

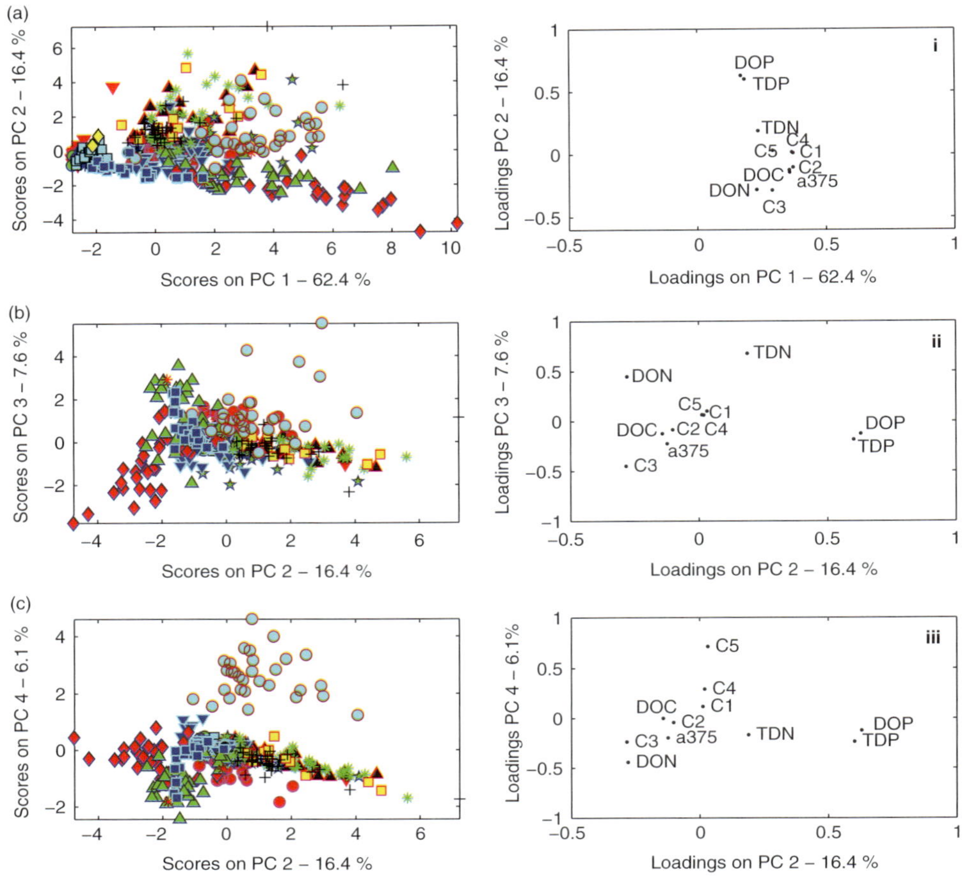

Plate 17. Principal components analysis of PARAFAC components and water quality parameters for the Horsens catchment data set. Score plots (a–c) showing variation among sites are on the left, loading plots (i–iii) showing correlations between measured variables are on the right.

conditions of the user's application is the best means to determine its fitness for a given purpose. There are many possible causes of stray light. All components in an optical system contribute to the problem, including baffles, apertures, partially reflecting surfaces, scattering from internal walls, and the fluorescence of optical materials. Ambient light from the room where the system is physically housed is also a source of stray light.

Scattered light in a monochromator is light that is neither diffracted nor absorbed by the grating. Such light can arise from imperfections in the spacing and shape of the diffraction grating grooves and also from the roughness of the grating surface. Three main processes can classify scattered light from a grating:

- *Diffuse scattered light* emanates into a hemisphere in front of the grating and is due to the micro-roughness of the grating surface. It is the primary scattering process for holographic gratings. The intensity of diffuse scattered light is higher near the diffraction orders for a particular wavelength than between the orders. Therefore the intensity of diffuse scattered light exiting a monochromator is proportional to the slit area and also to $1/\lambda^4$.
- *In-plane scattered light* is light that is not wanted in the dispersion plane of the monochromator and is due mainly to variations in groove spacing or depth of the grating. The intensity of diffuse scattered light exiting a monochromator is proportional to the slit area and also to $1/\lambda^2$.
- *Ghost light* is primarily a scattering effected in mechanically ruled gratings and is caused by periodic errors in the groove spacing at the time of ruling. It manifests as systematic periodic spikes on the background signal. Holographic gratings do not, in general, suffer from this potential scattering problem.

Stray light that is not caused by the scattered light from the grating is called instrumental stray light. Every monochromator will reflect light in the zero order and this must be trapped to minimize its contribution to the overall instrumental stray light. Similarly light from the other diffraction orders may also find its way to the exit slit and therefore contributes to the stray light. Careful instrument design, particularly with respect to baffles, along with correct illumination of the optics, minimization of sharp edges, and "nonoptical" reflective surfaces will all contribute to minimizing stray light problems.

The key performance indicator for any instrument is the overall SNR, which evaluates the ratio of diffracted light to unwanted light. Because this is an instrument function there is no clear rule-of-thumb that indicates what grating type, ruled or holographic, might provide the higher SNR. It is the SNR of an instrument that will determine the system linearity and dynamic range.

5.4.14 Cuvettes, Cleaning and Handling

There is a large range of optical cells available for spectroscopy measurements and they vary in materials, size, shape, and spectral transmission characteristics. The most commonly used sample holder in fluorescence spectroscopy is a 10 mm × 10 mm × 45 mm volume cuvette made from fused silica (for UV to near-IR operation), glass (visible), or a plastic that is often polycarbonate and disposable. Of course, it is essential that the cuvette chosen is suitable for the application and experiment and that it is clean, and handled with maximum care.

Checking the transmission of the cuvette and solvent in the wavelength range of excitation and emission is critical. This should be done using a quality absorption spectrophotometer. After all, if there is no transmission of light to the sample or the emission signal is blocked to the analyzing channel of the fluorimeter then there will be no fluorescence measurement. Plastic cuvettes are useful because they are inexpensive, disposable, do not require cleaning and are robust. Even so, such cuvettes are unsuitable for some applications such as:

- They demonstrate strong polarization effects and are therefore not suitable for any measurement involving polarizers.
- They can be dissolved by many organic solvents commonly used for fluorescence measurements.

In most cases, cuvettes should be used with stoppers, especially if the solvent is corrosive or volatile. This will help to minimize the possibility of spills and to maintain sample concentration. The optical windows that form the cuvette should not be touched with bare hands (powder-free gloves are often used), as human fingerprints exhibit fluorescence. In addition, careful and thorough cleaning of fused silica and glass cuvettes is fundamental to accurate measurements. For this, it is common to soak cuvettes for several hours in a nitric acid solution followed by a thorough rinse with deionized water. Before use, cuvettes should be rinsed several times with the solvent to be used in the experiment. An alternate method of cleaning glassware is to use a detergent solution but this has the risk that the detergent may have some fluorophore in it. Therefore, it is sometimes appropriate to make a fluorescence measurement on a dilute solution of any detergent that is used for cleaning purposes. At the same time, please note that there are whitening agents in many filter papers, clothes, and tissues and these can introduce contaminant fluorophores.

5.4.15 Solvents and Contaminants

The fluorescence emission from a fluorophore can be strongly dependent on its local environment. As such, choice of solvent and solvent purity are very important. Incorrect choice of solvent can cause spectral shifts and changes in the peak shapes and reduce samples emission. As a matter of good practice, high-purity solvents should be used and then checked before use by checking their absorption and fluorescence properties. Handling techniques should be employed to minimize the possibility of stock solutions becoming contaminated – even very small levels of contaminants may cause appreciable background signals. As such, solvents should not be stored in plastic containers and should be regularly screened for contaminations before use. Regular screening of the solvent using a fluorimeter will also indicate the position and magnitude of any Raman signal from the solvent or contaminants in the analysis spectral range. Contamination opportunities exist at each stage of sample preparation and handling. Trace levels or contaminants are enough to destroy measurement quality. Insufficient cleaning of the cuvette and related glassware, leaching from plastic parts in contact with samples, dirty pipettes, fingerprints, vacuum grease from degassing stations, impure solvents and solutes, and old stock solutions are just some of the opportunities for contaminated samples and all lead to erroneous fluorescence signals.

5.4.16 Background Signals: Rayleigh and Raman Scattering

Rayleigh and Raman scattering effects can be useful for providing reference signals to enable comparison between samples; however, they can often be the cause of signal interference too. Rayleigh scattering is associated with *elastic* scattering from small molecules or particles and can be seen by scanning the emission monochromator over the spectral region of the excitation wavelength. This invariably will produce intense optical signals and is normally to be avoided if possible, as such intense signals can "blind" photomultiplier detectors. Even so, there are situations where this signal will be present in both first and second orders of the analyzing monochromator, as is evident in excitation–emission matrices.

Raman scattering, on the other hand, is *inelastic* in nature (Raman and Krishnan, 1929), and observed signals are generally wavelength shifted to lower energies (longer wavelengths). In many solvents, the Raman scattered signal may overlap with observed fluorescence signals. Depending on one's application needs, the presence of these signals can be advantageous or a nuisance. For many applications this Raman signal can be used as a reference intensity to compare a fluorescence signal against, or to determine the day-to-day "stability" of an instrument (Mosier-Boss, 1995). In relation to aquatic fluorescence, the water Raman signal is often used as both a measure of instrument stability and as an internal normalization standard. It is usual to perform a water Raman test using deionized water in a sealed cuvette to minimize contaminants interfering with the measurement.

5.4.17 Spectral Irradiance of the Excitation Channel

The fluorescence emission signal is dependent on the excitation light intensity. In general, for dilute samples and assuming that photo-bleaching of the sample is not a problem, the fluorescence signal can be expressed as shown in Eq. (5.13):

$$F = 2.303kI(\lambda_{exc})\varepsilon cl = k'\phi_0 c \tag{5.13}$$

The excitation channel contains a light source in which intensity varies with wavelength, a monochromator where transmission efficiency varies with wavelength, and bandpass adjustment that can change the wavelength resolution. All of these effects, either singularly or in combination, introduce a different amount of light onto the sample. Because the fluorescence intensity is directly proportional to the incident light intensity it is therefore difficult to determine if a change in observed fluorescence signal is attributable to

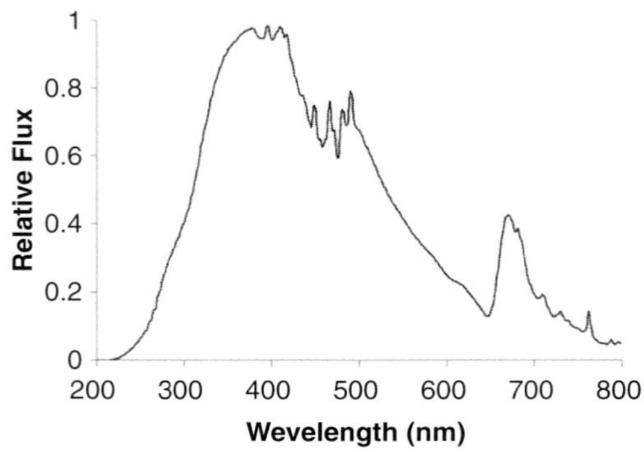

Figure 5.20. Relative intensity of the excitation channel of a fluorimeter as a function of wavelength.

changes in the light source or the sample. Therefore, it is necessary to characterize the excitation channel in terms of how much light, in relative terms, is incident upon the sample. For this, the excitation channel of a fluorimeter is calibrated to ensure that the excitation spectrum exhibits the correct spectra in terms of intensity and wavelength positions. Excitation intensity levels can change by more than two orders of magnitude and there are several dips and intensity spikes visible that will cause measurement errors. Figure 5.20 shows a typical *uncorrected* excitation intensity profile at the sample position of a fluorimeter.

Most modern fluorimeter instruments have a means to monitor the excitation intensity as a function of wavelength and over time. This task is usually performed using a beam splitter to separate a small proportion of the excitation light and record that signal using some form of photon detector such as a photodiode, photomultiplier, or a quantum counter. Originally, such reference detectors used a quantum counter approach in which a concentrated dye solution, often Rhodamine B, would absorb all photons incident upon it and whose emission spectrum and emission intensity are not excitation wavelength dependent. A wide range of quantum counter-designs have been proposed but none overcome the problems associated with polarization, geometry, concentration, photo-bleaching, and limited spectral range. Finally, these approaches fundamentally measure photon flux as opposed to optical power. Nearly all fluorimeter systems are equipped with a reference detector, usually a UV-sensitive photodiode, to correct the fluorescence signal both spectrally and temporally. The photodiode is usually calibrated against a known irradiance standard under the exact same conditions of illumination and spectral range as used in the instrument. This provides a known and traceable responsivity of the photodiode, $R_{pd}(\lambda)$. The measurement of the excitation light level can then be made by monitoring the signal output from the photodiode $S_{pd}(\lambda)$, and this can be used to correct resulting spectra.

5.4.18 Correcting Excitation Signal Channels

Correction can be divided into two types:

- To use a previously measured correction file, or
- To actively monitor the light intensity during the measurement and make the correction live

In the first method it is important to know the spectral response of the reference detector. This can be measured separately using a calibrated tungsten light source and the monochromator under test in the system if the spectral response of the reference detector is not already known. Normally, the *correction file*, $Ex\_Cor(\lambda)$, is normalized to give a range from 0 to 1, where the peak intensity is 1 and the minimum fluorescence intensity is 0. If the measured spectrum is $S_m(\lambda)$ then the *corrected sample file* $Sc\_Ex(\lambda) = S_m(\lambda) / Ex\_Cor(\lambda)$.

It is imperative that the $Ex\_Cor(\lambda)$ file should be determined under the same bandpass parameters as the sample measurement. More important is to ensure that the step size for each file is the same and that the $Ex\_Cor(\lambda)$ file covers at least the same spectral region as the measurement, $S_m(\lambda)$. If the step sizes are different between these two files then an interpolation needs to be ascertained in order to fill in the "missing data points." For example: if $Ex\_Cor(\lambda)$ is measured with 2 nm step size then there is a data point every 2 nm. If the sample $S_m(\lambda)$ is measured at 0.5 nm step size over the same spectral range then there are four times as many data points in the S_m file as there are in the Ex_Cor file, meaning that a simple division of one file with the other becomes problematic. This issue is negated by interpolating between the data points in the Ex_Cor file to make an equivalent number of data points of the same file length. Often this interpolation is a straight line between the existing points, but more complex functions are also sometimes used. This problem applies equally to using a reference detector and also to the emission channel correction.

In some ways the process of using the reference detector is quite similar to that of the correction file. The main difference is that we make the correction measurement "live" with every excitation scan. We need to know what the spectral response of the reference detector is so that we can correct for the reference detector itself. If the spectral response of the reference detector is, $Ref\_Det(\lambda)$, and the measured signal from the reference detector is $Ref\_Sig(\lambda)$, then the $Ex\_Cor(\lambda) = Ref\_Sig(\lambda) / Ref\_Det(\lambda)$. Thus the corrected excitation spectrum is:

$$Sc\_Ex(\lambda) = S_m(\lambda) / Ex\_Cor(\lambda) \, or \, Sc\_Ex(\lambda) = \left[S_m(\lambda) * Ref\_Det(\lambda) \right] / Ref\_Sig(\lambda) \tag{5.14}$$

Figure 5.21 provides examples of corrected and uncorrected excitation spectra.

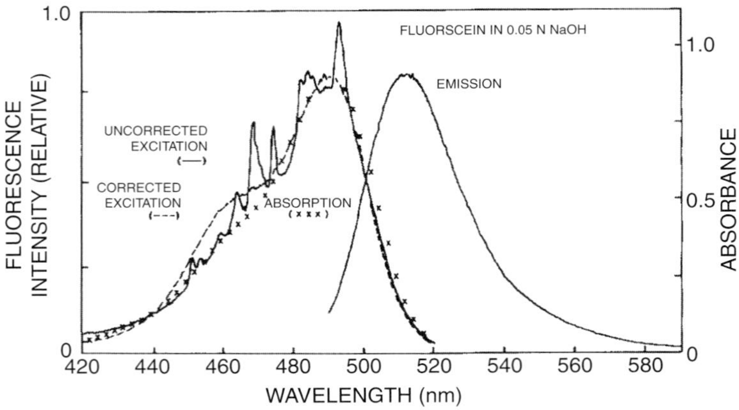

Figure 5.21. Example of corrected and uncorrected excitation spectra.

5.4.19 Correcting Emission Signal Channels

The spectral correction of the emission channel is necessary to provide the "true" emission spectrum of the sample because both the analyzing monochromator and detector each have a spectral response. The normal situation is to measure this correction function once and store the result in a file for use during a measurement. The correction file is measured using a calibrated tungsten source operated under very strict constant current conditions (often in conjunction with a calibrated deuterium source for UV correction). The tungsten source has a spectral emission file that corresponds to the light intensity at a certain distance from the source. The International Commission on Illumination (CIE) is responsible for publishing technical data of all the well known standard illuminants. The CIE publishes relative spectral power distribution data of such illuminants, and for a tungsten-filament source the data will cover 380 nm to 780 nm in increments of 5 nm (www.cie.co.at). However, because the emission spectrum is smooth, interpolating this data to provide higher resolution data is possible.

The spectral power distribution of the tungsten lamp is used to generate a spectral emission file, $W\_lamp(\lambda)$. If the measured signal through the emission channel is $Em\_Sig(\lambda)$, then the correction file $Em\_Cor(\lambda) = Em\_Sig(\lambda) \,/\, W\_lamp(\lambda)$. Note that the spectral emission file is usually normalized (from 0 to 1). Thus the corrected emission signal is equal to $Em\_Sig(\lambda) \,/\, Em\_Cor(\lambda)$.

5.4.20 Quantum Yield

The quantum yield (QY) is a fundamental property of a luminescent sample. Specifically, it refers to the ratio of the number of photons of light radiated from the photoluminescent material to the number of photons of light that the material absorbs. Practically, QYs are important as they allow the quantitative assessment of fluorescence from materials and the effects of interferences on the fluorescence properties. Such measurements can be made on

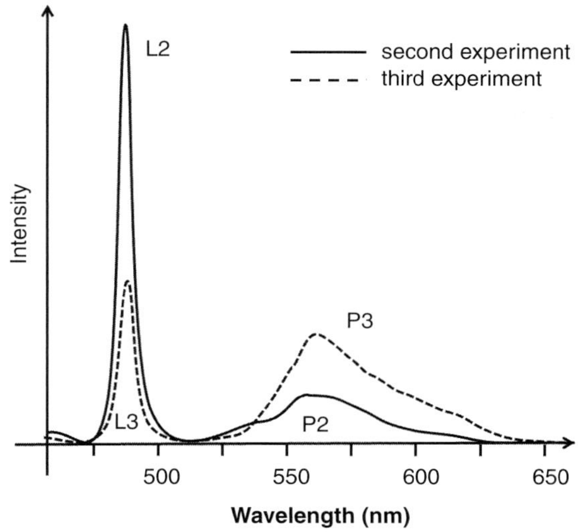

Figure 5.22. Graphic illustration of the three-measurement technique.

solutions, powders, and thin films and apply to a wide range of applications including, for example, fluorescent materials for whiteners and white lights, organic and inorganic light-emitting diode materials, biology, fluorescent probes and quantum dots, laser threshold requirements, determining suitability of wavelength shifters, and for studies of radiationless transitions in molecular systems. There are a variety of methods for measuring QYs, and many of these have been described in the literature (Resch-Genger et al., 2007). The measurement of absolute quantum yields is difficult owing to the range in experimental errors that need to be avoided or compensated for. Relative quantum yields are more commonly measured by comparing an unknown sample to that of a sample with a known QY in the same spectral region. In this case, the accuracy of the unknown QY is determined by that of the known reference sample. As in all fluorescence measurements, care is needed to minimize interactions occurring within the sample that could impact on the QY. Such factors include inner filter effects and factors affecting the fluorophore's microenvironment such as temperature, solvent, pH, presence of dissolved oxygen and other quenchers, polarity, viscosity, and fluorophore binding. All of these are capable of introducing significant errors to QY determination. Because the efficiency of the fluorescence process within a given sample is determined by the QY, this parameter is of major importance. The QY is essential for the calculation of quenching-rate constants, radiative and nonradiative rate constants, and energy transfer. In essence the QY is needed to help describe or define the samples photo-physical behavior (Fery-Forgues and Lavabre, 1999).

Measuring the true QY of any fluorophore is complex. One approach is to use an integrating sphere (IS). In this, the sphere is hollow, with entrance and exit ports, and the interior is coated with a diffuse reflective coating. Light scattered by the interior of the *IS* is uniform and evenly distributed over all angles. As a result the flux (total power) of any light source can be measured without errors caused by complex optical geometries and arrangements.

5.4.21 Measuring Quantum Yields: The Three-Measurement Technique

The three-measurement technique (Figure 5.22) is one of several methods of measuring the quantum yield of a sample. When used with an integrating sphere (IS) it is a method that provides compensation for the reabsorption effects in the measurements.

Measurement 1: Excitation Power – NO sample is in the IS.

- No sample is present in the *IS*.
- The excitation monochromator is set to the excitation wavelength.
- Emission scan across spectral range including the excitation peak.
- Determine the peak wavelength from the measurement.
- Correct the spectrum for the integration period to give the number of photons per second.
- Integrate the spectrum from the background (zero level) before the peak to the same background after the peak to give the total number of photons per second in the wavelength range of the peak, and therefore available to the sample ***L***1.
- Store the spectrum and the integrated photons per second signal rate.

Measurement 2: Re-Absorption Signal – Sample is present in the IS.

- Place the sample into the IS but in the OFF position, such that it is not directly illuminated by the excitation light.
- Excitation monochromator is set to the excitation wavelength.
- Emission scan across the spectral range, including the excitation peak.
- Determine the peak wavelength from the measurement.
- Correct the spectrum for the integration period to give the number of photons per second.
- Integrate two regions of the spectrum:
  - The excitation spectrum from the back ground (zero level) before the peak to the same background after the peak to give the total number of photons per second in the wavelength range of the peak, and therefore the number of excitation photons that are not reabsorbed by the sample. ***L***2
  - The PL emission spectrum from the background (zero level) before the peak to the same background after the peak to give the total number of photons per second in the wavelength range of the peak, and therefore the number of emission photons that are emitted by the sample. ***P***2
- Store the spectrum and the two integrated photons per second signal rates, ***L***2 and ***P***2.

Measurement 3, Reabsorption Signal – Sample is present in the IS.

- Place the sample into the IS but in the ON position such that it is directly illuminated by the excitation light.
- Excitation monochromator is set to the excitation wavelength.
- Emission scan across the spectral range, including the excitation peak.
- Determine the peak wavelength from the measurement.
- Correct the spectrum for the integration period to give the number of photons per second.

- Integrate two regions of the spectrum:
  - The excitation spectrum from the background (zero level) before the peak to the same background after the peak to give the total number of photons per second in the wavelength range of the peak, and therefore the number of excitation photons that are not absorbed by the sample. ***L*3**
  - The photoluminescence emission spectrum from the background (zero level) before the peak to the same background after the peak to give the total number of photons per second in the wavelength range of the peak, and therefore the number of emission photons that are emitted by the sample. ***P*3**
- Store the spectrum and the two integrated photons per second signal rates, ***L*3** and ***P*3**.

When the light hits the sample in measurement 3, some light is absorbed, and $\boldsymbol{A}$ and a fraction $1 - \boldsymbol{A}$ are reflected or transmitted. The light that is not absorbed is reflected by the integrating sphere inner surface and a fraction of this unabsorbed light is reabsorbed by the sample, $\boldsymbol{\mu}$. It is not possible to know exactly the behavior of the sample in terms of reflectance, transmission, and refractive index when it is illuminated so we can only evaluate the effect of secondary absorption in terms of the photons absorbed and photoluminescence photons emitted. This is the reason two measurements are taken, that is, to simulate the secondary absorption and assume that the samples and sphere have the same manner regardless of where the excitation light comes from. Therefore, measurements 2 and 3 need to be compared to determine the ratio of reabsorbed photons, $\boldsymbol{\mu}$. At the same time, some light, $1 - \boldsymbol{\mu}$, is not reabsorbed and this excitation light leaves the sphere in an identical manner to measurement 1. Thus, $\boldsymbol{LB} = \boldsymbol{LA}\ (1 - \mu)$.

The absorption ratio, $\boldsymbol{A}$, is determined from measurement 3. $\boldsymbol{LA}$ photons are in the excitation and $A \times LA$ are absorbed, while $(1 - A) \times LA$ are reflected and μ of these reflected photons are reabsorbed by the sample. Thus, $\boldsymbol{LC} = \boldsymbol{LA}(1 - A)(1 - \mu)$.

Assuming that absorption is independent of excitation wavelength (assuming a narrow excitation line), then from the preceding equations we can derive the following; $A = (1 - \mathbf{LC}/\mathbf{LA})$. In measurement 2 all the photoluminescence comes from the reabsorbed light whereas in measurement 3 it originates from the direct absorption of light and also the secondary reabsorption effect. This means that the total integrated spectrum (signal) for measurement 3: the excitation and emission comprises these two contributions. $\boldsymbol{X}$ and $\boldsymbol{Y}$. Thus, $\boldsymbol{LC} + \boldsymbol{PC} = \boldsymbol{X} + \boldsymbol{Y}$.

The unabsorbed part is determined by measurement 2 and hence $\boldsymbol{Y} = (1 - \mathbf{A})(\mathbf{LB} + \mathbf{PB})$. The directly absorbed light is re-emitted by the sample and hence $\boldsymbol{X} = \eta \mathbf{A}\,\mathbf{LA}$, where η is called the quantum efficiency. Thus, $\boldsymbol{LC} + \boldsymbol{PC} = \eta \mathbf{ALA} + (1 - \mathbf{A})(\mathbf{LB} + \mathbf{PB})$. Therefore the photoluminescence quantum efficiency, η, can be expressed as shown in Eq. (5.15):

$$\eta = \frac{\mathbf{PC} - (1 - \mathbf{A})\mathbf{PB}}{\mathbf{ALA}} \tag{5.15}$$

where $A = 1 - \dfrac{LC}{LB}$.

Because the spectral features are measured the values of **PC**, **PB**, **LC**, **LB**, and LA can be substituted from the integration analysis of the curves of measurements 1, 2, and 3. This means that the quantum efficiency can be expressed as: (5.16)

$$\eta = \frac{\mathbf{P}3 - (1 - \mathbf{A})\mathbf{P}2}{\mathbf{AL}1} \tag{5.16}$$

where $A = 1 - \frac{L3}{L2}$.

5.4.22 Fluorescence Units – What Are They?

Fluorescence units are often a term used to describe the intensity axis of the spectral plot (Resch-Genger, 2007). In reality, there appears to be no "standard" definition of this term, nor any means to quantify it in an absolute radiometric meaning. There are so many effects, such as optical, sample, and instrumentation related, involved in fluorescence measurements that without a full and strict radiometric calibration fluorescence units are no more than some *arbitrary* scale of intensity. Fluorescence units are therefore not directly comparable from instrument to instrument or from laboratory to laboratory. The situation becomes even more complex when thinking about excitation–emission matrices (EEMs) as the spectral position in excitation–emission space can be dramatically affected by the spectral correction used in both channels as well as the possibilities of signal saturations. These are important considerations to understand when reporting any fluorescence intensity data.

References

Arecchi, A.V., Messadi, T., and Koshel, R.J. (2007). *Field Guide to Illumination*. Bellingham, WA: SPIE Press,

Fery-Forgues, S. and Lavabre, D. (1999, September). Are fluorescence quantum yields so tricky to measure? A demonstration using familiar stationery products. *J. Chem. Educ.*, **76** (9), 1260–1264.

Fortin, G. (2008). Graphical representation of the diffraction grating equation. *Am. J. Phys.*, **76**(1), 43–47.

Long, D.A. (2002). *The Raman Effect: A Unified Treatment of the Theory of Raman Scattering by Molecules*. Hoboken, NJ: John Wiley & Sons.

Mosier-Boss, P.A., Lieberman, S.H., and Newbery, R. (1995). Fluorescence rejection in Raman spectroscopy by shifted-spectra, edge detection, and FFT filtering techniques. *Appl. Spectrosc.*, **49**, 630–638.

Niclass, C., Rochas, A., Besse, P-A., and Charbon, E. (2005). Design and characterization of a CMOS 3-D image sensor based on single photon avalanche diodes. *IEEE J. Solid-State Circuits*, **40**(9), September.

Parker, C.A. (1968). *Photoluminescence of Solutions* (pp. 220–221). New York: Elsevier.

Raman, C.V. and Krishnan, K.S. (1929). *Proc. Roy. Soc. London*, 122, 23.

Resch-Genger, U., Hoffmann, K., and Pfeifer, D. (2007). Simple calibration and validation standards for fluorometry. In C.D.P. Geddes (Ed.), *Reviews in Fluorescence* (pp. 1–31). New York: Springer Science+Business Media.

Skoog, D.A. and Leary, J.J. (1998). *Principle of Instrumental Analysis*, 5th ed. Philadelphia: Saunders College.

www.horiba.com (2012). Optics Tutorial. Retrieved from www.horiba.com/us/en/scientific/products/optics-tutorial.

www.newport.com (2012). *The Gratings Handbook*. Retrieved from http://gratings.newport.com/library/handbook/

6

Experimental Design and Quality Assurance: In Situ Fluorescence Instrumentation

ROBYN N. CONMY, CARLOS E. DEL CASTILLO, BRYAN D. DOWNING, AND ROBERT F. CHEN

6.1 Introduction

Both instrument design and capabilities of fluorescence spectroscopy have greatly advanced over the last several decades. Advancements include solid-state excitation sources, integration of fiber optic technology, highly sensitive multichannel detectors, rapid-scan monochromators, sensitive spectral correction techniques, and improved data manipulation software (Christian et al., 1981; Lochmuller and Saavedra, 1986; Cabaniss and Shuman, 1987; Lakowicz, 2006; Hudson et al., 2007). The cumulative effect of these improvements have pushed the limits and expanded the application of fluorescence techniques to numerous scientific research fields. One of the more powerful advancements is the ability to obtain in situ fluorescence measurements of natural waters (Moore, 1994).

The development of submersible fluorescence instruments has been made possible by component miniaturization and power reduction including advances in light source technologies (light-emitting diodes, xenon lamps, ultraviolet [UV] lasers) and the compatible integration of new optical instruments with various sampling platforms (Twardowski et al., 2005 and references therein). The development of robust field sensors skirts the need for cumbersome and or time-consuming filtration techniques, the potential artifacts associated with sample storage, and coarse sampling designs by increasing spatiotemporal resolution (Chen, 1999; Robinson and Glenn, 1999). The ability to obtain rapid, high-quality, highly sensitive measurements over steep gradients has revolutionized investigations of dissolved organic matter (DOM) optical properties, thereby enabling researchers to address novel biogeochemical questions regarding colored or chromophoric DOM (CDOM).

This chapter is dedicated to the origin, design, calibration, and use of in situ field fluorometers. It will serve as a review of considerations to be accounted for during the operation of fluorescence field sensors and call attention to areas of concern when making this type of measurement. Attention is also given to ways in which in-water fluorescence measurements have revolutionized biogeochemical studies of CDOM and how those measurements can be used in conjunction with remotely sensed satellite data to understand better the biogeochemistry of DOM in aquatic environments.

6.2 Historical Perspective of *In Situ* Sensors

The advent of field fluorometers was necessitated by a pressing need, particularly in ocean environments, to overcome difficulties in obtaining measurements from remote locales and over great depths. Limitations of discrete Niskin bottle sample collection prevented the capture of fine-scale measurements over sharp temporal and spatial gradients. This section is dedicated to the timeline of field sensor development beginning with the earliest instrument designs through the common designs of the present day. It offers an appreciation of previous efforts within the community to move past sampling limitations and describes how this work has improved the collective understanding of aquatic DOM.

6.2.1 Chlorophyll Field Sensors: Precursors to In Situ *DOM Fluorometers*

The first fluorescence field sensor was customized for chlorophyll measurements in open ocean environments (Lorenzen, 1966; Scripps Institute of Oceanography). A Turner III benchtop fluorometer was modified to extend the response of a standard photomultiplier tube (PMT) from 650 to 685 nm, thereby increasing sensitivity for chlorophyll by an order of magnitude. A blue fluorescent lamp was used to excite the sample, where the primary filter with a maximum transmission at 430 nm isolated the blue light close to the maximum excitation wavelength of chlorophyll (440 nm). A submersible pump was used to pump unfiltered seawater via a through-hull fitting (~2 m deep) into a flow-through cuvette. Results showed a strong linearity between discrete chlorophyll extractions on filter pads and *in vivo* chlorophyll fluorescence data; therefore discrete data were used to calibrate the sensor. Over a continuous 21-day collection (Lorenzen, 1966), it was demonstrated that flow-through fluorescence instrumentation was a viable means of collecting rapid measurements with improved temporal and spatial resolution. This work was also the first to address two critical environmental concerns with field instruments: (1) the interference of bubbles on flow-through measurements, where they implemented a bubble trap and subsequently gravity-fed the water to the instrument; and (2) biofouling, where daily cleaning with alcoholic KOH was used to prevent fouling.

A decade later, Herman and Denman (1976) developed a truly in situ method with the argument that the pumping of water is not "in situ" because it introduces some mixing of chlorophyll samples. To avoid this phenomenon, a Variosens Fluorometer (Impulsphysik GmbH, Hamburg, Germany) was mounted onto a Batfish towed instrument package (Figure 6.1) and data from this first in situ instrument was compared to a Turner Designs fluorometer that was fed by a towed submersible pump. Both were calibrated with discrete samples. Although the Variosens Fluorometer provided fluorescence values that were biased high, had slow response times for capturing signal change through sharp vertical gradients, and exhibited data interpretation challenges due to time lag with pumped water samples, the in situ profiler remained the optimal choice over the pumped system. Shortly after, Geiskes et al. (1978) mounted the same model Variosens Fluorometer on a rosette

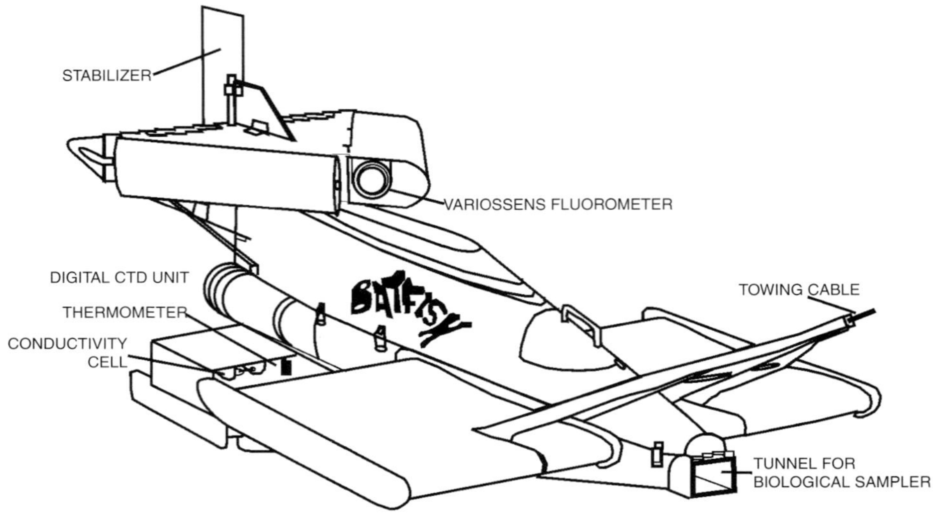

Figure 6.1. Variosens fluorometer mounted on batfish platform. This is the first truly in situ fluorometer to be deployed in water. (Redrawn from Herman & Denman, 1976.)

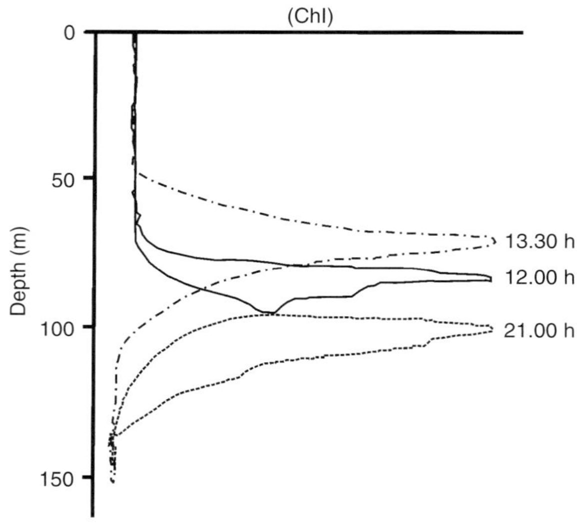

Figure 6.2. Position of deep chlorophyll max observed using a Variosens fluorometer in the Tropical Atlantic. (Redrawn from Geiskes et al., 1978.)

sampler, similar to the ones used today, to examine the existence of the deep chlorophyll maximum in the Tropical Atlantic Ocean. It was found that by using this sensor, narrow bands of increased algal concentrations were resolved and that vertical movement of these bands was observed on the order of hours (Figure 6.2).

6.2.2 Evolution of DOM Field Sensors

In the 1940s von Kurt Kalle (1949) discovered that seawater exhibited a blue fluorescence when irradiated by UV light. It wasn't until the 1990s that in situ fluorometers were modified to measure this fluorescence of organic matter in seawater. It is important here to clarify the pools of natural organic matter (NOM), which is comprised of particulate, dissolved, colored (POM, DOM, COM, respectively) organic matter, and noncolored constituents. The term "dissolved" indicates a filtrate that passes through a filter of some pore size (normally between 0.2 and 1.0 μm) and the term "colored" indicates the property of absorbing light. Field sensors are traditionally designed to collect measurements in unfiltered mode and include both the dissolved and particulate OM signatures, of which the sum has been termed COM. In many aquatic systems, the dissolved signal dominates over the particulate portion, so essentially CDOM ~ COM (Chen, 1999; Belzile et al., 2006). However, contribution by particles in some waters can be significant, as demonstrated by operating field sensors in unfiltered and filtered modes (Downing et al., 2009). Beyond just the presence or absence of particles, there is also a need to subdivide the colored pool because some fraction of COM also fluoresces and can be termed fluorescent organic matter (FOM). So it follows that all FOM is colored but not all COM fluoresces, and it is the former that in situ fluorometers are actually measuring. However, nomenclature used in the literature does not accurately reflect this because often "CDOM" is reported when in fact it is the nonfiltered fluorescent material that is truly measured. In some instances, FOM can be accurately reported as (1) FDOM if a filter was used to collect in situ measurements or (2) FDOM or CDOM if the sensor was calibrated to discrete samples of fluorescence intensity or absorption coefficients, respectively. For practical purposes when discussing cited literature within this chapter, nomenclature will be consistent with originally published literature. However, in an effort to avoid confusion about which pools are reported in the future, the authors recommended standardization of nomenclature within the community through the use of FOM, FDOM (when filtered in situ or calibrated to discrete fluorescence samples) and CDOM (when calibrated to discrete absorption samples) for in situ fluorescence data.

The first reports of in situ fluorescent DOM were published in 1991 (Coble and Gagosian, 1991; Coble et al., 1991). Investigations of CDOM in the Black Sea utilized a pumped profiling system, where water was pumped from depth through on-board continuous flow-through fluorometers customized for the fluorophore of interest (excitation/emission [ex/em] wavelengths described in Figure 6.3). The sampling design focused on measuring chlorophyll, DOM and flavins to search for material from photosynthetic bacteria. Before hydrocasts, the baseline of the DOM fluorometer was set to zero in the surface waters to observe relative changes throughout the water column, so these uncalibrated values are reported as relative fluorescence (Figure 6.3).

The first submersible sensor of DOM fluorescence was the Yellow Matter Meter developed by Sea Tech (Corvallis, OR). The instrument employed right angle detection and interference filters centered on ex/em = 330/450 nm with bandwidths of 80 and 65 nm,

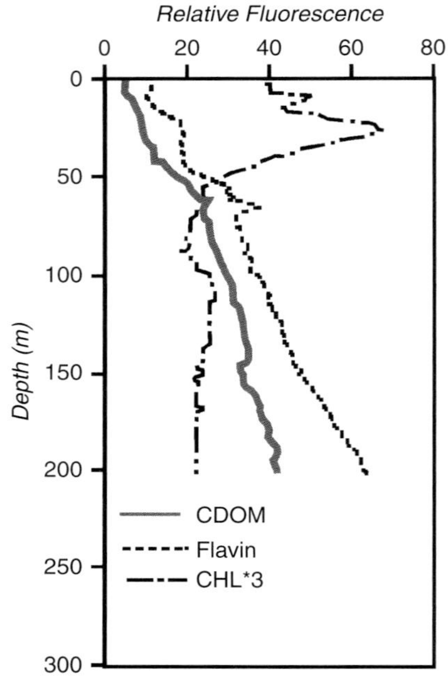

Figure 6.3. First published in situ DOM fluorescence profile from the Black Sea. DOM measured at ex/em = 320–390/475–530 nm, flavins at ex/em = 420–470/527 nm, and chlorophyll ex/em = 420–470/660 nm. (Redrawn from Coble et al., 1991.)

respectively. Meter output was in volts and was observed to be linear (using quinine sulfate dihydrate [QS] standards) over a wide range of concentrations due to operator selection among three gain settings. This sensor was tested and utilized by a number of researchers (Chen and Bada, 1992; Coble et al., 1998; Chen, 1999; and others) and demonstrated that it was a robust tool for tracking DOM optical properties in aquatic environments. The optical design is still used today and is discussed later in this chapter.

The following years gave rise to a number of instrument design types for various applications. In 1992, Lieberman et al. developed a fiber optic–based system to measure petroleum hydrocarbons in the San Diego Bay. It utilized a pulsed N_2 laser source (ex = 337 nm), measured fluorescence at 360 nm (50 nm bandpass), and possessed a sheath on the optic probe to protect the sensor from seaweed and provide constant viewing volume. Strong agreement was found between this design and flow-through measurements (Figure 6.4a), but with the advantage of rapid response times of measurements to align CDOM fluorescence data with CTD measurements. Another fiber optic design was introduced by Klinkhammer in 1994. The submersible profiling zero angle photon spectrometer (ZAPS) probe offered the sensitivity of PMT detection and utilized a xenon flash lamp and the versatility of fiber optics. It measured ex/em = 320/420 nm down to 6000 m. Data were calibrated using QS standards at several temperatures. Good agreement between a Sea

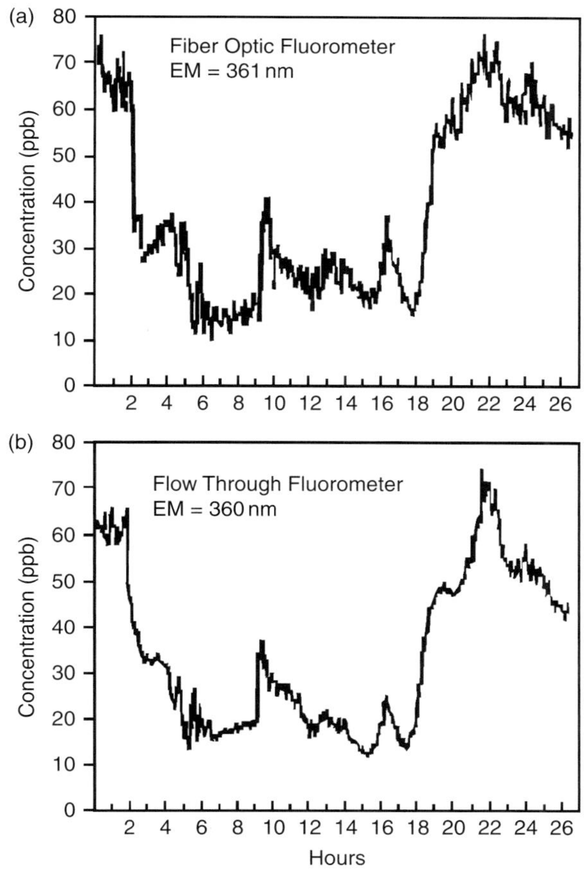

Figure 6.4. (a) Comparison of fiber-optic and flow-through methods for DOM fluorescence measurements. (Redrawn from Lieberman et al., 1992.) (b) Comparison of ZAPS (Klinkhammer, 1994) with discrete samples. (Redrawn from Chen and Bada, 1992.)

Tech in situ CDOM fluorometer was found with discrete samples taken in the NE Pacific (Figure 6.4b) (Chen and Bada, 1992). Work demonstrated that in situ profiles overcame uncertainties associated with preservation and filtration for deep ocean samples.

Sensor development by the mid-1990s saw an emphasis on multispectral and hyperspectral fluorometers. Heuermann et al. in 1995 reported on a new submersible instrument (ME Meerestechnik-Elektronik GmbH, Schleswig-Holstein, Germany) designed to measure at three excitation and nine emission wavelengths for CDOM, protein, and pigments. By employing multiple bands, fluorescence properties of two algal cultures could be discerned including 2 (short and long wavelength) humic bands (Figure 6.5). This unique instrument even included a spectral band to record the Raman peak. Shortly after, Desiderio et al. in 1997 published on the development of a fluorometer that could measure at 6 excitation and

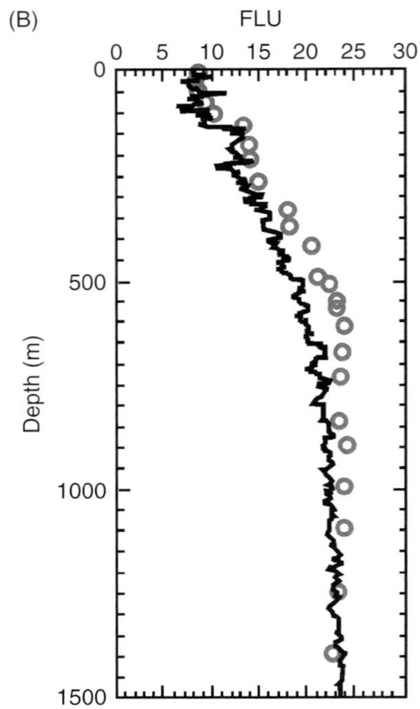

Figure 6.4. (*cont.*)

16 emission wavelengths (WetLabs Inc SAFIre – Spectral Absorption and Fluorescence Instrument) thereby producing coincident measurements of CDOM, protein, and chlorophyll fluorescence (Del Castillo et al., 2001; Conmy et al., 2004). These types of flash lamp, interference filter-based instruments were more affordable and less cumbersome than fiber optic designs. In addition, another unique design was developed by Physical Sciences, Inc (Mazel, 1997). It was a submersible, handheld and negatively buoyant design for diver operation. It utilized a spectrometer, measuring fluorescence between 250 and 750 nm using a halogen bulb, filters for selecting excitation wavelength, interference filters for emission wavelength selection, and a grating/charge-coupled device (CCD) detector combo. This sensor was optimized for fine-scale spatial resolution in ocean environments.

Laser-induced fluorescence (LIF) systems became popular in the late 1990s (Hoge et al., 1998). Systems like the LIF used by Chen (1999) utilized a UV N_2 laser at excitation = 337 nm, 30 m of fiber optic cable and a sensitive detector to measure bulk-integrated fluorescence between 350 and 550 nm with a 0 ns time delay. A probe could be placed directly in the ocean or in a flowing seawater system (water pumped from ~3 m deep in the ship's bow). This system had the advantage of carrying out time-resolved fluorescence measurements and detecting pyrene, a polycyclic aromatic hydrocarbon, with a detection limit of 5 parts per trillion in seawater (Rudnick et al., 1998). Later designs include a system

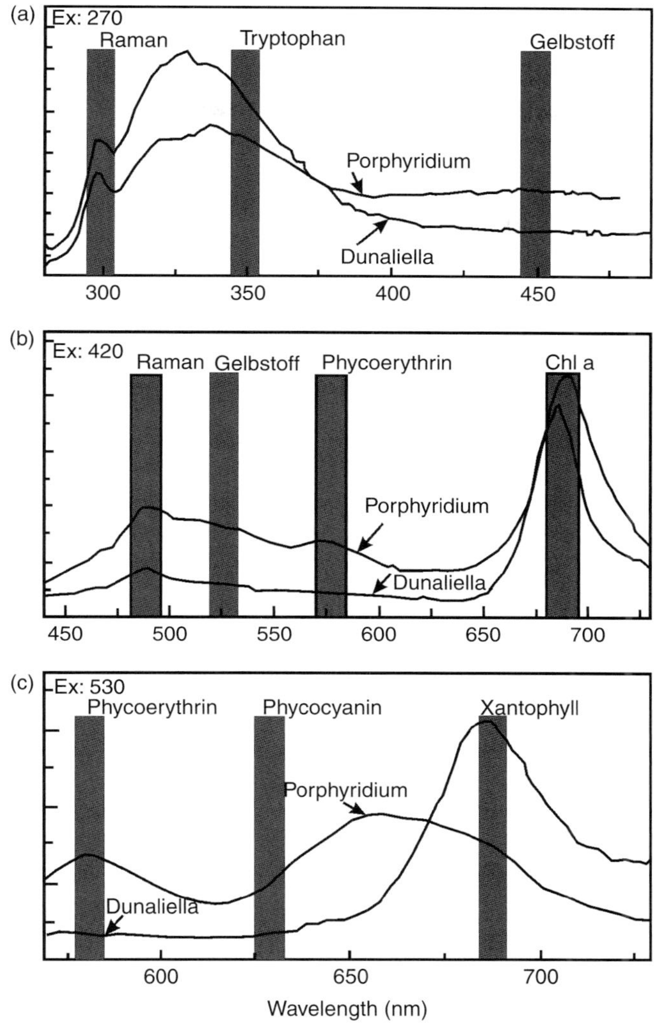

Figure 6.5 . Position of detection wavelengths collected using the submersible fluorometer designed by GmbH. (a) ex 270 nm, (b) ex 420 nm, and (c) ex 530 nm. (Redrawn from Heuermann et al., 1995.)

developed by Sivaprakasam et al. in 2003 for both NOM and bisphenol-A detection. It employed a 266 nm microchip laser and narrow bandwidth of filters (7.5–14 nm) resulting in an improved sensitivity limit to 0.005 ppb QS, rivaling benchtop spectrofluorometers. Shown in Figure 6.6 are the 13 detection wavelengths overlaid on top of the emission spectrum of a discrete sample of NOM (Figure 6.6a) and changes in the NOM spectra for a cruise tract from Tampa Bay, Florida out to Gulf of Mexico waters and returning to the

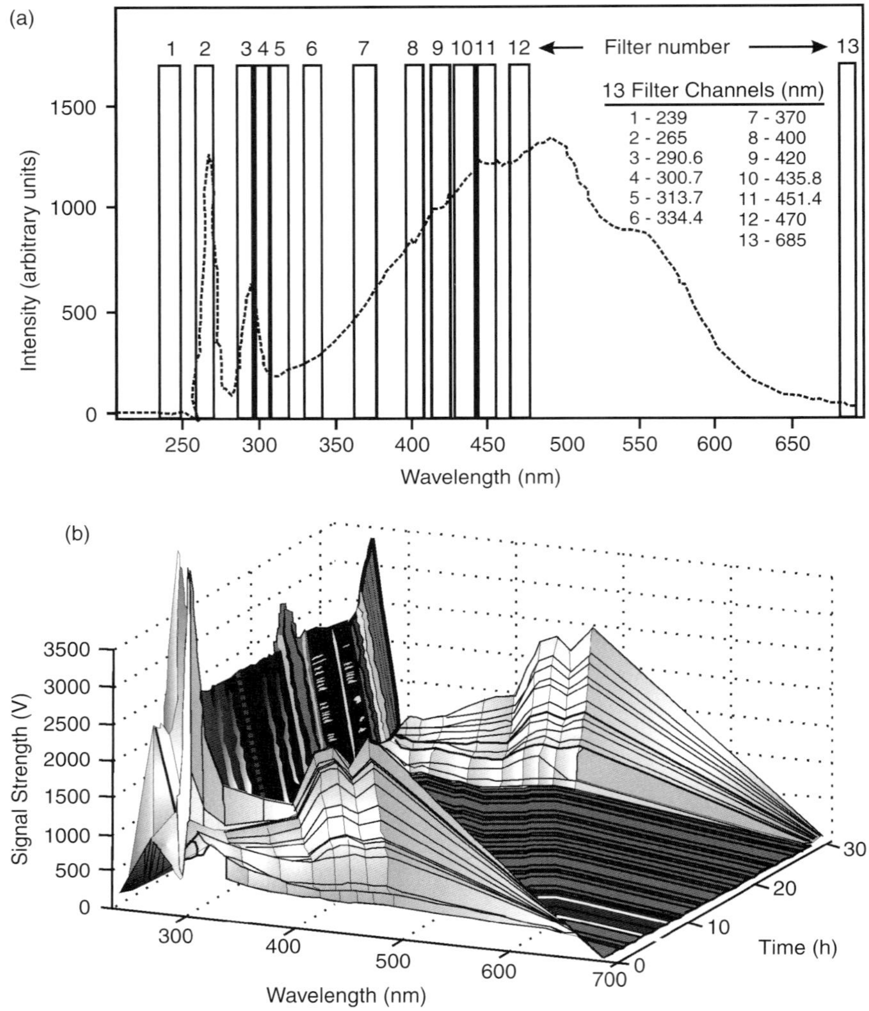

Figure 6.6 . (a) Wavelength positions of the interference filters used in a portable LIF system overlaid on top of a fluorescence spectrum of seawater. (b) Laser-induced fluorescence results for seawater as a function of time along a cruise tract. (Redrawn from Sivaprakasam et al., 2003.)

bay over a 32 hour period (Figure 6.6b). These types of instruments allow for sensitive multispectral fluorescence data with fine temporal and spatial resolution.

In addition to the transition of fluorescence techniques from the laboratory to submersible field measurements, there have also been great advances in the development of non-submersible, portable, handheld fluorometers. Most of this technology has been geared toward wastewater characterization during treatment processes, protein detection, and as a tool for water management issues. For these instruments, power requirements are less of a concern. One example is the portable SMF2 spectrophotometer (Safe Training Systems Ltd., Wokingham, UK), which utilizes a xenon flash lamp, bandpass and interference filters for peak excitation wavelength of 280 nm. The fluorescence signal of tryptophan is measured between 350 and 360 nm. Measurements from this sensor can be calibrated using diluted river or waste water samples to detect effluent in natural systems (Baker et al., 2004).

A great deal of effort has gone into transitioning fluorescence spectroscopy from the laboratory benchtop to in situ field measurements. In Section 6.3, we will focus on the most common submersible design types that are commercially available at the time of this book's publication that can offer analysts an appreciation of the options for their particular applications.

6.3 Instrument Design Types

6.3.1 Sensor Configurations

There exists a wide range of sensor designs used by researchers to collect in situ OM fluorescence measurements. The challenges of in situ measurements are the same – how to collect high-quality fluorescence data that is inexpensive and limited by low-power UV capabilities? Now we will focus on the most commonly designed, commercially available field sensor designs. Quite simply, an in situ fluorometer consists of five main optical components: a UV light source, optical hardware to bring excitation light to the sample volume, optical hardware to collect the emitted fluorescence from the sample, optical filters to separate the excitation and emission wavelengths of interest, and a photodetector (Figure 6.7). These components are housed within pressure housings of two types of sensor geometries: (1) the open-faced design, which can either be right-angle (Figure 6.7a) or flat-faced (optical backscatter type; Figure 6.7b) with a fixed cell geometry and volume and (2) the flow-through design, which may require a pump to pass the sample through a quartz cuvette flow tube (Figure 6.7c). With either geometry, sensors can be configured to collect single or multispectral measurements. Collection over multiple wavelengths greatly improves the ability to characterize FOM as well as provide a measurement of the bulk intensity or amount of FOM.

The flow-through geometry type tends to have greater efficiency in signal output compared to the open-faced design, in part due to the absorption of light by the plastic (or epoxy resin) facing of open-faced meters. Flow-through designs are an ideal choice for waters with low OM concentration and few particles as the sensor geometry (sample volume) is well defined as the width and length of the detection area is based primarily on the cuvette and the geometric area of the excitation and photo diode detector assembly. This geometry is also well suited for waters with high OM concentration as they have small pathlengths and are therefore less susceptible to alterations in the sample volume cone that can erroneously reduce fluorescence signals. Limitations of flow-through designs exist for deployment in waters that are not optically clear due to difficulty in maintaining cleanliness of the

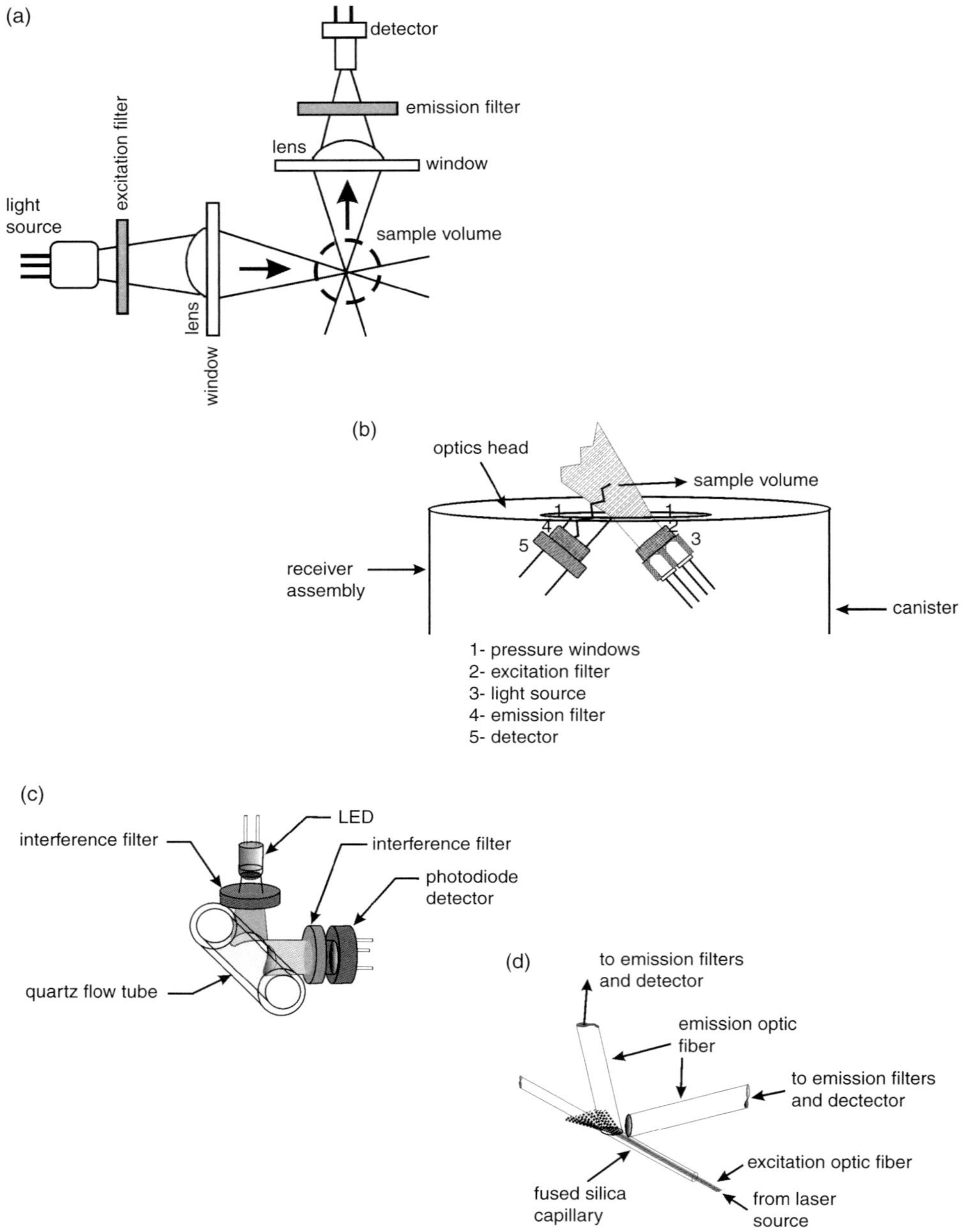

Figure 6.7. Conceptual diagrams of fluorescence detection hardware. (a) Right angle detection, (b) intersecting cones, (c) flow-through, and (d) fiber optic designs.

internal quartz tube and any tubing that connects the sensor to the pump that is required with this design type. In contrast, the flat-faced sensor utilizes two intersecting excitation and emission cones as its sample volume and is directly related to the optical aperture or geometric size of the excitation source and photodiode detector. These sensors are ideal for waters with high concentrations of DOM, where sensitivity is less of an issue and because cleaning of the optics head is easier than with the flow-through design. Analysts should be cautioned though that the presence of particles can affect the size of the viewable sample volume cone, altering signal intensity.

There is a fourth design, based on laser-induced fluorescence, which utilizes laser light sources, and fused silica fiber optics (Figure 6.7d). These systems exhibit high sensitivity and low detection limits, which are advantageous in environments with dilute concentrations. Internal quenching in highly fluorescent waters is virtually eliminated by normalizing to the Raman scattering signal of water (Rudnick and Chen, 1998). In flow through systems (e.g., Chen and Bada, 1990), a reduced sample volume is ideal for measurements that require small sample size (e.g., porewaters, rainwaters) or benefit from it (e.g., pumped flow-through sampling). One drawback however is the inability to submerge the entire sensor, therefore a fiber optic probe (up to 50 m in length) is deployed in situ or water is pumped to the sensor placed in a flow through cell.

6.3.2 Light Sources and Detectors

Original in situ sensor designs utilized xenon flash (or pulsed) lamps due to their high-energy output in the UV and visible regions and long lamp life. But these have been mostly replaced by light-emitting diodes (LEDs) as light sources, resulting in significant reduction in cost and power consumption, warm-up time and size of the sensor package. This transition was made possible due to advances in LED technology; specifically the ability of LEDs to produce nearly monochromatic excitation light with high signal even at lower UV wavelengths and nearly infinite lamp life. There are some applications that still require xenon lamps (measurement of protein fluorescence below 275 nm, low UV polycyclic aromatic hyrdocarbons and crude oil, and excitation over numerous wavelengths); however, LEDs can now be used to excite FOM from the low-300-nm up to the 700-nm region with narrow bandwidths and high spectral resolution. Alternatively, lasers can be used as light sources, as in LIF sensors and their corresponding applications, which can provide increased sensitivity of the measurements. Nitrogen and HeCd gas lasers excite at 337 and 325 nm, respectively; and solid state lasers such as the Nd:YAG at 266 and 405 nm.

As with advances in light source technology, improvements in detectors have yielded savings in instrument size and power consumption. PMTs were originally used in most fluorometers where the resulting current is proportional to light intensity, and the counting of individual photons is possible. Owing to the amplification of electrons per each photon, low light levels can be measured yielding highly sensitive detection. At present most photomultiplier detectors have been replaced by photodiode detectors within field sensors. These detectors are rugged and offer low cost, low power consumption, and small size. There is also the possibility of simultaneously measuring a full spectrum of wavelengths when employing a photo diode array (PDA) or CCD detector. So the popularity of photodiodes rests in the tradeoff between gaining the benefits stated above and sacrificing the high sensitivity of PMT detection.

6.3.3 Optical Filters

A variety of spectral filters are used to isolate the desired wavelengths of light along the excitation (incident) and the emitted light path of the resulting fluorescence. There are also neutral density filters that are used to reduce intensity without spectral discrimination. Spectral filters are of three general categories: bandpass, interference, and optical edge and are used to discriminate wavelengths. The former comprises absorptive and interference (dichroic) types. Absorptive filters are made of glass with the addition of inorganic and organic compounds to selectively absorb and transmit wavelengths of light. They are inexpensive and exhibit low peak transmittance and broad peak shape (Figure 6.8). Interference filters, however, consist of sequential layers of reflective cavities that resonate with desired wavelengths and destructively cancel or reflect nondesired wavelengths. Controlling the thickness and sequence of the layered coatings means that exact color ranges are possible. Interference filters have high peak transmittance and narrow peak shapes. Because of these properties and improvements in coating technology, interference filters are an ideal choice for precise scientific work requiring high spectral resolution (Macleod, 2001). Bandpass optical filters are described by critical features including peak transmittance, center wavelength (CWL), or nominal wavelength, and full width at half maximum (FWHM), or effective bandwidth (Figure 6.8). CWL is the average of two half-height wavelengths in a spectrum and FWHM is the wavelength range at half the max transmittance. Optical edge filters are referred to as cutoff or blocking filters and can be used to transmit much of the UV-visible regions of the spectrum, but reflect far UV and near infrared energy. UV protection filters also reduce photochemical degradation without damage to the components when the light output is intensified. Two or more cutoff filters can also be used to produce a peak similar to a bandpass filter (Figure 6.8).

6.3.4 Optical Configurations

Most submersible fluorometers are configured to utilize one or two LEDs, interference and cutoff filters, and photodiodes. Combinations of filters for effective fluorescence measurements can include the use of UV-violet blocking filters and an interference filter (centered on desired excitation wavelength(s)) to discriminate against the scattered UV incident light and reject small amounts of red light emitted by the blue LEDs. For emission detection a Schott (blue glass) glass cutoff filter (with optimal stopband or passband limits) and a bandpass filter (centered on desired emission wavelength(s)) can be used to select red fluorescence to be detected by a photodiode (WetLabs, Inc. fluorometers). The physical

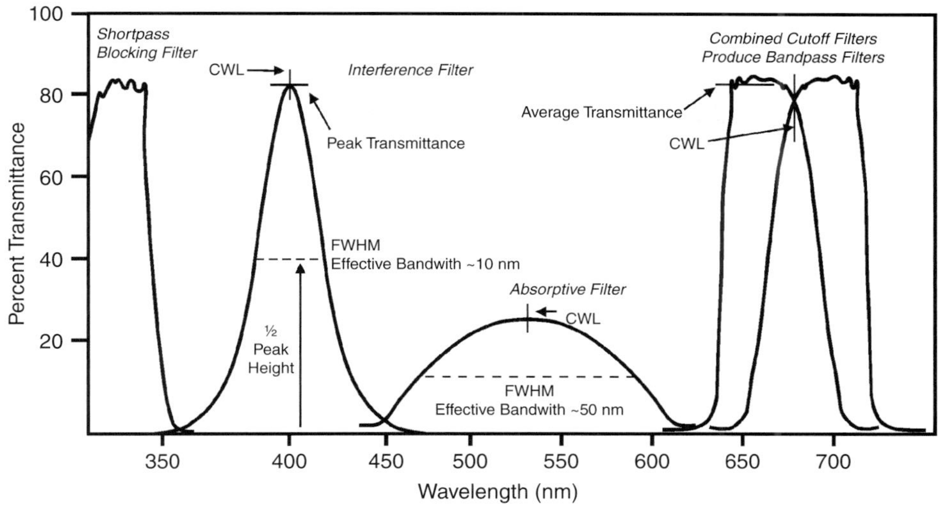

Figure 6.8. Conceptual diagram of optical filters and terminology.

dimension of optical apertures (some are small) vary amongst sensors and some optical designs also include internal reference measurement of the emitted light to compensate for aging and temperature dependences of the high-efficient LEDs (TriOS). Examples of specifications that manufacturers employ for NOM fluorometers are shown in Table 6.1. Note that the listing of instruments does not equate to an endorsement by the authors. This compilation merely serves as a reference of some commonly used fluorescence sensors.

6.3.5 Data Output

Beyond options of sensor configuration, there are also choices between passive or programmable sensors and between analog or digital data outputs. The output of any sensor should be thoroughly examined. If using an analog output, it is recommended that signal degradation be considered especially if the sensor is further than 10 m from the data logging device. High-quality communication-grade cable should always be used, and the resistivity of the cable product known. A good alternative is to choose an instrument with a digital readout which aids in preserving signal integrity, assuming an appropriate signal cable is utilized, and can be transmitted reliably over the distances required. Alternatively, internal logging of fluorescence measurements can also be conducted.

6.4 Calibration and Correction Procedures

This section is a compilation of regularly practiced calibration methods for field sensors found in the literature. Readers are encouraged to consult with the standard operating

Table 6.1 *Sensor specifications of commercially available submersible fluorometers*

Manufacturer	Instrument	Light source	Excitation λ (nm)	Emission λ (nm)	Detector	Dynamic range	Temperature reference	Design	Application
Chelsea Technologies Group	UV AQUAtracka	Xenon lamp	239, 26 FWHM	360, 70 FWHM	PMT	0.001–10 μgL^{-1} Carbazole	T probe available	B	L, R, M, W
	UV AQUAtracka	Xenon lamp	239, 26 FWHM	440, 110 FWHM	PMT	0.001–10 μgL^{-1} Perylene	T probe available	B	L, R, M, W
HOBI Labs	HydroScat-6	LED	370 or 395 CWL[a]	420 CWL[a]	Photodiode	determined by analyst[b]	Reference photodiode	O	M
Seapoint Sensors	SUVF	LED	370, 12 FWHM	440, 40 FWHM	Photodiode	0.1–1500 μgL^{-1} QS		B	L, R, M, W
TriOS, GmbH	MicroFlu-CDOM	LED	370 CWL	460, 100 FWHM	Photodiode	0.2–200 μgL^{-1} QS	Internal thermistor	B	L, R, M, W
TriOS, GmbH	EnviroFLU-HC, DS	Xenon lamp	254, 25 FWHM	360, 50 FWHM	Photodiode	0–5000 ppb Phenanthren	Internal thermistor	B	L, R, M, W
Turner Designs	Cyclops (CDOM)	LED	320, 130 BP	470, 60 BP	Photodiode	0–2500 ppb QS	T sensor equipped	B	L, R, M, W
	Cyclops (Crude oil)	LED	320, 130 BP	510, 180 BP	Photodiode	0–2700 ppb PTSA Salt	T sensor equipped	B	L, R, M, W
	Cyclops (Fine oil)	LED	254, 40 BP	350, 50 BP	Photodiode	0–10,000 ppb NDD Salt	T sensor equipped	B	L, R, M, W
WetLabs	WetStar	LED	370, 10 FWHM	460, 120 FWHM	Photodiode	0.100–1000 ppb QS		F	L, R, M
	ECO-FL3, triplet, puck	LED	370, 10 FWHM	460, 120 FWHM	Photodiode	0.01–500 ppb QS[c]	T sensor available	B	L, R, M

[a]Customizable.

[b]Manufacturer does not report dynamic range.

[c]Instrument-specific ranges between values listed.

Note: Wavelengths are listed as CWL (center wavelengths). Letters in Design column represent Open-faced (O), Flow-through (F), or Both (B). Letters in Application column represent Lakes (L), Marine (M), Rivers (R), and Wastewater (W) procedures (SOPs) from individual manufacturers regarding recommended protocols and instrument performance.

6.4.1 Temperature Correction

All fluorescence measurements are subject to alterations in signal as a function of temperature changes. This can arise either from self-heating by the instrument or from the ambient temperatures of the deployment environment. The degree to which temperature influences the fluorescence intensity through self-heating can be determined by laboratory characterization and will vary for each instrument. Many instruments are now equipped with thermistors and reference values to correct for temperature-related effects on fluorescence response. When characterizing these effects, attention must be paid to (1) the warm-up period where sharp increases in temperature and signal occur and (2) the gradual increase in temperature that may exist after the initial warm-up period. Both periods can exhibit an influence on fluorescence signal as evidenced from a laboratory experiment where a solution of standard was continually pumped through a flash lamp-based, flow-through sensor while submerged in a water bath of constant temperature (Figure 6.9). During the 8 hour experiment, fluorescence varied by 10% and internal temperature by 7°C. The extent of these effects are sensor dependent and manufacturers recommend a specific warm-up time, typically on the order of tens of minutes (but can be longer for flash lamp-based sensors), where signal should not be used for data collection. Although such recommendations are provided, analysts should repeat the characterization themselves to ensure appropriate warm-up times of their sensor. Beyond the warm-up period, fluorescence intensity may continue to increase with temperature, albeit at a smaller rate. If significant, data can be normalized to temperature to correct for this increase in signal.

6.4.2 Blank Subtraction

In the words of Cullen and Davis (2003: 29) "It is hard to imagine a topic that seems more boring and trivial than the measurement of nothing. In a sense, determination of an analytical blank is exactly that – measurement of the signal associated with the absence of the property being detected." However, in many cases determining an acceptable blank is a critical part of a calibration routine. Blank subtraction can be troublesome in some environments where the magnitude of the sample intensity is small relative to the blank, and analysts may choose to not apply this step for fear of underestimating the signal or in some cases, obtaining negative values (i.e., the oligotrophic ocean where the magnitude of the blank may be similar to the clearest ocean sample). Conversely, the fluorescence signal of a blank may be insignificant relative to sample water, and blank subtraction may be irrelevant (e.g., highly concentrated river systems). In either instance blank subtraction may not be conducted, and it is up to the operator to determine if this is a reasonable step for his or her application. In an examination of 14 peer-reviewed papers from 1998–2010

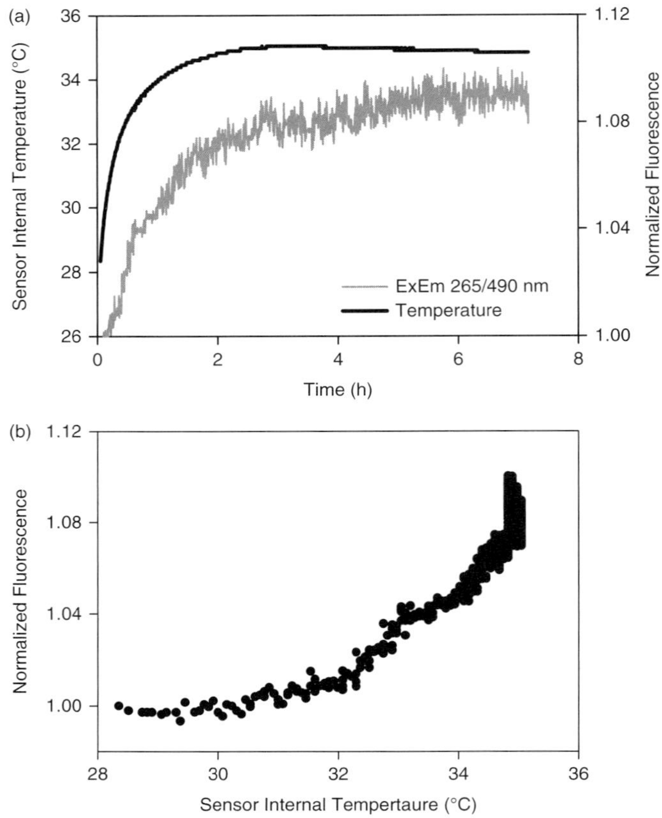

Figure 6.9. (a) Internal temperature and fluorescence signal increase during a laboratory experiment using a WetLabs SAFIre fluorometer. (b) Fluorescence as a function of instrument temperature.

that employed 10 different sensors in varying environments, only 4 performed blank subtraction (Table 6.2) for their applications. The environmental considerations discussed in Section 6.5 of this chapter may be of assistance in determining whether blank subtraction is needed. Selection of the blank type is of great importance. Published blanking methods have been routinely conducted with purified water, artificial seawater, and filtered seawater. Which blank is appropriate (if any) must be determined by the analyst and requires a series of experiments to establish the sensitivities and detection limits of the field sensors.

Once blank subtraction is deemed necessary, a blank may be acquired by either placing the instrument into a blank solution (open-face design canisters) or pumping water through the instrument (flow-through design canisters). In most cases the manufacturer's software can be used to acquire signal output for a blank. However, some sensors have the ability to measure the signal with a standard voltmeter. There are two accepted methods of applying the blank: (1) saving the blank as a separate file to be applied during post-processing of

Table 6.2. *Published calibration methods for fluorescence sensors for various applications*

Reference	Instrument	Deployment	Intensity calibration	Blank subtraction	Temperature correction	Environment
Coble et al. (1998)	Sea Tech, Yellow Matter Meter	Profiling	QS, 0.45-μm filtered samples	Yes	No	Arabian and Black Seas
Chen (1999)	Sea Tech, Yellow Matter Meter	Profiling	QS, discrete samples	No	No	San Diego Bay, Boston Harbor
Del Castillo et al. (2001)	WET Labs, SAFIre	Flow-through	Gf/F filtered samples	No	Yes	West Florida Shelf
Breves and Reuter (2001)	TriOS, GmbH	Profiling	relative units	No	No	Arabian Sea
Esser and Volpe (2002)	Chelsea, AQUAtracka	Flow-through	QS	No	No	N. Pacific Surface Water
Coble and Conmy (unpub. 2003)	WET Labs, WETStar	Towed undulating, profiling	Gf/F filtered samples	No	Yes	NW Australian Shelf
Chen and Gardner (2004)	Sea Tech and Seapoint, SUVF	Profiling, towed Undulating	Gf/F filtered samples	No	No	Louisiana Bight
Conmy et al. (2004)	WET Labs, SAFIre	Profiling, flow-through	0.45-μm and Gf/F filtered samples	No	Yes	MAB, WFS
Conmy et al. (2004)	WET Labs, SAFIre	Flow-through	Gf/F filtered samples	No	Yes	Louisiana Bight
Etheridge and Roesler (2004)	WET Labs, FlashPak	Time series	<0.2-μm, Gf/F filtered samples	No	No	Long Island Sound
Gardner et al. (2005)	Seapoint, SUVF	Towed undulating	QS, Gf/F filtered samples	No	No	Boston Harbor
Belzile et al. (2006)	WET Labs, WETStar	Profiling	QS, discrete samples	Yes	No	East Siberian Sea
Belzile et al. (2006)	WET Labs, WETStar	Profiling	0.2-μm filtered samples	No	No	Beaufort Sea
Downing et al. (2009)	WET Labs, WETStar	Moored time series	QS, filtered flowpath	Yes	No	San Francisco Bay
Tedetti et al. (2010)	TriOS, EnviroFlu-HC	Profiling	0.2-μm filtered samples	Yes	No	Mediterranean Sea
Conmy et al. (2014)	Chelsea, AQUATracka; WET Labs, ECO; Turner, Cyclops	Rack mounted in wave tank	Petroleum and water samples	Yes	No	Wave tank

sample data or (2) using an auto-blanking function found in some manufacturer's software, where the blank file is recorded before sample collection and applied to the data while they are being collected. The first method allows for the recording of raw fluorescence data. Readers should beware that if using the latter method, the data stream could be subject to complications if an ideal blank is not used, as it may be impossible to retrieve the nonblanked values once data collection is completed. Regardless of the method selected, blanks must be monitored carefully to ensure accuracy and to minimize systematic error.

6.4.3 Standards and Intensity Calibration

Throughout the literature there appear numerous findings of NOM fluorescence values reported in volts or relative fluorescence. Although the scientific community gains from these reports, the relative spatial and temporal trends in fluorescence values, it is impossible to compare absolute intensities without calibration to a known standard. Calibration of field sensor fluorescence intensity is routinely conducted using three standards. QS and natural organic matter (NOM) samples are utilized for humic- and fulvic-like fluorescence calibration, where NOM can be either humic acid standards or filtered sample water. The third standard is tryptophan, which is used for calibration of protein fluorescence.

QS is a standard reference material (SRM) that is a strong absorber of UV light with a fluorescence peak at ex/em = 347.5/450 nm. It has a high quantum fluorescent yield, particularly if diluted in weak acids and its peak fluorescence corresponds to regions of the spectrum where organic matter components absorb and fluoresce. Just as with laboratory instruments, a QS dilution series can establish the sensitivity and the linearity of a field sensor, and the range of concentration should be appropriate for the environment which is being sampled. Fluorescence intensity is converted to QSE (QS Equivalents) through the following equation:

$$[\text{sample}] = \frac{\left(C_{\text{sample}} - C_{\text{CWO}}\right)}{\text{QS Slope}} \tag{6.1}$$

where C_{sample} = sample raw counts, C_{cwo} = blank raw counts (if choosing to blank subtract the data), and QS Slope = linear regression slope of QS dilution series versus fluorescence counts. Analysts should also note that some manufacturer software automatically calculates the fluorescence data in QSE, so a dilution series need only to be conducted to confirm manufacturer scale factors and to verify that instrument is maintaining its calibration or the dynamic range of the sensor. Similarly, for sensors customized for protein detection, a dilution series of tryptophan standards prepared in purified water or seawater can be used for calibration (Tedetti et al., 2010).

NOM can also be used to calibrate field sensors and is used because no true standard for DOM exists. This method is conducted using either discrete filtered samples or commercially available humic acid standards. The latter method requires dissolution in purified water and the generation of a dilution series (Breves and Reuter, 2001), where the method

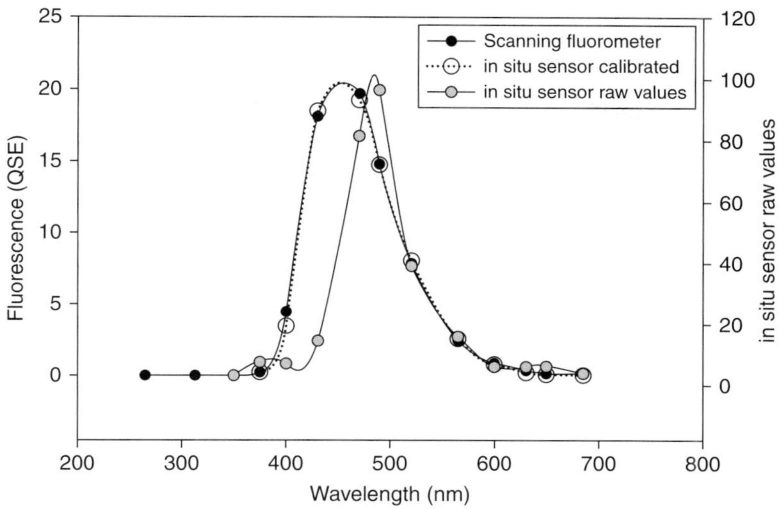

Figure 6.10. Emission spectra from a multispectral in situ sensor and a scanning fluorometer demonstrating pre- and post-calibration.

is then similar to the QSE approach. Most commonly, the NOM method utilizes discrete filtered samples that are collected coincident with fluorescence measurements from a field sensor (Del Castillo et al., 2001; Conmy et al., 2004). Samples are analyzed with a benchtop fluorometer, where data are fully corrected and calibrated to QSE. A scaling factor is applied to the in situ data through the equation,

$$\text{scaling factor} = \frac{FL_{\text{lab}}}{FL_{\text{in situ}}} \tag{6.2}$$

where FL_{lab} and $FL_{\text{in situ}}$ are fluorescence intensities for discrete samples and in situ data, respectively. This secondary standard method is commonly used for multichannel fluorometers, where calibration factors are calculated for each wavelength channel (Figure 6.10) (Del Castillo et al., 2001; Conmy et al., 2004). This method can also be used to convert field sensor measurements to Raman Equivalents (REs), which is an intensity normalization using the water Raman peak instead of QS, if the discrete data are calibrated in this manner. However, analysts should be aware that at the time of this chapter's writing, direct calibration to the water Raman peak cannot be conducted with commercial field sensors due to their wide filter bandpasses and filter wavelengths not being centered on Raman bands.

6.4.4 Correction for Inner Filter Effects

Field fluorometers are potentially subject to inner filter effects (IFEs) at high OM concentrations, just as are benchtop instruments. This effect is observed through an apparent

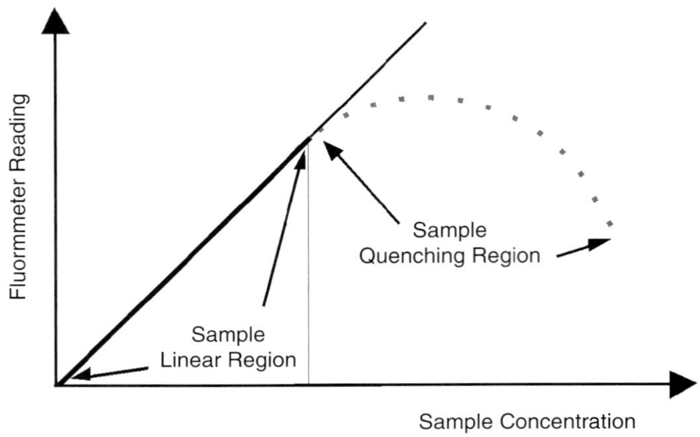

Figure 6.11. Conceptual diagram showing instrument response and saturation of fluorescence signal for a highly colored environment.

decrease in fluorescence intensity resulting from the reabsorption of emitted light within a solution. Previous studies have shown a 20% reduction in fluorescence response above 100 QSE (Gardner et al., 2005). To correct for this effect, the fluorescence of sample water and a series of diluted samples mixed with purified water were measured. A quadratic response curve (fluorescence vs. % whole water) was then calculated, where the linear portion of the fitted equation represented instrument response in the absence of the IFE. The correction factor was determined by dividing the linear portion of the fitted equation by the full equation, then applying it to the voltage output of the fluorometer. This estimated the response of the instrument in the absence of inner filtering.

6.4.5 Dynamic Range

Calibration curves are performed to identify the resolution limit, linear range of detection and saturation range of an instrument (Figure 6.11). Therefore dynamic range considerations are imperative when collecting fluorescence measurements. Many sensors are equipped with adjustable gain settings to maximize the linear range, and analysts can ensure that measurements do not approach saturation range by serially diluting hand-picked samples and quantifying any inner filter effects. This is critical for determining whether a sensor is appropriate for a specific application and is discussed further in Section 6.5.

6.5 Environmental Considerations

When deploying field sensors, analysts must be aware of environmental factors that can potentially influence fluorescence measurements. Attributes of a given environment can be used to decide the ideal optical and sampling designs to yield high-quality fluorescence data. Both the flow-through and the open-faced sensor designs can be used in most natural waters. However, differences amongst sensors (peak wavelengths, bandpass of interference filters, gain settings, flexibility of the sensor platform, and automated programmable internal dataloggers) should be taken into consideration when selecting a sensor for a specific application. For any deployment, it is recommended to perform manufacturer recommended cleaning and maintenance procedures and frequent intercalibration checks with a benchtop fluorometer or standard material. Fluorescence data should be verified with ancillary field measurements to avoid discovering voltage or temperature dependency on the data stream during final data analysis, and data should be recorded when possible to avoid loss of measurements due to unforeseen events.

6.5.1 *Factors of Concern*

6.5.1.1 Particles

The presence of inorganic and organic particles in natural waters can be problematic for any in situ optical measurement of dissolved species. Field fluorometers are subject to output bias from particle interferences through increased scatter of light within the sample volume, particularly in moderate to highly turbid systems. The result can either be inhibition or contribution to fluorescence signal. In a study by Saraceno et al. (2009), two WET Labs flow-through fluorometers were deployed in filtered (in-line 10 μm and 0.2 μm) and unfiltered modes within an agriculturally impacted California watershed, Willow Slough, demonstrating that the presence of particles in this system inhibited fluorescence during peak discharge events (~50% underestimation of dissolved species), but not during periods of little freshwater influx to the system (Figure 6.12). Conversely, the presence of particles can contribute to the fluorescence signal, such as bloom events containing phycobilipigments that fluoresce in the green portion of the spectrum, interfering with the humic fluorescence signal. It is important to note that a simple field proxy (e.g., turbidity) is not always adequate for particle prediction with respect to quenching of the sensor output. Particles may absorb excitation or emission light or in fact, reduce the apparent sample volume in front-faced instruments. FOM sensors in particular are sensitive to variations in turbidity and can easily be affected by particle quantity or quality, yet particle quality is an often overlooked parameter. Some manufacturers have bench-tested sensors to show turbidity rejection to levels as high as 400 NTU (nephelometric turbidity unit). The problem with this is that turbidity standards used (e.g., high-quality polystyrene beads) are typically monochromatic and absorb well below the excitation or emission bandpass of the sensor. Particles in natural systems, however, are not homogeneous and do not always behave in a predictable manner. There is much work to be done to understand fully the limitations and potential impacts of particle size and quality on in situ fluorescence measurements. To evaluate the contribution of particles on fluorescence measurements, analysts can operate two sensors simultaneously in filtered and unfiltered modes during field operations. If only one sensor is available, fluorescence data can be collected within

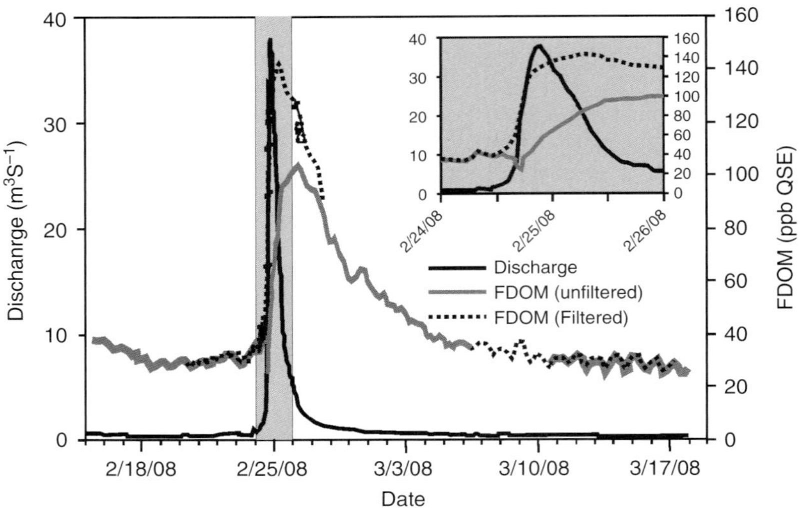

Figure 6.12. Filtered and unfiltered FDOM sensor output demonstrating interference of particles on fluorescence signal. (Redrawn from Saraceno et al., 2009.)

a laboratory whereby a sensor measures unfiltered and filtered water samples and the difference in signal is calculated.

6.5.1.2 Bubbles

Similar to particles, the presence of bubbles in the water column can also interfere with fluorescence measurements, resulting in an increase or decrease in signal output. Bubbles formed naturally due to wave processes in oceans and estuaries or due to flowing water within streams cannot be avoided when using open-faced sensors. To ensure that these effects are minimized, ancillary data for river flow, currents or sea surface roughness can be used to validate such interferences within a data stream. To minimize bubbles within flow-through sensors, deploying with degassing Y fittings (Twardowski et al., 1999), mounting of the sensor in a vertical orientation and monitoring the flow rate can all help to remove bubbles from in-line water flows. In addition, if using a vertical profiling platform, the sensor can be lowered to 1 m depth before data collection so as to equilibrate and clear bubbles from the flow-through line. Analysts should beware that bubbles can accumulate within sample flow tubes, or on optical windows in both flow-through and open-faced instruments, so instruments should be monitored within specific environments.

6.5.1.3 Dynamic Range

An understanding of DOM concentration range within natural waters is also critical when choosing a sensor, as systems can have a wide range in fluorescence intensity. To ensure high data quality, the dynamic range of a sensor must be representative of DOM concentrations found in the study area. Therefore a major consideration should be instrument gain. Many sensors are available with fixed gains and these settings may not be suitable for some environments. Undoubtedly a sensor optimized for linear, full-scale response in optically dilute waters is most likely not suitable for use in certain wetlands or vice versa. Some instruments use a logarithmic voltage response (e.g., Chelsea UV Aquatracka) to eliminate the need for multiple fixed gain settings. In environments with high DOM concentrations, internal quenching or absorption of light within the sample volume may prevent fluorescence signals from reaching detectors that results in nonlinear fluorescence intensities (Figure 6.11). A priori knowledge of measured optical properties in most study environments is critical in the application of modern NOM sensors. At present, there is no sensor that includes algorithms to optimize excitation or emission peaks, bandwidth or electronic gain setting. There are, however, some sensors that include automatic gain settings, albeit in fixed ranges.

6.5.1.4 Temperature Effects

All fluorescence measurements are subject to influences from temperature (Chen and Bada, 1992). In addition to the issue of self-heating discussed in Section 6.4, variations in the ambient water temperatures of the sampling environment can also impact fluorescence values. This can be monitored with some instruments with internal and external thermistors and some manufacturer's software offers temperature correction algorithms to apply to data streams. As discussed in Section 6.4, correction procedures for temperature effects can be conducted in the laboratory, and analysts should be aware that various deployment styles and environments can influence temperature effects differently. Consider the case where a vertically profiling sensor is deployed through a steep thermocline where rapidly changing external temperatures causes the internal temperature of the sensor to vary quickly. Depending on the difference between the external and internal temperatures, the effects on the sensor can be instantaneous or lag behind fluorescence measurements making corrections for temperature challenging. Conversely, sensors deployed on a mooring may undergo large fluctuations in environmental temperature, but over longer time scales (i.e., weekly, monthly, seasonally) where the sensor has time to reach thermal equilibrium. This can make for easier corrections to the data stream. Similarly, if operating a sensor where water is pumped to the meter while submerged in a water bath or a holding tank, the water bath can essentially serve to regulate the internal temperature of the sensor.

6.5.1.5 Biofouling

All sensors will biofoul if given enough time deployed in natural water systems. The extent of biofouling varies as a function of environment, so analysts are encouraged to determine how problematic this will be in their system, as fouling can severely interfere with measurements (Figure 6.13a) (Davis et al., 2000). Historically, substances containing tributyl tin (TBT) were used for long-term aquatic anti-fouling methods, but TBT has a negative environmental impact and is no longer in use today. Laboratory and field testing of TBT-based products, antifungal agents, and silicon-based compounds on glass and acrylic surfaces

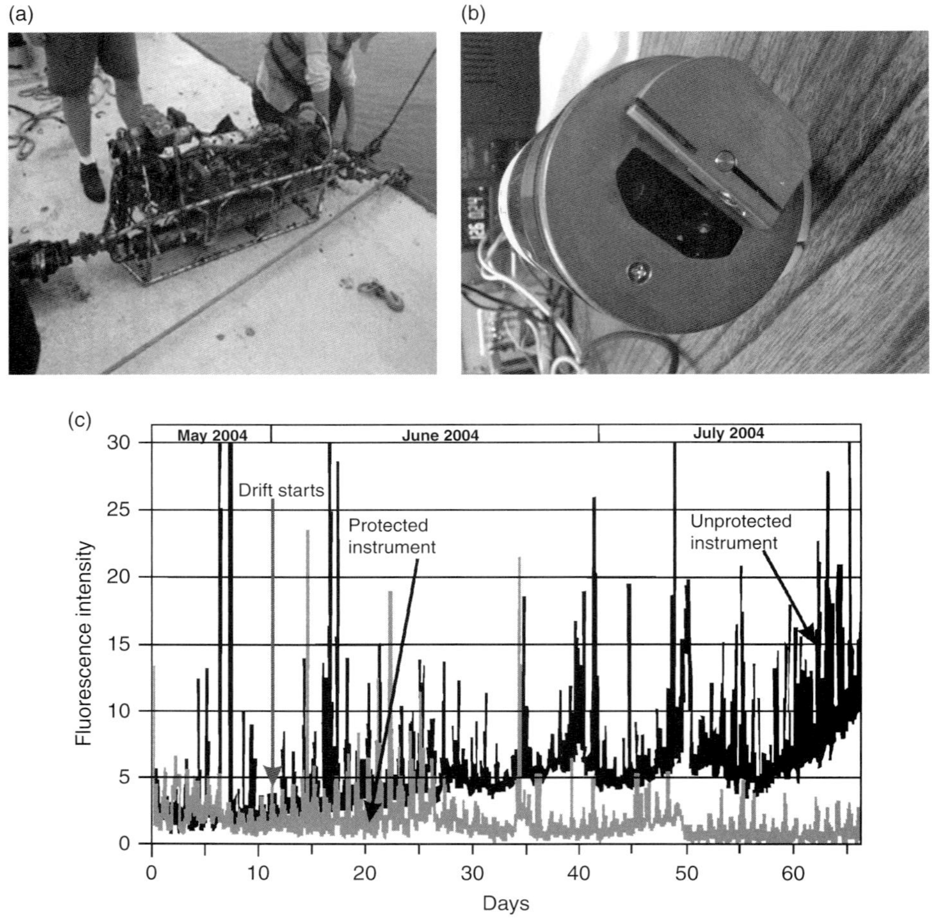

Figure 6.13. (a) Retrieval of an optical package after a deployment in the Penobscot River, Maine. (Photo courtesy of C. Roesler and A. Barnard.) (b) Example of a copper biofouling shutter on a sensor. (Photo courtesy of B. Downing.) (c) Drift of unprotected fluorometer due to biofouling. (Redrawn from Delauney et al., 2010.) (See Plate 9.)

found that the products were only marginally effective against algal growth (McLean et al., 1997). The same study reported that these coatings could ultimately serve as substrate for microfauna because it increased the surface roughness of optical windows, therefore it is best not apply them at all. Common anti-biofouling strategies used today include the use of copper shutters, plates and wipers for open-faced sensors (Manov et al., 2003; see example in Figure 6.13b). Copper tubing or tubing covered with foil or black tape to block light can be used for pumped instruments. Most recent anti-biofouling technologies include pressured air cleaning and the use of nanocoating technologies (Trios microFlu-CDOM, Trios Mess GmbH, Oldenberg, Germany). Also noteworthy is the introduction of nano-treated plastics into sensor production to minimize adhesion of biofouling. This is a particularly promising invention, as it removes common side effects of bio-wipers such as impedance of the sensor signal due to mechanical positioning on the sensor emitter/detector surface, interference due to wiper shading the signal, and wiper failure.

Biofouling can impact data by decreasing or increasing fluorescence signal (Delauney et al., 2010), the former due to physical blocking of optical path and the latter if the fouling material fluoresces at same wavelengths measured by the sensor (Figure 6.13c). In the event that biofouling impacts data quality, portions of a data stream can either be excluded, or attempts to correct the impact can be conducted. This is largely dependent on the degree, type, and duration of fouling. In a study in 2009 by Cetinic et al., measurements from a biofouled sensor were corrected by assuming a linear drift in instrument response as a function of increased biofouling. In the paper, analysts are warned that complications with this approach can arise because biofilm growth is exponential and undergoes a five-stage series of development, meaning that the response may not in fact be linear. However, given that in most deployments, calculating true biofilm growth rate is not possible, linear corrections may be the only reasonable option. In addition, the manner of sensor deployment is also a factor, where correcting data from a profiling mooring or autonomous underwater vehicles (AUVs) may be more challenging (due to sensor sampling multiple water types) than from a moored stationary sensor where biofouling rate may be calculated. In any case, analysts should note that if biofouling on any optical window is sufficiently developed to the point of swamping the ambient signal, even the best calibration procedure will not yield high data quality.

6.5.1.6 Understanding NOM Sources Within Environments

Commercially available sensors are often configured to measure at excitations and emissions which are representative of humic fluorescence peaks at longer wavelengths (e.g., peak C; Coble, 1996). Many studies have found peak C–based NOM fluorescence useful as surrogates for dissolved organic carbon (DOC) concentration in a myriad of pristine and anthropogenically influenced watersheds. Analysts should be forewarned that changes in fluorescence signal from single-channel sensors may not stem from a change in fluorescence intensity alone, but rather from a shift in the wavelengths of the fluorescence peak. These shifts result from source and compositional differences of organic material. Consider, for example, a case where a freshwater system has three dominant sources of CDOM: wastewater, plant leachates, and autochthonous production. Variations in sensor signal could arise from changes in concentrations of the material, but could also result from the sensor detecting one source over another. Sensors configured at a single fixed set of wavelengths cannot measure the shift and can compromise signal intensity. Before deployment of a peak C sensor, it is recommended that a preview of excitation–emission matrix spectroscopy (EEMS) from the study site to preestablish FDOM/DOC relationships that are to be considered. Review of EEMs and DOC will usually yield a sense of how reliable the sensor proxy will be for that system. In addition, ongoing collection of discrete (calibration) samples for EEMs analysis may also prove to be necessary to evaluate continued instrument performance as well.

6.5.2 Sensor Choices for Specific Environments

6.5.2.1 Optically Dilute Systems

Environments that are weakly absorbing with little scattering properties are considered optically dilute. Examples of these systems include open-ocean, oligotrophic lake, and groundwater environments. Biofouling and particle interference in these systems would be minimal, so there would be little complication with deploying either a flow-through or flat-faced sensor. The main advantage to the former is a higher signal-to-noise ratio, which becomes increasingly important in low-DOM waters.

6.5.2.2 Optically Thick Systems

Environments that are strongly absorbing due to either dissolved or particulate matter are considered optically thick. Examples of these systems include coastal ocean, estuary, river, eutrophic lake, and wastewater environments. The occurrence of self-shading can be an issue in these waters, where at high concentrations incident light is prevented from reaching organic material within the optical sampling volume, thereby decreasing the apparent fluorescence response. Signal-to-noise ratio is not an issue in these environments, but filtration and biofouling are. Applications where filtration is necessary may require a flow-through sensor, but note that some flat-faced sensors are also available with flow-through chambers. Analysts should make note that the sensitivity and sample control volume of the flow-through design is unequalled. A major consideration when using flow-through sensors is maintenance of the internal cuvette, which is prone to biofouling. This can be problematic in optically-thick systems, thereby making flat-faced sensors equipped with a bio-wiper to reduce fouling a better choice in systems with fewer particles.

6.5.2.3 Turbid Systems

Strongly scattering environments are a special case because all turbid waters are considered optically thick, but not all optically thick waters are turbid. In highly turbid environments (e.g., greater than 100 NTU), a flow-through sensor deployed with a microporous membrane particle filter may be the best choice to avoid interferences from particles (Belzile et al., 2006; Downing et al., 2009; Saraceno et al., 2009). Although flat-faced sensors are very attractive as they do not require the added power requirements of a pump and filter assembly, the effects of both quantity and quality of particle interferences on these sensors is not fully understood.

6.5.2.4 Energetically Flashy Environments

Some systems undergo extreme fluctuations in water flow, wave energy, tides, suspended particles, temperature, and anthropogenic discharge and can be termed "flashy" environments. These systems can also exhibit large DOM concentration gradients, and the presence of other fluorescing organic species (natural or anthropogenic) which can lead to interfering fluorescence signals. Flashy environments may also necessitate a sensor with a rapid response time, where the sensor optics and associated electronics are designed to react and recover from rapid and extreme biasing (interferences) causing photo-detector/sensor output saturation. Choosing the proper sensor design and deployment platform can be challenging in these systems. Questions must be raised regarding instrument gain, whether or not the sensor will be submerged during the entire deployment time (e.g., drought-prone or tidal rivers/estuaries), and if the optical system is strong enough to tolerate the physical energy of the environment (e.g., crashing waves or corrosion in acidified lakes). Analysts want to ensure not only high-data quality of fluorescence measurements, but also that their sensors will remain securely in place during extended deployments (e.g., moorings or on vertical profilers).

6.6 Revolutionizing NOM Studies via High-Resolution Fluorescence Measurements

The development of in situ fluorometers has led to the ability to examine NOM on spatial and temporal scales of traditional hydrographic parameters such as temperature and salinity. This has enhanced biogeochemical studies in a manner similar to how chlorophyll fluorometers have impacted biological oceanography (Geiskes et al., 1978). As a result, improved understanding of processes such as tidal flushing and DOM outwelling from mangroves and salt marshes; groundwater DOM sources to coastal and fresh waters; photochemical and biodegradation pathways that affect NOM distributions and quality; and benthic fluxes of sedimentary OM and hydrocarbon seeps have taken place. Continued understanding of OM cycling over varying spatial or temporal scales at land–ocean interfaces, sediment–water interfaces, or during episodic events will be made possible by deploying small, low-power, high-sensitivity, robust fluorescence sensors via a variety of sensor platforms.

6.6.1 Deployment Platforms

Scientists and engineers never cease to develop new and clever means of obtaining optical measurements, thereby increasing the ways field sensors are deployed in natural waters. Novel sensor platforms and deployment strategies have revolutionized the community's ability to obtain NOM measurements and understand the NOM pool (Twardowski et al., 2005). Just like with sensor choice, analysts must choose a platform that aligns well with sampling design. Hence consideration must be given to length of deployment, spatial and temporal variability of environment, power requirements, sensor size and stability, data logging methods, and optimal sampling rates. In addition, one must ensure that sensor placement and orientation on a platform does not interfere with its performance (e.g., objects within the optical path or altered water circulation near sensor).

6.6.1.1 Spatial Resolution

Common tethered platforms utilized from ship deck include vertical profilers, towed vehicles, and pumped flow-through systems, which can accommodate most commercially

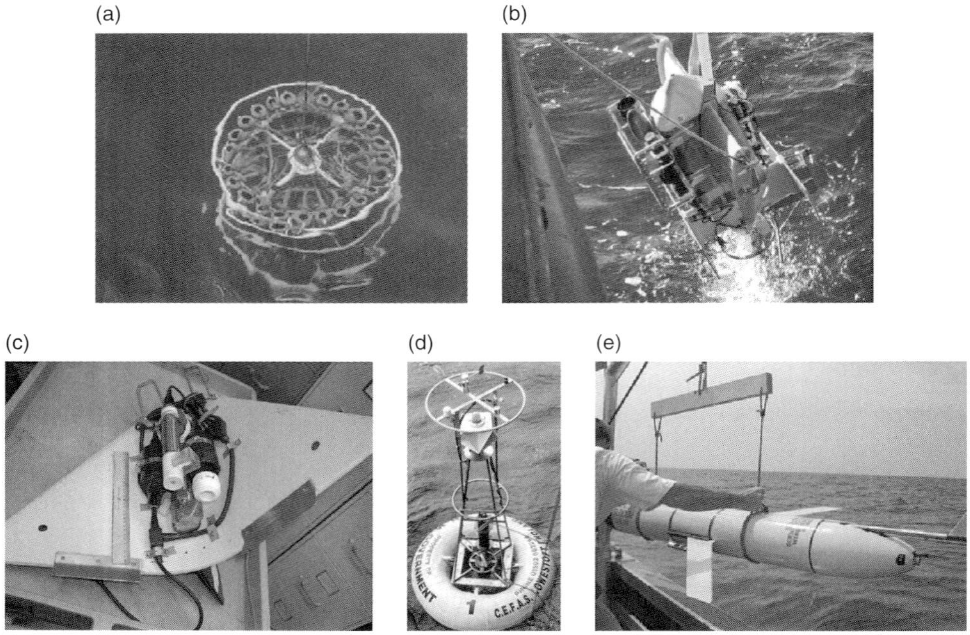

Figure 6.14. Fluorescence sensor deployment platforms. (a) CTD Rosette vertical profiler. Water was pumped in series through a fluorometer and other environmental sensors, where instruments were mounted at the bottom of the Rosette and CDOM fluorometer. (Courtesy of R. Conmy.) (b) Towed vehicle with various optical and chemical sensors. (Courtesy of R.F. Chen.) (c) Minishuttle tow-yo vehicle deployed with chlorophyll and NOM fluorometers, dissolved oxygen sensor, and CTD. (Courtesy of R.F. Chen.) (d, e) Buoys, moorings, and gliders are also platforms for optical and environmental sensors. (Courtesy of Cefas.) (See Plate 10.)

available sensors (Figure 6.14). Vertical profilers allow for depth-dependent fluorescence observations and the simultaneous collection of discrete water samples from Niskin bottles. Towed vehicles have been successfully deployed to study FOM in estuaries and river plumes and can operate either in undulating mode (computer-controlled wing determines depth pattern) or in tow-yo mode (wire out controls depth). These systems can also pump water from the platform surface through tubing to the deck of a boat for discrete bottle sampling (Chen and Gardner, 2004). Towed systems offer the added benefit of observations made in both the horizontal and the vertical directions, creating cross-sectional distributions of measurements.

AUVs and gliders can house the smallest fluorometers and allow autonomous deployment of instruments with preprogrammed missions or adjustable missions when the AUV surfaces, and AUVs and gliders have been used to survey coastal areas without ship support. Advantages include greater spatial coverage, the potential for use of a fleet of AUVs to obtain three-dimensional distributions and deployment in rough sea states. Disadvantages include less control over the location and direction of coverage, power limitations, and limited ability for near-real-time data. Gliders are slower than AUVs, but have longer duration capabilities.

6.6.1.2 Temporal Resolution

Fluorometers have been used on buoys in oceans, estuaries, and lakes; attached to docks and bridges; or at cabled observatories. CDOM fluorometers, due to new low power consuming light sources and low power detectors, have become a common addition to many aspects of coastal ocean observatories. With fluorescence, a measurement of NOM can be obtained at temporal resolutions of 1 second or less allowing time-varying distributions to be studied. Some moored platforms can even be of a profiling nature and provide both temporal and spatial data (e.g., bottom stationing ocean profiler [BSOP]; Langebrake et al., 2002).

6.6.2 Importance of Scale

The availability of sensitive, in situ sensors and appropriate deployment platforms allows for capturing NOM measurements over challenging scales. Thin layers of particles, plankton, or bacteria have been known to concentrate on density benches (regions where sharp changes in temperature and/or salinity occur in natural waters), especially during stratification events (Chen and Gardner, 2004). These distributions are difficult to study due to their ephemeral nature and small vertical distributions (they can be centimeters to meters thick). Conversely, river plumes can extend hundreds of kilometers (e.g., Amazon) with freshwater sitting upon higher density seawater, thus isolating the fresher river plume waters with all their nutrients, particles, contaminants, and organisms. In addition to these spatial measurement challenges, distributions of organic matter change temporally due to tides, winds, runoff, and episodic events such as storms or effluents. To capture these temporal variations, repeated measurements over time (e.g., via transects, buoys, or profilers) are required over the area of interest. The following three field experiments are provided as case studies for examining NOM at varying scales using in situ sensors on towed platforms.

6.6.2.1 Neponset River Estuary in Boston Harbor- Small Temporal and Spatial Scales

The Neponset River Estuary drains the second largest river into Boston Harbor, MA, and is a tidally dominated urban system fringed with tidal salt marshes. To examine tidal and seasonal variations in DOM sources and distributions within the estuary, surveys were conducted using a miniature towed instrument package called the Mini-Shuttle (Gardner et al., 2005). With all hydrographic and optical sensors mounted within 4 cm of each other, vertical salinity variations of 20 PSU over just 1–2 m depths can be delineated (Figure 6.15). This study demonstrated that a significant but seasonally varying input of CDOM within the estuary originates from the degradation of salt marsh organic material. This input yields

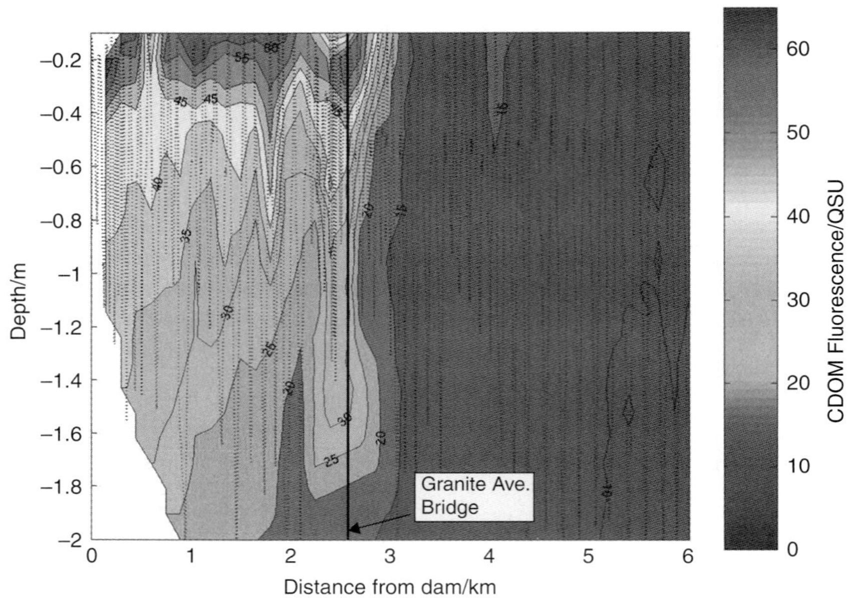

Figure 6.15. CDOM fluorescence in the Neponset Estuary at high tide. Dotted lines represent path of towed vehicle Bridge is located 2.6 km from dam. (See Plate 11.)

a nonconservative mixing curve of salinity with CDOM. River CDOM inputs were seen to fluctuate with discharge. Beyond source identification, the sensors on the towed platform illustrated a tidal restriction imposed by a local bridge (2.6 km from dam) that imposed a major influence on estuarine circulation. The effect of tidal restrictions on holding low salinity, high CDOM water from flowing out with the tide, the effect of bottom topography altering the circulation and thus CDOM variations at a single location, and temporally varying CDOM freshwater endmembers were all documented at this site, demonstrating the value of measurements at a centimeter scale.

6.6.2.2 Hudson River Estuary – Large Spatial and Small Temporal Scales

At a larger spatial scale, variability in the mixing of river and tributary sources can be examined. CDOM has been measured and modeled in the Hudson Estuary where multiple inputs (Hudson, Raritan, Hackensack, Passaic Rivers, and sewage outfalls) were differentiated using a towed, undulating platform, the ECOShuttle (Chen and Gardner, 2007; Figure 6.16). It was determined that the sewage effluent influenced CDOM distributions in the upper New York Bay. Tidal mixing of several effluent sources added significantly to a mid-estuary source of CDOM that is exported offshore with the Hudson River Plume. The detection and quantification of this temporally varying CDOM resulted in the incorporation of this phenomenon in the biogeochemical module of the physical circulation models for the region (Georgas and Blumberg, unpublished).

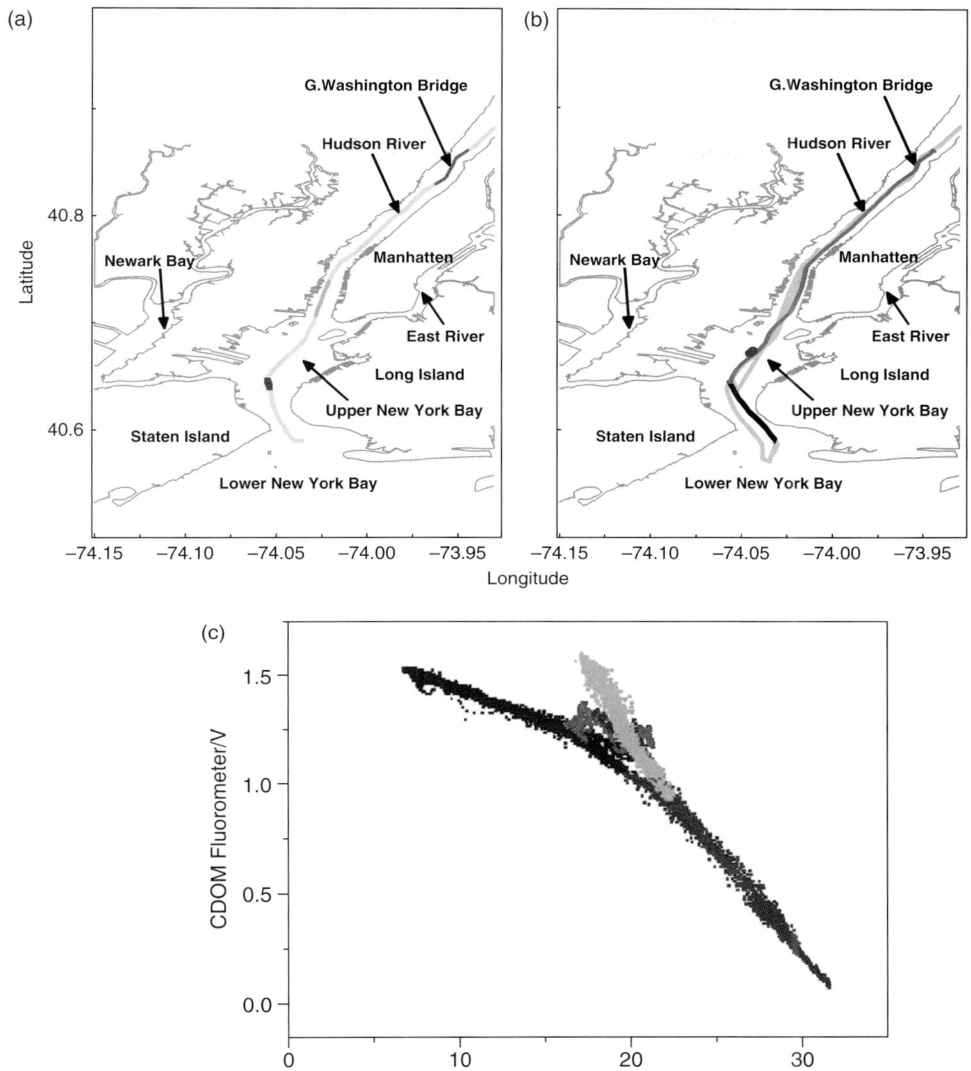

Figure 6.16. (a, b) CDOM as a function of salinity in the Hudson Estuary during ebb and flood tides, respectively. (c) Colors of data symbols correspond to colored portions of cruise tract on the map. Note the yellow CDOM contribution (presumably sewage-derived CDOM) during the ebb tide (a) that is not observed in the shipping channel during the flood tide (b). (See Plate 12.)

6.6.2.3 Mississippi Bight Region – Large Spatial and Temporal Scales

Examining the Mississippi River plume over large spatial and temporal scales using a towed, undulating vehicle has demonstrated time-varying freshwater endmembers and evidence of seasonal DOM photobleaching in this region (Chen and Gardner, 2004). This sampling strategy allowed for observing thin layers of high-CDOM waters below the seaward edge

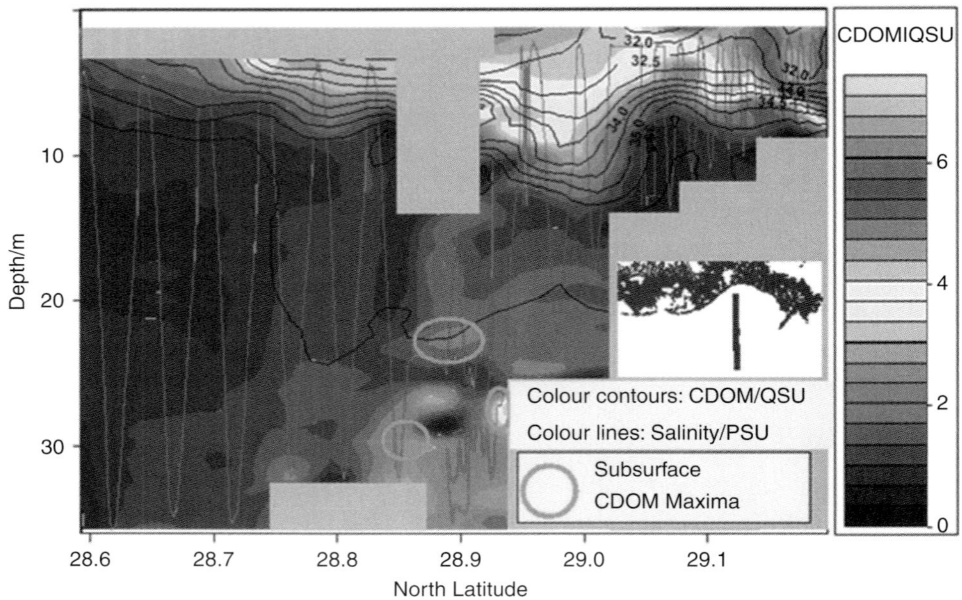

Figure 6.17. CDOM (grayscale colors) and salinity (black lines) contours across the Mississippi plume in June, 2000. Salinity is represented by the contour lines. Inset map shows the location of the survey transect in relation to the Mississippi Delta. Gray lines represent path of towed vehicle and gray boxes represent missing data. Bottom depths were approximately 5 m below the maximum depths of the fluorescence and salinity measurements. Subsurface CDOM maxima are highlighted with circles. (Redrawn from Chen and Gardner, 2004.) (See Plate 13.)

of the river plume. Thin layers of CDOM maxima at depth are not accompanied by lower salinities (Figure 6.17; circles) and are presumably formed by bacteria that are processing particles that sink from the river plume and concentrate on density benches as a result of stratification.

6.7 Remotely Sensed NOM Measurements

Fluorescence spectroscopy is a powerful tool in the study of organic matter in the ocean, yet most important geophysical variable related to CDOM is its wavelength-dependent light absorption coefficient ($a_g\lambda$). Because CDOM absorbs light, it may influence vertical distribution of phytoplankton, primary productivity, and water temperature. The color of CDOM can be detected using space and airborne sensors (e.g., NASA – Moderate Resolution Imaging Spectroradiometer [MODIS] and ESA – Medium Resolution Imaging Spectrometer [MERIS]). This allows for the study of processes controlling the distribution of CDOM in the oceans in ways that are not possible using field measurements. In this section we discuss the contribution of fluorescence spectroscopy to the validation of CDOM remote sensing algorithms, and comment on the use of Light Detection and Ranging (LIDAR) and passive measurements of natural CDOM fluorescence.

6.7.1 Fluorescent CDOM and Validation of Remote Sensing Products

Flow-through CDOM fluorometers are useful to validate remote sensing retrievals of a_g. Traditionally, remote sensing data are compared to field data collected within ±3 hours of a satellite overpass (Bailey and Werdell, 2006). Often the comparison is between one field value and a remote sensing retrieval that represents an average over the size of the pixel (from 250 m^2 to ~1 km^2 or more depending on sensor, viewing angle, and spatial binning used). This comparison assumes low spatial and temporal variability and may be correct in the open ocean. However, the assumption can be incorrect in coastal environments with high spatial variability (Yuan et al., 2005). For example, a perfect satellite retrieval of a_g (representing the average a_g in a pixel) could be different from any number of discrete samples collected within that pixel. This may give the impression of inaccurate remote sensing retrievals. Collecting a large number of discrete samples within the area of a typical pixel for analysis in a bench-top fluorometer or spectrophotometer is not a practical solution. However, flow-through CDOM fluorometers can be well calibrated using discrete samples to produce accurate estimates of CDOM absorption coefficient at wavelengths relevant to remote sensing (i.e., Ferrari and Tassan, 1991; Hoge et al., 1993; Green and Blough, 1994; Del Castillo et al., 1999). For example, Hoge et al. (1993) demonstrated the use of fluorescent CDOM (FCDOM) to retrieve values of a_g, and later applied this technique to the calibration of airborne LIDAR to retrieve a_g values (Hoge et al., 1995). Their linear regression curves between a_g and FCDOM had r^2 values between 0.89 and 0.98, but typically >0.98. Similar results have been reported elsewhere (Blough et al., 1993; Del Castillo et al., 1999).

Because the fluorescence spectrum of CDOM is broad, and its absorption spectrum monotonic, it is easy to establish relationships between FCDOM and a_g over a large range of wavelengths. Nevertheless, for the purpose of using this technique to validate remote sensing retrievals of CDOM, it is convenient to report values at wavelengths covered by satellite sensors, and to provide data on the absorption spectra used to calibrate the FCDOM to a_g including spectral slope S and how it was calculated. This facilitates comparison between published data.

The user must be aware of changes in S (Blough and Del Vecchio, 2002) and fluorescence efficiency (φ_f) (Blough and Del Vecchio, 2002 and references therein; Lakowicz, 2006) along salinity gradients. A constant S allows for reconstruction of a CDOM absorption spectrum based on measurement of a_g at any λ (within the λ range used to calculate S). A constant φ_f indicates that CDOM fluorescence is proportional to the amount of light absorbed by CDOM, allowing to calculate a_g (λ) from FCDOM. These are typically observed at salinities >30, when the marine CDOM end-member starts to influence spectral properties of CDOM. Although the changes in φ_f are typically small, the user should be aware of these changes and collect calibration samples along the salinity gradient. Sampling along buoyant river plumes is more problematic. A typical flowthrough water intake in a research vessel is located a few meters below the surface. Therefore, any instrument connected to the system may not sample the highly colored river plume waters that are mostly responsible for the water leaving radiance detected by the remote sensor. Under these conditions, it may be necessary to use a large number of discrete surface samples collected away from disturbances caused by the research vessel, or to use an undulating, towed sampling system (i.e., Chen and Gardner, 2004; Gardner et al., 2005). These instruments allow for very fine scale measurements along the water column and have the capability of pumping water onboard, so collection of multiple calibration samples is possible. The use of this instrumentation is a viable option for the validation of ocean color products in turbid waters.

6.7.2 Active Remote Sensors

LIDAR is a useful use of fluorescence in remote sensing of organic matter in the oceans. The use of LIDAR for detection of CDOM and chlorophyll fluorescence has been well demonstrated (Hogue, 2005; Hoge et al. 1981, 1983, 1995, 1998, 2005; Barbini et al., 2001; Drozdowska, 2007). One excellent examples of the application of LIDAR to CDOM detection (and other applications) is found in Hoge et al. (1995), who demonstrated that well-calibrated sensors are very stable and useful in coastal and oceanic waters. LIDAR is particularly useful in coastal waters because of its high spatial resolution and dynamic range. For example, the LIDAR system flown by Hoge et al. (1995) has a spatial resolution of ~130 m – superior to that of any satellite sensor. An additional advantage of airborne LIDAR is that it can be flown under clouds with obvious advantage over satellite sensors. The use of airborne LIDAR should be considered as an alternative to satellite imagery to study highly dynamic coastal regions and when cloud cover limits availability of satellite imagery. Airborne LIDARS can sample large area of the ocean during a satellite overpass, making it more productive for satellite validation than traditional sampling from a research vessel. These capabilities make LIDAR a possible option for validation of future spaceborne high spatial resolution ocean color radiometers.

LIDAR systems can also be deployed from the bow of ocean-going vessels (i.e., Barbini et al. 2001), and can provide simultaneous measurements of FCDOM (hence a_g) and chlorophyll. Barbini et al. (2001) used a system similar to Hoge et al. (1994) to validate remote sensing retrievals of Chl from SeaWiFS in the Southern Ocean demonstrating the usefulness of shipboard LIDAR to validate remote sensors. LIDAR, however, has its complications. The instrumentation and mission-planning capabilities are highly specialized and expensive. Flying a LIDAR system regularly is a luxury that only a well-funded program can afford. Moreover, LIDAR data, like most field data, are collected for specific campaigns and do not offer the global coverage or temporal resolution available from satellite sensors.

6.7.3 Fluorescence of CDOM and Passive Sensors

CDOM fluoresces under sunlight producing a broad emission spectrum that varies with the source of the CDOM and with the environmental conditions. This solar-induced fluorescence can be strong enough to contribute significantly to irradiance reflectance (*R*) and could affect the interpretation of remote sensing data. FCDOM may also fill Fraunhofer lines suggesting a tool for detection of CDOM by remote sensors.

Several authors have evaluated the contribution of FCDOM to water leaving radiance (L_W, radiance flux coming out of the water in W m^{-2} sr^{-1} nm^{-1}), and hence to *R* (ratio of irradiance reflected from a surface to irradiance incident on a surface) (i.e., Spitzer and Dirks, 1985; Peacock et al., 1990; Hawes et al., 1992; Lee et al., 1994; Vodacek et al., 1994; Haltrin et al., 1997; Zhou et al., 2009). Although these authors used different assumptions and different sources of CDOM in their studies, the consensus is that FCDOM only becomes an important contaminant to *R* when CDOM is very high. How high depends on the total optical properties of the water, and researchers should be cautions when working in low-sediment, high-CDOM waters. Complications to satellite remote sensing measurements are probably lower because high CDOM values required to influence *R* are typically restricted to near-shore or inland water bodies. Ocean color remote sensing images normally mask an area along the coast depending on sensor resolution (i.e., ~1 km for NASA-MODIS and SeaWiFS at nadir) to eliminate contamination from land reflectance. So pixels more likely to be contaminated with FCDOM are likely masked during image processing.

The broad fluorescence spectrum of CDOM fills Fraunhofer lines presenting the possibility of using very sensitive remote and in situ sensors to study CDOM and chlorophyll in ocean waters (Stoertz et al., 1969; Gee et al., 1993; Vodacek et al., 1994; Hu and Voss, 1998). Although under some conditions has shown promise (i.e., Vodacek et al., 1994), Natural FDOM is very faint ($_f$ ~ 1%) and the emission spectrum is to broad to be useful for passive remote sensing of organic matter.

6.7.4 Remote Sensing Summary

CDOM fluorescence measurements are a powerful tool to study dynamics of organic matter in the oceans. The application of fluorescence to remote sensing of CDOM is limited to the use of airborne LIDAR systems that can measure $a_g(\lambda)$ accurately. Contribution of FCDOM to *R* and filling of Fraunhofer lines can be important only during special conditions of high CDOM and low scatterers. FCDOM can be extremely useful in the validation of remote sensing estimates of $a_g(\lambda)$.

6.8 Summary

There are a number of commercially available fluorescence sensors and selecting the ideal sensor for a particular application, while understanding the capabilities and limitations of fluorometer design, is key to acquiring robust in situ CDOM fluorescence measurements.

Likewise, selecting deployment platforms and knowledge of study environment is essential to collecting measurements over appropriate temporal and spatial scales. Beyond gathering measurements is the need for analysts to adhere collectively to standardized methodologies for calibration of sensors. Whatever methods are conducted (e.g., blank subtraction, temperature correction, intensity standardization) need to be documented clearly in the literature. This will result in high-quality assurance of data throughout the community and the ability for intercomparison of data among laboratories.

Forging ahead, continuous advancements in sensor development and materials research are enabling measurements with LEDs and filters centered on shorter wavelengths, lower power consumption, and longer deployment durations. Emerging are reliable sensors with multiwavelength capabilities, sondes that include fluorescent NOM measurements for characterizing CDOM and "smart" FDOM sensors that include algorithms to utilize contemporaneous ancillary water quality measurements to calculate corrections in situ. These advancements lend promise to improved biogeochemical investigations, both qualitatively and quantitatively, of DOM in aquatic environments.

Acknowledgments

The authors wish to thank the science and engineering pioneers who forged ahead to develop and demonstrate the utility of in situ NOM fluorescence measurements. We greatly appreciate the reviewers of this chapter and all those who contributed pictures, data or diagrams to this chapter, particularly Paul Soderlind for redrawing of previously published figures. The information in this document has been funded in part by the U.S. Environmental Protection Agency. It has been subjected to review by the National Health and Environmental Effects Research Laboratory and approved for publication. Approval does not signify that the contents reflect the views of the Agency, nor does mention of trade names or commercial products constitute endorsement or recommendation for use. This is a contribution from the Gulf Ecology Division.

References

Bailey, S.W. and Werdell, P.J. (2006). A multi-sensor approach for the on-orbit validation of ocean color satellite data products. *Remote Sens. Environ.*, **102**, 12–23.

Baker, A., Ward, D., Lieten, S.H., Periera, R., Simpson, E.C., and Slater, M. (2004). Measurement of protein-like fluorescence in river and waste water using a handheld spectrophotometer. *Water Res.*, **38**, 2934–2938.

Barbini, R., Colao, F., Fantoni, R., Fiorani, L., and Palucci, A. (2001). Remote sensing of the Southern Ocean: Techniques and results. *J. Optoel. Adv. Mat.*, **3**, 817–830.

Belzile, C., Roesler, C.S., Christensen, J.P., Shakhova, N., and Semiletov, I. (2006). Fluorescence measured using the WETStar DOM fluorometer as a proxy for dissolved matter absorption. *Est. Coast. Shelf Sci.*, **67**, 441–449.

Blough, N.V., Zafiriou, O.C., and Bonilla, J. (1993). Optical absorption spectra of waters from the Orinoco River outflow: Terrestrial input of colored organic matter to the Caribbean. *J. Geophys. Res.*, **98**(C2), 2271–2278.

Breves, W. and Reuter, R. (2001). Bio-optical properties of gelbstoff in the Arabian Sea at the onset of the southwest monsoon. *Earth Planet. Sci.*, preprint.

Cabaniss, S.E. and Shuman, M.S. (1987). Synchronous fluorescence spectra of natural waters: Tracing sources of dissolved organic matter. *Mar. Chem.*, **21**, 37–50.

Cetinic, I., Toro-Farmer, G., Ragan, M., Oberg, C., and Jones, B.H. (2009). Calibration procedure for Slocum glider deployed optical instruments. *Opt. Express*, **17**, 15420–15430.

Chen R.F. (1999). In situ fluorescence measurements in coastal waters. *Org. Geochem.*, 397–409.

Chen, R.F. and Bada, J.L. (1990). A laser-based fluorometry system for investigations of seawater and porewater fluorescence. *Mar. Chem.*, **31**, 219–230.

Chen, R.F. and Bada, J.L. (1992). The fluorescence of dissolved organic matter in seawater. *Mar. Chem.*, **37**, 191–221.

Chen, R.F. and Gardner, G.B. (2004). High-resolution measurements of chromophoric dissolved organic matter in the Mississippi and Atchafalaya plume regions. *Mar. Chem.*, **89**, 103–125.

Chen R.F., Gardner, G.B., Zhang, Y., Vlahos, P., Wang, X., and Rudnick, S.M. (1999). Chromophoric dissolved organic matter (CDOM) in four US estuaries. *EOS*, **80**, 92.

Chen, R.F., Gardner, G.B., Huang, W., and Peri, F. (2007). Chromophoric dissolved organic matter (CDOM) dynamics in the Hudson River Plume. ASLO, Santa Fe, February 4–9, 2007.

Christian, G.D., Callis, J.B., and Davidson, E.R. (1981). Array detectors and excitation-emission matrices in multicomponent analysis. In E.L. Wehry (Ed.), *Modern Fluorescence Spectroscopy*, Vol. 4, pp. 111–165. New York: Plenum Press.

Coble, P.G. (1996). Characterization of marine and terrestrial DOM in seawater using excitation-emission matrix spectroscopy. *Mar. Chem.*, **51**, 325–346.

Coble, P.G. and Gagosian, R.B. (1991). The nature and distribution of fluorescent dissolved organic matter in the Black Sea and the Cariaco Trench. In E. Izdar and J.W. Murray (Eds.), *Black Sea Oceanography*, pp. 361–378. Dordrecht, the Netherlands: Kluwer Academic.

Coble, P.G., Gagosian, R.B., Codispoti, L.A., Friederich, G.E., and Christernsen, J.P. (1991). Dissolved fluorescence in the Black Sea. *Deep-Sea Res. Pt. I*, **38**, S985–S1001.

Coble, P.G., Del Castillo, C.E., and Avril, B. (1998). Distribution and optical properties of CDOM in the Arabian Sea during the 1995 Southwest Monsoon. *Deep-Sea Res. Pt. II*, **45**, 2195–2223.

Conmy R.N., Coble, P.G., and Del Castillo, C.E. (2004). Performance and calibration of the WetLabs SAFire in-situ fluorometer. *Cont. Shelf Res.*, **24**(3), 431–442.

Cullen, J.J. and Davis, R.F. (2003). The blank can make a big difference in oceanographic measurements. *Limnol. Oceanogr. Bull.*, **12** (2), 29–34.

Davis, R.F., Stabeno, P.J., and Cullen, J.J. (2000). Use of optical measurements from moorings to detect coccolithophore blooms in the Bering Sea. In *Proceedings of Ocean Optics XV Conference,* Monaco. CD-ROM.

Delauney, L., Compère, C., and Lehaitre, M. (2101). Biofouling protection for marine environmental sensors. *Ocean Sci.*, **6**, 503–511.

Del Castillo C.E., Coble, P.G., Morel, J.M., Lopez, J.M,. and Corredor, J.E. (1999). Analysis of the optical properties of the Orinoco River Plume by absorption and fluorescence spectroscopy. *Mar. Chem.*, **66**, 35–51.

Del Castillo C.E., Coble, P.G., Conmy, R.N., Muller-Karger, F.E., Vanderbloomen, L., and Vargo, G.A. (2001). Multispectral in-situ measurements of organic matter and chlorophyll fluorescence in seawater: Documenting the intrusion of the Mississippi River plume in the West Florida Shelf. *Limnol. Oceanogr.*, **46**(7), 1836–1843.

Desiderio, R.A., Moore, C., Lantz, C., and Cowles, T.J. (1997). Multiple excitation fluorometer for in situ oceanographic applications. *Appl. Optics*, **36**, 1289–1296.

Downing, B.D., Boss, E., Bergamaschi, B.A., Fleck, J.A., Lionberger, M.A., Ganju, N.K., Schoellhamer, H., and Fujii, R. (2009). Quantifying fluxes and characterizing compositional changes of dissolved organic matter in aquatic systems in situ using combined acoustic and optical measurements. *Limnol. Oceanogr. Methods*, **7**, 119–131.

Drozdowska, V. (2007). The LIDAR investigation of the upper water layer fluorescence spectra of the Baltic Sea. *Eur. Phys. J. Special Topics*, **144**, 141–145.

Ferrari, G.M. and Tassan, S. (1991). On the accuracy of determining light absorption by "yellow substance" through measurements of induced fluorescence. *Limnol. Oceanogr.*, **36**, 777–786.

Gardner, G.B., Chen, R.F., and Berry, A. (2005). High-resolution measurements of chromophoric dissolved organic matter (CDOM) in the Neponset River Estuary, Boston Harbor, MA. *Mar. Chem.*, **96**, 137–154.

Gee, Y., H.R. Gordon, and Voss, K.J. (1993). Simulation of inelastic scattering contributions to the irradiance field in the ocean: Variation in Fraunhofer line depths. *Appl. Optics*, **32**, 4028–4036.

Geiskes, W.W.C., Kraay, G.W., and Tijssen, S.B. (1978). Chlorophylls and their degradation products in the deep pigment maximum layer of the tropical north Atlantic. *Nether. J. Sea Res.*, **12**(2), 195–204.

Green, S.A. and Blough, N.V. (1994). Optical absorption and fluorescence properties of chromophoric dissolved organic matter in natural waters. *Limnol. Oceanogr.*, **39**, 1903–916.

Haltrin, V.I., Kattawar, G.W., and Weidemann, A.D. (1991). Modeling of elastic and inelastic scattering effects in oceanic optics. In S. G. Ackleson and R. Frouin (Eds.), *Proceedings of Ocean Optics XIII Conference*. SPIE 2963, 597–602.

Hawes S.K., Carder, K.L., and Harvey, G.R. (1992). Quantum fluorescence efficiencies of fulvic and humic acids: Effects on ocean color and fluorometric detection. In *Proceedings of Ocean Optics XI Conference*. SPIE 1750, 212–223

Herman, A.W. and Denman, K.L. (1976). Rapid underway profiling of chlorophyll with an in situ fluorometer mounted on a batfish vehicle. *Deep-Sea Res.*, **24**, 385–397.

Heuermann, R., Loquay, K.D., and Reuter, R. (1995). A multi-wavelength in situ fluorometer for hydrographic measurements. *Adv. Remote Sens.*, **3**, 71–77.

Hoge, F.E. (2005). Oceanic inherent optical properties: Proposed single laser lidar and retrieval theory. *Appl. Optics*, **44**, 7483–7486.

Hoge, F.E. and Swift, R.N. (1981). Airborne simultaneous spectroscopic detection of laser-induced water Raman backscatter and fluorescence from chlorophyll a and other naturally occurring pigments. *Appl. Optics*, **20**, 3197–3205.

Hoge, F.E. and Swift, R.N. (1983). Airborne detection of oceanic turbidity cell structure using depth-resolved laser-induced water Raman backscatter. *Appl. Optics*, **22**, 3778–3786.

Hoge, F.E., Vodacek, A., and Blough, N.V. (1993). Inherent optical properties of the ocean: Retrieval of the absorption coefficient of chromophoric dissolved organic matter from fluorescence measurements. *Limnol. Oceanogr.*, **38**, 1394–1402

Hoge, F. E., Vodacek, A., Swift, R.R., Yungel, J.K., and Blough, N. (1995). Inherent optical properties of the ocean: Retrieval of the absorption coefficient of chromophoric dissolved organic matter from airborne laser spectral fluorescence measurements. *App. Optics*, **34**, 7032–7038.

Hoge, F.E., Wright, C.W., Swift, R.N., and Yungel, J.K. (1998). Airborne laser-induced oceanic chlorophyll fluorescence: Solar-induced quenching corrections by use of concurrent downwelling irradiance measurements. *Appl. Optics*, **37**, 3222–3226.

Hoge, F.E., Lyon, E., Wright, P.E., Wayne, C., Swift, R.N., and Yungel, J.K. (2005). Chlorophyll biomass in the global oceans: Airborne lidar retrieval using fluorescence of both chlorophyll and chromophoric dissolved organic matter. *Appl. Optics*, **44**, 2857–2862.

Hu, C. and Voss, K.J. (1998). Measurement of solar-stimulated fluorescence in natural waters. *Limnol. Oceanogr.*, **43**, 1198–1206.

Hudson, N., Baker, A., and Reynolds, D. (2007). Fluorescence analysis of dissolved organic matter in natural, waste and polluted waters. *River Res. Appl.*, **23**, 631–649.

Kalle, K. (1949). Fluoreszenz und Gelbstoff im Bottnischen und Finnischen Meerbusen. *Dtsch. Hydrogr. Z.*, **2**, 117–124.

Klinkhammer, G.P. (1994). Fiber optic spectrometers for in-situ measurements in the oceans: The ZAPS probe. *Mar. Chem.*, **47**, 13–20.

Lakowicz, J.R. (2006). *Principles of Fluorescence Spectroscopy*, 3rd ed. New York: Springer Science+Business Media, 954 pp.

Langebrake, L.C., Lembke, C.E., Weisberg, R.H., Byrne, R.H., Russell, D.R., Tilbury, G., and Carr, R. (2002). Design and initial results of a bottom stationing ocean profiler. *IEEE*, **1**, 98–103.

Lee Z., Carder, K.L., Hawes, S.K., Steward, R.G., Peacock, T.G., and Davis, C.O. (1994). Model for the interpretation of hyperspectral remote-sensing reflectance. *Appl. Optics*, **33**(24), 5721–5732.

Lieberman, S.H., Inman, S.M., and Theriault, G.A. (1992). Laser-induced fluorescence over optical fibers for real-time in situ measurement of petroleum hydrocarbons in seawater. In *Proceedings of Oceans '91*, pp. 507–514. Oceanic Engineering Society of IEEE, 91CH 3063–5.

Lochmuller, C.H. and Saavedra, S.S. (1986). Conformational changes in a soil fulvic acid measured by time-dependent fluorescence depolarization. *Anal. Chem.*, **58**, 1978–1981.

Lorenzen, C.J. (1966). A method for the continuous measurement of in vivo chlorophyll concentration. *Deep-Sea Res.*, **1(2)**, 223–227.v

Macleod, H.A. (2001). *Thin-film optical filters*, 3rd ed. London: Institute of Physics Publishing, 668 pp.

Manov, D.V., Chang, G.C., and Dickey, T.D. (2003). Methods for reducing biofouling of moored optical sensors. *J. Atmos. Ocean. Technol.*, **21**, 958.

Mazel, C.H. (1997) Coral fluorescence characteristics: Excitation–emission spectra, fluorescence efficiencies, and contribution to apparent reflectance. In *Proceedings of Ocean Optics XIII Conference*, SPIE **2963**, 240–245.

McLean, S., Schofield, B., Zibordi, G., Lewis, M., Hooker, S., and Weidemann, A. (1997). Field evaluation of anti-biofouling compounds on optical instrumentation. *Proc. SPIE*, **2963**, 708–713.

Moore, C.M. (1994). In situ biochemical, oceanic optical meters: Spectral absorption, attenuation, fluorescence meters – a new window of opportunity for ocean scientists. *Sea Technol.*, **35**, 10–16.

Peacock, T.G., Carder, K.L., Davis, C.O., and Steward, R.G. (1990). Effects of fluorescence and water Raman scattering on models of remote-sensing reflectance. In R.W. Spinrad (Ed.), *Ocean Optics X, Proc. Soc. Photo-Opt. Instrum. Eng.* 1302, 303–319.

Robinson, A.R. and Glenn, S.M. (1999). Adaptive sampling for ocean forecasting. *Nav. Res. Rev.*, **51**, 26–38.

Rudnick, S.M., Chen, R.F., and Gardner, G.B. (1998). In situ, time-resolved fluorescence measurements in Boston Harbor. In *Ocean Sciences '98*, San Diego, February 1998.

Saraceno, J.F., Pellerin, B.A., Downing, B.D., Boss, E., Bachand, P.A.M., and Bergamaschi, B.A. (2009). High-frequency in situ optical measurements during a storm event: Assessing relationships between dissolved organic matter, sediment concentrations, and hydrologic processes. *J. Geophys. Res.*, **114**, doi:10.1029/2009JG000989.

Sivaprakasam, V., Shannon, R.F., Jr., Luo, C., Coble, P.G., Boehme, J.R., and Killinger, D.K. (2003). Development and initial calibration of a portable laser-induced fluorescence system used for in situ measurements of trace plastics and dissolved organic compounds in seawater and the Gulf of Mexico. *Appl. Optics*, **42**, 6747–6756.

Spitzer, D. and Dirks, R.W.J. (1985). Contamination of the reflectance of natural waters by solar-induced fluorescence of dissolved organic matter. *Appl. Optics*, **24**(4), 444–445.

Stoertz, G.E., Hemphill, W.R., and Markle, D.A. (1969). Airborne fluorometer applicable to marine and estuarine studies. *Mar. Technol. Sci. J.*, **3**, 11–26.

Tedetti, M., Guigue, C., and Goutx, M. (2010). Utilization of a submersible UV fluorometer for monitoring anthropogenic inputs in the Mediterranean coastal water. *Mar. Poll. Bull.*, **60**, 350–362.

Twardowski, M.S., Sullivan, J.M., Donaghay, P.L., and Zaneveld, J.R.V. (1999). Microscale quantification of the absorption by dissolved and particulate material in coastal waters with an ac-9. *J. Atmos. Ocean. Technol.*, **16**(12):691–707.

Twardowski, M.S., Lewis, M.R., Barnard, A.H., and Zaneveld, J.R.V. (2005). In-water instrumentation and platforms for ocean color remote sensing applications. In: R.L. Miller, C.E. Del Castillo, and B.A. McKee (Eds.), *Remote Sensing of Coastal Aquatic Environments*, pp. 69–93. Dordrecht, the Netherlands: Springer Science+Business Media.

Vodacek, A., Green, S.A., and Blough, N.V. (1994). An experimental model of the solar-induced fluorescence of chromophoric dissolved organic matter. *Limnol. Oceanogr.*, **39**, 1–11.

Yuan, J., Dagg, M., and Del Castillo, C.E. (2005). In pixel variations of chl a fluorescence in northern Gulf of Mexico and their implications for calibrating remotely sensed chl a and other products. *Cont. Shelf Res.*, **25**(15), 1894–1904

Zhou, J., Tonizzo, A., Ioannou, I., Hlaing, S., Gilerson, A., Gross, B., Moshary, F., and Ahmed, S. (2009). Evaluation of solar estimulated CDOM fluorescence and its impact on the closure of remote sensing reflectance. In R.J. Frouin (Ed.), *Ocean Remote Sensing Methods*. SPIE 7459, doi:10.1117/12.825374.

Part III

Environmental Effects

7

Physicochemical Effects on Dissolved Organic Matter Fluorescence in Natural Waters

CHRISTOPHER L. OSBURN, ROSSANA DEL VECCHIO, AND THOMAS J. BOYD

7.1 Introduction

Dissolved organic matter (DOM) fluorescence properties are strongly affected by physicochemical properties of natural waters – increasing, decreasing, or altering DOM fluorescence characteristics through a variety of mechanisms. Natural water physicochemical properties are often interdependent at many levels, and separating their impacts on DOM fluorescence is difficult. Natural water ionic strength varies widely due to changing concentration of dissolved solids from freshwater (<1000 mg L^{-1}) to seawater (avg. 35,000 mg L^{-1}) to saline lakes and brines (>50,000 mg L^{-1}), and much of these dissolved solids are metal salts which contribute to solution alkalinity. In estuarine environments, freshwater DOM in rivers mixes with seawater, dramatically changing the physicochemical environment in which freshwater DOM may be flocculated (Fox, 1983; Sholkovitz et al., 1976; Spencer et al., 2007a). However, such flocculation is not always observed and likely varies with the chemical composition, perhaps DOM humic content or dissolved metals concentration (Mantoura and Woodward, 1983; Spencer et al., 2007a) as much as the overall chemical milieu in which these components are mixed. Layered "on top" of these interactive chemistries are physical effects such as mixing and sunlight. Similarly, wastewater treatment streams, rich in DOM, can have very high ionic strengths. In all of these diverse environments, pH and the concentration of metals can vary widely, and metal–ligand complexation reactions can strongly affect DOM fluorescence. To study these effects experimentally, researchers often perturb one variable (e.g., lower or raise pH) and observe the resulting effects on DOM fluorescence.

Many reactions between fluorophores and their environment will reduce DOM fluorescence and are termed *quenching reactions*. Fluorescence is related primarily to various chemical components constituting DOM, which reflects DOM sources and molecular size and mass. Colloidal organic matter (COM), fractionated by a variety of ultrafiltration and flow fractionation methods, shows distinct patterns in fluorescence, but these patterns are not consistent among freshwater and seawater sources. In freshwater environments fluorescence is often found in the highest molecular size fractions, whereas in seawater environments, fluorescence is often found in lower molecular size fractions.

The wide variation in physical and chemical properties of natural waters imparts wide variation in the attendant effects on DOM fluorescence. Freezing and thawing may extensively modify DOM, especially if it becomes polymerized, but cyclical dehydration and rehydration appears to be more critical, making aridity a considerable influence on soil DOM fluorescence. Changes in DOM fluorescence can occur from changes to ionization potential in phenolic and carboxylic fluorophores, as well as intramolecular rearrangements. Therefore, solution pH exerts a major control on DOM fluorescence. In addition, metal–ligand complexation (primarily by salicylic or phenolic moieties within DOM) generally causes fluorescence quenching and is strongly influenced by pH. However, some fluorescence intensity increase can occur from Mg(II) and Al(III) metal ions when they are complexed with DOM. Changes to ionic strength appear to have a lesser effect on DOM fluorescence than pH and metal–ligand interactions. A major synergistic effect of increased ionic strength on DOM fluorescence could result from suppression in functional group ionization, coincident with the metal–ligand quenching previously described. Moreover, intramolecular rearrangements and ion suppression are likely mechanisms leading to fluorescence quenching in high ionic strength solutions. Photodegradation, or photobleaching, of DOM fluorescence occurs after exposure to sunlight and generally results in diminished emission intensity and blue shifting of fluorescence to shorter wavelengths, and to some degree is a function of molecular size, pH, and ionic strength.

This chapter explores the physical and chemical effects on DOM fluorescence in natural waters, with a focus on the changes in fluorescence induced by changes in pH, ionic strength, metal–DOM interactions, temperature, and particle size (colloidal versus non-colloidal). First, we discuss the topic of quenching as a mechanism with which to examine physicochemical effects on fluorescence. This is followed by an examination of the effect of molecular weight and size of fluorophores, prior to discussing, in turn, the effects of temperature, pH, metals, ionic strength, and particles. Finally, the interaction of these environmental effects and photobleaching is discussed. The aim is to provide the reader with an idea of how the physicochemical environment changes the fluorescent properties of DOM through either fluorescence quenching or enhancement. We have attempted to draw examples from the freshwater, marine, and soil DOM literature to connect these topics. A key literature review on DOM fluorescence and environmental effects studied by emission spectroscopy and synchronous fluorescence (SF) spectroscopy will be introduced and supplemented by current literature utilizing excitation–emission matrix (EEM) fluorescence analyses where possible. For this discussion, we use the convention of ex_λ/em_λ to denote excitation wavelength and emission wavelength, λ, respectively. For SF, we use SF_λ to denote excitation wavelength and indicate the offset value ($\Delta\lambda$) where appropriate. For the EEM discussion, we follow the typical B, T (protein); C and A (terrestrial humic); M (marine humic); and N (phytoplankton-derived) peak assignments developed elsewhere (Coble, 1996; Stedmon et al., 2003).

7.2 The Quenching of DOM Fluorescence

A consideration of environmental effects on fluorescence is centered on how the physicochemical environment either increases or decreases DOM fluorescence through a variety of mechanisms. DOM fluorescence is a photophysical process under which excited species expend absorbed light energy and return to their ground state; quantum yields are only ca. 1%. Reactions between fluorophores and their environment that inhibit a molecule's ability to release energy via photon emission at longer wavelength (i.e., at lower energy) will reduce DOM fluorescence. In the environmental effects considered here, the dominant effect will be the reduction in the intensity of fluorescence, termed here *quenching*. A number of processes can produce this effect at the molecular level.

Collisional quenching involves fluorophore (A) deactivation in the excited state (A*) on contact with another molecule, termed the quencher [Q] (Lakowicz 2006). The fluorescent energy that is quenched is released as heat and can be represented as:

$$A^* + Q \xrightarrow{k_q} A + Q + \text{heat}. \tag{7.1}$$

This is a diffusive process and so proximity of the fluorophore and the quencher are important, but no chemical reaction results. Probably the most effective collisional quencher in natural waters is molecular oxygen, which quenches nearly all fluorophores (Lakowicz, 2006). Molecular oxygen prevalence in surface waters means that DOM fluorescence will be modulated by its presence, and it should be expected that anaerobic or anoxic systems (e.g., hypolimnia of lakes and sediment pore waters) could exhibit higher fluorescence in situ than if a sample were taken and allowed to equilibrate with air (see Chapter 4 in this volume). Oxygen and paramagnetic metal ions (e.g. Cu^{2+}, Pb^{2+}, Mn^{2+}) quench DOM fluorescence by enhancing intersystem crossing for the excited singlet state to a triplet state and then deactivate the fluorophore to the ground state with loss of heat energy rather than photon emission. The triplet state itself is readily quenched by oxygen and other species in solution. In addition to intersystem crossing, several mechanisms, including charge transfer and electron exchange, may also contribute to overall fluorescence quenching, and discerning individual processes can be difficult (Lakowicz, 2006).

Quenching may also be caused by inorganic and organic compounds. Inorganic species in natural waters such as halides, with chloride and bromide being very effective collisional quenchers, may exert some control on DOM fluorescence as ionic strength of naturals waters increases. Like paramagnetic oxygen, halide (e.g., the heavy atom effect; Senesi, 1990) or amide quenching is most likely due to the enhancement of intersystem crossing from the excited singlet to the excited triplet state, followed by rapid decay to the ground state (Lakowicz, 2006). Organic compounds may also serve as collisional quenchers of DOM fluorescence. Aromatic and aliphatic amines, chlorinated hydrocarbons, and some olefins are known quenchers, and acrylamide and methyl viologen (a herbicide) have also been shown quenching effects (e.g., Milne and Zika, 1989).

Like collisional quenching, *resonant energy transfer* is an excited-state interaction in which the two participating molecules exist as a donor molecule that transfers the excitation energy (via electronic coupling) to an acceptor molecule rather than emitting photons. The result is an excited acceptor and a deactivated donor and altered absorption or fluorescence can detect the interaction, which is an important consideration for assessing DOM optical properties in natural waters (Del Vecchio and Blough, 2004).

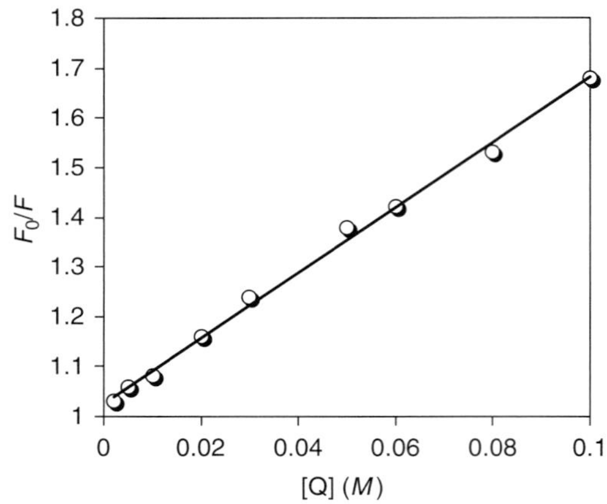

Figure 7.1. An idealized example of fluorescence quenching presented in a Stern–Volmer plot. F_0 is the initial fluorescence and F is the fluorescence after the quencher, Q, has been added.

Static quenching occurs when the fluorophore forms a stable, nonfluorescent complex with the quencher in the ground state. The newly formed complex may have a different absorption spectrum from the fluorophore alone, which can be used to detect the complex. In some cases, self-quenching occurs when the fluorophore quenches itself (in most cases, concentration dependent). Metal cations are very effective static quenchers because they form coordination complexes with the fluorescent ligands with the DOM pool (see Section 7.6). Similarly, protons can be effective quenchers (Senesi, 1990); both transition metals and protons are effective electron scavengers, likely underlying their quenching effect by forming coordination complexes.

DOM fluorescence quenching is traditionally observed using a Stern–Volmer plot, in which the fractional change in loss of fluorescence (ratio of initial fluorescence, F_0 to fluorescence after addition of the quencher, F) is plotted against the concentration of the quencher, Q (Figure 7.1). The slope of the line describing the linear fit is the quenching constant (K_{SV}) from the Stern–Volmer equation:

$$F_0 / F = 1 + K_{SV}[Q] = 1 + k_q \tau_0[Q] = \tau_0 / \tau \qquad (7.2)$$

where τ_0 and τ are the fluorescent lifetimes in the absence and presence of the quencher, Q, respectively, and k_q is the rate constant for the quenching process. Although they may have an impact on quenching within short time domains, fluorescence lifetimes are not discussed in detail in this chapter.

In studying humic substance electrostatic quenching properties, Green et al. (1992) found that both fulvic (FA) and humic (HA) acids' anionic components were the fluorophores that may be quenched. Presumably, this was due to the formation of stable complexes between metals and ligands, such as deprotonated carboxyl groups. They found higher K_{SV} for cationic quenchers than for neutral quenchers due to Coulombic cation to anion attraction in the FA and HA. Thus, as ionization of functional groups such as hydroxyls, phenolic hydroxyls, and, probably most importantly, carboxyl groups proceeds, the effect of static quenching with cations likely increases.

Work with amino acids like tryptophan (Trp) has shown that a quenching effect is determined by accessibility of the fluorophore to quenchers in solution. Trp residues on the surface of macromolecules (or by extension bound to surfaces of DOM or clays) will be exposed to the aqueous environment containing the quenchers and thus will be quenched to a much greater degree than Trp contained within the unexposed interior of the macromolecular structure (Lakowicz, 2006). For natural DOM, it is most likely the fluorophores on macromolecular surfaces that are most effectively quenched in aqueous solutions. This phenomenon has been used to determine molecular conformation by fluorescence (Eftink, 1991).

7.3 Effects of Molecular Weight and Fluorophore Size

DOM is operationally defined as organic molecules less than 0.2–0.7 μm in size (equivalently, 200–700 nm). Most fluorescence analyses have thus been measured on DOM because this size separation is easily achieved via filtration. Colloidal organic matter (COM) is generally defined as particles from 1 to 1000 nm in size (from measurements of hydrodynamic radius) and so DOM less than 200 nm and COM may operationally overlap. In relation to the ultrafiltration methods of separating COM, a 1 nm pore size approximates 1000 Daltons (Da, measurement of molecular weight) cutoff (Benner, 2002). Aquatic colloids have been isolated using cross-flow (or tangential flow) ultrafiltration, producing COM of sizes greater than 1000 Da (and roughly equivalent to 1 nm spherical diameter; Floge and Wells, 2007), though other size separations are achieved using different ultrafilters (Mopper et al., 1996; Guo and Santchi, 1997; Wells, 2002), flow-field-flow fractionation (FlFFF; e.g., Zanardi-Lamardo et al., 2002; Boehme and Wells, 2006), and split-flow thin cell fractionation (SPLITT; e.g., Lead et al., 2006). FlFFF uses a flow field to separate masses via diffusion (Zanardi-Lamardo et al., 2002), whereas SPLITT achieves separation via gravitational force. Both FlFFF and SPLITT are advantageous over other separation methods as they do not require ultrafiltration and can perhaps avoid artefacts caused by the aggregation of colloids, a possible interference in ultrafiltration methods (Lead et al., 2006).

Colloids can be a significant fraction (10–40%) of marine (Benner et al., 1992; Wells, 2002) and (>65%) of freshwater DOM (Liu et al., 2007). However, COM represents only a fraction of the DOM pool, and making interpretations on total DOM biogeochemical reactivity using COM fluorescence therefore requires some caution. Mopper et al. (1996) conducted a comprehensive survey of COM isolated by ultrafiltration (1 kDa filters) from marine samples and suggested that COM (>1 kDa) accounted for roughly 50% of the ambient DOM based on fluorescence quantum yields (fluorescence divided by absorption) data. However, coastal mixing of COM does not appear to be uniform across different regions, possibly due to varying COM sources. For example, Boyd and Osburn (2004) examined changes in the fluorescent properties of freshwater COM (>1 kDa) isolated from two estuaries, added to low-fluorescing ultrafiltration permeates to recreate a salinity gradient, and then exposed to estuarine and marine bacteria. They found different responses in freshwater COM fluorescence as a function of salinity from San Francisco Bay, which exhibited no predictable changes in fluorescence intensity when compared to Chesapeake Bay, in which humic fluorescence decreased as salinity increased. This result appeared to be related to the different riverine sources of the COM used in these experiments (i.e., Sacramento vs. Susquehanna River, Boyd and Osburn, 2004).

COM fluorescence properties, as investigated by excitation–emission matrix (EEM) spectra, show differences for natural water DOM based on source. As an example for discussion, Figure 7.2 compares separately surface water DOM from a freshwater riverine site in the Chesapeake Bay (Figures 7.2a–c) and from a coastal marine site in the Middle Atlantic Bight (Figures 7.2d–f) with their respective high molecular weight (HMW) retentate (>1 kDa) and low molecular weight (LMW) permeate DOM (<1 kDa) fractions (Osburn, unpublished data). To produce these separations, a 1000 Da molecular weight cutoff (MWCO) tangential flow (or cross-flow) ultrafiltration (TFF or CFUF) system was used after initial 0.2 μm filtration (Boyd and Osburn, 2004). To describe similarities and differences in EEM results here and elsewhere in this chapter, we use the convention of peak assignments developed by Coble (1998; Table 7.1) and we focus on terrestrial humic, terrestrial fulvic, marine humic, and protein peaks (A, C, M, and T, respectively).

For both water samples, and for each fluorescence peak, the concentrated retentate had the highest fluorescence (Figure 7.2). However, when normalized for the concentration factor, HMW fluorescence actually was less than the starting DOM fluorescence yet remained higher than for the LMW permeate fluorescence (Table 7.2). The differences among the three fractions were greater for the river sample than for the marine sample. Further, EEM fluorescence varied for different peaks, and recovery was greater for the marine sample than for the river sample. Recoveries are comparable with others reported (Wells, 2002; Boyd and Osburn, 2004; Liu et al., 2007). Despite the lower recovery for the river DOM (which could be due to sorption onto the filter' Mopper et al., 1996; Liu et al., 2007), the EEMs are similar in appearance among the surface DOM, retentate, and permeate fractions. In contrast, the marine surface DOM sample showed distinct A and T peaks (Figure 7.2d). A peak shoulder is noticeable at the intermediate (Int) region between the M and C peak regions, as described by Boyd et al. (2010a). After ultrafiltration, the Int peak is notably lacking in the marine permeate (Figure 7.2e) while the T peak is less noticeable in the marine retentate, potentially masked by the A peak fluorescence (Figure 7.2f). This result suggests that protein (T peak) fluorescence occurs primarily in the LMW fraction of marine DOM whereas the Int peak, intermediate between M and C, is primarily in the HMW fraction consistent with work by Liu et al. (2007). Further, the Int peak is prominent in the river CDOM, permeate, and retentate fractions (Figures 7.2a–c).

Because fluorescence intensity changed on molecular weight separation, peak ratios could be employed to understand how DOM sources are partitioned between size classes, based on their fluorescence (Table 7.2; Coble, 1998; Parlanti et al., 2000). As defined in

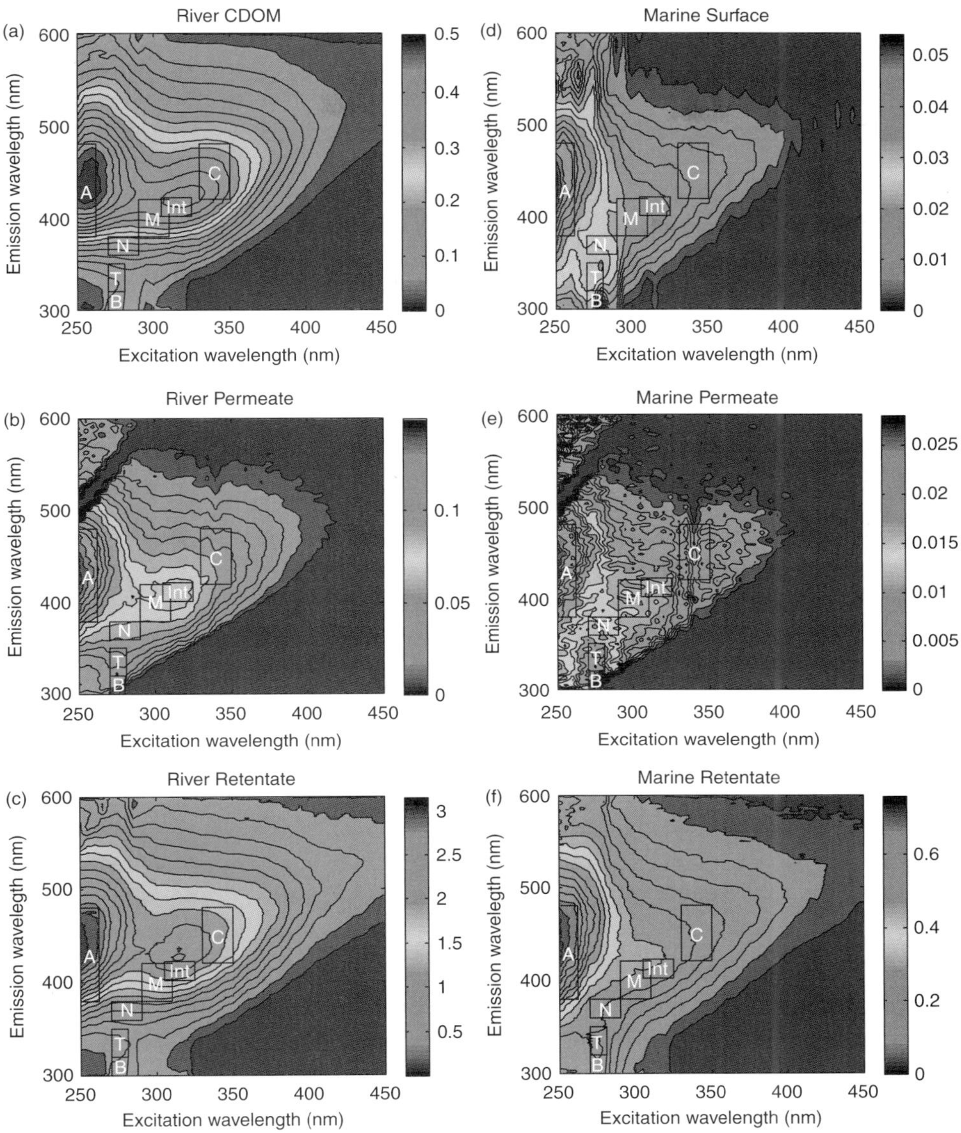

Figure 7.2. The variability in EEM fluorescence for whole DOM collected from surface waters compared to ultrafiltered LMW permeate (<1 kDa MWCO) and HMW retentate (>1 kDa MWCO) fractions. (a–c) Susquehanna River water collected at the head of the Chesapeake Bay. (d–e) Coastal seawater collected from the Middle Atlantic Bight. The boxes and letters on each EEM contour plot refer to the fluorescent peak region designations from Table 7.1.

Table 7.1. *EEM peak locations (excitation/emission wavelength ranges)*

Peak	Ex/Em (nm)	Description
A	260/380–460	Terrestrial humic-like
C	320–360/420–460	Terrestrial fulvic-like
M	290–310/370–410	Marine humic-like; microbial
N	280/370	Phytoplankton-derived; labile
T	270/340	Tryptophan; protein-like
B	270/305	Tyrosine; protein-like

Source: Reproduced from Coble (1998).

Table 7.2. *The differences in DOM and COM fluorescence (in calibrated Raman units) compared between river (fresh) and marine waters*

Peak or peak ratio	Surface	Permeate	Retentate	% Recovery
	River DOM			
A	0.517	0.124	0.313	76
C	0.365	0.059	0.218	68
M	0.304	0.077	0.166	72
T	0.097	0.048	0.058	98
M/C	0.83	1.31	0.76	n/a
T/A	0.19	0.39	0.19	n/a
	Marine DOM			
A	0.042	0.020	0.022	99
C	0.014	0.003	0.009	86
M	0.021	0.013	0.009	103
T	0.031	0.018	0.009	85
M/C	1.51	4.33	1.01	n/a
T/A	0.73	0.90	0.40	n/a

The Retentate fluorescence was normalized by its concentration factor for comparison to the Surface and Permeate fluorescence. Ratios of peak fluorescence are dimensionless. % Recovery is the volume-weighted mass balance of permeate fluorescence + retentate fluorescence (not normalized) divided by the surface DOM fluorescence.

Table 7.1, the M and T peak regions correspond to more freshly produced (autochthonous) DOM sources (microbial and protein, respectively), while the A and C peak regions correspond to more recalcitrant, terrestrial humified DOM sources (Coble, 1998). Even if these discrete peaks were not observable in EEM contour plots, the ratio of fluorescence in the M peak region to the C peak region (M/C), for example, could indicate relative abundance of microbial to terrestrial DOM sources. Similarly, the ratio of T/A describes the protein fluorescence relative to recalcitrant terrestrial humic fluorescence. Thus both ratios describe the relative fluorescence intensity of microbial, proteinaceous DOM to terrestrial humic DOM. For the DOM fluorescence results shown in Figure 7.2, a similar pattern in these ratios is apparent even between two DOM sources (Table 7.2). In both the river and marine samples, M/C decreases in the retentate relative to the surface DOM. This would indicate that the >1 kDa fraction of DOM is representative of the bulk, and exhibits a largely terrestrial character. By contrast, permeate fractions had larger M/C and T/A ratios than the surface DOM, indicating a relatively greater proportion of autochthonous DOM to terrestrial DOM in the <1kDa fraction. Further, T/A ratios were much lower in the retentate than in either the permeate or surface fractions. These results indicate that more terrestrial, and less autochthonous, proteinaceous, DOM is found in the >1 kDa fraction in coastal waters.

Liu et al. (2007) examined fluorescence on freshwater DOM and on a Trp standard using a 1 kDa MWCO filter. Through successive concentration of the Trp standard, fluorescence intensity remained constant in the retentate and increased asymptotically in the permeate – both fractions reached nearly the same value. Although a slight decrease in the retentate was attributed to sorption effects, the authors also showed that the ratio of permeate fluorescence intensity to retentate fluorescence intensity remained constant. They concluded that Trp-like fluorescence is primarily in the <1 kDa fraction in freshwater DOM.

Size fractionation of colloidal DOM has been improved by techniques such as field flow fractionation (FFF) and size-exclusion chromatography. These techniques have been used to collect fractions of DOM based on various size discriminations and then measure the fluorescence of each fraction. Employing FFF, colloids from aquatic environments have been separated into a continuum of molecular sizes (fractograms) and characterized by EEMs (Boehme and Wells, 2006; Lead et al., 2006; Battchelli et al., 2009; Huguet et al., 2010; Stolpe et al., 2010). The additional DOM size classes obtainable by this technique results in slightly different DOM fluorescence interpretations based on size, compared to static 1 kDa MWCO described previously, and provides additional insight into the relationship between DOM source and fluorescence properties.

Boehme and Wells (2006) separated colloids from marine DOM via FFF and their fractograms showed two major peaks indicating small (~1–5 kDa) and large (~15–150 kDa) size fractions that varied in abundance as well as in fluorescent properties, with the smaller sizes generally being seasonally variable, very labile, and probably related to phytoplankton blooms. Whereas smaller colloids exhibited a sharp peak (suggestive of molecules of rather narrowly defined size range), larger colloids exhibited broader and polydisperse peaks (indicative of a mixture of molecules in a continuum sizes).Their EEMs revealed a protein-like signature for the smallest colloids (~1–5 kDa fraction), whereas a humic-like fingerprint was observed for the larger colloids whose fluorescence emission was red shifted with increasing molecular size. The seaward edge of the Damariscotta River estuary where they conducted their study had phytoplankton blooms and illustrated an autochthonous-dominated system. Their 1–5 kDa fraction corresponded closely with chlorophyll measurements, leading to the conclusion that LMW (~1–5 kDa) material's fluorescence changed as a result of phytoplankton blooms, whereas HMW (~13–150 kDa) material's fluorescence remained unchanged during bloom conditions. The fractionation scheme offered greater size resolution than results shown in Figure 7.2, and elucidate an important size distinction between size classes and DOM sources. Boehme and Wells (2006) stated that T peak fluorescence dominated the "small, dynamic colloidal fraction," yet this fraction noticeably lacked marine or terrestrial humic and fulvic fluorescence signals. By contrast, and as observed in Figure 7.2e for the marine permeate (<1 kDa in size), both terrestrial and marine humic fluorescence was observed for a marine permeate sample that was collected from a river-dominated margin (the outflow of Chesapeake Bay) and not after a phytoplankton bloom.

In colloids, an increase in average molecular size may induce hydrophobic character, resulting in reduced quenching. This could occur if the hydrophobic character of the colloidal material inhibits the interaction of the quencher and fluorophore during the lifetime of its excited state (Lakowicz, 2006). Presumably, environmental effects that increase the hydrophobic nature of DOM could also reduce quenching effectiveness, limiting it to surface fluorophores. Lead et al. (2006) employed split-flow thin cell (SPLITT) fractionation to separate particles and colloids in lake water, an approach somewhat different than ultrafiltration and FFF, and found the fractions greater than 1 kDa had higher protein fluorescence than did humic fractions. They attribute this effect to fluorescence measurement of tryptophan (or tyrosine) residues on large proteins (or bound to humics or particles) rather than measurement of free protein. Moreover, they demonstrated that fulvic acid fluorescence (C peak) efficiency (normalized to absorption) was lower for their >1 nm (1 kDa) fraction than for their <1 nm fraction. This would imply some increased quenching of the C peak fluorescence in the higher molecular size class.

Similar results were obtained for FlFFF separations of DOM from riverine inputs to the coastal region of the Gulf of Mexico (Stolpe et al., 2010) and via ultrafiltration (ca. 12 kDa MWCO) before FlFFF separation for DOM from the humic River Thurso estuary (Batchelli et al., 2009). In fact, greater than 50% of the T peak fluorescence in the Atchafalaya River and Mississippi Bight occurred from DOM fractions greater than 20 nm in size (ca. 20 kDa: Stolpe et al., 2010). In the River Thurso estuary, the authors noted the appearance of T peak fluorescence in the retentate of their ultrafiltered DOM, but not in the permeate (Batchelli et al., 2009). These riverine and estuarine results are contradictory to the results of Boehme and Wells (2006), but the large amount of T peak fluorescence found in the higher molecular weight (and size) fractions in the freshwater studies could be attributable to effects of salinity on DOM fluorescence in estuarine systems (see below; Boyd and Osburn, 2004; Batchelli et al., 2009).

Boehme and Wells (2006) worked on DOM from a largely phytoplankton-dominated system, in which case terrestrial humic substances are much lower in abundance and consequently the protein fluorescence was most concentrated in the 1–5 kDa fractions. In contrast, systems that had substantial humic river influences (or were exclusively freshwater) exhibited protein fluorescence in the highest size fractions. While free amino acids such as Trp are smaller molecules, the dominant T peak fluorescence probably occurs from Trp residues on larger proteins (e.g., >1000 nm), or perhaps bound to other DOM (possibly humics). The so-called "visible humic-like" fluorescence at long wavelengths of excitation and emission (C peak) is due primarily to terrestrial (or at least freshwater) humic substances. Consequently, these humic substances tend to be smaller, being dominant in the 1–5 kDa (1–5 nm) fraction (e.g., Stolpe et al., 2010). Altogether these results indicate that colloidal fluorescence properties are strongly dominated by protein and various humic fluorophores, whose intensity reflects both the size of DOM and its source.

7.4 Effect of Temperature

Increasing a solution temperature increases the chance for molecules to interact and thus fluorophore collisional quenching is expected to be greater at higher temperatures. However, not all fluorophores are deactivated in the same manner. Thus thermal quenching of DOM fluorescence is a physical phenomenon that can be used to characterize DOM in natural waters (Baker, 2005; Seredynska-Sobecka et al., 2007) as much as it is a consideration for the environmental conditions DOM encounters (Spencer et al., 2007b).

Thermal quenching appears to increase the deactivation of the excited state by internal conversion (Senesi, 1990). Indeed this effect was observed from 10°C to 45°C by Baker (2005) on examining several riverine and wastewater DOM samples and humic standards. Interestingly, protein fluorescence appeared most susceptible to thermal quenching in several instances as compared to fulvic material fluorescence. As thermal quenching is sensitive to the fluorophores' exposure to the energy supplied by increasing temperature, these results imply that some species may be more easily perturbed than others and can provide information on DOM sources (Baker, 2005).

Freezing and thawing may extensively modify DOM, especially if it becomes polymerized. DOM concentration increases during the freezing process as ice crystallizes. In Arctic natural waters, as well as laboratory experiments, a temperature-driven size exclusion DOM fractionation produces a brine that is generally higher in molecular weight and highly fluorescent. Amon et al. (2003) found a subsurface fluorescence maximum in the Arctic halocline in the Nordic Seas that followed the East Greenland Current. Similarly, Gueguen et al. (2007) found a distinctive humic signal in DOM (C peak) in the Arctic halocline, fluorescence that was nearly twice that of the surface layer. Belzile et al. (2002) used synchronous fluorescence to study freeze up in polar lakes and rivers, finding that DOM exclusion factors were nearly twice those for inorganic solutes. Belzile et al. (2002) also found that freezing effectively size fractionates DOM owing to the inability of ice crystal structures to accommodate large compounds. This ultimately had the result of diminished humic acid fluorescence after separation into <30 kDa and <5 kDa fractions. Notably, some of these effects could be due to pH or metal ion concentrations changing with brine formation. Amon et al. (2003) was able to use DOM fluorescence to track brine formation in the Arctic Ocean during sea ice formation and determined this process to be a major mechanism for organic carbon transport (and transformation) in the Arctic. The large amount of terrigenous, highly fluorescent DOM delivered by high-latitude rivers to the Arctic Ocean is generally abundant in soil-derived humic substances high in molecular weight. Spencer et al. (2007b) examined the freeze–thaw cycles on river DOM fluorescence and found variable responses to the process. Both increases and decreases to peak positions (e.g., red or blue shifting) as well as increases and decreases in peak intensities were observed both for protein and for humic fluorescence peaks. No clear pattern emerged in terms of the more than 30 freshwater samples studied and, perhaps most importantly, no EEM fluorescent features of the DOM were able to predict responses to freeze–thaw cycles.

Similar to ice exclusion (leading to organic matter concentrating in the resultant solution), DOM drying (dehydration) can induce substantial changes to fluorescence, a process that becomes important in pore water and soil solutions in arid climates. Zsolnay et al. (1999) studied soil DOM and found that drying temperature or rate (air vs. oven) was important in determining desiccation effects to DOM fluorescence. Air drying produced a red shift indicating a slight enhancement of humic fluorescence, probably resulting from polymerization and condensation reactions. Oven drying (105°C), noted as artificial yet analogous to soil desiccation, resulted in a blue shift toward protein-type fluorescence. Zsolnay et al. (1999) attributed these observations to biomass lysis, but noted that changes to physical pore structure in soils may affect soil DOM leaching. Otero et al. (2007) made similar observations on pore water DOM fluorescence as modified by freezing sediments.

Freeze–thaw cycles that modify DOM fluorescence appear coupled to dehydration and rehydration cycles, though this effect has not been investigated widely. Hudson et al. (2009) offer some of the first controlled studies on DOM fluorescence due to cyclical dehydration and rehydration. After one cycle at neutral pH, the 13 freshwater samples from central England survey exhibited significant decreases in DOM fluorescence in the Trp fluorescence region (34% and 40% loss for freeze–thaw and dehydration–rehydration, respectively) that were substantially greater than humic fluorescence loss (7% and 18%). Comparing one to five cycles of these processes, the authors note that the two distinctive peaks found for Trp fluorescence (T1 at ex/em of 280/350 nm and T2 at 215–220/340 nm, respectively) behaved differently – with T2 apparently being most susceptible to dehydration. Similarly for the humic peaks, the C peak was apparently more labile than the A peak. Dehydration caused a greater decrease in DOM fluorescence than did freezing. This suggests some destruction of fluorophores but it is unclear whether collisional or static quenching is induced by dehydration processes.

7.5 Effect of pH

A change in solution pH is expected to cause variable and complex effects on DOM fluorescence. Due to hydroxyl and carboxyl group protonation at low pH and their ionization at high pH, DOM fluorophore electronic state is altered by variation in pH, resulting in an enhancement of fluorescence when pH increases (Figure 7.3a). Changes to the structure and bonding environments by lowering or raising pH or to extremes were observed to have an effect on DOM fluorescence which was most likely due to protonation and ionization charge effects (cf. Hosse and Wilkinson, 2001). Ghosh and Shnitzer (1980) observed an increase in averaged molecular weight and in molecular area when the pH of humic acids (HA) or fulvic acids (FA) was increased, an effect attributed to coiling or homolytic bonding. Such effects were observed when either neutral electrolyte (i.e., NaCl) or hydrogen ion concentration was high.

HA and FA extracts from soils, sediment, and natural waters have been studied by fluorescence under varying pH. Generally, fluorescence changes nonlinearly as a function of pH (Figure 7.3a; Smart et al., 1976; Laane, 1982; Willey and Atkinson, 1982; Willey, 1984). Laane (1982) and Senesi (1990) described both "acid" and "alkaline" regions of greatest impact (Figure 7.3a) and ascribed the increased fluorescence intensity to ionization of low pK_a moieties (e.g., carboxylic acids) in the acid region and high pK_a (e.g., phenols or amino groups) in the alkaline region. Senesi (1990) summarized that most FA fluorescence emission studies found either decreasing intensity with increasing pH or an emission peak between pH 5 and 7. Over pH ranges of 3–9, FA excitation intensity increased for aquatic, peat, and microbial FA, yet decreased for soil FA. Senesi (1990) postulated that these contrasting results occur from the wide variety of acidic functional group ionization constants and molecular rearrangements. Fluorophore deprotonation (phenolic hydroxyls), inter- and intramolecular hydrogen bond disruption, and proton quenching were suggested as mechanisms for excitation intensity increase and emission intensity reduction. Red shifts in emission wavelengths up to 10 nm were also noted. These effects support the supposition that acidic and basic functional group ionization has occurred with changing pH. Emission spectra from extracted HA and FA as well as DOM all respond similarly to pH changes.

In addition to emission and excitation scans, pH effects on DOM fluorescence has been extensively studied using synchronous-scan fluorescence (SF). At lower pH, soil fulvic acids exhibit a shoulder peak at 360 and more pronounced peak at 394 nm (18 nm synchronous-scan offset). As pH is increased, these peaks decrease and merge into a single relatively flat peak at 400 nm (pH 10). A peak at ~460 nm is reasonably pronounced at low pH (4.5) but decreases, flattens and blue shifts to ~440 nm at pH 10 (Senesi, 1990). Using excitation–emission matrices (EEMs) is another means used to assess humic and fulvic acid fluorescence shifts due to pH variation. Both soil humic and aquatic humic substances have been analyzed with this technique. Fluorescence intensity has been shown to increase as pH is increased for soil and aquatic fulvic and humic acids (Mobed et al., 1996). In longer wavelength regions (~450 nm), an emission peak was red shifted for soil and aquatic humic and fulvic acids as pH increased. Aquatic humic acids had a lower wavelength emission peak (~320 nm) which blue shifted as solution pH was increased. Two separate pH-dependent wavelength shift behaviors (short and long wavelength emission peaks) were attributed to two general processes: the two fluorescence maxima found in phenolic compounds and conformational changes that might expose different functional groups to the solvent.

Humic and fulvic acids (or their spectra within natural DOM) have been the primary focus of studies assessing optical property changes with respect to pH. Ma et al. (2010) found no significant changes in the optical properties of humic and fulvic material as pH was increased from 7 to 10. Several recent studies have assessed changes in amino acid

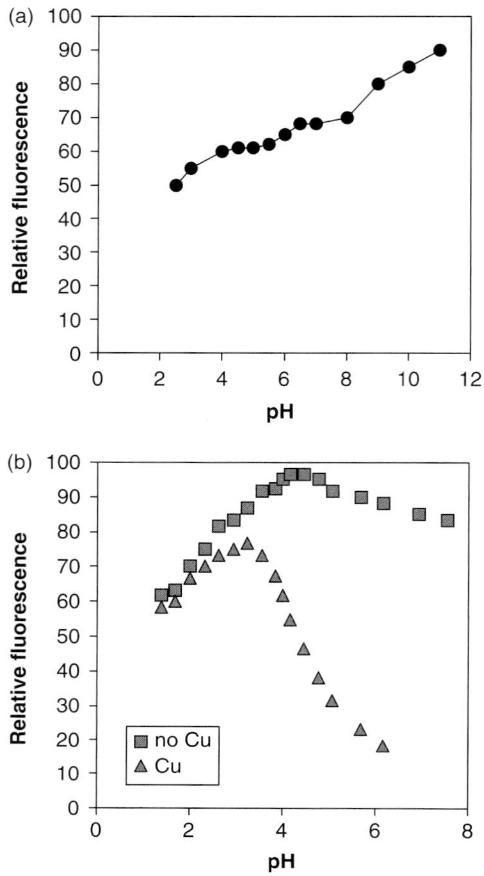

Figure 7.3. The effect of pH and Cu metal on DOM fluorescence. (a) The DOM fluorescence from the mouth of the Ems–Dollart estuary as a function of pH. (Reprinted from Laane, R.W.P.M. Influence of pH on the fluorescence of dissolved organic matter. *Mar. Chem.*, **11**, 395–401, 1982, with permission from Elsevier.). (b) The enhancement of soil fulvic acid fluorescence increases with pH, but the effect is mitigated by the addition of a transition metal such as Cu, which induces fluorescence quenching. (Adapted with permission from Saar, R.A. and Weber, J.H. Comparison of spectrofluorometry and ion-selective electrode potentiometry for determination of complexes between fulvic acid and heavy-metal ions, *Anal. Chem.*, **52**, 2095–2100. Copyright 1982 American Chemical Society.)

and protein fluorescence. Proteinaceous fluorescence emission intensities were observed to either show no variations with pH (Spencer et al., 2007b) or increasing signal with increasing pH (Baker et al., 2007). Several explanations have been given for the marked variability in results with proteinaceous materials. They are thought to be differentially impacted by colloidal material that may protect moieties from solution exposure, change tertiary structure in response to protonation–deprotonation, and perhaps change based on their acid–base content (Baker et al., 2007).

The increase in DOM fluorescence observed as pH increases can be partially attributed to alkaline hydrolysis. Kumke et al. (2001) investigated the changes to a brown water Lake Hochsee and to wastewater effluent upon hydrolysis with NaOH. An increase in fluorescence intensity monitored at 260 and at 330 nm excitation produced nearly a doubling of DOM fluorescence emission intensity near 450 nm. This increase in fluorescence was coincident with a decrease in molecular size. The smaller compounds produced by alkaline hydrolysis had stronger ultraviolet (UV) absorption and thus excitation, but the authors argue that hydrolysis may also disrupt the quenching activity.

More recent studies have focused on natural organic matter (NOM) rather than extracted humic substances. Using sequential tangential ultrafiltration (STUF), NOM from the Amazon basin was separated into particulate, colloidal, and dissolved fractions (Patel-Sorrentino et al., 2002). A and C peak fluorescence intensity both increased with increasing pH. The effect was similar for both "black water" and clear water Amazon basin rivers. As noted previously, phenolic moiety deprotonation, tertiary structural changes (coiling and uncoiling), and ionization were suggested as possible mechanisms for pH-induced variations.

Although no concrete picture for how spectral properties vary with pH, applicable to all environments and DOM substrates seems to be coming to light, it is clear that many properties do vary such that caution should be applied when comparing samples – particularly if pH ranges significantly. In natural waters, for example, an acidic river end member could potentially deliver DOM whose fluorescent properties could significantly change on mixing with the buffered seawater end member during estuarine mixing. Changes in pH along with buffering and divalent metal–organic matter interactions (cf. Willey, 1984) offer the potential to introduce nonconservative spectral signals during estuarine mixing. Several conceptual processes have been offered to explain DOM spectral changes with pH changes at the molecular and colloidal level (see earlier). In addition, humic substances could be a source for sorbing other fluorescent organic compounds (e.g., polycyclic aromatic hydrocarbons) in catchment areas with relatively low pH. This might allow estuarine transport via molecular protection, particularly if aggregate particles are formed. Then, as pH increases in estuaries, disaggregation could occur – releasing compounds and potentially impacting estuarine bulk DOM optical properties. Although this may be a concentration-dependent process (Avena and Wilkinson, 2002), such pH-transitional regions are most likely to exhibit the optical property changes described here.

7.6 Effect of Metals

Natural organic matter (NOM) interacts with metals in the environment primarily through complexation with organic ligands. For example, metals such as Al and Fe released via weathering reactions can be complexed to NOM and thus mobilized in surface waters, increasing their dispersion from the continental to the marine environment. Toxic metals (e.g., Cu, Cd, Hg, Pb) may also serve as complexing constituents for NOM in solution, though metal toxicity may also be mitigated by ligand complexation (Boyd et al.,

2005). Moreover, atmospheric deposition of metals in dust can greatly impact certain surface water ecosystems (e.g., Psenner, 1999). Water column effects may strongly control metal–ligand interactions. For example, Maloney et al. (2005) showed that reoxygenation of Fe-rich anoxic lake water can increase CDOM absorption within UV regions as Fe(II) is oxidized to Fe(III). DOM ligand complexation may lead to greater metal solubility in seawater than expected based on ionic composition alone (Liu and Millero, 2002). In addition, the micronutrient metals availability (e.g. Fe) and even macronutrients (e.g., P) can be strongly affected by metal–ligand complexes (Maranger and Pullin, 2002), some of which are highly photoreactive (e.g., Francko and Heath, 1982). Similarly, photochemistry can change metal solubility in seawater (Liu and Millero, 2002). Thus, ligand–metal binding is of clear biogeochemical importance.

Ligand complexation with metals will strongly affect DOM fluorescence owing to the static quenching induced by the formation of a coordination complex (ligands, L, around a metal, M). Collisional quenching is not as important for metals if their concentrations are sufficiently dilute (Lakowicz, 2006), though as seen earlier, increasing temperature can increase the importance of collisional quenching. Fluorescence quenching titration can be used to determine the stability of the metal–organic ligand complex. Metal salts are added to a natural water sample containing native or extracted DOM (e.g., humic or fulvic acids from natural waters, soils, or sediments) at a constant ionic strength, though ionic strength may be modified to assess any secondary effects on complexation. The fluorescence intensity (EEM, SF, or emission spectra) is recorded during stepwise addition of the metal salt (e.g., Figure 7.3b; Saar and Weber 1980). Any residual fluorescence must result from fluorophores that are not quenched over the concentration range of the metal added. Rayleigh scattering can be monitored to gauge the formation of precipitates as complexation proceeds (Ryan and Weber, 1982).

In contrast to systematically varied pH effects on DOM fluorescence where effects are most marked at the extremes, metal–ligand complexes that quench fluorescence intensity do not appear to alter general spectral shape. Early studies utilized fluorescence to examine metal–ligand interactions because it is a direct measure of free ligand concentration remaining in solution after introduction of metals (e.g., Levesque, 1972). This methodology was refined and compared to other quenching estimators (e.g., ion selective electrodes; Saar and Weber, 1980) which revealed fluorescence loss was directly related to the bound metal and not free metal in solution. Eventually, fluorescence quenching titrations lead to complexing capacity calculations and ligand stability constants (cf. Ryan and Weber 1982). Humic and fulvic acids, being the most abundant DOM in natural waters, were utilized in these and many subsequent studies. Many metals ranging in environmental relevance have been studied, with particular emphasis on divalent Cu, Hg, Pb, Fe, Mn, Ni, and trivalent Al. Of these, only Al(III) can both quench and enhance fluorescence – regulated by varying solution pH. Mg(II) and Ca(II) have also been shown to enhance DOM fluorescence. The enhancement appears to result from a newly fluorescent complex formed between the metal and the ligand.

Metal–ligand complexation is sensitive to the additional variable pH (Figure 7.3b). Saar and Weber (1980) added Cu(II) to a 5 × 10^{-5} *M* soil FA solution. No difference in the relative fluorescence (monitored ex = 350 nm and em = 445–450 nm) occurred until pH was studied as another variable. pH strongly controls fluorescence emission intensity via H^+ ion quenching. Increasing the pH decreases the H^+ ion concentration, which results in a reduced fluorescence quenching due to proton binding. Without an additional cation (i.e., in the Cu(II) free treatment), fluorescence generally increases to a maximum at pH ~5.0 (a commonly observed maxima) and then slightly decreases (see Section 7.5 on pH effects). After metal addition, the fluorescence change is similar to the FA without the metal from pH ~2 to 3.5, at which point the Cu(II) quenching becomes dominant. From pH 4 to 6, the fluorescence continues to be quenched by the metal. Of six metals examined (Ni, Mn, Co, Pb, Cu, and Cd), Cu(II) was the most effective quencher, indicating metal specificity for DOM ligand binding sites. Further, using ion selective electrode potentiometric titrations, Saar and Weber (1980) found that Cu(II) complexation by FA was nearly identical to Cu(II)–DOM fluorescence quenching, indicating that indeed the formation of the metal–ligand complex is quantitatively quenching the fluorescence. From these results, stability and binding coefficients can then be calculated.

Ligand complexation has been assumed to display a linear relationship between the complexing metal concentration and the change in fluorescence intensity. This assumption has been challenged by Cabaniss and Shuman (1986), who used SF to study Cu^{2+} binding with fulvic acid and found nonlinear relationships. Cabaniss (1992) expanded the use of differential synchronous fluorescence to identify metals that quench and also contribute to fulvic acid fluorescence. Presumably, the nonlinearity is caused by differing responses across wavelength regions and thus complexation capacity and ligand stability calculations depended on the wavelength pairs chosen for modeling (Luster et al., 1996).

Al(III) appears both to decrease and increase DOM fluorescence when forming a complex with fulvic acid (FA), isolated from soil (Ryan et al., 1996), river (Elkins and Nelson, 2002), and marine (da Silva and Machado, 1996) environments. The effect is presented in Figure 7.4 (redrawn from Elkins and Nelson, 2002), in which the change to relative fluorescence intensity of a river FA after addition of Al(III) is compared to additions of Cd(II) and of Cu(II). Immediately after addition of the Al(III) metal, fluorescence (ex = 344/em = 424) increased and reached a stable intensity at 5.0 × 10^{-5} *M* Al(III). By contrast, additions of up to 2 × 10^{-4} *M* Cd(II) did not change river FA fluorescence intensity. The addition of Cu(II) to the river FA solution produced fluorescence quenching, as shown in Figure 7.3b.

Pullin and Cabaniss (1997) were able to increase synchronous fluorescence at 380 nm excitation (SF380) by adding Al(III) to river water DOM at pH 5 but noted the effect was smaller than that found for fulvic acid. Lakshman et al. (1993) used SF to investigate the binding sites to a soil FA, separated into three size fractions by ultrafiltration, and identified changes to three SF peaks. Peaks I and II (315 and 370 nm excitation, respectively; $\Delta\lambda$ was not reported) corresponded to carboxyl or hydroxyl groups attached to simple aromatic rings, while peak III (excitation = 470 nm) corresponded to condensed polyaromatic rings

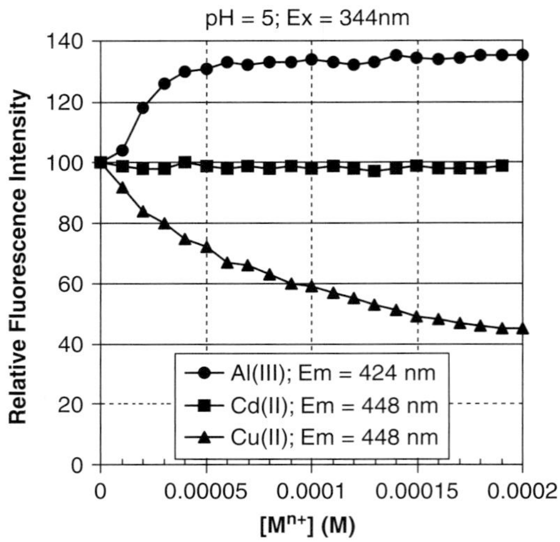

Figure 7.4. The change to fluorescence emission intensity (excitation of 344 nm) of fulvic acid after the addition of three different metals at pH = 5. As metal (M) concentration increased, Cd(II) showed no effect on fluorescence emission, whereas the addition of Al(III) increased fluorescence by 40% and Cu(II) quenched fluorescence by about 60%. (Redrawn from Elkins and Nelson, 2002.)

(e.g., flavenoids) having extended π-conjugation. The molecular weight separation dramatically modified SF features. On adding Al(III) concentrations at 4–400 μ*M*, SF increased for all three peaks identified and most strongly for peak III, which was most prevalent in the 500–1000 MW fraction. Titration curves eventually reached a saturation point and these were interpreted to suggest that deprotonated (at pH 3.5) carboxyl–Al complexes were inducing fluorescence.

Ryan et al. (1996), Elkins and Nelson (2001, 2002), and Zhao and Nelson (2005) all found that Al(III) strongly interacts with salicylic acid moieties in fulvic acids, and that Al(III)–salicylic acid complexes produced new and highly intense fluorescent peaks in EEMs. Salicyclic acid has both phenolic and carboxylic functionality and is a major component of FA. A common spectral change observed by these workers was a significant 20 nm red shift in the excitation peak maxima and a 20 nm blue shift in emission peak maxima, after addition of Al(III) to FA solutions. The same effect occurred for solutions of salicyclic acid complexed with Al(III) (Elkins and Nelson, 2001). However, Zhao and Nelson (2005) also observed that although Al(III) outcompeted other metals for available binding sites on FA and formed a new fluorescent complex with FA, Fe(III) complexation by FA actually quenched the new Al(III)–FA fluorescence. During complexation, metals bind to anionic functional groups via donation of electrons in a binding environment that may range from ionic to covalent. Zhao and Nelson (2005) noted that the uniquely low covalent and high ionic properties of ions such as Al(III) and Fe(III) alter the electronic structure of FA through complex formation and inducing these contrasting effects on FA fluorescence.

EEMs provide an estimate of how nearly every measureable fluorophore responds to metal additions, enabling the study of the metal–ligand fluorophores of DOM in greater detail. Luster et al. (1996) used EEMs to study fluorescence changes on Cu^{2+} addition to DOM leached from leaf litter at copper to carbon ratios, Cu/C less than 1000. Up to Cu/C ~1:250, quenching occurred in protein and (to a lesser extent) phenolic fluorescence regions. Highly conjugated fluorophores were quenched at Cu/C greater than 250. Thus, the amount of quencher added produced substantially different degrees of fluorescence loss in DOM discernible in "identified" EEM peak regions. Luster et al. (1996) also determined stability constants for Cu(II) and Al(III) in the leaf litter–extracted DOM. They were able to model three binding sites on this DOM using the major terrestrial humic peaks found in their EEM spectra and relating these fluorophores to functional group ligands such as phenolics and carboxylic acids (both properties of salicyclic acid moiety). Metal additions, especially Al(III), can cause flocculation of material from solution. Sharpless and McGown (1999) found that aggregation was caused by adding Al(III) reducing short and long wavelength fluorescence in aquatic HA, but reducing only longwave fluorescence in terrestrial (peat and soil) HA. They concluded that a decrease in fluorescence intensity of HA is caused by precipitation of Al(III)–HA complexes from solution rather than fluorescence quenching.

Ohno et al. (2008) expanded on the use of EEM fluorescence to investigate metal binding parameters on soil DOM leached separately from conifer and from deciduous tree litters. They used a parallel factor model (PARAFAC) model to decompose three terrestrial humic components that responded differently to additions of Fe(III) and Al(III). Their component 1 (peak at ex325/em450, similar to C peak) in the deciduous DOM was strongly quenched both by Al and Fe, while their component 1 for the coniferous DOM was quenched only by Al. Their component 3 (ex240/em400, similar to A peak) was not quenched by addition of Fe(III) yet slightly increased in signal by addition of Al(III). Yamashita and Jaffe (2008) have also used EEM-PARAFAC to model Cu(II) and Hg(II) binding to mangrove DOM in the Florida Coastal Everglades. They found that Cu(II) actually decreases then enhances protein fluorescence, while Hg(II) quenches the protein component fluorescence.

Metals have variable affinity for binding sites on DOM, which produces a range of effects on DOM fluorescence quenching – effects that depend on both the metal and the fluorophore. The competition for binding sites between highly ionic Al(III) and Fe(III) is a key example. The work by Zhao and Nelson (2005) suggested that either Al(III) cannot replace complexed Fe(III), or that Fe(III) displaced into solution dynamically quenches FA fluorescence. Apparently, then, not all fluorophores complex equally; from the examination of four different divalent metals, Antunes et al. (2007) found distinct changes in freshwater FA and a commercial HA EEMs fluorescence that varied with metal addition. This result is similar to that found by Yamashita and Jaffe (2008) but their work specifically addresses the fact that often-overlapping broad emission spectra must be deconvolved (they used principal component analysis [PCA] and multivariate curve resolution) into discrete fluorescent components, which can be chemically meaningful (Ohno et al., 2008). Thus, the application of multivariate statistical techniques such as PARAFAC to EEM data from metal–ligand complexing experiments should improve the determination of stability constants for complex formation.

Functional ligand group abundance associated with a given DOM source will be a primary control on metal–ligand associations, followed by the pH. Note that many functional groups also exert some local control on solution pH, so the effects are related and probably synergistic. For example, carboxyl groups may deprotonate, leaving the carboxylate anion (COO^-) to which a metal cation can bind, forming a complex and perhaps altering the DOM's optical properties. In the well characterized IHSS humic substances, the ratio of carboxyl groups to phenolic groups was found to be ~4:1 (Ritchie and Perdue, 2003), so the importance of carboxylic groups to metal–ligand fluorescence is clear. Carboxyls are most abundant in terrestrial FA and least abundant in terrestrial HA, meaning that DOM fluorescence should be strongly affected by the presence of carboxyls because FAs (soluble at all pHs) are the larger contribution to CDOM (Weishaar et al., 2003). Therefore, carboxylate should be considered a major complexing ligand for most metals at low pH, and the effect would be minimized at higher pHs where carboxylic acids are protonated.

Displacement of one metal by another can reduce the quenching effect. This was observed by Willey (1984) and modeled with synchronous fluorescence on fulvic acids by Cabaniss and Shuman (1987). Both studies showed an increase in fluorescence with Mg^{2+} additions. Many other reports have investigated alkaline earth metal disruption of paramagnetic metal complexation by DOM using fluorescence (Cabaniss and Shuman, 1988; Cabaniss, 1992). The alkaline earth metals bind differently to DOM causing this effect, which appears to result from conformational changes induced by the binding. Lu and Jaffe (2001) and Wu et al. (2004) studied the effects of other species on metal–ligand complexation, finding that the Cl^- concentration also plays a role. At high pH, OH^- becomes competitive with DOM for metal binding (Cao et al., 1995). When these competitions occur, the molecular conformation of DOM may change, as does the electrostatic environment by raising pH. Functional group ionization also influences the ability of DOM to complex metals, with an attendant effect on DOM fluorescence. The conformation changes caused by incremental addition of the alkaline earth metals when river water mixes with seawater also has implications for DOM biological and photochemical reactivity (see Section 7.7).

7.7 Effect of Salinity (Ionic Strength)

Changes to ionic strength appear to have a lesser effect on DOM fluorescence than pH and metal–ligand interactions. However, several workers have observed bathychromic shifts (blue shifts) and/or decreased fluorescence intensity of certain peaks in EEM spectra analyzed from samples across estuarine transitions where river water mixes with sea water (e.g., Del Castillo et al., 1999; Kowalczuk et al., 2003; Alberts et al., 2004; Kowalczuk et al., 2009), and during simulated mixing of freshwater and seawater (Boyd and Osburn, 2004). Moreover, salinization can decrease SF intensity in the FA extracted from soils (Cilenti et al., 2005; Provenzano et al., 2008). These observations imply that fluorescent moieties of DOM change configuration at higher ionic strengths, similar to proteins in biological systems (Boyd et al., 2010a), and quench or inhibit fluorescence.

Anions such as halides, hydroxyl, nitrite/nitrate, and carbonate, all prominent in natural waters, have been examined for their ability to quench the fluorescence of aromatic compounds by heavy atom effects and electron and charge transfer mechanisms (Watkins, 1974; Shizuka et al., 1980; Treinin et al., 1983; Mac, 1995). Thus, ionic strength effects on DOM fluorescence could result from quenching effects of these solutes. Ghosh and Schnitzer (1979) identified decreases in excitation spectra intensity of soil HA and FA as NaCl concentration was increased from 0.001 to 0.1 *M*. They suggested that increased molecular "coiling" (e.g., Conte and Piccolo, 1999) and/or decreased ionization of phenolic hydroxyl moieties (Senesi, 1990) caused the observed fluorescence quenching. Thus, a major synergistic effect of increased ionic strength on DOM fluorescence could result from suppression of functional group ionization, coincident with the metal–ligand quenching previously described. It is also possible that photochemically formed and short-lived radicals or excimers (an excited state dimer) or exciplexes (excited state complex), especially when aromatic compounds are the excited species (Mac et al., 1993) can modify, or interfere with, fluorescence (Senesi, 1990). Further, alkyl halides formed via the halogenation of DOM can also alter DOM fluorescence as halogen substitution could also produce an internal "heavy atom" effect (Senesi, 1990; Senesi and D'Orazio, 2005). This effect could explain the blue shift in hydrophobic HS fluorescence (including a narrowing of the emission bandwith) observed in chlorination reactions (Korshin et al., 1999). Similar mechanisms during water treatment activities probably reduce fluorescence as DOM is oxidized (e.g., Henderson et al., 2009).

Despite these examples, scant data exist to support a large effect of ionic strength on DOM fluorescence in laboratory studies. Mobed et al. (1996) examined EEMs of a peat-derived FA from 0 to 1 *M* KCl and found visual changes that were not statistically significant, based on a matrix correlation method. One way to evaluate an effect of ionic strength on DOM fluorescence (which would exclude mixing of water masses such as in the coastal ocean) is to examine saline lakes situated in arid regions and hydrologically isolated (Figure 7.5). Although the ionic compositions of saline lake water can differ markedly from seawater, these systems can serve as models for the progression of DOM properties during evaporation (though often their salinities can fluctuate seasonally).

Studies of DOM from saline lakes and wetlands suggest that autochthonous DOM is dominant in these systems, partially attributed to their eutrophic to mixotrophic nutrient states (Leenheer et al., 2004; Ortega-Retuerta et al., 2007). Recently, DOM fluorescence spectra from saline lakes in the Great Plains of the United States were analyzed with PARAFAC (Osburn et al., 2011). A component from their PARAFAC model resembled the prominent INT peak from Figure 7.5, a fluorescent signal also prominent in a saline lake from Antarctica dominated by autochthonous production (McKnight et al., 2001) and in saline ponds situated in the Brazilian Patanal wetland (Mariot et al., 2007). Compiling results from several studies, Osburn et al. (2011) also found a consistent slope value (0.534 ± 0.127) for log–log regressions of DOC on conductivity for a range of inland

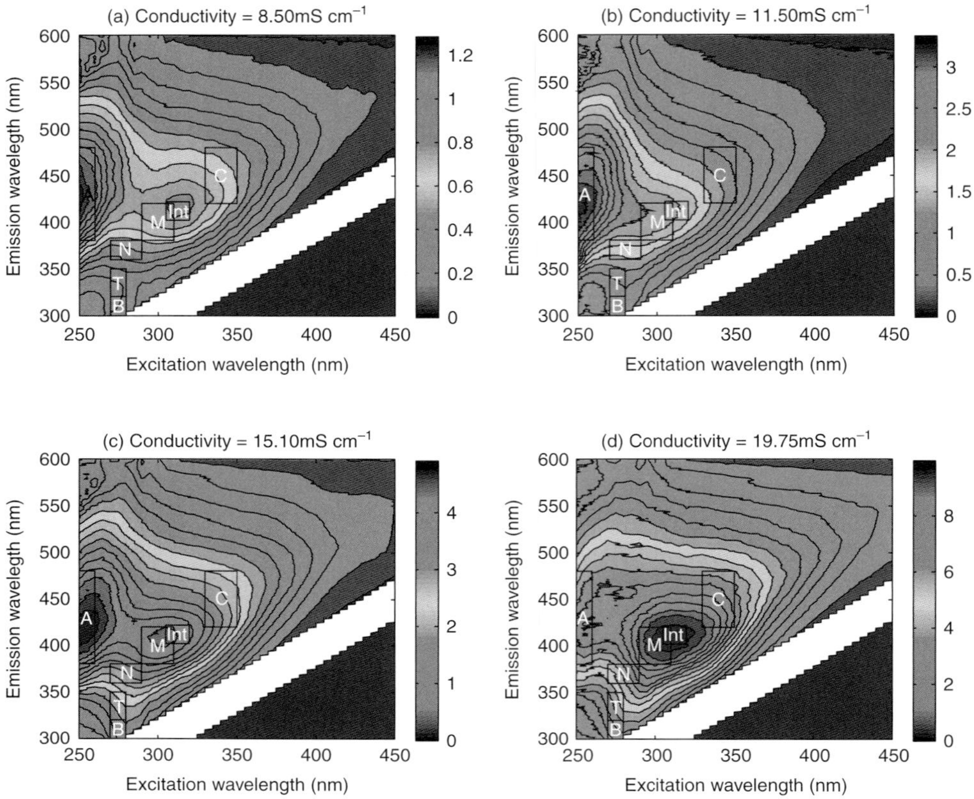

Figure 7.5. The change in EEM fluorescence of Alkaline Lake CDOM (North Dakota). (a) June 2001. (b) May 2004. (c) August 2004. (d) August 2005. The lake water conductivity is indicated for each measurement and the boxes and letter codes describe peak regions as for Figure 7.2. A decided enhancement of the Int peak intermediate between the microbial M peak region and the terrestrial C peak region is apparent.

saline lakes (Arts et al., 2000; Anderson and Stedmon, 2007; Mariot et al., 2007; Ortega-Retuerta et al., 2007). Although evapoconcentration appears to be a forcing factor on DOM fluorescence as salts, and hence ionic strength, increase along with organic matter, Osburn et al. (2011) also suggested that microbial processing of DOM in saline lakes exerts an influence on fluorescence properties in these ecosystems, rather than a strict concentration of allochthonous fluorescent DOM.

A saline lake (Lake Alkali) in North Dakota, USA, sampled during summer in 2001, 2004, and 2005 showed marked changes to EEM fluorescence of DOM as conductivity increased in the lake, from 8.50 mS cm^{-1} to 19.75 mS cm^{-1} (Figure 7.5; Osburn et al., 2011). Note that the intensity of fluorescence in these EEMs increases, which suggests that photodegradation was not a strong forcing factor on this DOM. Although seasonal inputs of DOM from runoff and precipitation may contribute to the DOM pool, it is clear from Figures 7.5a–d that evaporation is concentrating fluorescent DOM. Blue shifting in DOM fluorescence might occur if the increase in salinity is also causing a metal-quenching effect, *or* if increased salinity is causing conformational changes in DOM, as suggested by a number of studies (e.g., Lochmuller and Saavedra, 1986; Reche et al., 1999; Boyd and Osburn, 2004; Batchelli et al., 2009; Provenzano et al., 2010). However, the opposite effect was observed. The EEMs show an enhancement in a peak centered at 315 nm excitation and 400–420 nm emission (labeled INT on Figures 7.5a–d), intermediate between the region of the terrestrial humic C peak and the marine humic (or microbial) M peak (Boyd et al., 2010a). This peak increased with conductivity (Figure 7.6a), but the pH did not change appreciably in this lake from 2001 to 2005 (data not shown). However, Mg^{2+} concentration in Alkali Lake increased from May to August 2004 (2.46 to 3.41 mmol Mg^{2+}, respectively). It is possible that increasing Mg^{2+}concentrations in the lake water displaced quenching metals, similar to results found by Willey (1984) and Cabaniss (1992). Interestingly, the INT peak coordinates are very similar to ex/em maxima for the 3-hydroxybenzoic acid and salicylic acid moieties proposed to constitute humic fluorescence (Senesi et al., 2005), as well as amino sugars (Biers et al., 2007) and evidence for phytoplankton production of fluorescent DOM (Romera-Castillo et al., 2010). This latter point suggests that microbial processing of DOM, rather than ion displacement, created or transformed fluorescent DOM in Alkali Lake (Boyd and Osburn, 2004; Osburn et al., 2011).

The results from this saline lake are in contrast to the results from Provenzano et al. (2008) for hydrophilic (HI) and hydrophobic (HO) humic substances isolated from three increasingly saline soils (Figure 7.6b). This study showed a clear decrease in fluorescence intensity for both DOM fractions, though at different ex/em maxima. Provenzano et al. (2008) indicated that the suppression of ionization was likely responsible for this effect. Exchangeable Na^+ potential increased in each soil type with salinity, but information on Mg^{2+} ion was not available in this study. More work is needed on these systems to elucidate the roles of metal–ligand complexation and suppression of ionization in the DOM chemistry of saline lake and salinized soil environments.

EEMs fluorescence changes occurred when DOM was added to solutions of increasing ionic strength – while maintaining the natural in situ milieu – in experiments meant to simulate coastal mixing and to investigate biological and photochemical degradation. Boyd and Osburn (2004) found that mixing of COM isolated from the Susquehanna River by ultrafiltration into solutions of increasing salinity quenched the fluorescence intensity of terrestrial and marine humic peaks (C, A, M) but not protein peaks (T, B). In subsequent work, in which LMW organic matter fluorescence from several estuaries was studied, Boyd et al. (2010a) observed a relatively linear mixing relationship for freshwater-derived LMW DOM (<1000 nm) but found mid-estuarine LMW DOM (~16 salinity) became more fluorescent when mixed toward the freshwater end member. HMW DOM (>1000 nm) collected in freshwater and mid-estuarine regions in the same estuaries showed EEM peaks, peak ratios, and PCA and PARAFAC modeled components which generally varied (had lower signal) at lower salinities and increased toward the ocean end member. The B peak,

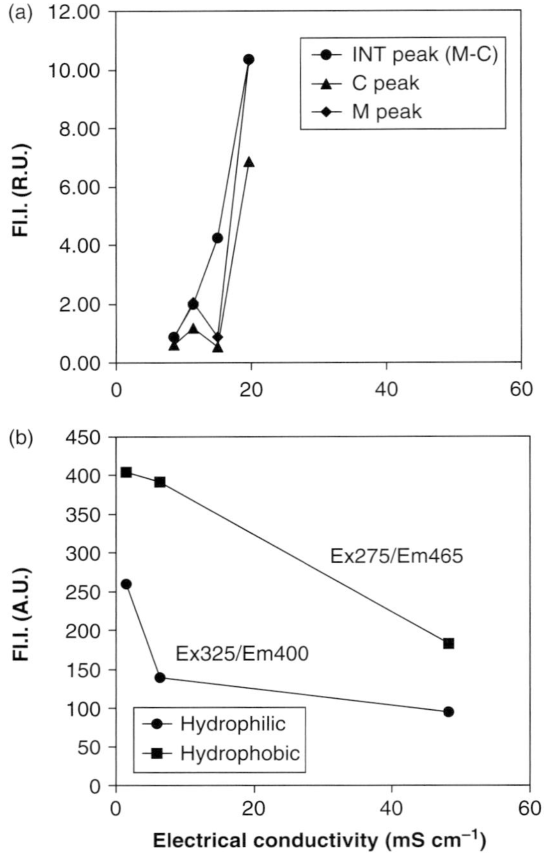

Figure 7.6. Differing effects of ionic strength measured as conductivity for (a) DOM collected from Alkali Lake, North Dakota and (b) hydrophilic and hydrophobic acids isolated from soils sampled in the west coast of Sicily. (Adapted from Provenzano et al. Spectroscopic investigation on hydrophobic and hydrophilic fractions of dissolved organic matter extracted from soils at different salinity. *Clean* **36**(9), 748–753, 2008. Used with permission of John Wiley & Sons.) DOM fluorescence of the Int peak increases fivefold as conductivity increased from 8.5 to 20 mS cm^{-1}. By contrast, both hydrophilic and hydrophobic acid isolates of DOM from salinized soils showed dramatic reductions in fluorescence in similar regions of fluorescence.

however, showed almost no variability in peak or modeled data. The B peak component modeled in river to ocean transects by Yamashita et al. (2008) also showed similar behavior in which no clear trend was observed with salinity.

Alberts et al. (2004) studied DOM fluorescence effects of a CDOM-rich river fractionated into three molecular weight size classes (<10 kDa, 10–50 kDa, and >50 kDa) and mixed at increasing salinity. The dominant EEM peak studied was the C-peak (ex 355/em 450) and its behavior was observed during mixing for each size fraction. Blue shifting from 450 to 440 occurred from 0 to 10 salinity for the 10–50 and >50 kDa fractions, similar to the Chesapeake Bay results from Boyd and Osburn (2004), then remained stable at salinity from 10 to 33. No change in emission peak position was found for the <10 kDa size fraction. Only slight changes were observed in fluorescence intensity, yet fluorescence efficiency decreased markedly in each size class from ~25 to 33 salinity.

Moreover, salinity appears to have contrasting effects on DOM removal mechanisms, especially photodegradation by sunlight. Minor et al. (2006) observed no effect on DOM photomineralization as a function of salinity, yet Osburn et al. (2009a) and Grebel et al. (2009) found distinct effects of salinity and halides, respectively, on DOM photobleaching. Grebel et al. (2009) specifically found a halide ion effect on Suwannee River DOM absorbance photobleaching but not on DOM fluorescence photobleaching. Halide ion abundance may become a factor in natural systems such as estuaries, where halide-rich seawater mixes with freshwater, and in regions where halide-rich groundwater mixes with surface waters. Halide ion effects due to waste treatment chlorination have been shown to narrow DOM fluorescence emission bands (termed "contraction") and to blue-shift fluorescence maxima (Korshin et al., 1999).

7.8 Effect of Particles

Particles such as Fe- and Al-hydroxides and clay minerals have the capacity to sorb organic components, perhaps even more so when "coated" with humic materials (Zhou and Rowland, 1997). Particle association can influence DOM fluorescence in soils and aquatic sediments (Kaiser and Guggenberger, 2000). Two mechanisms are likely at play: first, through humic sorption to clays (which does not appear to alter their physicochemical properties; Zhou et al., 1994) and second, by proteinaceous DOM sorption. Humic materials exist as macro-ions contributing varying sorptive capacity to clay minerals. In estuaries, acidic functional groups are impacted by conditions such as pH (decreasing sorption with increasing pH) and salinity changes (increasing sorption with increasing salinity) (Zhou et al., 1994; Specht et al., 2000). Although OM removal in estuaries has been postulated to occur by "salting out" (solubility decreased due to increasing salinity), most studies have demonstrated an increased sorptive capacity for minerals as salinity increases (Means, 1995; Zhou and Rowland, 1997). Mechanistically, as DOM becomes less soluble, it becomes more surface-active and preferentially sorbs to soil or aquatic particles. In terrestrial environments, especially in agricultural regions, salinization could cause similar effects in soil DOM (Cilenti et al., 2005; Provenzano et al., 2008, 2010).

Differential humic or proteinaceous material sorption to particles may impact fluorescence. Many studies have assessed the impact of cations at various concentrations – to simulate estuarine mixing – on natural DOM fluorescence (cf. Antunes et al., 2007). In the Alberts et al. (2004) study highlighted earlier in Section 7.7, the authors speculated humic material coagulation caused a decrease in absorbance at higher salinities. The salinity range causing the effect was similar to that reported (above) having maximal impact on clay-DOM sorption. If humic materials were preferentially lost to particles during this mixing experiment, a change in humic peaks (A and C) might reflect the phenomenon. Indeed, a peak shift and change in relative fluorescence intensity were observed for the C peak in response to increasing salinity.

Only recently have optical properties been linked to coagulation and flocculation associated with estuarine particulates. In a recent study, extracted aquatic humic acids were used to assess partitioning on different sized estuarine particles (Sun et al., 2009). Particles were collected from a natural estuary and thus likely contained previously sorbed organic matter (i.e., they were not pure clay minerals). Differential partitioning was observed between fluorescence components by relating partitioning coefficients to the ratio of A to C peak fluorescence intensity. Increased partitioning coincided with larger peak A to peak C intensity ratios. This implies that larger, perhaps more aromatic humic fractions preferentially sorb to estuarine particles. In another recent study (related to water treatment efforts), aluminum sulfate was used as an artificial coagulant (Gone et al., 2009). Below and above pH 5, coagulation was most pronounced (based on DOC measurements). At higher pH, more net negative charge was offered as an explanation for lower DOM sorption. Coagulation or natural DOM sorption in estuaries could produce a loss of fluorescence through increased DOM removal via sorption and flocculation. Lead et al. (2006) found such an effect on separation of freshwater DOM (SPLITT; see earlier) by molecular weight, in which ca. 40% of the T peak fluorescence was found in the >1 μm fraction – particles by operation definition. They also found no significant changes to A and C peak fluorescence with molecular size and postulated that humic and fulvic DOM moieties existed in bound form to "clays, biological cells, etc." rather than in free form. In summary, modifications to DOM fluorescence via particle sorption phenomena may be important in environments undergoing substantial changes in ionic strength (as in estuaries and coastal waters) as well as in freshwater systems.

7.9 Effect of Sunlight

Photodegradation, or photobleaching, of DOM fluorescence occurs after exposure to sunlight and generally results in diminished emission intensity and blue shifting (Coble, 1996). In stratified water bodies (e.g., lakes, estuaries, and coastal waters, especially those influenced by riverine discharge), the potential for DOM fluorescence bleaching is high owing to the delivery of photoreactive fluorescent material and a residence time long enough for photochemical bleaching reactions to occur. Light attenuation within the water column will be important to determining overall rates (Miller, 1998), as will light absorption by DOM itself (cf. Osburn et al., 2001; Del Vecchio and Blough, 2002; Osburn and Morris, 2003)

Natural sunlight is polychromatic. Under polychromatic light exposure, emission loss is broad and unstructured and extends throughout the entire spectral range, with the bleaching always being more pronounced in the spectral region passed by the cutoff filter. Examples of fluorescence photobleaching indicate a general nonspecific decrease in broad emission spectra (Koussai and Zika, 1990; Kieber et al., 1990; Vodacek et al., 1997). However, SF spectra from river DOM show some structure after photobleaching indicating faster photobleaching at longer wavelengths. For example, synchronous fluorescence at ex = 375 nm (SF375) decreases more rapidly than synchronous fluorescence at ex = 350 nm (SF350) (Pullin and Cabaniss, 1997). Similar results were found by Tzortziou et al. (2007) for the Rhode River estuary and by Osburn et al. (2009b) for the Mackenzie River in which the longer excitation wavelength fluorescence bleached faster than did the shorter excitation wavelength fluorescence. Few EEM measurements have been made of DOM after photobleaching. Mayer et al. (1999) showed that tryptophan protein fluorescence is more photo-labile than was tyrosine fluorescence. Using estuarine DOM, Moran et al. (2000) was one of the first reports exploring both photochemical and biological DOM degradation via fluorescence. They reported greater bleaching at humic (A, C, M) peaks than at protein peaks, though bleaching of the latter did occur. Coupled with this was blue shifting of the DOM fluorescence ex/em maxima due to the preferential loss of emission at longer wavelengths. Typically, DOM photobleaching is quite rapid initially, followed by a slower degradation stage (Koussai and Zika, 1990; Pullin and Cabaniss, 1997; Moran et al., 2000; Del Vecchio and Blough, 2002; Osburn et al., 2009b).

On monochromatic light exposure, the fluorescence signal is lost (bleached) as a function of the irradiation wavelengths. Along the excitation axis, monochromatic light exposure induces the greatest fluorescence loss nearest the irradiation wavelength (as observed for absorption) (Patsayeva et al., 1991; Boehme and Coble, 2000; Del Vecchio and Blough, 2002, 2004). However, losses also occur above and below the irradiation wavelength. A possible explanation for the loss at the excitation wavelength is direct photochemical destruction of fluorophores all absorbing at this wavelength. However the broad emission loss (extended across the entire range investigated) cannot arise from destruction of non interacting DOM components all excited at the irradiation wavelength and showing a continuum of red-shifted emission (very unlikely) but could instead arise from a more complex model of interacting species (Del Vecchio and Blough, 2004; Goldstone et al., 2004). The secondary loss (away from the irradiation wavelength) could be due (a) to the destruction (via indirect photochemistry) of a different class of fluorophores only excited at long wavelengths; or (b) to the direct photochemical destruction of a subset class of fluorophores excited at both wavelengths (or somehow coupled with fluorophores excited at short wavelengths). The continuous red shift of secondary loss with increasing irradiation wavelengths does not favor the superposition of a large number of fluorophores all overlapping the lowest excited state band and showing a continuous red shift of the highest excited band. Instead, Del Vecchio and Blough (2004) argue for a more complex model of interacting species originating the long wavelength emission.

The net effect on DOM fluorescence from this complex photochemistry is a blue shift in excitation after irradiation, but red shifting may also occur within emission spectra. An example of DOM fluorescence photobleaching showing a subtle red shift in emission is presented in Figure 7.7 for a Sphagnum-dominated bog (Figure 7.7a) and for DOM generated from an algae culture (Figure 7.7b) (cf. Osburn et al., 2001). For each type of DOM, fluorescence emission spectra (400–600 nm, ex370) decreased significantly. After photo-exposure, the bog DOM peak emission shifted slightly from 462 to 469 nm; the algae

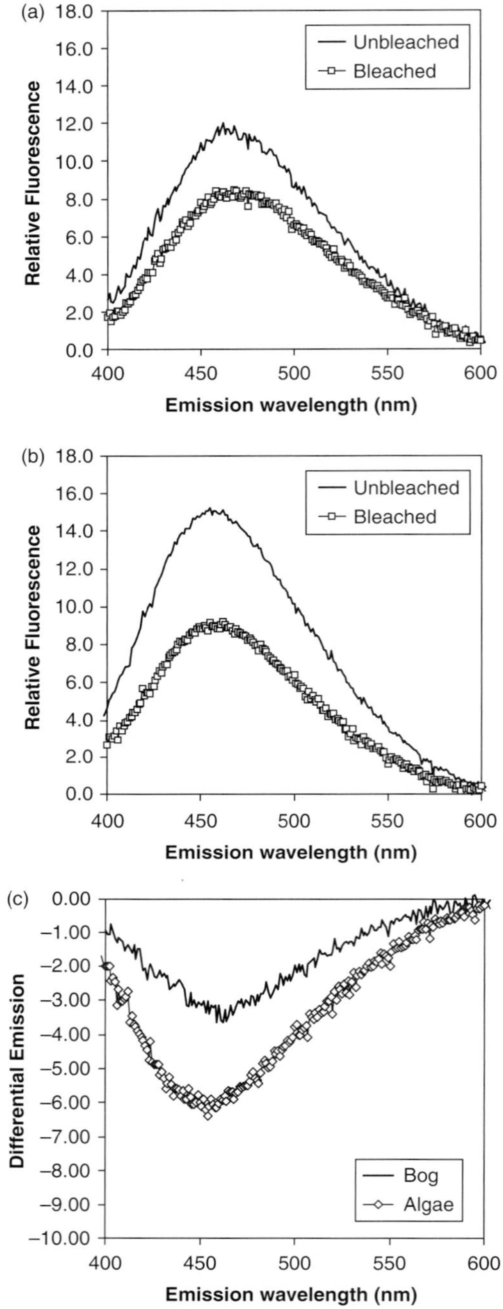

Figure 7.7. The effect of photobleaching on DOM emission spectra (ex370) for two sources of DOM. (a) Groundwater from a Sphagnum dominated bog. (b) An algae culture. The difference between unbleached and bleached spectra (c) is also shown.

DOM showed a similar red shift from 455 to 462 nm. The differential emission spectra for each DOM type show that the majority of photobleaching occurs at the peak emission wavelength (Figure 7.7c). The difference in emission peaks between these samples likely reflects their different chemical compositions (terrestrial vs. aquatic DOM). The common degree of red shifting (ca. 6 nm) could be due to partial oxidation of this organic matter. Increasing the amount of carboxyl and hydroxyl groups on each DOM type would shift fluorescence to longer wavelengths (Senesi and D'Orazio, 2005).

Fluorescence lifetime measurements provide additional information on DOM and humic materials' optical properties and photoreactivity. Many workers have studied decay rates (τ, in nanoseconds) and have modeled multicomponent lifetimes at less than 1, 2–5, and 6–14 ns (Clark et al., 2002 and references therein). These studies suggest three broad fluorophore components based on fluorescence lifetime range. Clark et al. (2002) photobleached humic Shark River (Florida) water using light at 280 nm, and found a significant decrease in the two shorter lifetime component (at less than 1 and 2–5 ns). Interestingly, after photobleaching Shark River DOM at a longer wavelength of 334 nm, Clark et al. observed that the shortest lifetime component was significantly more reduced (the second lifetime component's bleaching remain unchanged). Thus photobleaching results in measurably shorter lifetimes in the resultant DOM. It remains to be seen if these lifetimes correlate to specific humic peaks, though the fluorescence lifetime detection for Clark et al.'s study was ex337/em430, which falls in the C peak region (Coble, 1996).

Environments where ample sunlight exposure to DOM can occur will promote the greatest fluorescence photobleaching. A clear example is the seasonal stratification that occurs in surface waters. Gibson et al. (2001) observed a loss of DOM fluorescence in lakes in the Canadian Arctic that occurred seasonally during periods of surface water stratification and mixing that constantly bleaches the upper water column of stratified lakes. Vodacek et al. (1997) and Del Vecchio and Blough (2002) showed substantial loss of fluorescence in stratified coastal waters of the Middle Atlantic Bight. Similarly, Ma and Green (2004) examined the EEM fluorescence effects of photobleaching in Lake Superior, finding that, while humic fluorescence decreased with sunlight exposure, blue-shifted increases in DOM fluorescence were observed. They suggested that new chromophores were being formed, an effect that was also observed in seawater by Biers et al. (2007).

In coastal environments, mixing is important for diluting a "concentrated" DOM solution and for increasing the sunlight exposure as terrestrial runoff essentially mixes and spreads out in a thin layer atop a denser seawater bottom layer. This creates ideal conditions for substantial DOM photodegradation. Del Vecchio and Blough (2002) have shown this to be an effective removal mechanism for DOM fluorescence, noting that the photobleaching depth to vertical mixed depth ratio will provide an ultimate index to photobleaching extent. Plotting shifts in emission peak intensity at fixed excitation wavelengths can indicate whether substantial blue shifting or red shifting has occurred. These results may then be used to infer if DOM photobleaching or dilution has occurred (Coble, 2007; Conmy et al., 2009). Ultimately, these modeling constraints can then be applied to coastal observations of DOM fluorescence and incorporated into mixing models for terrigenous DOM transport (Blough et al., 1993; Vodacek et al., 1997; Del Castillo et al., 1999; Conmy et al., 2009).

Photobleaching of DOM fluorescence also occurs in rainwater. Kieber et al. (2007) photobleached coastal rainwater and found losses to EEM peaks A, C, and T similar to freshwater UDOM. They remark that the origin of the rain event (continental versus marine precipitation) likely exerts a large control over photobleaching. Freshwater rain can contribute to surface waters DOM and provide an environment for photoreaction in the atmosphere (Graber and Rudich, 2006). In fact, the molecular properties of DOM in atmospheric water suggest some degree of photobleaching has occurred. Rainwater is often acidic ($pH < 5$), which can enhance photobleaching potential within the atmosphere as well as in surface waters (Gennings et al., 2001).

Humic-regions (e.g., the C and A peaks from Coble, 1996) appear to be most photolabile. The degree of susceptibility to photobleaching depends both on exposure duration and the sunlight spectral quality. An example of these quantitative and qualitative controls on DOM photobleaching is shown for the Mackenzie River in Arctic Canada by Osburn et al. (2009b), where marked changes to SF spectra occurred at peaks centered on 350 nm and 380 nm excitation ($\Delta\lambda = 14$ nm) after sunlight exposure. SF emission intensity at these wavelengths generally corresponds to highly conjugated humic substances representative of terrestrial DOM (Senesi, 1990). In a kinetic experiment (Figure 7.8a), more DOM photobleaching occurred for the SF350 than SF380 nm after 72 hours of sunlight exposure. By contrast, Figure 7.8b shows the effect on these peaks when sunlight was modified by cutoff filters that selectively remove portions of the solar spectrum during the sunlight exposure. In the 335 nm filter treatment, for example, more photobleaching occurred for the 380 nm peak than for the 350 nm peak. The 314 nm treatment included nearly all environmentally relevant UV radiation and produced photobleaching results not statistically different from the treatment without a cutoff filter. These results would indicate that not all fluorophores are bleaching in the same manner and thus respond differently to sunlight exposure.

Specific patterns in DOM fluorescence photobleaching should identify molecular weight changes. Longwave emission should decrease as a function of light exposure as polyaromatic compounds are broken apart, reducing the extension of the π-electron system (Senesi and D'Orazio, 2005). Aromatic rings within HS are opened after photooxidation, disrupting charge transfer reactions and giving rise to long-wavelength absorption and fluorescence phenomena (Del Vecchio and Blough, 2004). Polyaromatic structures may generate reactive oxygen species (ROS) that photooxidize DOM. O'Sullivan et al. (2005) found a strong correlation between SF350 and hydrogen peroxide production from natural and simulated sunlight exposures of DOM from river and coastal waters. They inferred that the destruction of these polyaromatic structures (which have extensive π-electron systems) decreases the longer wavelength (red shifted) fluorescence) concomitant with a decrease in the generation of ROS. Singlet oxygen, for example, is known to be causative agent of DOM oxidation (e.g., Cory et al., 2010).

An additional consideration is the degree to which photobleaching can modify terrestrial DOM fluorescence and the similarity of that modified fluorescent signature to marine

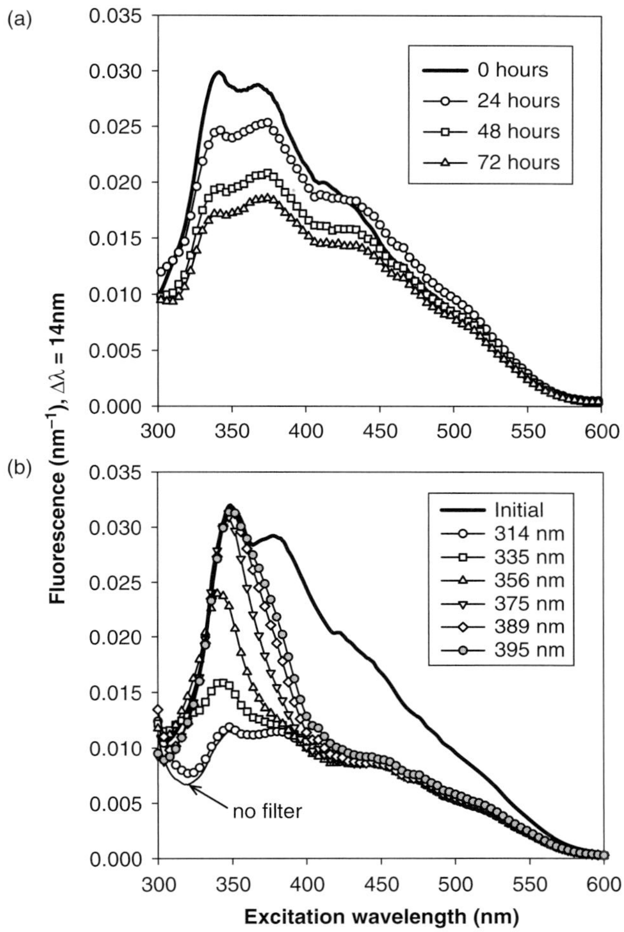

Figure 7.8. (a) The change in SF spectra (Δλ = 14 nm) due to the photobleaching of Mackenzie River DOM after exposure to sunlight for 3 days. (b) The change in SF spectra due to photobleaching, showing the effect of removing increasingly short wave portions of incident irradiance through the use of optical cutoff filters. In both experiments, the loss in SF intensity modified the spectral shape of the DOM fluorescence, making river water DOM fluorescence similar in appearance to DOM collected from the Mackenzie shelf region of the Beaufort Sea, western Canadian Arctic Ocean. (Reprinted from Osburn, C.L., Retamal, L., and Vincent, W.F., Photoreactivity of chromophoric dissolved organic matter transported by the Mackenzie River to the Beaufort Sea. *Mar. Chem.*, **115**, 10–20, 2009, with permission from Elsevier.)

DOM fluorescence. This consideration is important for interpreting the change to DOM fluorescence spectra in estuaries and in coastal waters both of large lakes and of the ocean. Osburn et al. (2009b) conducted this experiment with Mackenzie River DOM, reproduced in Figure 7.8. They found that substantial modifications to SF spectra occur that ultimately remove the humic fluorescence (SF 350 and SF380) but that SF280 fluorescence remained unchanged. To capture the spectral match of photobleached Mackenzie River DOM fluorescence to Arctic Ocean DOM fluorescence, Pearson correlation coefficients were calculated between SF spectra of river samples before and after photobleaching and SF spectra of Arctic Ocean DOM. In nearly all cases, photobleaching improved the correlation, often from $r = 0.8$ to $r = 0.9$. In fact, their results also indicated that the UV-A and blue region wavelengths (>360 nm) were most important for modifying Mackenzie River DOM to more closely match the spectral signature of Arctic Ocean DOM. A similar modeling approach by Pullin and Cabaniss (1997) determined that Cuyahoga River water DOM that was photolabile on the order of 3–7 days could obscure the use of SF as a conservative mixing tracer, producing spectra more similar to those of the Detroit River DOM.

Interactive solution chemistry effects (pH, ionic strength) on DOM fluorescence bleaching are obvious, but trends across environmental systems and with respect to DOM sources are inconclusive. Reche et al. (1999) showed high CDOM photobleaching rates increasing with alkalinity, yet acidification was shown to facilitate photobleaching in boreal lakes (Gennings et al., 2001). Although neither Minor et al. (2006) nor Hefner et al. (2006) found an effect of salinity on highly absorptive estuarine DOM and Suwannee River humic acid (SRHA), respectively, Osburn et al. (2009a) showed that longwave photobleaching of CDOM absorption increased with salinity for ultrafiltered DOM and for SRHA. The Osburn et al. (2009a) results were corroborated by Grebel et al. (2009) for CDOM absorption, but not for DOM fluorescence, so it appears that DOM fluorescence emission might respond differently to photobleaching than would DOM absorption.

However, in the SRHA experiment from Osburn et al. (2009a), EEM peak ratio data do suggest an effect of ionic strength (Figure 7.9, Osburn, unpublished results). The data show the percent change in A/T, A/C, and A/M peak ratios for SRHA mixed separately into freshwater, estuarine, and marine permeates and exposed to sunlight. Relative to proteinaceous material, humic material appeared more photolabile (A/T decreased) in lower salinity water than in seawater. The longer wavelength fluorescent humic material appeared most photoreactive. The A/C ratio increased with salinity to a greater degree than did the A/M ratio. This would indicate that the removal of the C peak is sensitive to changes in salinity, though perhaps for metal–humic interactions or efficiency of peroxide generation (O'Sullivan et al., 2005). Recently, it has been shown that both LMW DOM A and C peak fluorescence may be preferentially bleached at lower salinities while HMW DOM A and C peaks appeared to photobleach more at mid to high salinities (Boyd et al., 2010b). A four-dimensional PARAFAC model, representing excitation, emission, salinity, pre- and post-photobleaching showed that T-peak fluorescence increased after exposure while the A, C, and M peaks were reduced. The changes to peak ratios in Figure 7.9 support the findings of Boyd et al. (2010b) and suggest that salinity may increase photobleaching of DOM

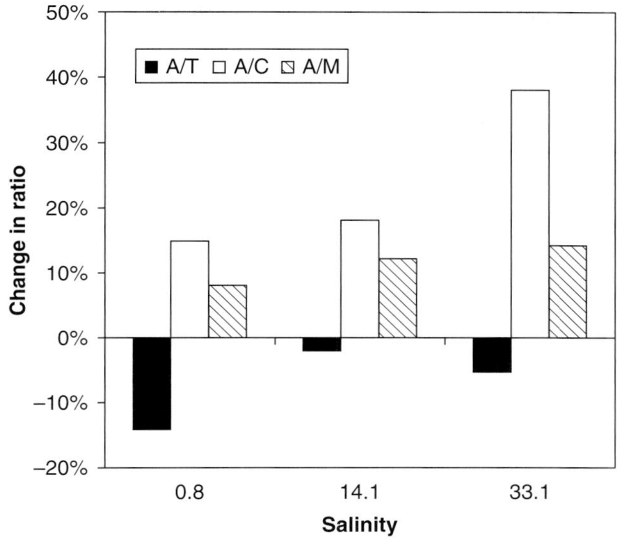

Figure 7.9. The change to EEM fluorescence peak ratios after Suwannee River humic acid (SRHA) was exposed to sunlight in solutions of increasing salinity. The ratio of A/T decreased in general without a consistent effect of salinity. Ratios of A/C and A/M after sunlight exposure increased with salinity, indicating an enhanced loss of C and M peak fluorescence as salinity increased. (Osburn, unpublished results.)

fluorescence and ultimately contribute to changes in DOM fluorescent properties seen in coastal waters.

The results from DOM fluorescence photobleaching studies suggest substantial molecular changes to DOM after it is exposed to sunlight. Dissolved inorganic carbon (CO_2) photoproduction, attributed to decarboxylation reactions, represents complete DOM degradation (mineralization) and can be used to assess carbon cycling implications for photodegradation (e.g., Miller and Zepp, 1995). Carboxylic acid oxidation should enhance fluorescence emission if a metal–ligand bond is disrupted once the carboxyl group is oxidized. Thus, models of DOM fluorescence changes coupled with DOM molecular changes will improve the predictability of carbon cycling in surface waters by means of fluorescence measurements.

7.10 Summary and Future Directions

Changing environmental conditions in terrestrial and aquatic ecosystems can likely impact the fluorescence signatures of DOM generated, transported, and transformed within and between these systems. A first-order control on the source of DOM fluorescence in natural waters will be the relative distribution of DOM molecular size and weight. The coupling of fluorescence with size fraction technologies and other means of colloid investigation have provided valuable insight into the differences in DOM fluorescence properties between freshwater and marine systems. However, we suggest that three environmental processes will exert the largest effect on DOM fluorescent properties in surface waters. First, surface water acidification that causes substantial decreases in pH below about 5 could induce diminished fluorescence and blue shifting seen by Laane (1982), Mobed et al. (1996), and Spencer et al. (2007b) among others. Evidence that decreasing pH or increasing pH (Reche et al., 1999), as well as modifying ionic strength (Osburn et al., 2009a; Boyd et al., 2010b), can promote photochemical reactions should also be noted, even if the effects to DOM fluorescence specifically are not apparent (Grebel et al., 2009). Whereas pH can change spectral shapes and lead to diminished fluorescence, it appears that metal quenching solely leads to diminished fluorescence intensity. The exception is Al(III)–OM complexes, which amplify DOM fluorescence in humic regions, therefore increasing fluorescence intensity in addition to imposing red shifting in emission spectra.

Second, photobleaching can permanently alter DOM fluorescence by degrading fluorophores. The photobleaching behavior under monochromatic and polychromatic light exposure cannot be explained by a simple superposition of non interacting fluorophores. Therefore, the interactivity of photobleaching on the molecular properties of DOM as it relates to bulk DOM fluorescence properties should be addressed. Advanced mass spectroscopy techniques such as Fourier transform-ion cyclotron resonance mass spectrometry (FT-ICR-MS) are currently being applied to these problems (e.g., Kujawinski et al., 2004; Mopper et al., 2007; Gonsior et al., 2009) and should provide substantial information regarding the molecular changes DOM undergoes as it is degraded. The information from this technique could provide insight into the molecular classes in given fluorophores.

Third, surface water metal speciation will offer multivalent elements that complex with available DOM and either quench or enhance DOM fluorescence. As erosion in upland catchments and aeolian rain and dust inputs increase, so should the metal content of surface waters. DOM fluorescence will respond accordingly. A possible future direction of study would be relating the regional mineralogy of a watershed (inland or coastal) to the DOM fluorescence properties in its streams and catchments. Gradual changes in DOM characteristics over time may therefore be related to the concentration and abundance of metals released into solution via weathering or anthropogenic inputs.

Ionic strength effects are more difficult to resolve. Although salinity was shown to increase DOM fluorescence to some degree, its overall quenching effect seems minimal, especially in coastal waters where freshwater and seawater continuously mix. The saline lake data provided a similar finding for DOM yet in a different chemical milieu than estuarine waters. Evapoconcentration of lake water does appear to alter DOM fluorescence properties as ionic strength increases. By contrast, the salinization of soils dramatically increased the ionic strength of soil solution and decreased the fluorescence as soil pore water DOM structures coil (possibly inducing steric hindrance) and functional groups ionize (Provenzano et al., 2008). Thus, DOM fluorescence offers the potential for new investigations in relation to climate and land use change.

Results of ionic strength on DOM photochemistry remain unclear. Experiments have not shown evidence supporting the possibility that increasing salinity enhances the photodegradation of humic DOM (Minor et al., 2006) and anthropogenic DOM (Kong and Ferry, 2003), yet studies do show an enhancement of DOM absorption photobleaching (Osburn et al., 2009a; Grebel et al., 2009), though the latter study did not find an effect on DOM fluorescence either in FA or in an algogenic DOM preparations. A multiestuary estuarine mixing study coupled with photobleaching showed an increase in T and B peak fluorescence indicating possible proteinaceous material photoproduction (Boyd et al., 2010b).

Similarly, a burgeoning literature on DOM fluorescence in aerosols is describing the abundance of atmospheric DOM with chemical properties similar to humic substances. The atmosphere also contains substantial amounts of metals in dust and photochemistry is important. Understanding the sources and reactivities of atmospheric DOM via fluorescence is important because of the many metal–ligand reactions that likely occur, some of which is clearly related to DOM photochemistry in the atmosphere.

In terms of environmental effects on DOM fluorescence, it appears that the most dramatic and permanent effect would result from photochemistry or from dehydration caused by evaporation. Largely, this is because pH, metal, and ionic strength effects appear to be reversible, whereas the effects of dehydration and photochemistry are not. Thus, DOM fluorescence ought to be most sensitive to global change conditions that promote photochemistry (pH and metals are linked to this) or dehydration.

Linkages between DOM in soils, sediments, surface waters and the atmosphere are clearly in need of further study because these systems are responsive to changing climate. Taking a system approach to understanding the role of pH, salinity, and metals in terms of DOM fluorescence will be particularly important now that multivariate statistical modeling (e.g., PARAFAC and other methods) can be used to determine similar fluorophore (or fluorescing components) among these linked environments. The effects of soil salinization on DOM properties using fluorescence has been investigated (Cilenti et al., 2005) and dehydration effects of a drier climate clearly alter the fluorescence and perhaps quantity of DOM in natural waters (Hudson et al., 2009). This understanding of extractable and pore water DOM fluorescence properties in relation to pH and salinity changes – perhaps even to temperature changes – affords the opportunity to examine these recorders of environmental change for information regarding the global C cycle. The utility of fluorescence for understanding the transport and cycling of DOM between the terrestrial, aquatic, and atmospheric environments has been demonstrated, and we suggest the future discovery will be utilizing DOM fluorescence further to link the chemistries of each environment with the larger picture of environmental change.

Acknowledgments

Braden Giordano is thanked for help in preparation of this manuscript. Jasmine Saros kindly provided conductivity data for Alkaline Lake. The Office of Naval Research (Work Unit Nos. N0001401WX20072 and N0001403WX20946) supported field collection and analysis of DOM fluorescence from the Chesapeake Bay and the Mackenzie River Estuary. The National Science Foundation (Division of Environmental Biology Award 0315665) supported field collection and analysis of DOM fluorescence from Alkaline Lake.

References

Alberts, J.J., Takacs, M., and Schalles, J. (2004). Ultraviolet-visible and fluorescence spectral evidence of natural organic matter (NOM) changes along an estuarine salinity gradient. *Estuaries*, **27**(2), 296–310.

Amon, R.M.W., Budeus, G., and Meon, B. (2003). Dissolved organic carbon distribution and origin in the Nordic Seas: Exchanges with the Arctic Ocean and the North Atlantic. *J. Geophys. Res. Oceans*, **108**, 3221, doi:10.1029/2002JC001594.

Anderson, N.J. and Stedmon, C.A. (2007). The effect of evapoconcentration on dissolved organic carbon concentration and quality in lakes of SW Greenland. *Freshwater Biol.*, **52**, 280–289.

Antunes, M.C.G., Pereira, C.C.C., and Esteves da Silva, J.C.G. (2007). MCR of the quenching of the EEM of fluorescence of dissolved organic matter by metal ions. *Anal. Chim. Acta*, **595**(1–2), 9–18.

Arts, M.T., Robarts, R.D., Kasai, F., Waiser, M.J., Tumber, V.P., Plante, A.J., Rai, H., and de Lange, H.J. (2000). The attenuation of ultraviolet radiation in high dissolved organic carbon waters of wetlands and lakes on the northern Great Plains. *Limnol. Oceanogr.*, **45**, 292–299.

Avena, M.J. and Wilkinson, K.J. (2002). Disaggregation kinetics of a peat humic acid: Mechanism and pH effects. *Environ. Sci. Technol.*, **36**(23), 5100–5105.

Baalousha, M., Motelica- Heino, M., and Le Coustumer, P. (2006). Conformation and size of humic substances: Effects of major cation concentration and type, pH, salinity, and residence time. *Colloids Surf. A: Physicochem. Eng. Aspects*, **272**(1–2), 48–55.

Baker, A. (2005). Thermal fluorescence quenching properties of dissolved organic matter. *Water Res.*, **39**(18), 4405–4412.

Baker, A. and Spencer, R.G.M. (2004). Characterization of dissolved organic matter from source to sea using fluorescence and absorbance spectroscopy. *Sci. Total Environ.*, **333**(1–3), 217–232.

Baker, A., Elliott, S., and Lead, J.R. (2007). Effects of filtration and pH perturbation on freshwater organic matter fluorescence. *Chemosphere*, **67**(10), 2035–2043.

Baker, A., Tipping, E., Thacker, S.A., and Gondar, D. (2008). Relating dissolved organic matter fluorescence and functional properties. *Chemosphere*, **73**(11), 1765–1772.

Banaitis, M.R. Waldrip-Dail, H., Diehl, M.S., Holmes, B.C., Hunt, J.F., Lynch, R.P., and Ohno, T. (2006). Investigating sorption-driven dissolved organic matter fractionation by multidimensional fluorescence spectroscopy and PARAFAC. *J. Colloid Interf. Sci.*, **304**(1), 271–276.

Batchelli, S., Muller, F.L.L., Baalousha, M., and Lead, J.R. (2009). Size fractionation and optical properties of colloids in an organic-rich estuary (Thurso, UK). *Mar. Chem.*, **113**(3–4), 227–237.

Belzile, C. and Guo, L.D. (2006). Optical properties of low molecular weight and colloidal organic matter: Application of the ultrafiltration permeation model to DOM absorption and fluorescence. *Mar. Chem.*, **98**(2–4), 183–196.

Belzile, C., Johannessen, S.C., Gosselin, M., Demers, S., and Miller, W.L. (2000). Ultraviolet attenuation by dissolved and particulate constituents of first-year ice during late spring in an Arctic polynya. *Limnol. Oceanogr.*, **45**(6), 1265–1273.

Belzile, C., Gibson, J.A.E., and Vincent, W.F. (2002). Colored dissolved organic matter and dissolved organic carbon exclusion from lake ice: Implications for irradiance transmission and carbon cycling. *Limnol. Oceanogr.*, **47**(5), 1283–1293.

Benner, R. (2002). Chemical composition and reactivity. In D.A. Hansell and C.A. Carlson (Eds.), *Biogeochemistry of Marine Dissolved Organic Matter* (pp. 59–90). San Diego: Academic Press.

Benner, R., Pakulski, J.D., McCarthy, M., Hedges, J.I., and Hatcher, P.G. (1992). Bulk chemical characteristics of dissolved organic matter in the ocean. *Science*, **255**(5051), 1561–1564.

Biers, E.J., Zepp, R.G., and Moran, M.A. (2007). The role of nitrogen in chromophoric and fluorescent dissolved organic matter formation. *Mar. Chem.*, **103**(1–2), 46–60.

Blough, N.V., Zafiriou, O.C., and Bonilla, J. (1993). Optical absorption spectra of waters from the Orinoco River outflow: Terrestrial input of colored organic matter to the Caribbean. *J. Geophys. Res.*, **98**(C2), 2271–2278.

Boehme, J. and Wells, M. (2006). Fluorescence variability of marine and terrestrial colloids: Examining size fractions of chromophoric dissolved organic matter in the Damariscotta River estuary. *Mar. Chem.*, **101**(1–2), 95–103.

Boehme, J.R. and Coble, P.G. (2000). Characterization of colored dissolved organic matter using high-energy laser fragmentation. *Environ. Sci. Technol.*, **34**(15), 3283–3290.

Boyd, T.J. and Osburn, C.L. (2004). Changes in CDOM fluorescence from allochthonous and autochthonous sources during tidal mixing and bacterial degradation in two coastal estuaries. *Mar. Chem.*, **89**(1–4), 189–210.

Boyd, T.J., Wolgast, D.M., Rivera- Duarte, I., Holm- Hansen, O., Hewes, C.D., Zirino, A., and Chadwick, D.B. (2005). Effects of dissolved and complexed copper on heterotrophic bacterial production in San Diego Bay. *Microb. Ecol.*, **49**(3), 353–366.

Boyd, T.J., Barham, B.P., Hall, G.J., and Osburn, C.L. (2010a). Variation in ultrafiltered and LMW organic matter fluorescence properties under simulated estuarine mixing transects: 1. Mixing alone. *J. Geophys. Res.*, **115**, G00F13.

Boyd, T.J., Barham, B.P., Hall, G.J., Schumann, B.S., Paerl, R.W., and Osburn, C.L. (2010b). Variation in ultrafiltered and LMW organic matter fluorescence properties under simulated estuarine mixing transects: 2. Mixing with photoexposure. *J. Geophys. Res.*, **115**, G00F14.

Buesseler, K.O. Bauer, J.E., Chen, R.F., Eglinton, T.I., Gustafsson, O., Landing, W., Mopper, K., Moran, S.B., Santschi, P.H., Vernon-Clark, R. and Wells, M.L. (1996). An intercomparison of cross-flow filtration techniques used for sampling marine colloids: Overview and organic carbon results. *Mar. Chem.*, **55**(1–2), 1–31.

Cabaniss, S.E. and Shuman, M.S. (1986). Combined ion-selective electrode and fluorescence quenching detection for copper-dissolved organic matter titrations. *Anal. Chem.*, **58**(2), 398–401.

Cabaniss, S.E. and Shuman, M.S. (1987). Synchronous fluorescence spectra of natural waters: Tracing sources of dissolved organic matter. *Mar. Chem.*, **21**(1), 37–50.

Cabaniss, S.E. and Shuman, M.S. (1988). Copper binding by dissolved organic matter: I. Suwannee River fulvic acid equilibria. *Geochim. Cosmochim. Acta*, **52**(1), 185–193.

Cabaniss, S.E. (1992). Synchronous fluorescence spectra of metal-fulvic acid complexes. *Environ. Sci. Technol.*, **26**(6), 1133–1139.

Cao, Y., Conklin, M., and Betterton, E. (1995). Competitive complexation of trace metals with dissolved humic acid. *Environ. Health Perspect.*, **103**, 29–32.

Cilenti, A., Provenzano, M.R., and Senesi, N. (2005). Characterization of dissolved organic matter from saline soils by fluorescence spectroscopy. *Environ. Chem. Lett.*, **3**(2), 53–56.

Clark, C.D., Jimenez- Morais, J., Jones, G., Zanardi- Lamardo, E., Moore, C.A., and Zika, R.G. (2002). A time-resolved fluorescence study of dissolved organic matter in a riverine to marine transition zone. *Mar. Chem.*, **78**(2–3), 121–135.

Clark, C.D., Litz, L.P., and Grant, S.B. (2008). Salt marshes as a source of chromophoric dissolved organic matter (CDOM) to Southern California coastal waters. *Limnol. Oceanogr.*, **53**(5), 1923–1933.

Coble, P.G. (1996). Characterization of marine and terrestrial DOM in seawater using excitation emission matrix spectroscopy. *Mar. Chem.*, **51**(4), 325–346.

Coble, P.G. (2007). Marine optical biogeochemistry: The chemistry of ocean color. *Chem. Rev.*, **107**(2), 402–418.

Coble, P.G., Del Castillo, C.E., and Avril, B. (1998). Distribution and optical properties of CDOM in the Arabian Sea during the 1995 Southwest Monsoon. *Deep-Sea Res. Pt. Ii*, **45**(10–11), 2195–2223.

Conmy, R.N., Coble, P.G., Chen, R.F., and Gardner, G.B. (2004). Optical properties of colored dissolved organic matter in the Northern Gulf of Mexico. *Mar. Chem.*, **89**(1–4), 127–144.

Conmy, R.N., Coble, P.G., Cannizzaro, J.P., and Heil, C.A. (2009). Influence of extreme storm events on West Florida Shelf CDOM distributions. *J. Geophys. Res. Biogeosci.*, 114, G00F04, doi:10.1029/2009JG000981

Conte, P. and Piccolo, A. (1999). Conformational arrangement of dissolved humic substances. Influence of solution composition on association of humic molecules. *Environ. Sci. Technol.*, **33**(10), 1682–1690.

Cory, R.M., McNeill, K., Cotner, J.P., Amado, A., Purcell, J.M., and Marshall, A.G. (2010). Singlet Oxygen in the coupled photochemical and biochemical oxidation of dissolved organic matter. *Environ. Sci. Technol.*, **44**(10), 3683–3689.

da Silva, J. and Tauler, R. (2006). Multivariate curve resolution of synchronous fluorescence spectra matrices of fulvic acids obtained as a function of pH. *Appl. Spectrosc.*, **60**(11), 1315–1321.

daSilva, J. and Machado, A. (1996). Characterization of the binding sites for Al(III) and Be(II) in a sample of marine fulvic acids. *Mar. Chem.*, **54**(3–4), 293–302.

da Silva, J.C.G., Machado, A., and Oliveira, C.J.S. (1998). Effect of pH on complexation of Fe(III) with fulvic acids. *Environ. Toxicol. Chem.*, **17**(7), 1268–1273.

Del Castillo, C.E., Coble, P.G., Morell, J.M., Lopez, J.M., and Corredor, J.E. (1999). Analysis of the optical properties of the Orinoco River plume by absorption and fluorescence spectroscopy. *Mar. Chem.*, **66**(1–2), 35–51.

Del Vecchio, R. and Blough, N.V. (2002). Photobleaching of chromophoric dissolved organic matter in natural waters: Kinetics and modeling. *Mar. Chem.*, **78**(4), 231–253.

Del Vecchio, R. and Blough, N.V. (2004). Spatial and seasonal distribution of chromophoric dissolved organic matter and dissolved organic carbon in the Middle Atlantic Bight. *Mar. Chem.*, **89**(1–4), 169–187.

Eftink, M. (1991). Fluorescence techniques for studying protein structure. In C.H. Suelter (Ed.), *Methods of Biochemical Analysis: Protein Structure Determination*, Vol. 35 (pp. 127–205). Hoboken, NJ: John Wiley & Sons.

Elkins, K.M. and Nelson, D.J. (2001). Fluorescence and FT-IR spectroscopic studies of Suwannee river fulvic acid complexation with aluminum, terbium and calcium. *J. Inorg. Biochem.*, **87**(1–2), 81–96.
Elkins, K.M. and Nelson, D.J. (2002). Spectroscopic approaches to the study of the interaction of aluminum with humic substances. *Coord, Chem. Rev.*, **228**(2), 205–225.
Esteves, V.I., Santos, E.B.H., and Duarte, A.C. (1999). Study of the effect of pH, salinity and DOC on fluorescence of synthetic mixtures of freshwater and marine salts. *J. Environ. Monit.*, **1**(3), 251–254.
Floge, S.A. and Wells, M.L. (2007). Variation in colloidal chromophoric dissolved organic matter in the Damariscotta Estuary, Maine. *Limnol. Oceanogr.*, **52**(1), 32–45.
Fox, L.E. (1983). The removal of dissolved humic acid during estuarine mixing. *Estuar. Coast. Shelf Sci.*, **16**(4), 431–440.
Francko, D.A. and Heath, R.T. (1982). UV-sensitive complex phosphorus: Association with dissolved humic material and iron in a bog lake. *Limnol. Oceanogr.*, **27**(3), 564–569.
Fu, P.Q., Wu, F.C., Liu, C.Q., Wang, F.Y., Li, W., Yue, L.X., and Guo, Q.J. (2007). Fluorescence characterization of dissolved organic matter in an urban river and its complexation with Hg(II). *Appl. Geochem.*, **22**(8), 1668–1679.
Gennings, C., Molot, L.A., and Dillon, P.J. (2001). Enhanced photochemical loss of organic carbon in acidic waters. *Biogeochemistry*, **52**(3), 339–354.
Ghosh, K. and Schnitzer, M. (1979). UV and visible absorption spectroscopic investigations in relation to macromolecular characteristics of humic substances. *J. Soil Sci.*, **30**(4), 735–745.
Ghosh, K. and Schnitzer, M. (1980). Macromolecular structures of humic substances. *Soil Sci.*, **129**(5), 266–276.
Gibson, J.A.E., Vincent, W.F., and Pienitz, R. (2001). Hydrologic control and diurnal photobleaching of CDOM in a subarctic lake. *Arch. Hydrobiol.*, **152**(1), 143–159.
Goldstone, J.V., Del Vecchio, R., Blough, N.V., and Voelker, B.M. (2004). A multicomponent model of chromophoric dissolved organic matter photobleaching. *Photochem. Photobiol.*, **80**(1), 52–60.
Gone, D.L., Seidel, J.L., Batiot, C., Bamory, K., Ligban, R., and Biemi, J. (2009). Using fluorescence spectroscopy EEM to evaluate the efficiency of organic matter removal during coagulation-flocculation of a tropical surface water (Agbo reservoir). *J. Hazard. Mater.*, **172**(2–3), 693–699.
Gonsior, M., Peake, B.M., Cooper, W.T., Podgorski, D., D'Andrilli, J., and Cooper, W.J. (2009). Photochemically induced changes in dissolved organic matter identified by ultrahigh resolution Fourier transform ion cyclotron resonance mass spectrometry. *Environ. Sci. Technol.*, **43**(3), 698–703.
Graber, E.R. and Rudich, Y. (2006). Atmospheric HULIS: How humic-like are they? A comprehensive and critical review. *Atmos. Chem. Phys.*, **6**, 729–753.
Grebel, J.E., Pignatello, J.J., Song, W.H., Cooper, W.J., and Mitch, W.A. (2009). Impact of halides on the photobleaching of dissolved organic matter. *Mar. Chem.*, **115**(1–2), 134–144.
Green, S.A., Morel, F.M.M., and Blough, N.V. (1992). Investigation of the electrostatic properties of humic substances by fluorescence quenching. *Environ. Sci. Technol.*, **26**(2), 294–302.
Gueguen, C., Guo, L.D., Yamamoto-Kawai, M., and Tanaka, N. (2007). Colored dissolved organic matter dynamics across the shelf-basin interface in the western Arctic Ocean. *J. Geophys. Res. Oceans*, **112**, C05038, doi:10.1029/2006JC003584.

Guo, L.D. and Santschi, P.H. (1997). Composition and cycling of colloids in marine environments. *Rev. Geophys.*, **35**(1), 17–40.

Hefner, K.H., Fisher, J.M., and Ferry, J.L. (2006). A multifactor exploration of the photobleaching of Suwannee River dissolved organic matter across the freshwatersaltwater interface. *Environ. Sci. Technol.*, **40**(12), 3717–3722.

Henderson, R.K., Baker, A., Murphy, K.R., Hamblya, A., Stuetz, R.M., and Khan, S.J. (2009). Fluorescence as a potential monitoring tool for recycled water systems: A review. *Water Res.*, **43**(4), 863–881.

Hosse, M. and Wilkinson, K.J. (2001). Determination of electrophoretic mobilities and hydrodynamic radii of three humic substances as a function of pH and ionic strength. *Environ. Sci. Technol.*, **35**(21), 4301–4306.

Hudson, N., Baker, A., Reynolds, D.M., Carliell- Marquet, C., and Ward, D. (2009). Changes in freshwater organic matter fluorescence intensity with freezing/thawing and dehydration/rehydration. *J. Geophys. Res. Biogeosci.*, 114, G00F08, doi:10.1029/2008JG000915

Huguet, A., Vacher, L., Saubusse, S., Etcheber, H., Abril, G., Relexans, S., Ibalot, F., and Parlanti, E. (2010). New insights into the size distribution of fluorescent dissolved organic matter in estuarine waters. *Org. Geochem.*, **41**(6), 595–610.

Kaiser, K. and Guggenberger, G. (2000). The role of DOM sorption to mineral surfaces in the preservation of organic matter in soils. *Org. Geochem.*, **31**(7–8), 711–725.

Kieber, R.J., Zhou, X.L., and Mopper, K. (1990). Formation of carbonyl compounds from UV-induced photodegradation of humic substances in natural waters: Fate of riverine carbon in the sea. *Limnol. Oceanogr.*, **35**(7), 1503–1515.

Kieber, R.J., Hydro, L.H., and Seaton, P.J. (1997). Photooxidation of triglycerides and fatty acids in seawater: Implication toward the formation of marine humic substances. *Limnol. Oceanogr.*, **42**, 1454–1462.

Kieber, R.J., Willey, J.D., Whitehead, R.F., and Reid, S.N. (2007). Photobleaching of chromophoric dissolved organic matter (CDOM) in rainwater. *J. Atmos. Chem.*, **58**(3), 219–235.

Kong, L. and Ferry, J.L. (2003). Effect of salinity on the photolysis of chrysene adsorbed to a smectite clay. *Environ. Sci. Technol.*, **37**(21), 4894–4900.

Korshin, G.V., Kumke, M.U., Li, C.W., and Frimmel, F.H. (1999). Influence of chlorination on chromophores and fluorophores in humic substances. *Environ. Sci. Technol.*, **33**(8), 1207–1212.

Koussai, A.M. and Zika, R.G. (1990). Light-induced alteration of the photophysical properties of dissolved organic matter in seawater. 1. Photoreversible properties of natural water fluorescence. *Nether. J. Sea Res.*, **27**(1), 25–32.

Kowalczuk, P., Cooper, W.J., Whitehead, R.F., Durako, M.J., and Sheldon, W. (2003). Characterization of CDOM in an organic-rich river and surrounding coastal ocean in the South Atlantic Bight. *Aquat. Sci.*, **65**(4), 384–401.

Kowalczuk, P., Durako, M.J., Young, H., Kahn, A.E., Cooper, W.J., and Gonsior, M. (2009). Characterization of dissolved organic matter fluorescence in the South Atlantic Bight with use of PARAFAC model: Interannual variability. *Mar. Chem.*, **113**(3–4), 182–196.

Kujawinski, E.B., Del Vecchio, R., Blough, N.V., Klein, G.C., and Marshall, A.G. (2004). Probing molecular-level transformations of dissolved organic matter: Insights on photochemical degradation and protozoan modification of DOM from electrospray ionization Fourier transform ion cyclotron resonance mass spectrometry. *Mar. Chem.*, **92**(1–4), 23–37.

Kumke, M.U., Specht, C.H., Brinkmann, T., and Frimmel, F.H. (2001). Alkaline hydrolysis of humic substances – spectroscopic and chromatographic investigations. *Chemosphere*, **45**(6–7), 1023–1031.
Laane, R. (1982). Influences of pH on the fluorescence of dissolved organic matter. *Mar. Chem.*, **11**(4), 395–401.
Lakowicz, J. (2006). *Principles of Fluorescence Spectroscopy*. New York: Springer Science+Business Media.
Lakshman, S., Mills, R., Patterson, H., and Cronan, C. (1993). Apparent differences in binding site distributions and aluminum(III) complexation for three molecular weight fractions of a coniferous soil fulvic acid. *Anal. Chim. Acta*, **282**(1), 101–108.
Lead, J.R., De Momi, A., Goula, G., and Baker, A. (2006). Fractionation of freshwater colloids and particles by SPLITT: Analysis by electron microscopy and 3D excitation-emission matrix fluorescence. *Anal. Chem.*, **78**(11), 3609–3615.
Leenheer, J.A., Noyes, T.I., Rostad, C.E., and Davisson, M.L. (2004). Characterization and origin of polar dissolved organic matter from the Great Salt Lake. *Biogeochemistry*, **69**(1), 125–141.
Levesque, M. (1972). Fluorescence and gel filtration of humic compounds. *Soil Sci.*, **113**(5), 346–353.
Liu, X. and Millero, F.J. (2002). The solubility of iron in seawater. *Mar. Chem.*, **77**(1), 43–54.
Liu, R.X., Lead, J.R., and Baker, A. (2007). Fluorescence characterization of cross flow ultrafiltration derived freshwater colloidal and dissolved organic matter. *Chemosphere*, **68**(7), 1304–1311.
Lochmuller, C.H. and Saavedra, S.S. (1986). Conformational changes in soil fulvic acid measured by time-dependent fluorescence depolarization. *Anal. Chem.*, **58**(9), 1978–1981.
Lu, X.Q. and Jaffe, R. (2001). Interaction between Hg(II) and natural dissolved organic matter: A fluorescence spectroscopy based study. *Water Res.*, **35**(7), 1793–1803.
Luster, J., Lloyd, T., Sposito, G., and Fry, I.V. (1996). Multi-wavelength molecular fluorescence spectrometry for quantitative characterization of copper(II) and aluminum(III) complexation by dissolved organic matter. *Environ. Sci. Technol.*, **30**(5), 1565–1574.
Ma, J.H., Del Vecchio, R., Golanoski, K.S., Boyle, E.S., and Blough, N.V. (2010). Optical properties of humic substances and CDOM: Effects of borohydride reduction. *Environ. Sci. Technol.*, **44**(14), 5395–5402.
Ma, X.D. and Green, S.A. (2004). Photochemical transformation of dissolved organic carbon in Lake Superior – An in-situ experiment. *J. Great Lakes Res.*, **30**, 97–112.
Mac, M. and Wirz, J. (1993). Deriving intrinsic electron-transfer rates from nonlinear Stern-Volmer dependencies for fluorescence quenching of aromatic molecules by inorganic anions in acetonitrile. *Chem. Phys. Lett.*, **211**(1), 20–26.
Mac, M. (1995). Fluorescence quenching of aromatic molecules by inorganic anions in polar solvents. *J. Luminesc.*, **65**(3), 143–151.
Maloney, K.O., Morris, D.P., Moses, C.O., and Osburn, C.L. (2005). The role of iron and dissolved organic carbon in the absorption of ultraviolet radiation in humic lake water. *Biogeochemistry*, **75**(3), 393–407.
Mantoura, R.F.C. and Woodward, E.M.S. (1983). Conservative behavior of riverine dissolved organic carbon in the Severn Estuary – Chemical and geochemical implications. *Geochim. Cosmochim. Acta*, **47**(7), 1293–1309.
Maranger, R. and Pullin, M.J. (2002). Elemental complexation by dissolved organic matter in lakes: Implications for Fe speciation and the bioavailability of Fe and P. In

S. Findlay and R. Sinsabaugh (Eds.), *Dissolved Organic Matter in Aquatic Ecosystems* (pp. 186–217). San Diego: Academic Press.

Mariot, M., Dudal, Y., Furian, S., Sakamoto, A., Valles, V., Fort, M., and Barbiero, L. (2007). Dissolved organic matter fluorescence as a water-flow tracer in the tropical wetland of Pantanal of Nhecolandia, Brazil. *Sci. Total Environ.*, **388**, 184–193.

Mayer, L.M., Schick, L.L., and Loder, T.C. (1999). Dissolved protein fluorescence in two Maine estuaries. *Mar. Chem.*, **64**(3), 171–179.

McKnight, D.M., Boyer, E.W., Westerhoff, P.K., Doran, P.T., Kulbe, T., and Andersen, D.T. (2001). Spectrofluorometric characterization of dissolved organic matter for indication of precursor organic material and aromaticity. *Limnol. Oceanogr.*, **46**, 38–48.

Means, J.C. (1995). Influence of salinity upon sediment-water partitioning of aromatic hydrocarbons. *Mar. Chem.*, **51**(1), 3–16.

Miller, W.L. (1998). Effects of UV radiation on aquatic humus: Photochemical principles and experimental considerations. In D.O. Hessen and L.J. Tranvik (Eds.), *Aquatic Humic Substances: Ecology and Biogeochemistry* (pp. 125–141). Berlin: Springer-Verlag.

Miller, W.L. and Zepp, R.G. (1995). Photochemical production of dissolved inorganic carbon from terrestrial organic matter: Significance to the oceanic organic carbon cycle. *Geophys. Res. Lett.*, **22**(4), 417–420.

Milne, P.J. and Zika, R.G. (1989). Luminescence quenching of dissolved of dissolved organic matter in seawater. *Mar. Chem.*, **27**(3–4), 147–164.

Minor, E.C., Pothen, J., Dalzell, B.J., Abdulla, H., and Mopper, K. (2006). Effects of salinity changes on the photodegradation and ultraviolet-visible absorbance of terrestrial dissolved organic matter. *Limnol. Oceanogr.*, **51**(5), 2181–2186.

Mobed, J.J., Hemmingsen, S.L., Autry, J.L., and McGown, L.B. (1996). Fluorescence characterization of IHSS humic substances: Total luminescence spectra with absorbance correction. *Environ. Sci. Technol.*, **30**(10), 3061–3065.

Mopper, K., Feng, Z.M., Bentjen, S.B., and Chen, R.F. (1996). Effects of cross-flow filtration on the absorption and fluorescence properties of seawater. *Mar. Chem.*, **55**(1–2), 53–74.

Mopper, K., Stubbins, A., Ritchie, J.D., Bialk, H.M., and Hatcher, P.G. (2007). Advanced instrumental approaches for characterization of marine dissolved organic matter: Extraction techniques, mass spectrometry, and nuclear magnetic resonance spectroscopy. *Chem. Rev.*, **107**(2), 419–442.

Moran, M. A., Sheldon, W.M., and Zepp, R.G. (2000). Carbon loss and optical property changes during long-term photochemical and biological degradation of estuarine dissolved organic matter. *Limnol. Oceanogr.*, **45**(6), 1254–1264.

Ohno, T., Amirbahman, A., and Bro, R. (2008). Parallel factor analysis of excitation-emission matrix fluorescence spectra of water soluble soil organic matter as basis for the determination of conditional metal binding parameters. *Environ. Sci. Technol.*, **42**(1), 186–192.

Ortega-Retuerta, E., Pulido-Villena, E., and Reche, I. (2007). Effects of dissolved organic matter photoproducts and mineral nutrient supply on bacterial growth in mediterranean inland waters. *Microb. Ecol.*, **54**, 161–169.

Osburn, C.L., and Morris, D.P. (2003). Photochemistry of chromophoric dissolved organic matter in natural waters. In E.W. Hebling and H.E. Zagarese (Eds.), *UV Effects in Aquatic Organisms and Ecosystems* (pp. 185–217). London: Royal Society of Chemistry.

Osburn, C.L., Zagarese, H.E., Morris, D.P., Hargreaves, B.R., and Cravero, W.E. (2001). Calculation of spectral weighting functions for the solar photobleaching of chromophoric dissolved organic matter in temperate lakes. *Limnol. Oceanogr.*, **46**(6), 1455–1467.

Osburn, C.L., O'Sullivan, D.W., and Boyd, T.J. (2009a). Increases in the longwave photobleaching of chromophoric dissolved organic matter in coastal waters. *Limnol. Oceanogr.*, **54**(1), 145–159.

Osburn, C.L., Retamal, L., and Vincent, W.F. (2009b). Photoreactivity of chromophoric dissolved organic matter transported by the Mackenzie River to the Beaufort Sea. *Mar. Chem.*, **115**(1–2), 10–20.

Osburn, C.L., Wigdahl, C.R., Fritz, S.C., and Saros, J.E. (2011). Dissolved organic matter composition and photoreactivity in prairie lakes of the US Great Plains. *Limnol. Oceanogr.*, **56**(6), 2371–2390.

O'Sullivan, D.W., Neale, P.J., Coffin, R.B., Boyd, T.J., and Osburn, S.L. (2005). Photochemical production of hydrogen peroxide and methylhydroperoxide in coastal waters. *Mar. Chem.*, **97**(1–2), 14–33.

Otero, M., Mendonca, A., Valega, M., Santos, E.B.H., Pereira, E., Esteves, V.I., and Duarte, A. (2007). Fluorescence and DOC contents of estuarine pore waters from colonized and non-colonized sediments: Effects of sampling preservation. *Chemosphere*, **67**(2), 211–220.

Parlanti, E., Wörz, K., Geoffroy, L., and Lamotte, M. (2000). Dissolved organic matter fluorescence spectroscopy as a tool to estimate biological activity in a coastal zone submitted to anthropogenic inputs. *Org. Geochem.*, **31**(12), 1765–1781.

Patel-Sorrentino, N., Mounier, S., and Benaim, J.Y. (2002). Excitation–emission fluorescence matrix to study pH influence on organic matter fluorescence in the Amazon basin rivers. *Water Res.*, **36**(10), 2571–2581.

Patsayeva, S.V., Fadeev, V.V., Filippova, E.M., Chubarov, V.V., and Yuzhakov, V.I. (1991). Temperature and laser ultraviolet radiation influence on luminescence spectra of dissolved organic matter. *Vest. Moskov. Universit. Ser. 3 Fizik. Astronom.*, **32**(6), 71–75.

Provenzano, M.R., Cilenti, A., Gigliotti, G., and Senesi, N. (2008). Spectroscopic investigation on hydrophobic and hydrophilic fractions of dissolved organic matter extracted from soils at different salinities. *Clean Soil Air Water*, **36**(9), 748–753.

Provenzano, M.R., Caricasole, P., Brunetti, G., and Senesi, N. (2010). Dissolved organic matter extracted with water and a saline solution from different soil profiles. *Soil Sci.*, **175**(6), 255–262.

Psenner, R. (1999). Living in a dusty world: Airborne dust as a key factor for alpine lakes. *Water Air Soil Pollut.*, **112**(3–4), 217–227.

Pullin, M.J. and Cabaniss, S.E. (1997). Physicochemical variations in DOM-synchronous fluorescence: Implications for mixing studies. *Limnol. Oceanogr.*, **42**(8), 1766–1773.

Reche, I., Pace, M.L., and Cole, J.J. (1999). Relationship of trophic and chemical conditions to photobleaching of dissolved organic matter in lake ecosystems. *Biogeochemistry*, **44**(3), 259–280.

Ritchie, J.D. and Perdue, E.M. (2003). Proton-binding study of standard and reference fulvic acids, humic acids, and natural organic matter. *Geochim. Cosmochim. Acta*, **67**(1), 85–96.

Romera-Castillo, C., Sarmento, H., Alvarez- Salgado, X.A., Gasol, J.M., and Marrase, C. (2010). Production of chromophoric dissolved organic matter by marine phytoplankton. *Limnol. Oceanogr.*, **55**, 446–454.

Ryan, D.K. and Weber, J.H. (1982). Fluorescence quenching titration for determination of complexing capacities and stability constants of fulvic acid. *Anal. Chem.*, **54**(6), 986–990.

Ryan, D.K., Shia, C.P., and Oconnor, D.V. (1996). Fluorescence spectroscopic studies of Al-fulvic acid complexation in acidic solutions. *Humic Fulv. Acids Isolat., Struct., Environ. Role*, **651**, 125–139.

Saar, R.A. and Weber, J.H. (1980). Comparison of spectrofluorometry and ion-selective electrode potentiometry for determination of complexes between fulvic acid and heavy-metal ions, *Anal. Chem.*, **52**, 2095–2100.

Senesi, N. (1990). Molecular and quantitative aspects of the chemistry of fulvic acid and its interactions with metal ions and organic chemicals. 2. The fluorescence spectroscopy approach. *Anal. Chim. Acta*, **232**(1), 77–106.

Senesi, N. and D'Orazio, V. (2005). Fluorescence spectroscopy. In D. Hillel (Ed.), *Encyclopedia of Soils in the Environment* (pp. 35–52). Amsterdam: Academic Press.

Seredynska-Sobecka, B., Baker, A., and Lead, J.R. (2007). Characterisation of colloidal and particulate organic carbon in freshwaters by thermal fluorescence quenching. *Water Res.*, **41**(14), 3069–3076.

Sharpless, C.M. and McGown, L.B. (1999). Effects of aluminum-induced aggregation on the fluorescence of humic substances. *Environ. Sci. Technol.*, **33**(18), 3264–3270.

Shaw, P.J., Jones, R.I., and De Haan, H. (2000). The influence of humic substances on the molecular weight distributions of phosphate and iron in epilimnetic lake waters. *Freshwater Biol.*, **45**(4), 383–393.

Shizuka, H., Nakamura, M., and Morita, T. (1980). Anion-induced fluorescence quenching of aromatic molecules. *J. Phys. Chem.*, **84**(9), 989–994.

Sholkovitz, E.R. (1976). Flocculation of dissolved organic and inorganic matter during mixing of river water and seawater. *Geochim. Cosmochim. Acta*, **40**(7), 831–845.

Sierra, M.M.D., Donard, O.F.X., and Lamotte, M. (1997). Spectral identification and behaviour of dissolved organic fluorescent material during estuarine mixing processes. *Mar. Chem.*, **58**(1–2), 51–58.

Smart, P.L., Finlayson, B.L., Rylands, W.D., and Ball, C.M. (1976). Relation of fluorescence to dissolved organic carbon in surface waters. *Water Res.*, **10**(9), 805–811.

Specht, C.H., Kumke, M.U., and Frimmel, F.H. (2000). Characterization of NOM adsorption to clay minerals by size exclusion chromatography. *Water Res.*, **34**(16), 4063–4069.

Spencer, R.G.M., Ahad, J.M.E., Baker, A., Cowie, G.L., Ganeshram, R., Upstill-Goddard, R.C., and Uher, G. (2007a). The estuarine mixing behaviour of peatland derived dissolved organic carbon and its relationship to chromophoric dissolved organic matter in two North Sea estuaries (UK). *Estuar. Coast. Shelf Sci.*, **74**(1–2), 131–144.

Spencer, R.G.M., Bolton, L., and Baker, A. (2007b). Freeze/thaw and pH effects on freshwater dissolved organic matter fluorescence and absorbance properties from a number of UK locations. *Water Res.*, **41**(13), 2941–2950.

Stedmon, C.A., Markager, S., and Bro, R. (2003). Tracing dissolved organic matter in aquatic environments using a new approach to fluorescence spectroscopy. *Mar. Chem.*, **82**(3–4), 239–254.

Stolpe, B., Guo, L.D., Shiller, A.M., and Hassellov, M. (2010). Size and composition of colloidal organic matter and trace elements in the Mississippi River, Pearl River and the northern Gulf of Mexico, as characterized by flow field-flow fractionation. *Mar. Chem.*, **118**(3–4), 119–128.

Sun, L.Y., Sun, W.L., and Ni, J.R. (2009). Partitioning of water soluble organic carbon in three sediment size fractions: Effect of the humic substances. *J. Environ. Sci. China*, **21**(1), 113–119.

Treinin, A., Loeff, I., Hurley, J.K., and Linschitz, H. (1983). Charge transfer interactions of excited molecules with inorganic anions – The role of spin orbit coupling in controlling net electron transfer. *Chem. Phys. Lett.*, **95**(4–5), 333–338.

Tzortziou, M., Osburn, C.L., and Neale, P.J. (2007). Photobleaching of dissolved organic material from a tidal marsh-estuarine system of the Chesapeake Bay. *Photochem. Photobiol.*, **83**(4), 782–792.

Vahatalo, A.V., Salonen, K., Salkinoja- Salonen, M., and Hatakka, A. (1999). Photochemical mineralization of synthetic lignin in lake water indicates enhanced turnover of aromatic organic matter under solar radiation. *Biodegradation*, **10**(6), 415–420.

Vodacek, A., Blough, N.V., DeGrandpre, M.D., Peltzer, E.T., and Nelson, R.K. (1997). Seasonal variation of CDOM and DOC in the Middle Atlantic Bight: Terrestrial inputs and photooxidation. *Limnol. Oceanogr.*, **42**(4), 674–686.

Watkins, A.R. (1974). Kinetics of fluorescence quenching by inorganic anions. *J. Phys. Chem.*, **78**(25), 2555–2558.

Weishaar, J.L., Aiken, G.R., Bergamaschi, B.A., Fram, M.S., Fujii, R., and Mopper, K. (2003). Evaluation of specific ultraviolet absorbance as an indicator of the chemical composition and reactivity of dissolved organic carbon. *Environ. Sci. Technol.*, **37**(20), 4702–4708.

Wells, M.L. (2002). Marine colloids and trace metals. In D.A. Hansell and C.A. Carlson (Eds.), *Biogeochemistry of Marine Dissolved Organic Matter* (pp. 367–404). San Diego: Academic Press.

Willey, J.D. (1984). The effect of seawater magnesium on natural fluorescence during estuarine mixing and implications for tracer applications. *Mar. Chem.*, **15**(1), 19–45.

Wu, F.C., Mills, R.B., Evans, R.D., and Dillon, P.J. (2004). Kinetics of metal-fulvic acid complexation using a stopped-flow technique and three-dimensional excitation emission fluorescence spectrophotometer. *Anal. Chem.*, **76**(1), 110–113.

Yamashita, Y. and Jaffe, R. (2008). Characterizing the interactions between trace metals and dissolved organic matter using excitation-emission matrix and parallel factor analysis. *Environ. Sci. Technol.*, **42**(19), 7374–7379.

Zanardi-Lamardo, E., Clark, C.D., Moore, C.A., and Zika, R.G. (2002). Comparison of the molecular mass and optical properties of colored dissolved organic material in two rivers and coastal waters by flow field-flow fractionation. *Environ. Sci. Tech.*, **36**(13), 2806–2814.

Zhao, J. and Nelson, D.J. (2005). Fluorescence study of the interaction of Suwannee River fulvic acid with metal ions and Al^{3+}-metal ion competition. *J. Inorg. Biochem.*, **99**(2), 383–396.

Zhou, J.L. and Rowland, S.J. (1997). Evaluation of the interactions between hydrophobic organic pollutants and suspended particles in estuarine waters. *Water Res.*, **31**(7), 1708–1718.

Zhou, J.L., Rowland, S., Mantoura, R.F.C., and Braven, J. (1994). The formulation of humic coatings on mineral particles under simulated estuarine conditions – A mechanistic study. *Water Res.*, **28**(3), 571–579.

Zhou, J.L., Rowland, S.J., Mantoura, R.F.C., and Lane, M.C.G. (1997). Desorption of tefluthrin insecticide from soil in simulated rainfall runoff systems – Kinetic studies and modelling. *Water Res.*, **31**(1), 75–84.

Zsolnay, A., Baigar, E., Jimenez, M., Steinweg, B., and Saccomandi, F. (1999). Differentiating with fluorescence spectroscopy the sources of dissolved organic matter in soils subjected to drying. *Chemosphere*, **38**(1), 45–50.

8

Biological Origins and Fate of Fluorescent Dissolved Organic Matter in Aquatic Environments

COLIN A. STEDMON AND ROSE M. CORY

8.1 Introduction

All natural waters on Earth contain dissolved organic matter (DOM). The ubiquitous presence of organic life in all environments has resulted in its widespread distribution, from the depths of the oceans, to trapped within polar ice sheets, to aerosol droplets in the upper atmosphere. Despite the ability of microbes to evolve to degrade all types of complex structures, an organic residue always remains, owing to it being either energetically unfavorable, at too low a concentration, or preserved as a result of some other external limiting factor such as auxiliary nutrients.

Large pools of organic carbon exist on land as living biomass (500 Pg C) and soils (2300 Pg C) (Jobbágy and Jackson, 2000; Houghton, 2007). It has recently been estimated that 2.9 Pg C yr^{-1} are leached from land to inland waters, with 0.9 Pg C being consequently exported to the ocean (Tranvik et al., 2009). In the oceans the majority of the organic carbon (663 Pg C) is found as DOM (Hansell et al., 2009). Ocean productivity results in a net accumulation of 2 Pg C yr^{-1} as semilabile DOM (Hansell & Carlson, 1998) that is gradually remineralized during the mixing time of the oceans, whereas there is thought to be a more or less constant background of oceanic refractory DOM (Hansell et al., 2009). These carbon reservoir sizes and fluxes are considerable and an essential link in the global carbon cycle that controls climate. To put these numbers into context, the reservoir of carbon in the atmosphere as CO_2 is approximately 800 Pg (Houghton, 2007) and fossil fuel emissions are currently estimated at 7.2 Pg C yr^{-1} (Canadell et al., 2007). It is therefore important that we obtain a better understanding of the production, turnover, and fate of DOM in aquatic environments, if we are to attempt to understand its role in the global carbon cycle and the potential feedbacks to and change as a result of climate change (Jiao et al., 2010). To accomplish this, a battery of chemical techniques is required to quantify and characterize DOM and follow how its two major sinks, microbial and photochemical degradation, modify and ultimately remineralize organic carbon.

A fraction of the organic compounds present in dissolved organic matter absorbs light and a subfraction of these also fluoresce (Figure 8.1). For several decades, the optical (absorption and fluorescence) properties of DOM have been used to study both the distribution of DOM and its characteristics in aquatic environments. The major advantage of the approach is that it requires very small sample volumes and minimal preparation before

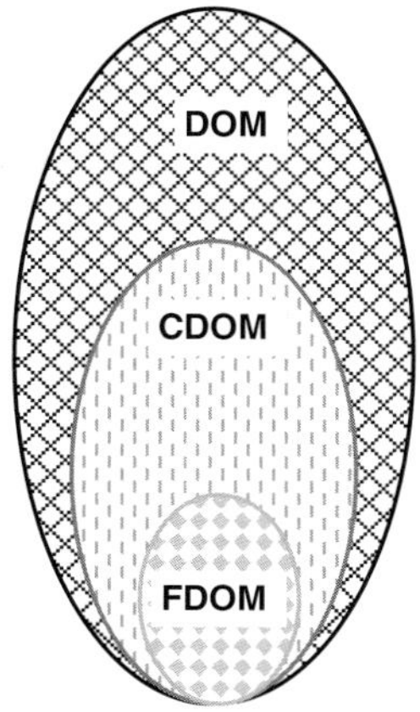

Figure 8.1. Schematic of the overlap between dissolved organic matter (DOM), colored dissolved organic matter (CDOM), and fluorescent dissolved organic matter (FDOM).

analysis. In addition, it is well suited for in situ or remote sensing measurement platforms, which can offer high spatial resolution measurements. The major pitfall of this approach is that the actual compounds responsible for the signal in general remain unknown. The absorption and fluorescence signals of DOM represent the sum of all optically active compounds present combined with any intramolecular charge transfer interactions between them (Del Vecchio and Blough, 2004). This complicates the interpretation of fluorescence spectra to some extent as the changes occurring in fluorescence intensity and spectra are due not only to the presence or absence of a specific fluorophore but also to the occurrence or absence of additional interactions, such as quenching or charge transfer (Del Vecchio and Blough 2004; Stedmon and Bro, 2008; Boyle et al., 2009). Despite this, some progress has been made in identifying likely fluorophores present in fluorescent dissolved organic matter (FDOM) samples.

In this chapter we briefly discuss the potential of the microbial food web as a source and sink for FDOM and examine how the fluorescence characteristics change as a result of microbial processing. In addition, the combined effects of microbial and photochemical degradation on FDOM characteristics are summarized. The literature on the subject is vastly expanding and it is not our aim to provide an exhaustive review but rather to highlight specific studies as examples of how fluorescence spectroscopy is being applied to studying the microbial turnover of DOM in natural aquatic environments.

8.2. Sources

8.2.1 Allochthonous versus autochthonous

Irrespective of which aquatic system is studied, the sources of FDOM can be classified into two categories: allochthonous and autochthonous. Allochthonous FDOM is material that is produced outside the system being studied and transported to it. For example, in lakes this consists of organic matter supplied by rivers that is often a mixture of material derived from soils and aquatic production upstream. Allochthonous inputs to an ocean basin can consist of precipitation and exchange of water with neighboring seas. In contrast, autochthonous FDOM is that which is released by organisms living within the system being studied. These definitions at first appear relatively simple and robust; however, it is useful to clarify their use in specific studies or systems, as they are often used interchangeably when referring to fixed carbon originating from either from terrestrial or aquatic environments. But as can be seen in the preceding example, aquatic systems are linked and it can be very difficult to distinguish between aquatic organic matter produced upstream or within a system.

8.2.2 Terrestrial Organic Matter

In soils, microbes degrade and transform organic matter from living and decaying plants and animals. Low molecular weight compounds either directly released or enzymatically cleaved from larger polymeric structures such as cellulose and lignin and the remaining material transformed to higher molecular weight compounds in a process commonly referred to as humification (Stevenson, 1982). During this process the chemical properties are altered, among them the number of carboxyl functional groups and C/H ratio increases. Fluorescence has for many years been used to characterize soil organic matter (Senesi et al., 1991) and its degree of humification (Zsolnay et al., 1999). With increased humification the emission spectrum shifts toward longer wavelengths (Figure 8.2). This shift can be characterized using the humification index (HIX), which is the emission intensity from 435–480 divided by the emission from 300–345 nm (Zsolnay et al., 1999).

Soil-derived DOM is a dominant source of FDOM to many fresh and coastal waters. Exceptions to this are hydrologically isolated systems surrounded by very little vegetation such as the Antarctic Dry Valley lakes (McKnight et al., 1991) and high-altitude mountain lakes (McKnight et al., 1997). The amount and quality of FDOM leached from soils can be expected to vary as a function of climate, soil hydrology, and catchment characteristics and slope, as seen for other DOM characteristics (Aitkenhead-Peterson et al., 2003; Mulholland, 2003; Stedmon et al., 2006; Fellman et al., 2009a). The concentration of FDOM in soil solutions and streams in natural catchments vary seasonally. For example, in a temperate Danish stream the highest concentrations are found in summer and lowest concentrations in winter (Stedmon and Markager, 2005a). This trend follows what is also observed for DOC in similar temperate systems (Tipping et al., 1999; Kalbitz et al., 2000) and is explained by a combination of reduced water throughput and greater soil microbial activity in the surface organic-rich layers (McDowell and Wood, 1984; Guggenberger et al., 1998; Kalbitz et al., 2000). In conjunction with this seasonal change in concentrations clear seasonal changes are also seen in the fluorescence characteristics of the leached FDOM. For example, for a small Danish forest stream the relative humic fluorescence at 412 nm and 504 nm varied considerably across season (components 3 and 2 respectively in

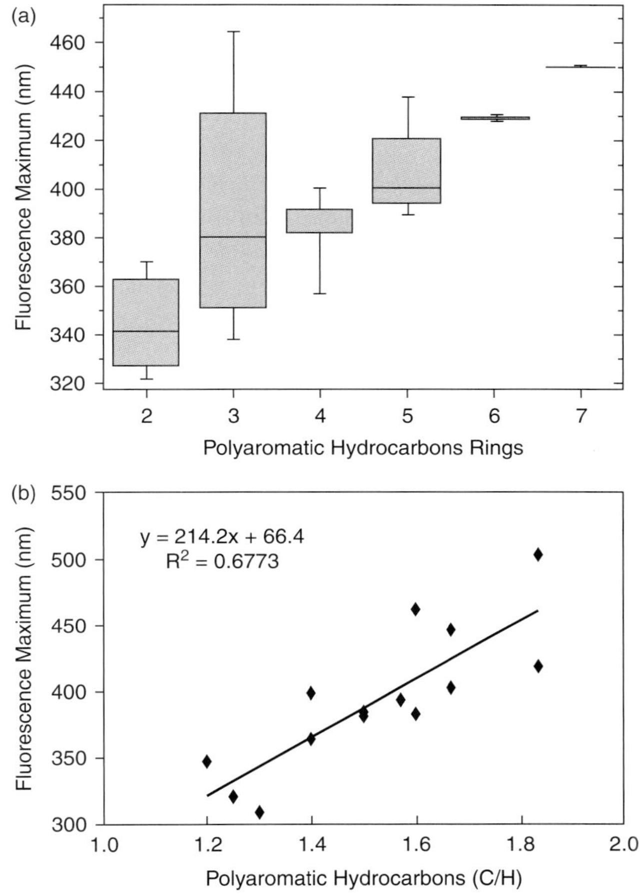

Figure 8.2. Example of how the fluorescence characteristics (emission maximum) of soil organic matter change with increasing aromaticity. (Reprinted from Zsolnay, 2003, with permission from Elsevier.)

Stedmon and Markager, 2005a). During colder temperatures the fluorescence signals were comparable but during the summer months the fluorescence at 412 nm was much greater.

In contrast, more anthropogenically influenced catchments, such as those dominated by agriculture, the mineral content and drainage time of the soils are notably higher. Subsequently the FDOM leached from these types of catchments differs from natural catchments. Seasonal changes in FDOM characteristics are less apparent and leaching of FDOM is largely driven by rapid drainage of precipitation. A recent study by Fellman et al. (2009a) has also shown that changes in the characteristics of FDOM exported from soils comparable to those seen across season in natural catchments can also occur over very short time periods, such as during a storm event. These compositional changes again reflect the dominant soil horizon flow path of the rainwater with recent labile surface organic matter being leached during storms compared to that during low-flow conditions. It is clear the quantity and characteristics of FDOM exported from catchments is largely controlled by a combination of soil microbial activity and the overall contact time that precipitation has in soils.

Lignin is a biopolymer unique to terrestrial plants (Sarkanen and Ludwig, 1971) and as a result useful for tracing the flux of organic matter from terrestrial ecosystems to rivers and eventually the open ocean (Ertel et al., 1986; Opsahl and Benner, 1997). Current techniques for quantifying and characterizing lignin content of DOM involve solid phase extraction and subsequent chemical oxidation to a suite of lignin phenol derivatives, which are detected using gas chromatography and mass spectrometry (e.g., Louchouarn et al., 2000). The approach is labor intensive and far from suitable for either routine or intensive sampling programs. As a result correlations between the lignin phenol content and characteristics and DOM ultraviolet (UV)-visible spectroscopic properties are being investigated (Del Vecchio and Blough, 2004; Boyle et al., 2009; Hernes et al., 2009; Spencer et al., 2009). Hernes et al. (2009) found that the region of fluorescence that gave best predictability of lignin concentrations and characteristics was the region with excitation below 300 nm and emission between 300 and 350 nm. These findings contradict previous understanding and current models that link the lignin fraction with humic material with longer wavelength fluorescence (Ertel et al., 1986; Lochmuller and Saavedra, 1986; Del Vecchio and Blough, 2004). For instance, strong similarities were observed between the spectral dependence of fluorescence quantum yields, emission peak maxima, and fluorescence lifetimes among extracted lignin and FDOM, suggesting that lignin and FDOM exhibit common photophysical and structural properties (Del Vecchio and Blough, 2004; Boyle et al., 2009). However, in support of the Hernes et al. (2009) finding, the fluorescence maxima of two breakdown products of lignin, vanillic and syringic acid, are at 326 and 338 nm respectively (Figure 8.3). Similarly, the results of Maie et al. (2007) also support this. They found that tryptophan-like fluorescence peak from DOM from the Florida Coastal Everglades could be chromatographically split using size-exclusion chromatography into two fractions; one correlated with organic nitrogen content (i.e., proteins) and another correlated to humic fluorescence signals.

Developments in data analysis now allow us to separate the fluorescence signal using advanced data analysis techniques, into underlying independent signals (Stedmon et al., 2003). This greatly simplifies distinguishing between different sources and processes acting on FDOM (e.g., Stedmon and Markager, 2005b), although much work is still required on understanding the actual chemical origins of these fluorescence signals, albeit specific fluorophores versus signals arising from complex interactions. In Figure 8.4 the fluorescence characteristics of components identified in some early studies are compared to those of organic fluorophores that can be expected to be found in aquatic environments. Ferulic acid and coumaric acid are two oxidation products from lignin. Soil fungi are known to produce enzymes (phenol oxidases) that degrade lignin to a range of polyhydroxy carboxylic acids. The similarity of the fluorescence spectra of these fractions with the standards

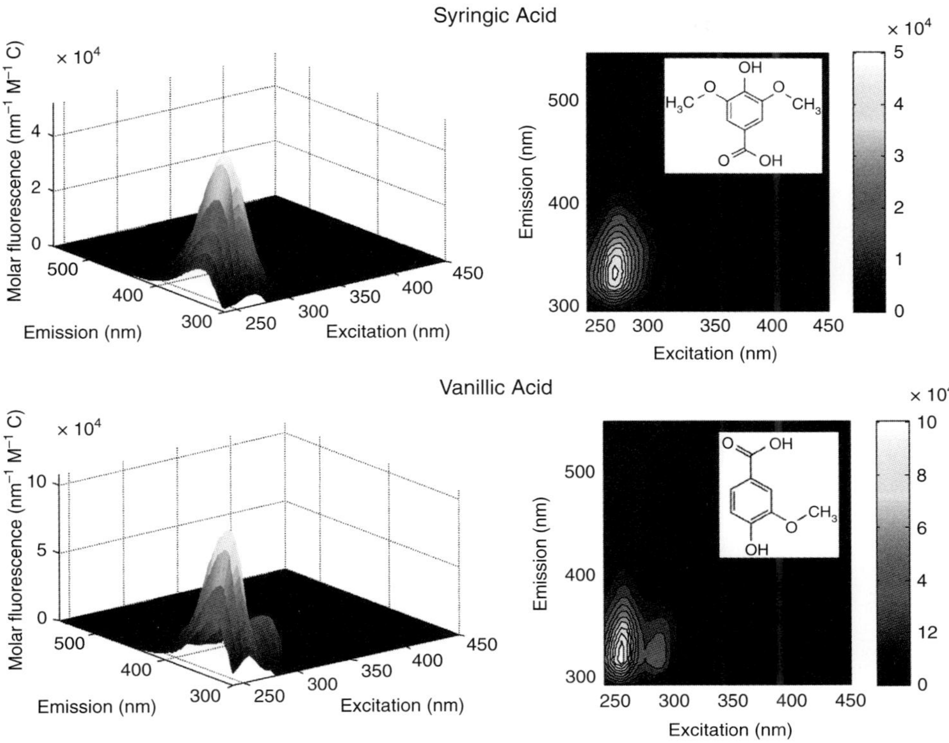

Figure 8.3. Molar fluorescence properties of Syringic (Sigma S 6881) and vanillic acid (Aldrich H3,600–1) in MilliQ water at room temperature. (Stedmon, unpublished data.) (See Plate 14.)

suggests that structures similar to these are present. However, this does not necessarily imply that lignin is the sole source of this fluorescence. A fluorescence signal similar to ferulic acid was originally labeled as marine humic-like material (M-peak) in early studies (e.g., Coble, 1996; Coble et al., 1998). Since then it has become increasingly clear that this signal is a product of both marine and terrestrially derived DOM (Stedmon et al., 2003; Murphy et al., 2006, 2008). Similarly, components strongly correlated with soil-derived organic matter in most freshwater environments have been observed in FDOM from Antarctic lakes (Fulton et al., 2004; Cory and McKnight, 2005), despite the fact that the watersheds draining these lakes are devoid of lignin (Spencer, unpublished data).

8.2.3 Aquatic Organic Matter

As with their soil counterparts, aquatic microbes process organic carbon continually modifying its characteristics. DOM is produced at all levels of the aquatic food web (Figure 8.5). Bacteria have been shown to be capable of rapidly releasing complex DOM despite

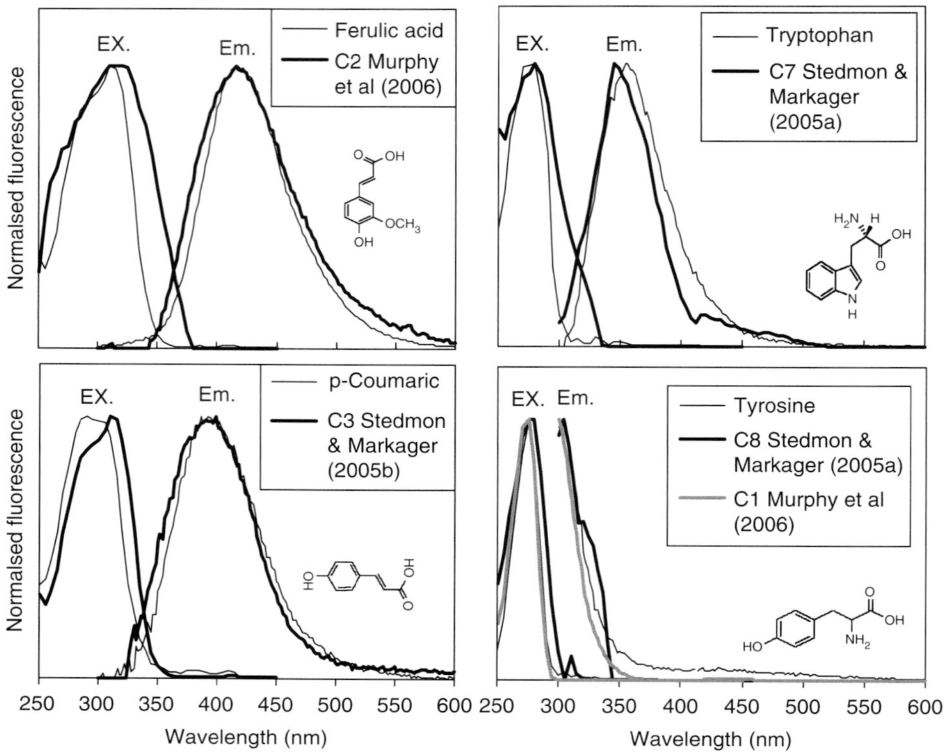

Figure 8.4. The fluorescence characteristics of four organic compounds compared to the characteristics of fluorescent components identified in earlier studies. Excitation and emission spectra are labelled as Ex. and Em, respectively.

growing on simple substrates (Ogawa et al., 2001). Bacteria can release between 14% and 31% of their carbon demand during growth (Kawasaki and Benner, 2006). In addition, grazing of bacteria by protozoans and cell lysis by viral infection also produce DOM (Strom et al., 1997; Middelboe and Lyck, 2002). Similarly, phytoplankton also release DOM during growth and as a result of grazing by zooplankton (Nagata, 2000). However, it is currently unclear to what extent FDOM is directly produced at each level in the food web. To date, studies have focused primarily on bacteria, phytoplankton, and zooplankton and some examples are given here. Although there are a range of organisms that possess fluorescent proteins and pigments, the discussion is focused on those fluorescent signals that are present in high enough concentrations and persist for long enough to be detected in natural DOM samples.

Determann et al. (1998) examined the fluorescence properties of several marine bacteria and phytoplankton species. The fluorescence signal from living bacteria was found to be remarkably constant and dominated by a blue-shifted tryptophan fluorescence. The fluorescence characteristics of the phytoplankton species investigated were slightly more

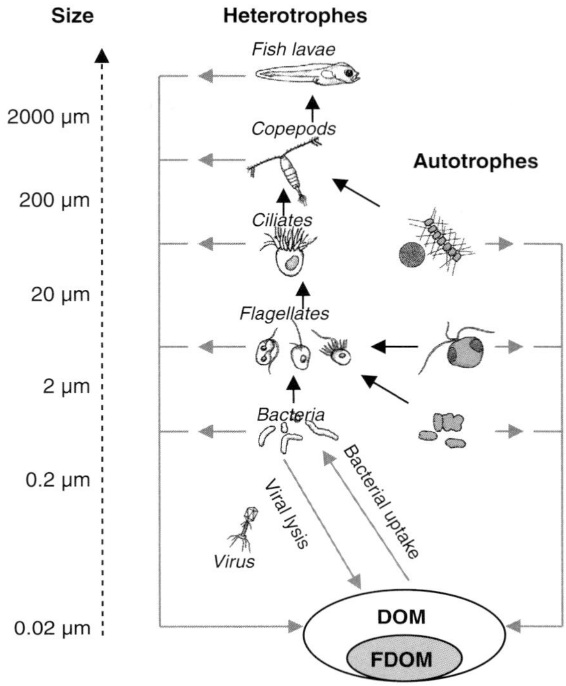

Figure 8.5. A simplified example of the lower marine food web with vira as the smallest organisms and fish lavae as the largest. Grazing is depicted as black arrows and DOM production from each trophic level is depicted with gray arrows. The production of FDOM has not been documented for all levels and it seems still unclear whether it is directly released or forms abiotically from compounds released by organisms.

variable but overall resembled the combined fluorescence of tyrosine and tryptophan (again blue shifted). The blue-shifted tryptophan signal is due to the fact that the amino acids where bound in proteins. In addition, some of the degradation products of these two amino acids were also found to fluorescence with emission maxima between 300 and 450 nm (Figure 8.6). Fluorescence signatures with these characteristics are often found in natural waters and these results clearly reveal that the decay of bacteria and phytoplankton or release of cell DOM are a likely source of FDOM in aquatic environments. In support of these results, Yamashita and Tanoue (2003) found strong relationships between FDOM fluorescence centered around 340 and 300 nm with concentrations of tryptophan and tyrosine measured using high-performance liquid chromatography (HPLC).

The origins of humic-like fluorescence at longer wavelengths remains to a certain extent illusive. Correlations between the apparent oxygen utilization (AOU) and humic-like fluorescence in the deep Pacific Ocean suggest that FDOM is generated from the oxidation and remineralization of organic particles in the dark ocean (Hayase and Shinozuka, 1995; Yamashita and Tanoue, 2008). Recently this has been found to be a global phenomenon

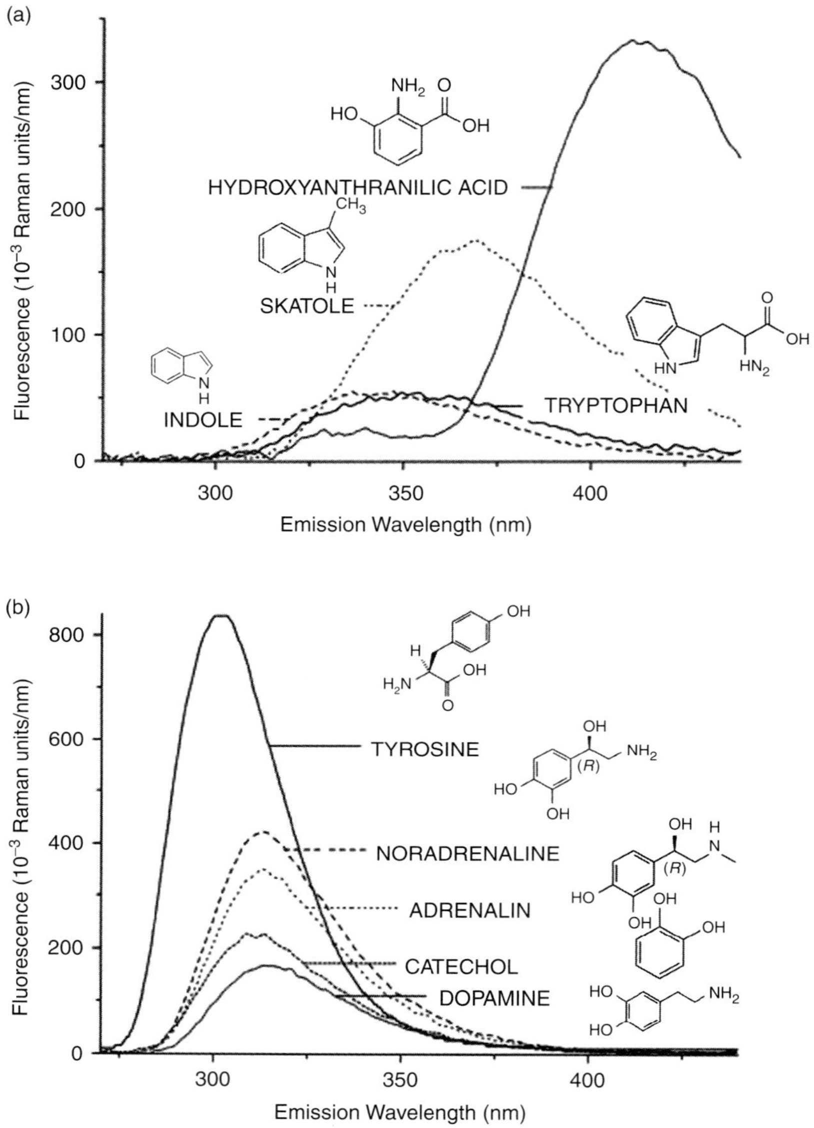

Figure 8.6. Fluorescence properties and structures of (a) tryptophan and (b) tyrosine compared with their respective degradation products. (Modified from Determann et al., 1998.)

(Jørgensen et al., 2011) and to also apply for CDOM absorption (Nelson et al., 2010). Earlier experimental work also supports these findings. Rochelle-Newall and Fisher (2002) and Stedmon and Markager (2005b) have shown how humic-like fluorescence is not evidently produced by phytoplankton directly, but by microbial activity on otherwise colorless DOM. In contrast, amino acid-like fluorescence appeared to be produced during the

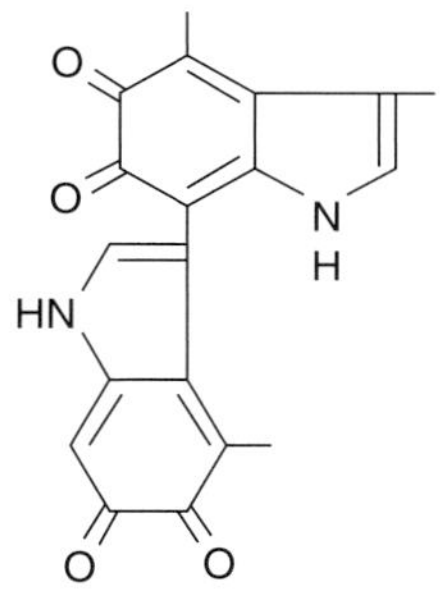

Figure 8.7. Melanin type structure resulting from the Maillard reaction of carbohydrates and amino acids.

exponential growth phase of phytoplankton (Stedmon and Markager, 2005b). In addition, the experimental evidence revealed that there is a close interplay between photochemical exposure and microbial degradation. The production of some of the humic fluorescence signals during these dark incubations was dependent on pre-exposure of the samples to photodegradation.

Recent work by Romero-Castillo et al. (2010) with axenic phytoplankton cultures appears to refute the suggestion that phytoplankton do not directly produce humic fluorescence. Their results show that both amino acid–like and humic-like FDOM signals are generated. The production of amino acid–like fluorescence agrees with earlier field and experimental observations (e.g., Stedmon & Markager, 2005b). Humic-like fluorescence, as mentioned previously, has otherwise been attributed to microbial turnover of otherwise colorless DOC released. So it would appear that this issue warrants further study. In a series of studies, Urban-Rich et al. (2004, 2006) have also shown that zooplankton grazing and excretion are a source of both humic- and amino acid–like fluorescence. Again the latter is to be expected as proteins of most aquatic organisms will to some extent contain tryptophan and tyrosine (e.g., Determann et al., 1998). The generation of humic-like fluorescence, however, is more intriguing and slightly counterintuitive. These broad fluorescence peaks are often associated with larger more complicated conjugated structures, so an alternate hypothesis could be that abiotic condensation processes may also be responsible.

In some of the earliest work on DOM fluorescence and marine humic matter summarized in Duursma (1965) it is suggested that a reaction similar to the Maillard reaction, well known as being responsible for the browning of food, may be responsible for producing this humic fluorescence signal. In this reaction carbohydrates and amino acids condensate to produce compounds with a melanin structure (Hedges, 1978) (Figure 8.7). Potentially this type of reaction can occur, as both these groups of compounds are known to be released by phytoplankton (Nagata, 2000) and are often detected in seawater DOM (Benner, 2002). The fact that they are labile to bacteria can explain why these relatively simple and ubiquitously produced structures represent so little of the DOC. Free concentrations in seawater are very low, so if this type of reaction is responsible it must be occurring rapidly in or around decaying cells, where concentrations can be expected to be high (Yamamoto and Ishiwatari, 1989; Ishiwatari, 1992). An alternative pathway for the formation of marine humic material and its associated fluorescence signal has been proposed and demonstrated by Harvey et al. (1983, 1984). In this pathway unsaturated lipids released by organisms undergo autoxidative cross-linking to form material that has characteristics very similar to those of marine humic isolates. Irrespective of the actual formation pathway, this rapid abiotic "aquatic humification" may be an important step in the marine carbon cycle, generating semi-labile and refractory DOM that has a humic-like fluorescence and that is less bioavailable than its precursors, thereby introducing a time delay between carbon fixation and remineralization back to CO_2 (Jiao et al., 2010).

8.3 Microbial Degradation of Fluorescent Dissolved Organic Matter

The fate of DOM is its mineralization via either heterotrophic respiration or photochemical reactions. The speed and efficiency with which this oxidation occurs is a major control on the global carbon cycle and therefore a field of intensive research. As well as remineralizing carbon, nitrogen, and phosphorus bound in organic matter, these two processes also alter the characteristics of the remaining DOM and thereby its function in aquatic ecosystems. In addition, there are considerable interactions between these two processes. Photochemical degradation can enhance, retard, and even compete with bacterial degradation (Benner and Biddanda, 1998; Moran et al., 2000; Tranvik and Bertilsson, 2001; Stedmon and Markager, 2005b). However, in combination these processes can result in near complete mineralization of the initial DOM pool, much faster than either process acting alone (Vähätalo and Wetzel, 2004, 2008).

Fluorescence characteristics of DOM are often used either to trace the transformations occurring in the DOM pool or to estimate susceptibility of the pool to photochemical and microbial degradation (Stedmon and Markager, 2005b; Cory et al., 2007; Fellman et al., 2009b). The different fluorescent fractions identified in DOM have been found to respond differently to degradation, resulting in some patterns of DOM fluorescence consistently related to DOM character, bio- and photo-lability, and state of degradation; these patterns are discussed in the text that follows.

8.3.1 Bioavailability of FDOM

8.3.1.1 Amino Acid–like Fluorescence

Amino acid–like fluorescence is often associated with autochthonous FDOM, and its presence has been consistently related to elevated biological activity in the water column from freshwater to marine systems, including pristine and polluted waters. This fluorescence signal has been positively correlated with chemical characteristics of DOM indicative of microbial derived precursor material, including the bulk $\delta^{15}N$ signature, the total organic N and aliphatic C content of the DOM (Cory et al., 2007), and the free amino acid or protein concentrations (e.g., Yamashita and Tanoue, 2003). Proteins and their degradation products are thought to comprise a biodegradable fraction of DOM preferred by bacteria. Thus, it is widely assumed that among a sample set, the relative amount of amino acid–like FDOM is an indication of the presence and proportion of labile organic carbon and nitrogen present (Hudson et al., 2008 and references therein; Fellman et al., 2009 a,b; Hood et al., 2009).

However, although amino acid–like FDOM may serve as a proxy measurement for bioavailable DOM, we currently do not understand the relationships among bacterial metabolic activity, bacterial degradation of DOM, and the presence of amino acid–like FDOM signals (e.g., Cammack et al., 2004). For example, laboratory incubation and field mesocosm studies have shown that bacterial degradation can be a source and a sink for amino acid–like FDOM (Moran et al., 2000; Yamashita and Tanoue, 2003; Boyd and Osburn, 2004; Cammack et al., 2004; Stedmon and Markager, 2005b, Nieto-Cid et al., 2006), which has contributed to the difficulty in unambiguously relating amino acid–like FDOM signals to DOM bioavailability. In some bacterial incubation studies, this fluorescence has been shown to increase as a function of incubation time (Moran et al., 2000; Cammack et al., 2004; Boyd and Osburn, 2004), whereas in other studies with a different source of DOM, it decreases as a function of incubation time (Yamashita and Tanoue, 2004; Stedmon and Markager, 2005b; Nieto-Cid et al., 2006).

There are several possible mechanisms for the production and removal of amino acid–like FDOM observed during bacterial degradation of DOM. In one case, this material is simply a byproduct of DOM degradation by bacteria. For example, it is possible that bacterial degradation of protein residues may lead to conformational changes (denaturation) that increase fluorescence from tryptophan and tyrosine residues (Determann et al., 1998; Lakowicz, 2006). This scenario would lead to an increase in amino acid–like FDOM signals, which does not necessarily reflect a direct product of bacterial degradation. Here the accumulation of amino acid–like FDOM as a function of incubation time indicates either that this material is not being consumed by bacteria, or that the bacterial uptake rate is less than the production rate (positive net production).

An alternative explanation for the production of amino acid–like FDOM as a function of incubation time is that bacteria directly produce the material during biosynthesis and growth. For example, Cammack et al. (2004) hypothesized that amino acid–like FDOM was a product of bacterial growth, based on strong, positive correlations between bacterial growth rates and fluorescence. However, the formation pathways and form of fluorescent moieties produced directly by bacteria during degradation of DOM have not been determined and remain unclear; for example, would these constituents leak from bacterial cells and if so under what conditions? In addition, fluorescent amino acids are energetically costly to produce (Akashi et al., 2002), so it is less likely that fluorescent amino acids would accumulate over time unless bacteria are experiencing little or no limitation to growth.

In either the first or second case, the presence of amino acid–like FDOM is likely an indication of the presence of bioavailable DOM. In the first case, accumulation of this fluorescence as a function of incubation time indicates that bacteria preferentially consume other labile DOM fractions, while concurrently (and indirectly) increasing amino acid–like

FDOM signals by degradation of DOM. These amino acid–like FDOM moieties accumulate because bacteria have other preferential sources of carbon and nitrogen available for growth. In the second case, production of energetically costly fluorescent amino acids indicates an abundance of labile carbon and nutrients to support bacterial growth.

These proposed scenarios are consistent with conclusions from the literature on studies aimed at identifying conditions under which bacterial degradation is a sink or source for amino acid–like FDOM (Cammack et al., 2004; Stedmon and Markager, 2005b, Nieto-Cid et al., 2006; Biers et al., 2007). Both Cammack et al. (2004) and Nieto-Cid et al. (2006) proposed scenarios under which amino acid–like FDOM may accumulate and be subsequently consumed, as a function of nutrient availability. Consistently, Biers et al. (2007) showed production of amino acid–like FDOM during bacterial incubation in seawater when labile sources of nitrogen were added. The results from these studies can explain some of the conflicting trends that have been reported.

In support of the link between amino acid–like FDOM and DOM bioavailability, Fellman et al. (2009b) found that protein-like fluorescence correlated strongly to the percent of biodegradable organic carbon (BDOC) in the DOM pool from a range of Alaskan freshwaters. A similarly strong relationship was found across a range of coastal watersheds in the Gulf of Alaska that vary widely in glacial coverage (Hood et al., 2009). Further, amino acid–like FDOM correlated positively with DOM mineralization in springs, streams, and permafrost soils (Balcarczyk et al., 2009), and with uptake rates of dissolved organic nitrogen (DON) in temperate streams (Fellman et al., 2009b), providing additional evidence that this fluorescence signal correlates with the presence of labile carbon and nitrogen. Stedmon and Markager (2005b) showed that one amino acid–like component of FDOM was removed by bacterial degradation and thus was clearly a proxy for bioavailable material.

In wastewaters and waters receiving wastewater effluent, amino acid–like FDOM was correlated positively with biological oxygen demand, a measure of bioavailable DOM (Hudson et al., 2008). Wastewaters are enriched in this fluorescence relative to fresh and marine waters (Reynolds and Ahmad, 1997), which provides additional support for the link between elevated amino acid–like FDOM and high rates of biological activity. Together, these studies strongly support the relationship between amino acid–like FDOM and biological activity and indicate that it is produced either directly or indirectly during conditions of high bacterial growth; however, because it is widespread and often persists (e.g., Vähätalo and Wetzel, 2008), the relationship between bacterial activity and the consumption or degradation of amino acid–like FDOM is likely more complicated or at least less well understood at this time.

8.3.1.2 Humic-like Fluorescence

Humic-like fluorescence is predominately but not solely associated with allochthonous organic matter derived from degrading soil and plant matter. Relative to microbial derived material enriched in amino acid–like FDOM, humic-like FDOM is not expected to be preferentially degraded by bacteria, and thus considered to be a proxy for slowly degrading or recalcitrant DOM. Consistently, studies have shown that bacterial degradation is more often a source rather than a sink of humic-like fluorescence (Moran et al., 2000; Stedmon and Markager, 2005b). In cases where bacterial degradation was found to be a sink, consumption of humic-like FDOM was less than that seen for amino acid–like fluorescent material. For example, Moran et al. (2000) showed little (~11–12%) or no loss of humic-like fluorescence over the course of a 51-day bacterial degradation experiment with estuarine water. Stedmon and Markager (2005b) identified five fluorescent humic-like fractions that were all produced during dark microbial incubations. Similarly, Yamashita and Tanoue (2004), Kramer and Herndl (2004), and Nieto-Cid et al. (2006) found that bacterial degradation of marine DOM increased humic-like fluorescence as a function of incubation time. Boyd and Osburn (2004) showed that bacterial degradation could be both a source and a sink for humic-like FDOM depending on the source of the coastal estuarine waters evaluated in their study. Amado et al. (2007) reported increased humic-like fluorescence in tropical freshwaters when DOM was degraded by bacteria. In addition, humic-like FDOM was found to behave conservatively with distance downstream on slug additions of allochthonous DOM to temperate streams in southeastern Alaska (Fellman et al., 2009b).

8.3.2 Interactions between Photochemical and Microbial Degradation

In sunlit surface waters, photochemical degradation of DOM strongly influences its bioavailability to bacteria (Moran et al., 2000; Tranvik and Bertilsson, 2001; Obernosterer and Benner, 2004). The current paradigm is that the effect of photodegradation on DOM bioavailability depends on the source of the DOM – for example, freshwater and marine autochthonous material becomes less bioavailable after light exposure (Benner and Biddanda, 1998; Tranvik and Kokalj, 1998), while DOM derived predominantly from terrestrial precursor matter (e.g., fresh or estuarine waters with high inputs of allochthonous material) becomes more labile after exposure (Moran et al., 2000; Tranvik and Bertilsson, 2001). In sunlit surface waters, tightly coupled photo- and biochemical degradation is the most important sink for DOM, and thus it is desirable to understand how these processes interact together to affect FDOM signals.

In a laboratory study where DOM derived from leaf leachate was exposed to long-term concurrent photochemical and bacterial degradation, FDOM associated with humic material composed only 0.1–0.3% of the total DOM remaining in the dark control, whereas protein-like FDOM was 4.3% of the dark control after 420 days (Vähätalo and Wetzel, 2008). These results show that in a closed system with enough time, combined photochemical and bacterial processes can remove nearly all the initial FDOM in addition to any FDOM produced during degradation. Counterintuitive to these results is that the amino acid–like FDOM, commonly associated with biologically labile DOM, was more persistent compared with humic-like FDOM, which in contrast is associated with refractory DOM. This is not to say that that in open systems, humic-like FDOM is not associated with slowly degrading DOM, or that protein-like DOM is refractory. Rather, these results exemplify their different susceptibility to photochemical and microbial degradation, and provide further evidence that bacterial degradation of DOM can be a sink and a source of amino acid–like FDOM.

It is understood that photodegradation is a more important sink for humic-like FDOM (e.g., Granéli et al., 1996; Moran et al., 2000) compared to amino acid–like FDOM. However the latter can be susceptible to photodegradation (Moran et al., 2000; Cory et al., 2007). Although exposure to light generally results in loss of FDOM, one study showed that photochemical exposure of seawater DOM resulted in the production of humic- and amino acid–like fluorescence when tryptophan was added (Biers et al., 2007).

A general pattern that emerges across studies is that bacterial processing of photodegraded DOM results in opposite shifts in FDOM compared to shifts caused by photodegradation (e.g., production of humic- and amino acid–like FDOM rather than loss; Moran et al., 2000; Kramer and Herndl, 2004; Stedmon and Markager, 2005b; Nieto-Cid et al., 2006; Amado et al., 2007). Photodegradation tends to remove humic-like fluorescence (e.g., FDOM with emission >400 nm) while it is produced by microbial degradation in the dark (Moran et al., 2000; Stedmon and Markager, 2005b). Because the effects of microbial growth and activity on FDOM after photodegradation are generally smaller than the effects of photodegradation, photochemical influences on FDOM dominate over bacterial influences during sequential photo- followed by biodegradation in laboratory studies, particularly for humic FDOM (Moran et al. 2000) (Figure 8.8).

A second emerging pattern from the literature is that photodegradation is an important sink for autochthonous humic and amino acid–like FDOM, that is, FDOM produced from phytoplankton growth and decay or FDOM produced as DOM is degraded by bacteria (see earlier). For example, Stedmon and Markager (2005b) found that photodegradation was an important sink for microbial produced humic and protein-like FDOM. In addition, in subsequent incubations of riverine and coastal FDOM, Nieto-Cid et al. (2006) observed a significant correlation between the bacterial production and the photochemical consumption of humic-like FDOM. They interpreted these results to mean that autochthonous humic material produced in the dark, either in the aphotic layer or during the night, is rapidly photo-degraded in the light. Similarly, Amado et al. (2007) observed that FDOM produced in the dark autochthonously in humic and eutrophic lagoons was rapidly degraded by sunlight.

Despite these general patterns describing the interrelations of photo- and biodegradation and production of FDOM, many questions remain to be addressed. Do photochemical and microbial processes impart distinct signatures on the FDOM that remain after degradation? Can those signatures be used to trace the FDOM in a given system to provide an unambiguous understanding of the relative importance of each process on the fate of DOM? Is the stimulation of microbial activity after photodegradation of DOM evident in the FDOM signature after photodegradation? Only careful studies that help to uniquely separate the different, and often competing, mechanisms of DOM alterations will help us answer these questions.

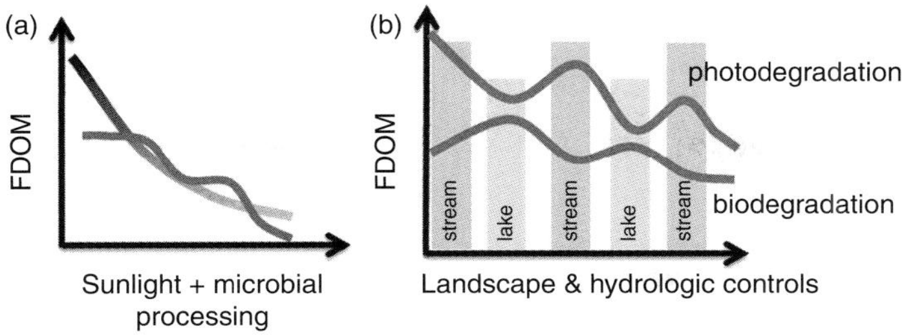

Figure 8.8. (a) Expected influences on FDOM from combined photo + microbial processing of DOM: phototransformations (red-yellow curve) are an important sink for FDOM while microbial processing (green curve) can be a sink and a source of FDOM. Because phototransformations of DOM generally stimulate its turnover by microorganisms, loss of FDOM by photodegradation may be rapidly followed by microbial production of FDOM. Microbial influences on FDOM signature are typically smaller than photochemical influences in laboratory studies conducted over short time scales (e.g., days to weeks). (b) In real systems, the net effect of combined photo + bio processes on FDOM will reflect the complex interplay between DOM source and its photo and biolability which will vary with landscape and hydrology controls (as shown for a hypothetical system of hydrologically connected lakes and streams, for example).

8.4 Future Research

Recent studies have put forward the hypothesis that photochemical and bacterial degradation act in a complementary way through formation of substrates available to the other process (Amado et al., 2007; Vähätalo and Wetzel, 2008). This hypothesis appears to be reflected in shifts in the FDOM pool during combined and/or subsequent photo and biochemical degradation of DOM, as summarized earlier, suggesting that FDOM is a good proxy for processes influencing the wider (non-fluorescent + fluorescent) DOM pool. However, one major limitation to using FDOM as a proxy to understand changes to the DOM pool is the lack of a quantitative relationship between FDOM components/fractions and DOM concentration. FDOM is a sensitive tracer that has provided new insight on how rapidly some fractions of the DOM pool respond to coupled photo-biochemical processes in the environment, often without detectable changes in the DOM pool measured on a carbon basis (e.g., Nieto-Cid et al., 2006). These findings may suggest that although in some systems FDOM may be a small fraction of the total DOM pool, some fractions of the DOM pool may be more dynamic than previously assessed based solely on spatial and/or temporal variation in DOC concentrations. Although challenging, it would be a major breakthrough if we could link FDOM components with carbon concentrations.

Although much work is needed to develop a quantitative approach to FDOM as a proxy for DOM processing in the environment, this chapter highlights some of the advances in the qualitative understanding of photochemical and microbial sources and sinks of FDOM. However, much more work is needed to unambiguously decipher the FDOM signature such that we can discern the relative importance of each process acting on the DOM pool in a given system. For example, although amino acid–like FDOM is pretty clearly associated with microbial processing of the DOM pool, the interpretation of amino acid–like components in the FDOM pool is complicated by the fact that these fluorophores are both produced and removed by microbial activity. Future work should be aimed at determining differences between recently produced amino acid–like FDOM from microbial degradation of colorless DOM versus amino acid–like FDOM associated with humic material that may represent an important bioavailable component of humic DOM.

One can hypothesize that the degree to which DOM has been exposed to sunlight and microorganisms (e.g., its diagenetic state) is generally inversely related to the rate at which it can be further degraded by photo and biochemical degradation. In other words, as the substrates most labile to photo and biochemical degradation are preferentially removed, the remaining DOM is degraded more slowly. One question is whether this hypothesis is reflected in the FDOM signature, for example, whether the diagenetic state of DOM is reflected either in the FDOM signature or perhaps from photo or bioassays based on changes to FDOM during photo- and biochemical incubations.

Why is understanding this link important? The reason is that in several key environments, there is evidence for the current or predicted increase in DOM in surface waters. Climate change–induced alteration in the hydrology of catchments are increasing the loading of terrestrial organic material to surface waters, such as the observed increased DOC concentrations in European and North American surface waters (Monteith et al., 2007) and the expected increase in DOC concentrations from permafrost thawing in Arctic regions (Rember and Trefry, 2004, Frey and Smith, 2005). The likely fate and impact of this additional DOM supply on aquatic ecosystems are unknown. The bioavailability of the organic matter associated with this increased supply has great influence on the net result, as an increased supply of labile carbon either directly or indirectly via photochemical process will favor heterotrophic bacteria in competition with phytoplankton for mineral nutrients (Thingstad et al., 2008). If we are able to link the FDOM signature of "new" DOM in a given system to its susceptibility to photo and biochemical degradation, it would significantly improve our ability follow and attempt to predict its fate and impact within the context of the regional and global carbon cycling.

Acknowledgments

C.A. Stedmon was funded by the Danish Research Council Grant no. 272–07–0485 and Carlsberg Foundation Grant no. 2007–01–0124. We also thank two anonymous reviewers for their constructive comments on the text.

References

Aitkenhead-Peterson, J.A., McDOwell, W.H., and Neff, J.C. (2003). Sources, production and regulation of allochthonous dissolved organic matter inputs to surface waters.

In S.E.G. Findlay and R.L. Sinsabaugh (Eds.), *Aquatic Ecosystems: Interactivity of Dissolved Organic Matter* (pp. 25–70). Aquatic Ecology Series. San Diego: Academic Press.

Akashi, H. and Gojobori, T. (2002). Metabolic efficiency and amino acid composition in the proteomes of *Escherichia coli* and *Bacillus subtilis*. *Proc. Natl. Acad. Sci. USA*, **99**(6), 3695–3700.

Amado, A.M., Cotner, J.B., Suhett, A.L., Esteves, F.D., Bozelli, R.L., and Farjalla, V.F. (2007). Contrasting interactions mediate dissolved organic matter decomposition in tropical aquatic ecosystems. *Aquat. Microb. Ecol.*, **49**, 25–34.

Balcarczyk, K.L., Jones, J.B., Jr., Jaffe, R., and Maie, N. (2009). Stream dissolved organic matter bioavailability and composition in watersheds underlain with discontinuous permafrost. *Biogeochemistry*, **94**, 255–270.

Benner, R. (2002). Chemical composition and reactivity. In D.A. Hansell and C.A. Carlson (Eds.), *Biogeochemistry of Marine Dissolved Organic Matter* (pp. 59–90). San Diego: Academic Press.

Benner, R. and Biddanda, B. (1998). Photochemical transformations of surface and deep marine dissolved organic matter: Effects on bacterial growth. *Limnol. Oceanogr.*, **43**, 1373–1378.

Biers E.J., Zepp R.G., and Moran M.A. (2007). The role of nitrogen in chromophoric and fluorescent dissolved organic matter formation. *Mar. Chem.*, **103**, 46–60

Boyd, T.J. and Osburn, C.L. (2004). Changes in CDOM fluorescence from allochthonous and autochthonous sources during tidal mixing and bacterial degradation in two coastal estuaries. *Mar. Chem.*, **89**, 189–210.

Boyle E.S., Guerriero, N., Thiallet, A., Del Vecchio, R., and Blough, N.V. (2009). Optical properties of humic substances and CDOM: Relation to structure. *Environ. Sci. Technol.*, **43**, 2262–2268.

Cammack, W.K., Kalff, J., Prairie, Y.T., and Smith, E.M. (2004). Fluorescent dissolved organic matter in lakes: Relationships with het- erotrophic metabolism. *Limnol. Oceanogr.*, **49**, 2034–2045.

Canadell, J.G., Le Quéré, C., Raupach, M.R., Ciais, P., Conway, T.J, et al. (2007). Recent acceleration in CO_2 emissions and the response of the global carbon cycle. *Proc. Natl. Acad. Sci. USA*, **104**, 18866–18870.

Coble, P.G. (1996). Characterization of marine and terrestrial DOM in seawater using excitation-emission matrix spectroscopy. *Mar. Chem.*, **51**, 325–346.

Coble, P.G., Del Castillio, C.E., and Avril, B. (1998). Distribution and optical properties of CDOM in the Arabian Sea during the 1995 Southwest Monsoon. *Deep-Sea Res. Pt. II*, **45**, 2195–2223.

Cory, R.M., McKnight, D.M., Chin, Y.P., Miller, P., and Jaros, C.L. (2007). Chemical characteristics of fulvic acids from Arctic surface waters: Microbial contributions and photochemical transformations. *J. Geophys. Res.*, 112, G04S51, doi:10.1029/2006JG000343 ER.

Del Vecchio, R. and Blough, N.V. (2004). On the origin of the optical properties of humic substances. *Environ. Sci. Technol.*, **38**, 3885–3891.

Determann, S., Lobbes, J.M., Reuter, R., and Rullköter, J. (1998). Ultraviolet fluorescence excitation and emission spectroscopy of marine algae and bacteria. *Mar. Chem.*, **62**, 137–156.

Duursma, E.K. (1965). Dissolved organic constituents of sea water. In P. Riley and G. Skirrow (Eds.), *Chemical Oceanography*, Vol, 1 (pp. 433–477). London: Academic Press.

Ertel, J.R., Hedges, J.I., Devol, A.H., Richey, J.E., and de Nazare Goes Ribeiro, M. (1986). Dissolved humic substances of the Amazon River system. *Limnol. Oceanogr.*, **31**, 739–754.

Fellman, J.B., Hood, E., Edwards, R.T., and D'Amore, D.V. (2009a). Changes in the concentration, biodegradability, and fluorescent properties of dissolved organic matter during stormflows in coastal temperate watersheds. *J. Geophys. Res.*, **114**, G01021.

Fellman, J.B., Hood, E. D'Amore, D.V., Edwards, R.T., and White, D. (2009b). Seasonal changes in the chemical quality and biodegradability of dissolved organic matter exported from soils to streams in coastal temperate rainforest watersheds. *Biogeochemistry*, **95**, 277–293.

Frey, K.E. and Smith, L.C. (2005). Amplified carbon release from vast West Siberian peatlands by 2100. *Geophys. Res. Lett.*, **32**, L09401.

Fulton, J.R., McKnight, D.M., Foreman, C., Cory, R., Stedmon, C., and Blunt, E. (2004). Changes in fulvic acid redox state through the oxycline of a permanently ice-covered Antarctic lake. *Aquat. Sci.*, **66**, 27–46.

Granéli, W., Lindell, M., and Tranvik, L. (1996). Photo-oxidative production of dissolved inorganic carbon in lakes of different humic content. *Limnol. Oceanogr.*, **41**, 698–706.

Guggenberger, G., Kaiser, K., and Zech, W. (1998). Mobilization and immobilization of dissolved organic matter in forest soils. *J. Plant Nutr. Soil Sci.*, **161**, 401–408.

Hansell, D.A. and Carlson, C.A. (1998). Net community production of dissolved organic carbon. *Global Biogeochem. Cycles*, **12**, 443–453.

Hansell, D.A., Carlson, C.A., Repeta, D.J., and Schlitzer, R. (2009). Dissolved organic matter in the ocean: New insights stimulated by a controversy. *Oceanography*, **22**, 52–61.

Harvey, G.R., Boran, D.A., Chesal, L.A., and Tokar, J.M. (1983). The structure of marine fulvic and humic acids. *Mar. Chem.*, **12**, 119–132.

Harvey, G.R., Boran, D.A., Piotrowicz, S.R., and Weisel, C.P. (1984). Synthesis of marine humic substances from unsaturated lipids. *Nature*, **309**, 244–246.

Hayase, K. and Shinozuka, N. (1995). Vertical distribution of fluorescent organic matter along with AOU and nutrients in the Equatorial Pacific. *Mar. Chem.*, **48**, 283–290.

Hedges J.I. (1978). Formation and clay mineral reactions of melanoidins. *Geochim. Cosmochim. Acta*, **42**, 69–76.

Hernes, P.J., Bergamaschi, B.A., Eckard, R.S., and Spencer, R.G.M. (2009). Fluorescence-based proxies for lignin in freshwater dissolved organic matter. *J. Geophys. Res.*, **114**, G00F03.

Hood, E., Fellman, J., Spencer, R.G.M., Hernes, P.J., Edwards, R., D'Amore, D., and Scott, D. (2009). Glaciers as a source of ancient and labile organic matter to the marine environment. *Nature*, **462**., 1044–1047.

Houghton, R.A. (2007). Balancing the global carbon budget. *Annu. Rev. Earth Planet. Sci.*, **35**, 313–347.

Hudson, N., Baker, A., Ward, D., Reynolds, D.M., Brunsdon, C., Carliell- Marquet, C., and Browning, S. (2008). Can fluorescence spectrometry be used as a surrogate for the Biochemical Oxygen Demand (BOD) test in water quality assessment? An example from South West England. *Sci. Total Environ.*, **391**(1), 149–158.

Ishiwatari, R. (1992). Macromolecular material (humic substance) in the water column and sediments. *Mar. Chem.*, **39**, 151–166

Jiao, N., Herndl, G.J., Hansell, D.A., Benner, R., Kattner, G., Wilhelm, S.W., Kirchman, D.L., Weinbauer, M.G., Luo T., Chen, F., and Azam, F. (2010). Microbial production of recalcitrant dissolved organic matter: Long-term carbon storage in the global ocean. *Nature Rev. Microbiol.*, 8, 593–599.

Jobbágy, E.G. and Jackson, R.B. (2000). The vertical distribution of soil organic carbon and its relation to climate and vegetation. *Ecol. Applicat.*, **10**, 423–436.

Jørgensen, L., Stedmon, C.A., Kragh, T., Markager, S., Middelboe, M., and Søndergaard, M. (2011). Global trends in the fluorescence characteristics and distribution of marine dissolved organic matter. *Mar. Chem.*, **126**, 139–148.

Kalbitz, K., Solinger, S., Park, J.H., Michalzik, B., and Matzner, E. (2000). Controls on the dynamics of dissolved organic matter in soils: A review. *Soil Sci.*, **165**, 277–304.

Kawasaki, N. and Benner, R. (2007). Bacterial release of dissolved organic matter during cell growth and decline: Molecular origin and composition. *Limnol. Oceanogr.*, **51**(5),2170–2180.

Kramer, G.D. and Herndl, G.J. (2004). Photo- and bioreactivity of chromophoric dissolved organic matter produced by marine bacterioplankton. *Aquat. Microb. Ecol.*, **36**(3), 239–246.

Lakowicz, J.R. (2006). *Principles of Fluorescence Spectroscopy*, 3rd ed. New York: Springer Science+Business Media.

Lochmuller, C.H. and Saavedra, S.S. (1986). Conformational changes in a soil fulvic acid measured by time dependent fluorescence depolarization. *Anal. Chem.*, **38**, 1978–1981.

Louchouarn, P., Opsahl, S., and Benner, R. (2000). Isolation and quantification of dissolved lignin from natural waters using solid-phase extraction (SPE) and GC/MS selected ion monitoring (sim). *Anal. Chem.*, **13**, 2780 –2787.

Maie, N., Scully, N., Pisani, O., and Jaffé R. (2007). Composition of a protein-like fluorophore of dissolved organic matter in coastal wetland and estuarine ecosystems, *Water Res.*, **41**(3), 563–570.

McDowell, W.H. and Wood, T. (1984). Podzolization-Soilo processes control dissolved organic carbon concentrations in stream water. *Soil Sci.*, **137**, 23–32.

McKnight, D.M., Aiken, G.R., and Smith, R.L. (1991). Aquatic fulvic acids in microbially based ecosystems: Results from two desert lakes in Antarctica. *Limnol. Oceanogr.*, **36**, 998–1006.

McKnight, D.M., Harnish, R., Wershaw, R.L., Baron, J.S., and Schiff, S. (1997). Chemical characteristics of particulate, colloidal, and dissolved organic material in Loch Vale Watershed. *Rocky Mt. Natl. Park Biogeochem.*, **36**, 99–124.

Middelboe, M. and Lyck, P.G. (2002). Regeneration of dissolved organic matter by viral lysis in marine microbial communities. *Aquat. Microb. Ecol.*, **27**, 187–194.

Monteith, D.T., Stoddard, J.L., Evans, C.D., de Wit, H.A., Forsius, M., Hogasen, T., Wilander, A., Skjelkvale, B.L., Jeffries, D.S., Vuorenmaa, J., Keller, B., Kopacek, J., and Vesely, J. (2007). Dissolved organic carbon trends resulting from changes in atmospheric deposition chemistry. *Nature*, **450**, 537–541.

Moran, M.A., Sheldon, W.M., and Zepp, R.G. (2000). Carbon loss and optical property changes during long-term photochemical and biological degradation of estuarine dissolved organic matter. *Limnol. Oceanogr.*, **45**, 1254–1264.

Mulholland, P.J. (2003). Large-scale patterns in dissolved organic carbon concentration, flux and sources. In S.E.G. Findlay and R.L. Sinsabaugh (Eds.), *Aquatic Ecosystems: Interactivity of Dissolved Organic Matter* (pp. 139–160). Aquatic Ecology Series. San Diego: Academic Press.

Murphy, K.R., Ruiz, G.M., Dunsmuir, W.T.M, and Waite, T.D. (2006). Optimized parameters for rapid fluorescence-based verification of ballast water exchange by ships. *Environ. Sci. Technol.*, **40**, 2357–2362.

Murphy, K.R., Stedmon, C.A., Waite, T.D., and Ruiz, G.M. (2008). Distinguishing between terrestrial and autochthonous organic matter sources in marine environments using fluorescence spectroscopy. *Mar. Chem.*, **108**, 40–58.

Nagata, T. (2000). Production mechanisms of dissolved organic matter. In D.L. Kirchman (Ed.), *Microbial Ecology of the Oceans* (pp. 121–152). Wiley Series in Ecological and Applied Microbiology. Hoboken, NJ: Wiley-Liss.

Nelson, N.B., Siegel, D.A., Carlson, C.A., and Swan, C. (2010). Tracing global biogeochemical cycles and meridional overturning circulation using chromophoric dissolved organic matter. *Geophys. Res. Lett.*, **37**, L03610.

Nieto-Cid, M., Alvarez-Salgado, X.A., and Perez, F.F. (2006). Microbial and photochemical reactivity of fluorescent dissolved organic matter in a coastal upwelling system. *Limnol. Oceanogr.*, **51**(3), 1391–1400

Obernosterer, I. and Benner, R. (2004). Competition between biological and photochemical processes in the mineralization of dissolved organic carbon. *Limnol. Oceanogr.*, **49**, 117–124.

Ogawa, H., Amagi, Y., Koike, I., Kaiser, K., and Benner, R. (2001). Production of refractory dissolved organic matter by bacteria. *Science*, **292**, 917–920.

Opsahl, S. and Benner, R. (1997). Distribution and cycling of terrigenous dissolved organic matter in the ocean. *Nature*, **386**, 480–482.

Rember, R.D. and Trefry, J.H. (2004). Increased concentrations of dissolved trace metals and organic carbon during snowmelt in rivers of the Alaskan Arctic. *Geochim. Cosmochim. Acta*, **68**, 477–489.

Reynolds, D.M. and Ahmad, S.R.A. (1997). Rapid and direct determination of wastewater BOD values using a fluorescence technique. *Water Res.*, **31**(8), 2012–2018.

Rochelle-Newall, E.J. and Fisher, T.R. (2002). Production of chromophoric dissolved organic matter fluorescence in marine and estuarine environments: An investigation into the role of phytoplankton. *Mar. Chem.*, **77**, 7–21.

Romera-Castillo, C., Sarmento, H., Álvarez-Salgado, X.A., Gasol, J.M., and Marrase C. Production of chromophoric dissolved organic matter by marine phytoplankton. *Limnol. Oceanogr.*, **55**(1), 2010, 446–454.

Sarkanen, K.V. and Ludwig, C.H. Lignins, Eds. (1971). *Occurrences, Formation, Structure and Reactions*. New York: Wiley-Interscience.

Senesi, N., Miano, T.M., Provenzano, M.R., and Brunetti, G. (1991). Characterization, differentiation and classification of humic substances by fluorescence spectroscopy. *Soil Sci.*, **152**, 259–271.

Spencer, R.G.M., Aiken, G.R., Butler, K.D., Dornblaser, M.M., Striegl, R.G., and Hernes, P. (2009). Utilizing chromophoric dissolved organic matter measurements to derive export and reactivity of dissolved organic carbon exported to the arctic ocean: A case study of the Yukon river, Alaska, Geophys. *Res. Lett.*, **36**, L06401.

Stedmon, C.A. and Markager, S.S. (2005a). Resolving the variability in dissolved organic matter fluorescence in a temperate estuary and its catchment using PARAFAC analysis. *Limnol. Oceanogr.*, **50**(2), 686–697.

Stedmon, C.A. and Markager, S.S. (2005b). Tracing the production and degradation of autochthonous fractions of dissolved organic matter by fluorescence analysis. *Limnol. Oceanogr.*, **50**(5), 1415–1426.

Stedmon, C.A. and Bro, R. (2008). Characterizing dissolved organic matter fluorescence with parallel factor analysis: A tutorial. *Limnol. Oceanogr. Methods*, **6**, 572–579.

Stedmon, C.A., Markager, S., and Bro, R. (2003). Tracing dissolved organic matter in aquatic environments using a new approach to fluorescence spectroscopy. *Mar. Chem.*, **82**, 239–254.

Stedmon C.A, Markager, S., Søndergaard, M., Vang, T., Laubel, A., Borch, N.H., and Windelin, A. (2006). Dissolved organic matter (DOM) export to a temperate estuary: Seasonal variations and implications of land use. *Estuaries Coasts*, **29**, 388–400.

Stevensen, F.J. (1982). *Biochemistry of the Formation of Humic Substances*. In F.J. Stevensen (Ed.), *Humus Chemistry* (pp. 195–220). New York: John Wiley & Sons.

Strom, S.L., Benner, R., Ziegler, S., and Dagg, M.J. (1997). Planktonic grazers are a potentially important source of marine dissolved organic carbon. *Limnol. Oceanogr.*, **42**, 1364–1374.

Tipping, E., Woof, C., Rigg, E., Harrison, A.F., Ineson, P., Taylor, K., Benham, D., Poskitt, J., Rowland, A.P., Bol, R., and Harkness, D.D. (1999). Climatic influences on the leaching of dissolved organic matter from upland UK moorland soils, investigated by a field manipulation experiment. *Environ. Int.*, **25**, 83–95.

Tranvik, L. and Kokalj, S. (1998). Decreased biodegradability of algal DOC due to interactive effects of UV radiation and humic matter. *Aquat. Microb. Ecol.*, **14**, 301–307.

Tranvik, L.J. and Bertilsson, S. (2001). Contrasting effects of solar UV radiation on dissolved organic sources for bacterial growth. *Ecol. Lett.*, **4**, 458–463.

Tranvik, L.J., Downing, J., and Cotner, J. (2009). Lakes and reservoirs as regulators of carbon cycling and climate. *Limnol. Oceanogr.*, **54**(1), 2298–2314.

Urban-Rich, J., McCarty, J.T., and Shailer, M. (2004). Effects of food concentration and diet on chromophoric dissolved organic matter accumulation and fluorescent composition during grazing experiments with the copepod, *Calanus finmarchicus*. *ICES J. Mar. Sci.*, **61**, 542–551.

Vähätalo, A.V. and Wetzel, R.G.. (2004). Photochemical and microbial decomposition of chromophoric dissolved organic matter during long (months-years) exposures. *Mar. Chem.*, **89**, 313–326.

Vähätalo, A.V. and Wetzel, R.G. (2008). Long-term photochemical and microbial decomposition of wetland-derived dissolved organic matter with alteration of C-13: C-12 mass ratio. *Limnol. Oceanogr.*, **53**, 1387–1392.

Yamamoto, S. and Ishiwatari, R. (1989). A study of the formation mechanism of sedimentary humic substances-II. Protein-based melanoidin model. *Org. Geochem.*, **14**, 479–489.

Yamashita, Y. and Tanoue, E. (2003). Chemical characterization of protein-like fluorophores in DOM in relation to aromatic amino acids. *Mar. Chem.*, **82**, 255–271.

Yamashita, Y. and Tanoue, E. (2004). In situ production of chromophoric dissolved organic matter in coastal environments. *Geophys. Res. Lett.*, doi: 10.1029/2004GL019734.

Yamashita, Y. and Tanoue, E. (2008). Production of bio-refractory fluorescent dissolved organic matter in the ocean interior. *Nature Geosci.*, 1, 579–582, doi:10.1038/ngeo279.

Zsolnay, A. (2003). Dissolved organic matter: Artefacts, definitions, and functions. *Geoderma*, **113**, 187–209.

Zsolnay, A., Baigar, E., Jimenez, M., Steinweg, B., and Saccomandi, F. (1999). Differentiating with fluorescence spectroscopy the sources of dissolved organic matter in soils subjected to drying. *Chemosphere*, **38**, 45–50.

Part IV

Interpretation and Classification

9

Fluorescence Indices and Their Interpretation

RACHEL S. GABOR, ANDY BAKER, DIANE M. MCKNIGHT,
AND MATTHEW P. MILLER

9.1 Introduction

In aquatic ecosystems and soil interstitial waters, dissolved organic material (DOM) is comprised of a heterogeneous mixture of organic compounds derived from plant and soil decomposition and leachate, degradation of microbial biomass, and microbial exudates. Organic material present in the solid phase of soils and sediments (SOM) is similarly complex, with variations not only in chemical composition, but also in the extent to which organic molecules are bound to minerals. The chemical characteristics and distribution of classes of organic compounds within DOM can be dynamic. For example, DOM quality can change seasonally with hydrologic drivers, such as snowmelt, or due to summertime algal blooms in lakes and streams. In contrast, changes in SOM may occur more slowly as soils and sediments age, a process historically referred to as humification.

Important fractions of DOM and SOM include chromophoric (light absorbing) and fluorophoric (light absorbing and light emitting) compounds. Humic material is a major chromophoric fraction of DOM and SOM, having a yellow to brown color associated with aromatic carbon moieties derived from plant pigments, lignin, and other precursor materials. In surface waters, for example, dissolved humics typically account for a substantial fraction of DOM and are the main light-absorbing constituents in the water column (McKnight et al., 2003). Humics derived from plant litter and soil are generally more chromophoric than humics derived from microbial biomass, reflecting their greater content of aromatic carbon moieties. A fraction of both plant/soil-derived and microbially derived humic molecules fluoresces with characteristic broad fluorophores occurring in the useful analytical range of ex 240–370 nm, em 350–550 nm. Another important chromophoric fraction is proteinaceous material associated with extracellular microbial products and plant litter degradation products, which often include fluorescent amino acids, specifically tryptophan and tyrosine. These amino acid-like fluorophores occur at lower emission wavelength ranges than typically observed for humic fluorophores. For these and other fluorescent classes of organic compounds present in DOM and SOM, more chemical information can be obtained by knowing the wavelength ranges at which the molecules both absorb and emit light than by absorbance information alone. In addition, absorbance in a natural water sample can be influenced by inorganic solutes, such as nitrate or iron. As a result, fluorescence is a more sensitive indicator of organic material source and chemical quality than absorbance and there is less potential for interference from other light absorbing compounds. For these reasons, and the observation of ubiquitous fluorescent organic materials in natural waters, soils, and sediments, fluorescence spectroscopy has proven to be a robust technique for studying the source and chemical composition of organic matter (e.g., Fellman et al., 2010). As explained in Chapter 2, 2-D fluorescence scans can be either a line scan (typically a fixed excitation wavelength and scan of emitted fluorescence over a range of wavelengths) or synchronous scan (scanning both excitation and emission wavelengths with a constant wavelength offset between the two). Three-dimensional fluorescence scans, referred to as excitation–emission matrices (EEMs), are acquired for a range of emission wavelengths when excited at multiple wavelengths.

Spectroscopic (absorbance and fluorescence) indices have been used to characterize natural organic material for many decades. For example, the determination of platinum color units is a chromophoric index that was employed in the 1950s (American Public Health Association, 1965) by limnologists and was based on comparison of a water sample to a standard series of platinum solutions. This absorbance index was used as a proxy for dissolved organic carbon (DOC) concentration. Although color units are no longer routinely measured, another chromophoric index, specific ultraviolet absorbance (SUVA), is commonly used. $SUVA_{254}$, reported in units of L mg^{-1} m^{-1}, is the absorbance of a water sample at 254 nm divided by the concentration of DOC and is associated with bulk aromaticity (Weishaar et al., 2003). Soil scientists employ an absorbance index called the E4/E6 ratio (the ratio of absorbance of a solution of extracted humic materials at 400 nm and 600 nm), which is used as an index of humification (Chen et al., 1977). However, in humic-poor aquatic samples, absorption at 600 nm is minimal, and this ratio is not often useful. Fluorescence indices were developed subsequently and can be generally defined as the ratio of fluorescence intensity measured at two different points or regions in optical space. These indices can be thought of as a sub-sample of the information contained in EEMs (Figure 9.1) of organic matter. Beginning several decades ago, indices were developed in association with the commonly available line scan or synchronous scan methodologies. The various indices were targeted to address several specific questions about the nature of organic material in a broad range of systems, such as soils, groundwater (Kalbitz et al., 1999), lakes, and streams (McKnight et al., 2001). These indices focus on different portions of the EEM and the utility of an index must be considered in the context of the questions to be addressed. Indices have been applied to understand a wide range of ecosystem processes including SOM associated with land cover and the changes in DOM that may occur on decadal time scales under a changing climate.

Currently, fluorescence indices are frequently developed in the context of data contained in full EEMs, which can now be obtained with high resolution in a reasonable time period (minutes) in a laboratory setting. It is also now possible to measure DOM fluorescence in situ continuously at one excitation wavelength with a laser source, and one, or possibly several, emission wavelengths; in situ methods are considered further in Chapter 6. Indices have proven to be a valuable tool for remote sensing applications owing to the current

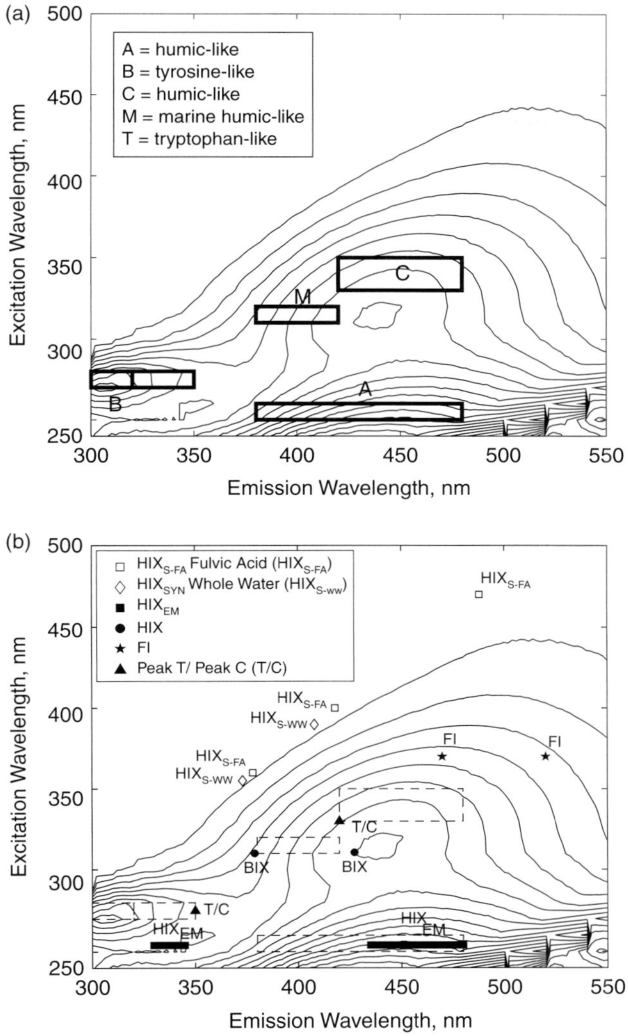

Figure 9.1. Identified regions of a typical representative corrected excitation–emission matrix (EEM) in Raman units (a), and measurement locations of fluorescence indices (b) discussed in this chapter. Where an area is used by Zsolnay et al. (1999), the emission wavelength range is shown by a line. Information plotted here can be found in Tables 9.1 and 9.2.

inability of in situ detectors to measure multiple wavelengths. It can be anticipated that in the future, in situ fluorometers will be developed that have the capabilities to measure a full EEM with reasonable spectral resolution. In either case, these fluorescence measurements result in large quantities of spectral data requiring analysis to be interpreted in a meaningful manner. In laboratory measurements with a modern scanning fluorometer, one EEM typically represents a large amount (typically 2000–3000 wavelength pairs) of data. Similarly, in situ measurements can be made at 10- to 15-minute intervals with probes deployed for weeks to months. For these reasons, fluorescence indices remain widely used to either (1) simplify the large amounts of data present in an EEM, or (2) design an in situ probe to obtain a quantified parameter that can be compared across sites or followed over time in a field study.

This chapter reviews commonly used indices that have been developed in the analysis of organic matter, along with their interpretation and potential problems. The chapter concludes with an overview of the use of fluorescence indices, some recommendations and ideas about future directions in their use, and some important spectroscopic considerations for collecting fluorescence measurements. For further reading on the use of fluorescence spectroscopy to understand DOM ecosystem dynamics, both Chapter 8 (this volume) and the review paper by Fellman et al. (2010) highlight additional ecological uses of fluorescence spectroscopy and indices. It is important to note that there are spectroscopic difficulties in the application of fluorescence methods as detailed in Chapters 2 and 5 (this volume). These are related to instrument-dependent characteristics of the fluorescence spectra, inner-filter effects associated with the absorbance of emitted light, and fluorescence quenching by metal complexation and pH effects. All of these need to be addressed to meaningfully compare results from different laboratories or in a single system over time (Cory et al., 2010). Despite these potential difficulties, the use of indices, along with the availability of standard reference materials, has allowed for comparisons across studies done by different investigators in different study sites to build a broader understanding of the dynamics of DOM and SOM extracts.

9.2 Overview of Common Fluorescence Indices

Table 9.1 summarizes the common fluorescence indices that have been employed in the literature and are discussed in this chapter. Fluorescence indices were first used in the late 1990s, building on the improved interpretation of soil organic matter extracts using synchronous scan techniques. Work such as that by Nicola Senesi and collaborators (e.g., Senesi et al., 1989, 1991; Miano and Senesi, 1992) formed the foundations of our understanding of how fluorescence spectroscopy could be used to classify and distinguish humic-like substances from various soils, composts, and sludges. Using synchronous scan approaches and humic and fulvic extracts analyzed as water solutions at relatively high concentrations and fixed pH, several consistent peaks in fluorescence intensity were observed between 300 and 500 nm emission (Senesi et al., 1991). Comparison of fluorescence spectra with ^{13}C-nuclear magnetic resonance (^{13}C-NMR), ultraviolet (UV) absorbance, and Fourier transform infrared (FTIR) spectroscopy demonstrated that fluorescence spectra were related to the number of highly substituted aromatic nuclei and/or conjugated unsaturated systems capable of a high degree of resonance (Senesi et al., 1989; Miano and Senesi, 1992). They found that a shift in maximum fluorescence intensity from shorter to longer wavelengths due to the degree of polycondensation of the organic matter could be captured and quantified by use of an index. Later Kalbitz et al. (1999) developed a "humification index" using

Table 9.1. *Fluorescence indices highlighted in this chapter, with relevant papers, and their definitions*

Fluorescence Index	Parameters	Comments
Humification Index (HIX_{SYN}) Kalbitz et al. (1999, 2000)	Calculated using synchronous scans with 18 nm offset. (ex 470 nm/em 488 nm)/(ex 360 nm/em 378 nm) or (ex 400 nm/em 418 nm)/(ex 360 nm/em 378 nm) for fulvic acids. (ex 390 nm/em 408 nm)/(ex 355 nm/em 373 nm) for whole water.	Correlated with C/N ratio, aromaticity, and polycondensation for soil extracted and groundwater OM. Higher numbers indicate more polycondensation and lower C/N.
Humification Index (HIX_{EM}) Zsolnay et al. (1999) Ohno (2002a)	At ex 254 nm, area of peak under em 435–480 nm divided by peak area under em 300–345 nm. Ohno later suggested modification for concentrated samples of area under em 435–480 nm divided by area under em 300–345 nm + 435–480 nm.	Indicates degree of humification of soil extracts, with higher numbers indicative of lower H/C ratios shifting the emission to longer wavelengths, attributed to a greater degree of humification.
Freshness Index (β/α) (BIX) Parlanti et al. (2000) Huguet et al. (2009) Wilson and Xenopoulos (2009)	Originally calculated as the ratio of the β peak (max intensity within ex 310–320 nm/em 380–420 nm) and α peak (max intensity within ex 330–350 nm/em 420–480 nm). Modified to intensity at em 380 nm divided by max intensity between em 420 nm and em 435 nm at ex 310 nm.	Indicates proportion of recently produced DOM. The β peak represents recently created (likely microbial) OM while the α peak is older, more decomposed OM. Developed for and mostly used in estuarine environments.
Fluorescence Index (FI) McKnight et al. (2001) Cory and McKnight (2005)	Originally calculated as the ratio of em 450 nm and em 500 nm at ex 370 nm for non-instrument corrected spectra. Modified, for instrument-corrected spectra, to em 470 nm/em 520 nm at ex 370 nm.	Indicates if precursor material for DOM is of a more microbial (FI ~1.8) in nature or more terrestrially derived (FI ~1.2).
Peak T/Peak C ratio Baker (2001)	Ratio of maximum fluorescence at ex 275 nm/em 350 nm (peak T) to max intensity of the region ex 320–340 nm/em 410–430 nm (peak C).	Used to identify impact of sewage effluent on a river. Indicates biochemical oxygen demand relative to dissolved organic carbon.
Redox Index (RI) Cory and McKnight (2005) Miller et al. (2006)	Sum of reduced quinone-like input over total quinone-like input from components of Cory-McKnight PARAFAC model.	Indicates whether quinone-like components within DOM are more reduced (closer to 1) or oxidized (closer to 0) in character. Typical values are around 0.42.

Note: em = emission wavelengths, ex = excitation wavelengths

Table 9.2. *Common labels for identified regions of observed fluorescence peaks in an excitation–emission spectrum of aquatic DOM*

Component	ex (nm)	em (nm)	Coble (1996)	Parlanti et al. (2000)
Humic-like	250–260	380–480	A	α'
Tyrosine-like	270–280	300–320	B	γ
Humic-like	330–350	420–480	C	α
Marine humic-like	310–320	380–420	M	β
Tryptophan-like	270–280	320–350	T	δ

Note: Note that peak M has subsequently been observed in nonmarine environments.

synchronous fluorescence spectra. At the same time, Zsolnay et al. (1999) developed a "humification index" using line scan spectroscopy.

Aquatic scientists also explored the application of fluorescence spectroscopy to study the chemical quality and source of DOM. For example, Stewart and Wetzel (1980, 1981) showed that in a freshwater lake the larger molecular weight aquatic humic fractions had a greater absorbance but lower fluorescence than smaller molecular weight fractions and that the higher molecular weight humic fractions were removed in calcium-rich waters. In marine systems, the fluorescence characteristics of DOM were shown to provide a useful means for its characterization despite low DOC concentrations (Coble, 1996). Coble identified common fluorophores present in marine and coastal waters. Five component peaks were identified as either protein-like or humic-like and their regions and labels can be found in Figure 9.1a and Table 9.2. Based on this work, fluorescence indices were developed by aquatic scientists as a means to understand the variations in DOM quality in natural waters. Parlanti et al. (2000) developed the "freshness index" (the β/α index, later modified to the "BIX" index) to identify microbial influence on marine DOM. McKnight et al. (2001) put forward a "fluorescence index" (FI) to examine differences in precursor organic materials for aquatic humics. Later, Miller et al. (2006) proposed a "redox index" (RI) as an indicator of the oxidation state of quinone-like moieties in DOM. Indices were also developed to quantify information in more specialized studies. Proctor et al. (2000) applied an index to fluorescent organic matter preserved in a cave stalagmite. Perrette et al. (2005) also developed a similar index for use in measuring fluorescent organic matter preserved in stalagmites, using a laser excitation source for high spatial resolution analyses.

Figure 9.1b shows the locations of the wavelength pairs measured in the fluorescence indices tabulated in Table 9.1, presented together with the regions of interest typically associated with identified peaks indicated in Table 9.2. Most of the indices focus on variations in the fluorescence intensity associated with what has historically been referred to as the "humic-like" peak in natural organic matter, sometimes also comparing it with what has been called the "protein-like" peak, although it is now clear that fluorescence in these regions can be attributed to more than one class of fluorescent organic matter. In all cases, the interpretation of the index is likely to vary with the organic matter source, matrix, and environmental conditions, such as changing salinity, pH, or sunlight. These processes are considered in detail in Chapters 7 and 8 (this volume) on chemical and biological effects on OM fluorescence, and are reviewed briefly in the context of fluorescence indices in Section 9.4. In the rest of this section we will consider each commonly used index and its interpretation separately, along with a brief explanation of how and why it was developed.

9.2.1 A "Humification Index" to Track Chemical Properties Developed by Kalbitz and Colleagues (HIX_{SYN})

Kalbitz et al. (1999) utilized spectroscopic techniques to distinguish differences in dissolved humic substances due to variations in land use in a fen area in Germany. Groundwater samples as well as topsoil aqueous extracts were collected from regions of arable land, grassland, and woodland in various stages of succession. UV, fluorescence, and IR spectroscopic properties were tested for their suitability as an indicator of land use and modification. They measured the fluorescent signature using synchronous scan fluorescence with an excitation range from 260 to 560 nm and an offset of 18 nm. Scans of resin-extracted humic and fulvic acids at 10 mg L^{-1} DOC were normalized to carbon content and characteristic peaks were identified at ex 360/em 378 nm and ex 400/em 418 nm excitation/emission wavelength pairs, and a shoulder at ex 470/em 488 nm. Kalbitz et al. (2000) compared the peak locations in the original whole water samples to the peak locations in the fulvic acid fractions and found that the peak locations in the whole water samples had shifted to shorter wavelengths after isolation, which was associated with fewer highly substituted aromatic nuclei.

Kalbitz et al. (1999) used fluorescence intensity ratios to calculate two humification (HIX_{syn}) indices for the extracted humic material. Each index ratios the fluorescence emission intensity at ex 360/em 378 nm against the emission intensity at ex 400/em 418 nm or ex 470/em 488 nm. The (ex 400/em 418)/(ex 360/em 378) and (ex 470/em 488)/(ex 360/em 378) indices were shown to correlate closely with each other ($r = 0.91$, $n = 46$) and could be reproducibly measured because they are independent of total fluorescence intensity. Kalbitz et al. (1999) argued that these indices are humification ratios, whereby a higher index value indicates a higher degree of humification, inferred from a higher degree of polycondensation and a shift to longer fluorescence wavelengths. The higher wavelength (ex 470/em 488)/(ex 360/em 378) humification index correlated inversely with the C/N ratio of the samples ($r = -0.63$) and positively with the infrared spectra at 1620 cm^{-1} (which is related to absorbance from aromatic C=C structures; $r = 0.71$). It also correlated inversely with the infrared absorbance ratio at 1725 cm^{-1}/1620 cm^{-1} (which is related to the ratio of COO– to C=C structures; $r = -0.69$), strongly suggestive that this index is indeed a measure of polycondensation. For the whole water samples Kalbitz et al. (2000) calculated an index with a (ex 390/em 408)/(ex 355/em 373) ratio to account for the shift to shorter wavelengths in the spectrum. As shown in Figure 9.2, this was found to strongly correlate to the index for the associated humic material ($r = 0.89$, $n = 46$, $\alpha < 0.001$), the FTIR absorbance at 1620 cm^{-1} ($r = 0.66$, $n = 44$, $\alpha < 0.001$) and the C/N ratio ($r = -0.48$, $n = 46$, $\alpha < 0.01$), indicating

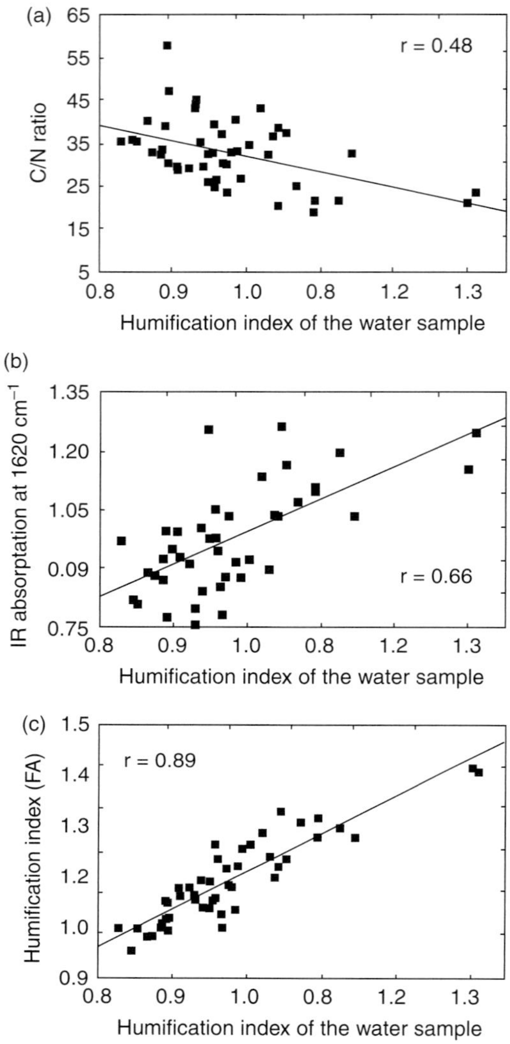

Figure 9.2. Correlations of humification index (HIX_{SYN}) for whole water (*x*-axis) with the (a) C/N ratio of the sample, (b) humification index of the fulvic acid fraction, and (c) IR absorption at 1620 cm^{-1}. The whole water HIX_{SYN} is calculated as (ex 390/em 408)/(ex 355/em 373) and the fulvic acid HIX_{SYN} graphed here is calculated as (ex 400/em 418)/(ex 360/ em 378) using synchronous scans with an 18 nm offset. (From Kalbitz et al. [2000] with minor modifications.)

the potential to use spectroscopy on the original water sample to determine the chemical character of the humic material.

Kalbitz and Geyer (2001) further tested the humification index for the effect of DOC concentration and spectrometer type (a LS 50B luminescence spectrometer by Perkin Elmer and a SFM 25 spectrometer by BIO-TEK Instruments). They confirmed that inter-instrument comparison of calculated humification indices was possible but required application of an appropriate correction curve specific to the compared instruments. This correction curve would require running the same set of standards on each instrument. They also found the index would be highly prone to inner-filter effects if samples were too concentrated, resulting in artificially high estimates of the degree of aromaticity (or conjugated ring structures) from the humification index. A sample concentration of approximately 10 mg L^{-1} DOC was recommended to minimize concentration and inner-filter effects.

Figure 9.1 demonstrates that the humification index of Kalbitz et al. (1999) is essentially measuring the intensity of fluorescence associated with what is now referred to as peak C. Rather than tracking the change in maximum fluorescence of the peak, the chosen wavelength pairs are at longer excitation wavelengths. By capturing the fluorescence at these longer wavelengths, the index identifies changes in the shape of the peak which may relate to aromaticity and other bulk properties of the sample.

9.2.2 *Zsolnay's Humification Index to Identify Soil Organic Matter Properties (HIX_{EM})*

Zsolnay et al. (1999) proposed a "humification index" (HIX_{EM}) around the same time as Kalbitz et al. (1999). Zsolnay et al. (1999) investigated the effects of drying on soil OM, seeking to identify the source of the observed increase of extractable DOM due to drying processes. Aqueous extracts (using 4 m*M* $CaCl_2$) of DOM from multiple soil types were separated into fulvic acid and material released by cell lysis. The fluorescence signature of the DOM and of each fraction was analyzed using a fixed excitation wavelength of 254 nm and usable emission was measured between 300 and 480 nm. Samples were run at a pH of 2 to negate pH effects in comparisons and diluted to an absorbance no greater than 0.1 cm^{-1}.

The HIX_{EM} was determined by dividing the area of fluorescence intensity between 435 and 480 nm by that between 300 and 345 nm (mistakenly given as 300–445 nm in the original text, but correct in Figure 9.1 in the paper, Figure 9.3 in this chapter). Zsolnay et al. (1999) used theoretical concepts similar to those of Kalbitz et al. (1999), proposing that an increase in the HIX_{EM} is associated with an increase in emission wavelength due to increased polycondensation (lower H/C ratios). Because humification is also associated with a decrease in the H/C ratio, they suggested the index represented the degree of humification of the organic matter. Figure 9.4 presents results showing the effects of air drying, oven drying, and fumigation on two contrasting soil OM samples in comparison to a fulvic acid standard, showing how oven drying decreased the HIX_{EM} and standard relative fluorescence (fluorescence per gram of carbon in quinine sulphate dihydrate equivalents). Because humified material is generally more condensed than its precursor material, and because the HIX_{EM} decreased during oven drying, Zsolnay et al. (1999) concluded that most of the organic material released is from non-humified sources such as cell lysis.

Ohno (2002a) investigated concentration and inner filter effects on the humification index. Field corn residue extract, water-extractable soil organic matter, and soil fulvic acid

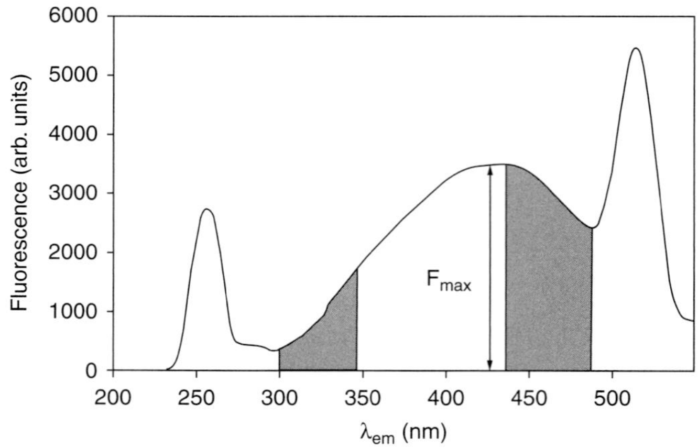

Figure 9.3. Fluorescence emission spectrum of an aqueous extract of soil at an excitation wavelength of 254 nm. The broad humic peak (300–500 nm) can be seen between the two sharp Rayleigh scattering peaks. The shaded regions represent the areas used to calculate the HIX_{EM}. The index is calculated by dividing the area under the upper region (435–480 nm) by the area under the lower region (330–345 nm) and is associated with the degree of humification in the sample. F_{max} indicates the peak emission intensity. (From Zsolnay et al., 1999.)

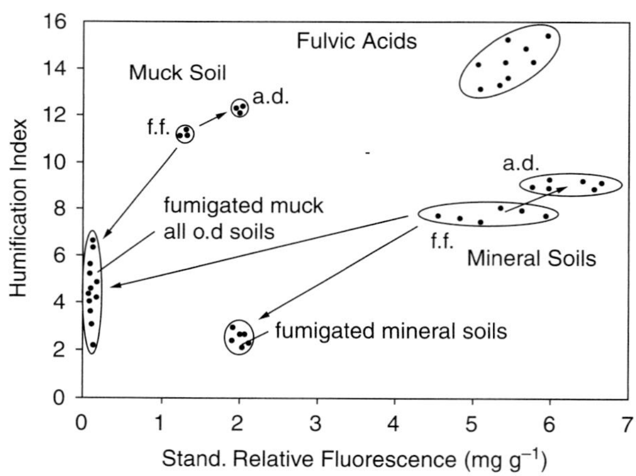

Figure 9.4. Comparison of the standard relative fluorescence intensity (in units of quinine sulfate dihydrate equivalent) to HIX_{EM}. Muck soils and mineral soils were treated by air drying, oven drying, and fumigation and then extracted aqueously for fluorescence analysis. The fulvic acid fraction of each soil was obtained with NaOH pyrophosphate. This plot shows the different signals from the field fresh (f.f), air dried (a.d.), oven dried (o.d.), and fumigated samples, indicating drying processes release different fractions of organic matter. (From Zsolnay et al., 1999.)

were measured at increasing concentrations and the HIX_{EM} was found to increase with concentration in all three samples. This was due to the greater absorption, and thus greater sensitivity to concentration effects, in the 300–345 nm region. Ohno (2002a) made two recommendations for avoiding the inner filter effect in this situation. The first was to dilute all solutions to a fixed absorbance (<0.3 cm^{-1}) and the second was to calculate the HIX_{EM} as a ratio of the area of the fluorescence intensity under 435–480 nm divided by the sum of the areas under 300–345 nm and 435–480 nm). Zsolnay (2002) commented that the concentration error in HIX_{EM} would be on the order of only 10% in dilute samples and Ohno (2002b) further emphasized that reducing concentration to limit the inner-filter effect allows a more accurate measure of DOM quality. Because HIX_{EM} is a ratio of two regions that are not equally affected by concentration, it should be used with caution, and concentration and inner-filter effects could have a large impact if not corrected. The original calculation proposed by Zsolnay et al. (1999) is most commonly used in the literature, and can be accurately used to compare samples. This is provided they are diluted to an absorbance no greater than 0.3 cm^{-1} at 254 nm to minimize inner-filter effects due to concentration and allow for inner-filter corrections to be effective (Ohno, 2002a).

9.2.3 Freshness Index to Identify Microbial Material in Marine DOM (the "β/α" and "BIX" Index)

Parlanti et al. (2000) reported fluorescent components and EEMs in filtered water samples collected from marine and freshwater end members in a coastal shelf environment. Peaks C, A, M, B, and T of Coble (1996) were renamed in Parlanti et al. (2000) where peak C is α, peak M is β, and peak B is γ (see Table 9.2). The α peak (max intensity within ex 330–350 nm, em 420–480 nm) is associated with humic-like components, the β peak (max intensity within ex 310–320 nm, em 380–420 nm) with marine humic-like components, and the γ peak with tyrosine-like, protein-like components. Observations of the β peak co-occurring with protein-like peaks supported the hypothesis of it originating from biological activity (Figure 9.5). The ratios of the fluorescence intensity of these three peaks were reported for both the marine and freshwater surface samples, and for a laboratory degradation experiment with blue-green algae. Parlanti et al. (2000) demonstrated that peaks γ and β both originate from recent biological activity and could be used as markers of biological activity in coastal zones. They found the ratio of these peaks to peak α (humic-like) provided an index of biologically produced to terrestrially derived fluorescent organic matter in coastal environments. They named this β/α ratio the freshness index, as it was indicative of the amount of organic matter recently produced from biological activity.

Huguet et al. (2009) further developed the β/α ratio, renamed BIX. This index accounts for the presence of the biological activity– based β peak causing a broadening of the humic portion of the emission spectrum. BIX is the ratio of the emitted fluorescence intensity at 380 nm (max of the β band) to the emitted fluorescence intensity at 430 nm (max of the α band) when excited at 310 nm. This is essentially a single line scan version of the β/α ratio, where an increase in BIX is related to an increase in the intensity of fluorophore β. Huguet

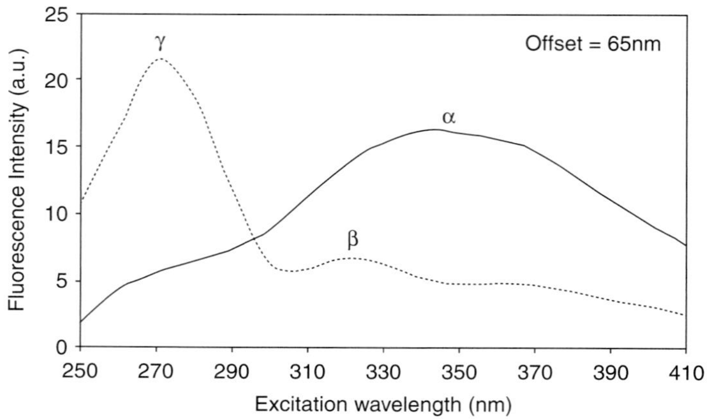

Figure 9.5. Synchronous fluorescence spectra with an offset of 65 nm of a freshwater sample (solid line) and marine water sample (dotted line). Locations of the α (humic-like), β (marine humic-like), and γ (tyrosine-like) peaks are identified. The marine humic-like peak has been associated with recently created humic material due to algal activity and is commonly seen with a corresponding protein peak. The β and α peaks are used to calculate the freshness index, which is considered an indicator of the degree of recently produced humic material in coastal systems. (From Parlanti et al. [2000], with minor adaptations.)

et al. (2009) investigated filtered estuarine samples, and after collecting EEMs, calculated the BIX. Low ratios were interpreted in this environment as containing dissolved organic matter with a low autochthonous component. Increasing ratios were correlated to a high autochthonous component of greater aquatic microbial origin.

Wilson and Xenopoulos (2009) suggested another slight variation on the β/α ratio. Riverine dissolved organic matter from 34 watersheds was compared to determine the impact of agricultural land use on DOM characteristics. Like Huguet et al. (2009) the 310 nm excitation wavelength was used. The β/α index was calculated by dividing the emission intensity at 380 nm (representing the β peak) with the maximum emission intensity observed between 420 and 435 nm (representing the α peak). Wilson and Xenopoulos (2009) found an increase in the β/α ratio with an increase in percent cropland coverage, corresponding with an increase in total dissolved nitrogen, a nutrient that often increases biological activity. These results reinforced the association of the freshness index with recently produced DOM from microbial sources.

9.2.4 Fluorescence Index to identify Precursor Material in Freshwater DOM (FI)

While studying humics found in lakes and streams, McKnight et al. (2001) investigated the relative importance of plant/soil-derived organic material compared to organic material derived from microbial biomass as precursor material. They used ^{13}C-NMR, ^{1}H-NMR, IR, and elemental analysis to study the chemistry of purely microbially derived fulvic acids isolated from lakes and streams in the McMurdo Dry Valleys, a plant-free desert oasis near Ross Island in Antarctica. Results showed that these microbially derived fulvic acids had a lower carbon to nitrogen ratio and lower aromaticity than plant/soil OM, reflecting the absence of lignin in the precursor pool of organic molecules (McKnight et al. 1994; Aiken et al. 1996). An opportunity to compare Dry Valley lake samples with lake samples from the Bunger Hills, another desert oasis in Antarctica, motivated a search for a fluorescence index that could be used on small-volume water samples.

Based on the work of Ewald and Belin (1987), who examined differences in fluorescence between marine samples at an excitation wavelength of 370 nm in the peak C region, a trial-and-error approach was used to identify changes in the fluorescence signal of various fulvic acids. This study confirmed that at an excitation wavelength of 370 nm, there was a significant difference in the emission line scan between an extracted fulvic acid from a dry valley lake and the International Humic Substances Society (IHSS) reference fulvic acid from the Suwannee River, which drains the Okefenokee Swamp, a large wetland in the southeastern United States. Whole waters from the Antarctic lakes and whole waters and isolated fulvic acids from North American streams and rivers were also analyzed. The spectra were inner-filter corrected and a fluorescence index (FI) was developed based on the ratio of the emission at 450 nm to 500 nm (ex = 370 nm) to quantify the sharper and narrower peaks of the Antarctic lake fulvic acids compared to the broader peak of the Suwannee River fulvic acid and other plant/soil-derived samples. The 450 nm wavelength was chosen because it was close to the mean of the maximum intensity for the dataset, and 500 nm because it was close to 50% of the peak emission for the microbially derived fulvic acids. The samples from the Antarctic lakes had the highest FI (1.7–2.0), corresponding to DOM derived from a microbial source associated predominantly with phytoplankton productivity. In contrast, the North American river samples had a lower FI (1.3–1.4), resulting from their origins in plant litter and soils. EEMs obtained for a few samples confirmed that increases in this index corresponded to a shift of the main peak C fluorophore to a lower emission wavelength. As expected, for the isolated fulvic acid samples, a good correlation was observed between the FI and the percent aromaticity as determined by ^{13}C-NMR (Figure 9.6). However, in a set of stream fulvic acid samples, removal of more aromatic fulvic acid by sorption onto streambed iron oxides caused a large decrease in aromaticity but only a small change in the index. This can be seen in Figure 9.6 where the Snake River (SR), which is fed by Deer Creek (DC), has a lower aromaticity but comparable fluorescence index. This observation indicates that the index reflects primarily organic precursor material rather than aromaticity. McKnight et al. (2001) commented that the FI would potentially be useful in screening samples to identify changes in DOM source and quality that could be explored using other more detailed measurements. McKnight et al. (2001) also noted that the range of values of the index would be instrument dependent.

Subsequent work by Cory and McKnight (2005) refined the FI based on instrument corrections applicable to all fluorometers. The refined FI is calculated as the ratio of emission intensity at 470 nm to that of 520 nm at an excitation of 370 nm. This study analyzed

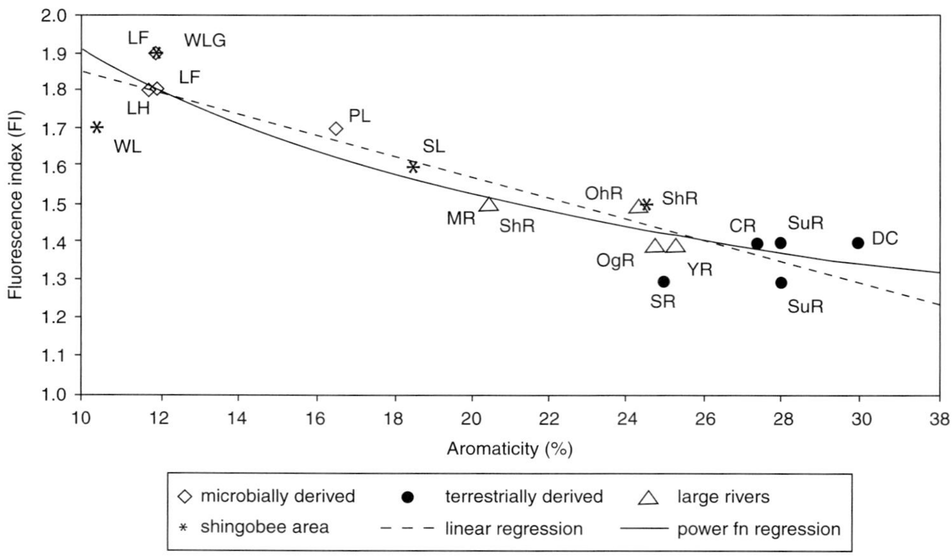

Figure 9.6. Plot of McKnight et al. (2001) fluorescence index (FI) vs. % aromaticity for a variety of isolated fulvic acid samples. The % aromaticity was calculated using ^{13}C-NMR as the ratio of the area of the aromatic carbon region to the total area of the spectrum. There is a strong correlation between FI and % aromaticity (linear regression equation is $y = -0.027x + 2.1$ ($R^2 = 0.85$), and the power function regression is $y = 3.94\,x^{-0.316}$ where x = % aromaticity and y = FI). Note that while Deer Creek (DC) flows into Snake River (SR), the aromaticity drops due to the aromatic fulvic acids sorbing to iron oxide in the streambed, but there is little change in the FI, indicating a robustness of the index.

Site codes: DC = Deer Creek, LF = Lake Fryxell, LH = Lake Hoare, MR = Missouri River, OgR = Ogeechee River, OhR = Ohio River, PL = Pony Lake, SL = Shingobee Lake, ShR = Shingobee River, SR = Snake River, SuR = Suwanee River, WL = Williams Lake, WLg = Williams Lake groundwater, YR = Yakima River. (From McKnight et al. [2001] where site descriptions are presented.)

a data set of EEMs from 379 samples that included fulvic acid extracts and whole waters from oxic and anoxic zones of diverse lakes, streams, and wetlands, including those in the Dry Valleys, alpine watersheds in Colorado, and an Arctic tundra in Alaska. Cory and McKnight (2005) used parallel factor analysis (PARAFAC) to develop a 13-component model that included 7 components with absorption and emission spectra that were similar to those of various model quinone compounds. (See Chapter 10 for details of PARAFAC analysis.) Three components were identified as representing oxidized quinone-like moieties (Q1, Q2, and Q3), three components represent semiquinone-like moieties (SQ1, SQ2, and SQ3), and one component represents a reduced hydroquione-like moiety (HQ). They identified a strong correlation between the FI and a relationship between two of these components (SQ1 and SQ2 [Figure 9.7]). These components had dual absorption peaks, one of which was centered on ~370 nm with emission peaks in the peak C range. The SQ2 component was resolved in a simpler PARAFAC model based only on the EEMs for the Antarctic lakes and was relatively more important in samples with higher FIs. Cory et al.

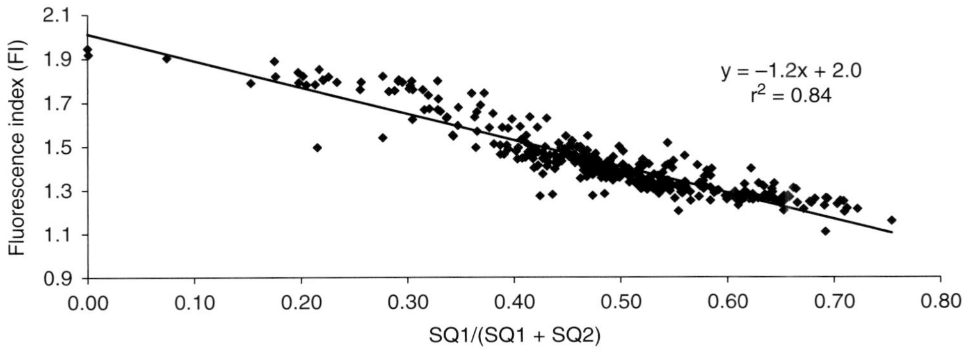

Figure 9.7. Two components of the Cory and McKnight (2005) PARAFAC model that have spectral properties similar to semiquinones (SQ1 and SQ2) explain the variation in the McKnight et al. (2001) fluorescence index (FI). The SQ2 component is resolved in a model using only samples from Antarctic lakes, indicating it is likely microbial in origin. As the portion of SQ2 to the total of SQ1 + SQ2 falls, so does the FI, and a lower FI is associated with less microbially derived material. (From Cory and McKnight, 2005.)

(2007) further investigated the effect of photodegradation on Arctic samples and the FI, demonstrating that for these samples the FI was insensitive to photodegradation. These results provide evidence that FI is tied to underlying chemistry based on precursor material. Finally, Cory et al. (2010) compared fluorometers and instrument correction factors, and showed that if an instrument has sufficient sensitivity, application of correction factors can allow for direct comparison of FI values in different studies. For example, Jaffé et al. (2008) analyzed, at two different laboratories, a set of diverse samples ranging from marine surface waters to soil interstitial waters in the Arctic tundra. The FI values from the two laboratories were highly correlated ($r^2 = 0.87$), with a near zero intercept once proper instrument correction factors were applied.

9.2.5 The "Peak T/Peak C Ratio" to Identify Sewage Impact on Rivers

Baker (2001) presented fluorescence EEMs from a selection of small UK rivers impacted by sewage effluents. Samples from the effluents themselves and from the rivers upstream and downstream of the sewage discharge demonstrated a change in fluorescence signature due to sewage input, indicating the potential use of fluorescence to analyze the water quality of sewage-impacted rivers. Tryptophan-like peak T (called peak A in the paper) was measured at 275 nm excitation, 350 nm emission. Fulvic-like peak C (called peak B in the paper) was measured at the wavelengths of maximum intensity within 320 –340 nm excitation, 410–430 nm emission. The ratio of peak T to peak C was used to differentiate upstream and downstream water samples. Upstream of the discharge sites, the unimpacted rivers had a protein-like/fulvic-like fluorescence intensity ratio of around 0.6 due to the peak C intensity being greater than the peak T intensity. The sewage effluents had a

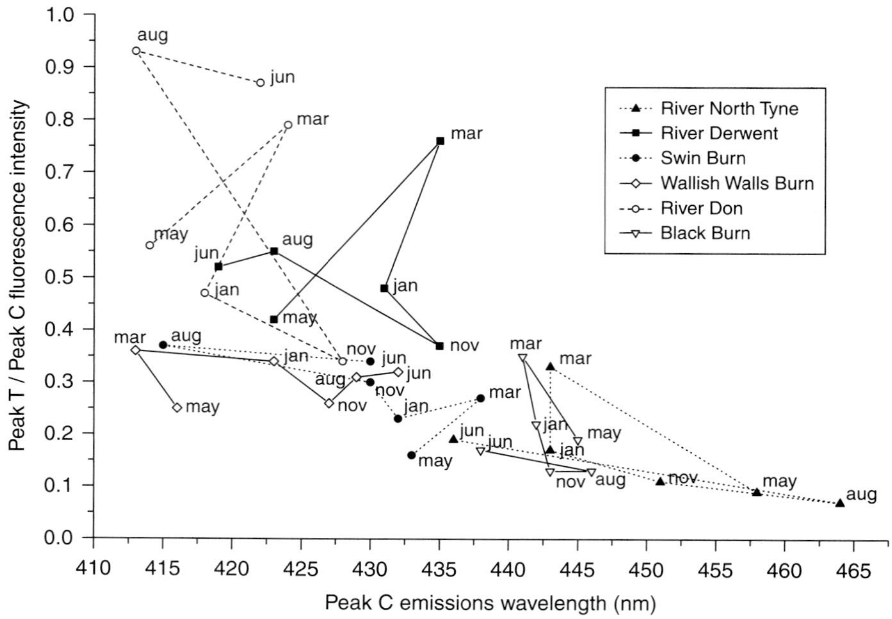

Figure 9.8. Example of the use of the peak T/peak C ratio to characterize river water samples, here graphed against the wavelength of maximum peak C emission. Samples are from the Tyne catchment, NE England over an annual sampling campaign 2002–2003. The North Tyne and Black Burn, upland peat-rich catchments, are characterized by a low peak T/peak C ratio and high peak C emission wavelength, considered to be a natural water signal. The more impacted lowland catchments – the Wallish Walls Burn, with intensive agriculture, and the Rivers Don and Derwent, two urbanized catchments – have lower peak C emission wavelengths and a greater range of peak T/peak C, though the ratio is generally higher than in un-impacted upland rivers. This shows the utility of the peak T/peak C index for identifying the influence of wastewater on rivers. (From Hudson et al. [2007], with minor adjustments.)

peak T/peak C ratio of around 1.0, showing a distinctive fluorescence EEM with high protein-like and fulvic-like peaks in approximately equal ratio. Downstream of the sewage discharge the EEMs continued to show the discharge signal, with the ratio slowly decreasing due to dilution effects.

Subsequent research has demonstrated that peak T fluorescence intensity is strongly correlated with biological oxygen demand in both river waters and final treated sewage effluents (Hudson et al., 2008). Together with the observation that peak C fluorescence intensity correlates with dissolved organic matter concentration for individual rivers, the peak T/peak C ratio can be conceptualized as a BOD/DOC ratio in terrestrial waters, and therefore related to water quality. Figure 9.8 presents the peak T/peak C ratio for some sub-catchments of the River Tyne, UK (Hudson et al., 2007) showing a contrast between upland catchments of good water quality and lowland, urban catchments of poor quality. The microbial sources of peak T fluorescence need not be limited to sewage-derived organic matter, with agricultural organic matter sources also potentially significant (Baker, 2002). The peak T/peak C ratio has also been used as an indicator of slurry run-off from fields (Naden et al., 2010). In marine and estuarine samples, however, a pollutant interpretation for peak T fluorescence intensity would not be valid, with peak T fluorescence attributed to marine biological activity. Conceptually, the peak T/peak C ratio is the freshwater equivalent of the freshness index (BIX) used for marine and estuarine waters. In both cases it presents the ratio of the intensity of newer, microbially derived material to more complex humic material.

9.2.6 Redox Index as an Indicator of the Oxidation State of Quinone-Like Moieties

Shortly after the development of a fluorescence index for characterizing the sources of aquatic humics, McKnight and others explored the use of fluorescence to characterize the involvement of humics in electron shuttling (redox) reactions in the environment. Research by Lovley et al. (1996) demonstrated that microbes can use humics as electron shuttles or as electron acceptors in a manner similar to the use of anthraquinone 2,6-disulfonate (AQDS), a model quinone. Thus, it was inferred that quinone-like moieties in humic material play an important role in redox reactions. Evidence for electron transfer to quinone moieties in humic molecules was obtained by Scott et al. (1998) using electron spin resonance (ESR) where an increase in the semi-quinone radical was documented following microbial electron transfer. This change was interpreted as representing a shift towards a preponderance of the reduced dihydroquinone species, similar to anthrahydroquinone 2,6-disulfonate (AHDS). The ESR measurements, however, required high concentrations of humics (2000 mg L^{-1}) that do not occur in natural systems. Klapper et al. (2002) explored the potential for fluorescence spectroscopy to characterize humic redox state. In experiments studying the microbial reduction of humics extracted from marine sediments, Klapper et al. (2002) showed that there was a shift of the main humic fluorophores to higher emission wavelengths following electron transfer, which was analogous to the wavelength shift of the AHDS fluorophores compared to the fluorophores of AQDS. A similar change in the EEMs across the oxycline of one of the Antarctic lakes was observed by Fulton et al. (2004).

In a field experiment conducted in an alpine stream–wetland ecosystem, Miller et al. (2006) investigated the role of hyporheic exchange in controlling the rates of redox reactions involving DOM and nitrogen in a short (100 m) reach of an alpine stream. In this study, whole water EEMs were fit to the Cory and McKnight (2005) PARAFAC model, and the loadings of the seven quinone-like components were used to develop the redox index (RI) as a measure of the redox state of quinone-like moieties in humic DOM. The RI is defined as the ratio of the sum of the loadings of the four reduced quinone-like components (SQ1, SQ2, SQ3, and HQ) to the sum of the loadings of all seven of the quinone-like components identified by the Cory and McKnight (2005) PARAFAC model. High RI values (~0.5–0.6) are representative of predominantly reduced quinone-like moieties and low RI values (<0.4) are representative of more oxidized quinone-like moieties. Miller et al. (2006) used RI values in concert with a reactive transport model to demonstrate that reduced quinone-like moieties (RI = 0.51) are transported from the hyporheic zone to the main stream channel, where they are rapidly oxidized (RI = 0.39). The study highlighted the usefulness in biogeochemical studies of fitting EEMs to an existing PARAFAC model constructed based on EEMs from a range of DOM sources and redox conditions, such as the Cory and McKnight (2005) model, in this case necessary to calculate RI. A discussion of the potential advantages and disadvantages to building a new PARAFAC model versus fitting EEMs to an existing PARAFAC model is provided by Fellman et al. (2009) and PARAFAC modeling is considered in detail in Chapter 10 (this volume).

The RI has been shown to be useful in understanding DOM–metal interactions across redox gradients in groundwater systems. Mladenov et al. (2010) calculated the RI for surface and groundwater samples collected in Bangladesh in the context of investigating the role of DOM source and redox reactivity in controlling dissolved iron and arsenic mobilization. In surface waters, the quinone-like moieties were oxidized (RI = 0.34–0.46) and had low dissolved iron (<0.1 mg L^{-1}) and arsenic (~5 μg L^{-1}) concentrations. In contrast, shallow groundwater samples had more reduced quinone-like moieties (RI = 0.47–0.48) and higher dissolved iron (6–10 mg L^{-1}) and arsenic (>200 μg L^{-1}) concentrations. Based on these results and results from a sediment incubation experiment, a pathway for microbially mediated electron shuttling via humic DOM resulting in mobilization of dissolved iron and arsenic in shallow groundwater was hypothesized (Figure 9.9) and applied to a similar system in the Okavango Delta (Mladenov et al., 2007, 2008). The RI has also been used to characterize DOM redox state in oxidizing environments such as in the water column in alpine lakes (Miller et al., 2009; Mladenov et al., 2009), and in chlorination during drinking water treatment (Beggs et al., 2009).

9.3 Applications of Fluorescence Indices

Table 9.1 and Figure 9.1 demonstrate that many of the fluorescence indices may be related, especially those that identify variations in the location of maximum intensity of the humic peak (the Kalbitz et al. [1999] and the McKnight et al. [2001] FI) or those which are ratios of microbiologically derived organic matter to soil-derived organic matter (BIX and the peak T/peak C ratio). Here we highlight several studies and show examples of how fluorescence indices have been used to further ecosystem understanding.

9.3.1 Using Fluorescence Indices to Identify Environmental Controls on Soil Organic Matter

As was discussed earlier in this chapter, much of the initial work using fluorescence and fluorescence indices to study and characterize organic matter was developed to understand changes in soil organic matter due to land use or environmental changes. Kalbitz et al. (1999) developed the HIX_{SYN} from the observation that aqueous extracts of topsoil from arable and intensively used sites demonstrated a higher C/N and smaller degree of

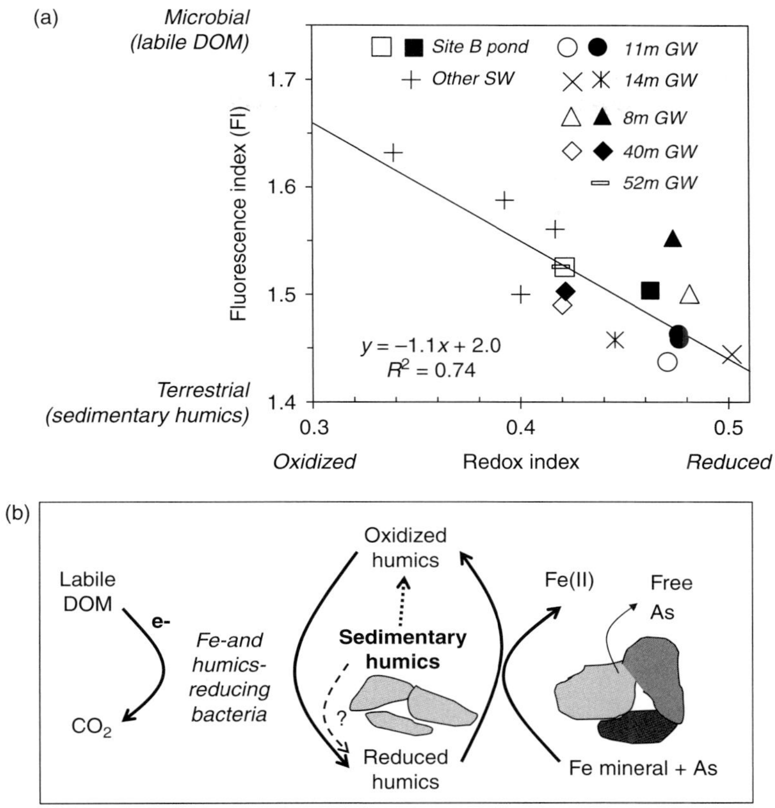

Figure 9.9. Figures showing the relationship between the McKnight et al. (2001) FI and (a) redox index in surface water (SW) and groundwater (GW) samples from Bangladesh. Samples with a more oxidized RI are from an oxidizing surface water environment containing low dissolved arsenic while the deeper samples are more reduced and come from high arsenic environments. (b) A proposed mechanism for microbially mediated electron shuttling cascade in which humic material acts as an electron shuttle between labile DOM in the substrate and iron oxide, ultimately resulting in mobilized iron and arsenic and a change in the redox index of the humic material. [Modified from Mladenov et al. (2010), where more details of the figure symbols can also be found.]

polycondensation than soil extracts from extensively used or unused land. Zsolnay et al. (1999) proposed the HIX_{EM} to measure the effects of soil drying on DOM quality, demonstrating that drying released a greater degree of non-humified material, and Agaki et al. (2007) utilized HIX_{EM} to assess the effects of two different soil sample treatments.

Several studies have used and compared both humification indices (i.e., both HIX_{SYN} and HIX_{EM}). Kalbitz et al. (2003) investigated the impact of biodegradation on the properties of dissolved organic matter extracted from 13 different soil samples. With increasing biodegradation both HIX_{EM} and HIX_{SYN} decreased, indicative of a smaller degree of humification. This decrease in humification was also associated with a decrease in aromatic

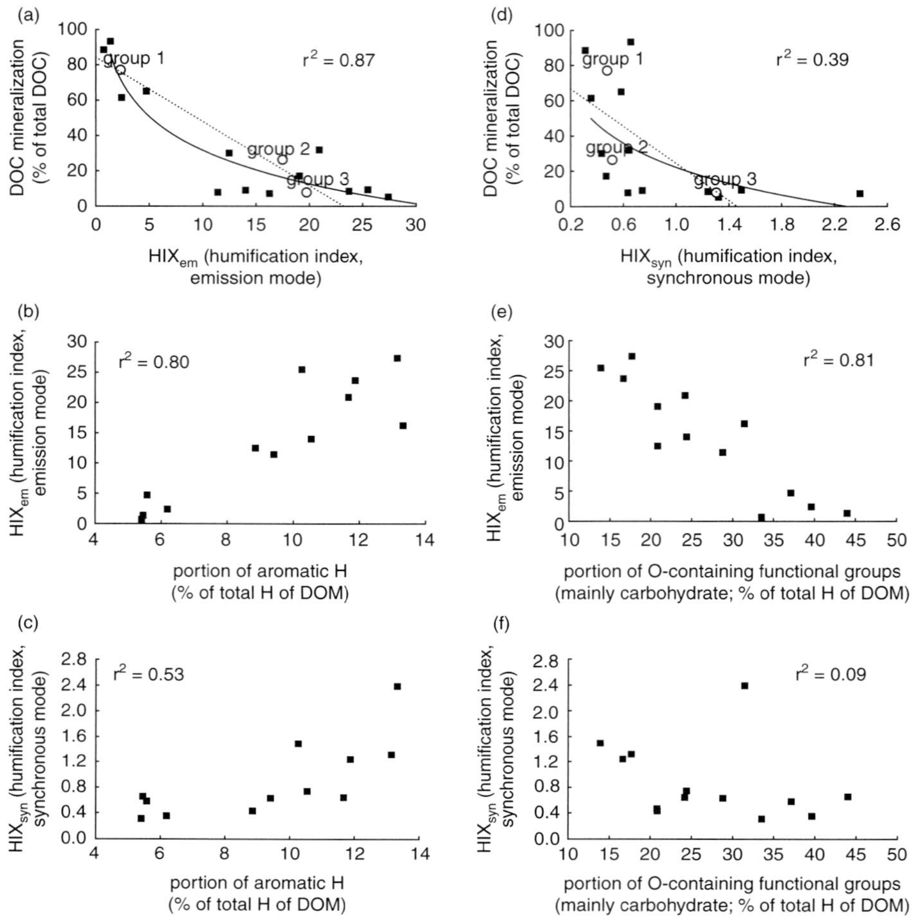

Figure 9.10. Comparison of the HIX_{SYN} and HIX_{EM} for a set of samples extracted from maize straw, forest floors, peats, and agricultural soils. Samples were incubated for 90-days to determine the impact of biodegradation on the DOM character and fluorescence signature. Group 1 samples were considered to have high biodegradation (measured as mineralization to CO_2) after 90 days, group 2 intermediate biodegradation, and group 3 low biodegradation. Graphs (a) and (b) show the logarithmic (solid line, r^2) as well as linear (dashed line) regressions using the average value from each group, showing the decrease in both HIXs with increased biodegradation. HIX_{EM} (c, d) and HIX_{SYN} (e, f) are also plotted vs. portion of aromatic H- (c, e) and O-containing functional groups (d, f). In both cases the HIX_{EM} has a stronger linear relationship. (From Kalbitz et al., 2003.)

H and an increase in O-containing functional groups (Figure 9.10). The HIX_{EM} showed a stronger correlation to these changes in chemical characteristic than the HIX_{SYN}. Bu et al. (2010) also saw a correlation between the two indices when investigating the water-extractable organic matter from soils taken from different vegetation areas along an elevational gradient. They found that surface soil OM had a greater degree of humification than deeper soil OM and conifer forest soil was more humified than other vegetation types, which possibly indicated slower C mineralization. Continuing with the biodegradation studies, Cannavo et al. (2004) studied the change in humification index in response to precipitation over time. They found a strong response of HIX_{EM} to rainfall events, hypothesizing that smaller molecules were more easily washed away, as well as a correlation with microbial activity. They also observed that a change in the index could occur without a corresponding change in total carbon, indicating the index measured a change in the chemical composition of the organic matter that concentration measurements could not detect.

Numerous studies have also observed the humification index (HIX_{EM}) decreasing with depth in the soil profile. Bu et al. (2010) found a noticeable decrease in all four soil types. Corvasce et al. (2006) and Hassouna et al. (2010) made similar observations to those of Cannavo et al. (2004) showing a decrease in aromaticity correlating to the decrease in HIX_{EM}. This decrease has been hypothesized to indicate that larger, more humified molecules are retained on mineral surfaces and the smaller, more mobile molecules can be transported to deeper soil layers. In a study on the affect of acidification on soil DOM, Ohno et al. (2007) also observed a decrease in HIX_{EM} with depth. They found that the decrease in HIX_{EM} with depth was much more pronounced in soil samples in deciduous than coniferous forests. In addition, in both forest types samples from an acidified watershed had higher HIX_{EM} values than samples from the reference watershed.

9.3.2 The Development of a Fluorescence Index to Measure Organic Matter Humification Preserved in Cave Stalagmites and Create Long-Term Records

Preservation of the chemical signature of organic matter has the potential to provide information about past environmental conditions. Groundwater hydrogeologists have observed that drip waters that form cave stalagmites preserve their fluorescent organic matter as they infiltrate through overlying soils and vegetation. To quantify the changes in this preserved organic matter, fluorescence indices were developed that were related to the emission in the humic region, corresponding to a simplification of the Kalbitz approach. Baker et al. (1998) conducted a calibration study of modern stalagmite samples and demonstrated that the emission wavelength of peak C provided useful information on the overlying soil type and extent of humification. Baker et al. (1998) found that an increase in wavelength of the maximum peak C intensity corresponded to an increase in polycondensation and aromaticity. Subsequently a summary of stalagmite fluorescence index data was presented by Baker and Bolton (2000). Based on these results, Proctor et al. (2000) determined a 1000-year record of peat humification from the shifts in the wavelength of peak C fluorescence intensity in a cave stalagmite. Proctor et al. (2000) used a standard fluorescence spectrophotometer coupled to a fiber optic probe to measure fluorescence emitted from a polished stalagmite surface. Using the ratio of fluorescence at ex 350/em 420 nm to that at ex 390/em 470 nm, Proctor et al. (2000) were essentially tracking the regions within peak C that are observed in peat-rich, colored terrestrial waters (Spencer et al., 2007). Figure 9.11 shows this 1000-year fluorescence ratio record, which was validated against other proxy climate archives

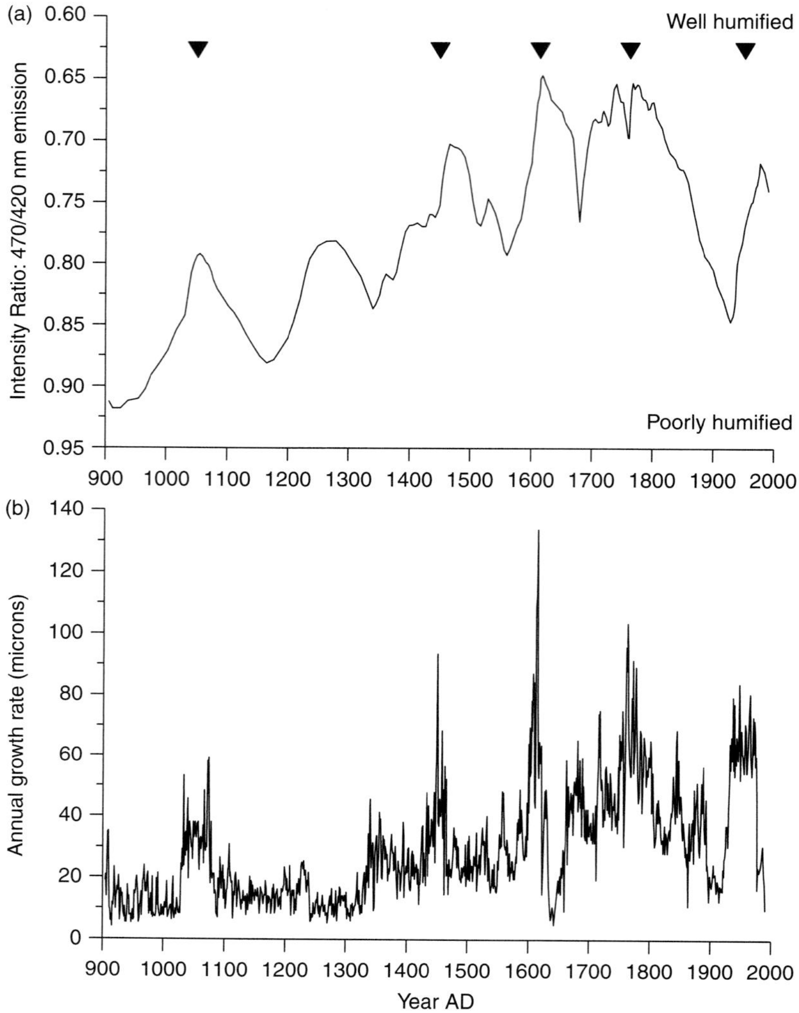

Figure 9.11. A 1000-year record of organic matter humification preserved in a cave stalagmite. (a) The annual band width (annual accumulation rate) of the stalagmite as determined by counting annual bands formed from fluorescent organic matter. (b) The fluorescence ratio as determined by measuring the intensity of emitted fluorescence at 420 and 470 nm. Low ratios (note the inverse scale) correlate with periods of faster stalagmite growth due to a decrease in water table in the overlying peat and increased CO_2 production and limestone dissolution. Low ratios therefore correspond to periods of high humification. (Adapted from Proctor et al., 2000.)

such as the annual lamina width to be a record of water table variations in the overlying peat and the extent of organic matter humification. The same stalagmite data have recently been combined with tree ring records of water availability to produce a 1000-year long record of the North Atlantic Oscillation (Trouet et al., 2009).

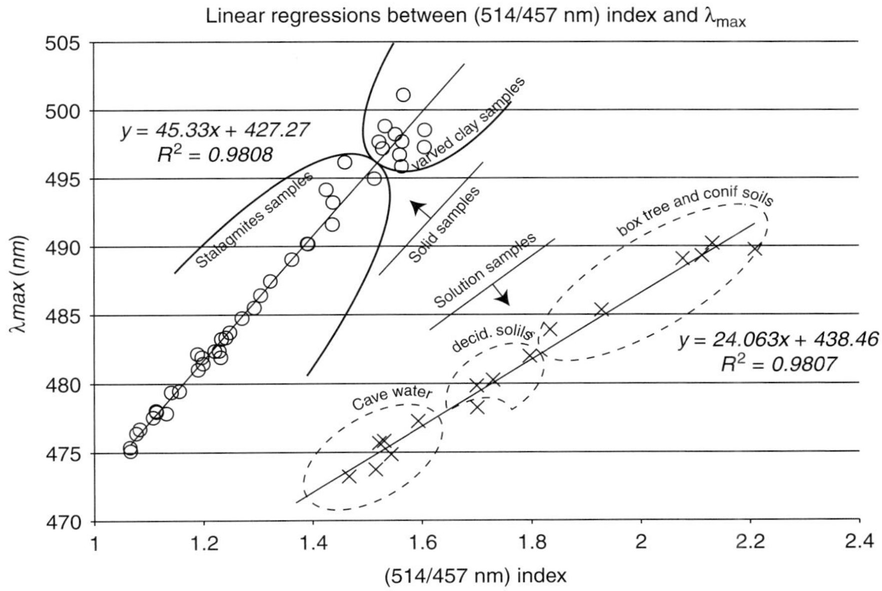

Figure 9.12. Linear regression between Perrette et al. (2005) fluorescence index (the ratio of emitted fluorescence at 514–457 nm) and the wavelength of maximum fluorescence intensity λ_{max}. Excitation wavelength is 364 nm using a laser light source. Lower regression is for aqueous samples from soil and groundwater, grey shaded regression is for geological samples. (From Perrette et al., 2005.)

An important advance in this approach has been the use of a laser excitation, which permits micron-scale excitation of organic matter in the stalagmite samples, equivalent to a sub-annual time resolution (Perrette et al., 2005). Thus, laser excitation has the possibility of high-resolution reconstructions of past organic matter fluxes. Perrette et al. (2005) developed an index based on excitation at 364 nm for use with a laser light source. Conceptually, the Perrette et al. (2005) fluorescence index is almost identical (although inverse) to the McKnight et al. (2001) FI and the index used by Proctor et al. (2000). Emitted fluorescence was detected at 514 and 457 nm, and the ratio of these values used to obtain a fluorescence index. Groundwaters, soil extracts, and cave stalagmites were analyzed, and the index compared to the emission wavelength of maximum fluorescence. Figure 9.12 shows the very low fluorescence index of ground water samples compared to soil water, indicative of the relatively hydrophilic nature of dissolved organic matter in groundwaters, as well as a difference in fluorescence indices between water and calcite matrices. The observation of very low fluorescence indices in the groundwater samples matches the observations of shorter wavelengths of maximum peak C emission observed by Baker and Genty (1999). If this is typical of groundwaters as a whole, it suggests that peak C–related fluorescence indices in surface waters may represent the mixing of groundwater and surface waters in rivers which intercept the water table.

9.3.3 Understanding Controls on DOM Source and Quality in Surface Waters

One initial application of the McKnight et al. FI was to understand the controls on DOM concentration in a system of alpine and subalpine lakes in the Rocky Mountains that are studied as part of the Niwot Ridge Long Term Ecological Research (LTER) project (Hood et al., 2003). In all the lakes an increase in DOC concentration occurred during snowmelt and this DOC peak was associated with a high percentage of humics and low FIs, indicative of a plant/soil source of organic material. During the summer algal bloom, the percent humics decreased and the FI increased. This increase in FI was greater in the alpine lakes than in the sub-alpine lakes, reflecting the greater initial input of terrestrial DOM from the subalpine forest compared to the alpine tundra and talus fields. A seasonal change in FI was not observed in the stream draining an adjacent watershed that does not contain lakes. The seasonal trend has been observed to occur every spring and summer, and the greater aromaticity of fulvic acid from the alpine lake in spring snowmelt compared to the summer was confirmed by chemical characterization of isolated samples by ^{13}C-NMR (Hood et al., 2005). Further, Miller et al. (2009) observed that an unusual 3-day summer rainstorm caused the FI in the alpine lake to decrease abruptly due to additional runoff, and then the FI rebounded to higher values when the algal populations increased in abundance (Figure 9.13).

Fluorescence indices have been applied to understand controls on DOM quality associated with land cover and land use for 34 temperate watersheds in Canada where the percentage of riparian cropland varied from 0 to 45% (Wilson and Xenopoulos, 2009). These investigators examined filtered water samples collected bimonthly over 2 years and the spectral data were inner filter corrected before analysis. In the study catchments there was a correlation between HIX_{EM} and moisture conditions, indicating a climate effect on DOM character. However, the FI and BIX strongly correlated with the amount of riparian cropland and total dissolved nitrogen respectively, and showed no relationship to moisture conditions, indicating that the chemical signal for source material was independent of moisture (Figure 9.14).

The correlation of the FI with continuous cropland demonstrated, for the first time, a possible link between land cover, DOM character, and aromaticity. The BIX index, originally developed to compare peaks M and C in marine and estuarine environments, was applied to a terrestrial environment and was shown to correlate with total dissolved nitrogen. This is likely to be a co-relationship, with increased microbial productivity leading to increased fluorescence in the peak M and peak T regions, increasing the value of this index due to eutrophication.

The McKnight et al. (2001) FI has also been used by a number of researchers to interpret spatial and seasonal trends, and often has been combined with a "microbial/autochthonous" versus "terrestrial/allochthonous" interpretation based on the calibration data set of surface waters (McKnight et al., 2001). In a comparison across a diverse set of long-term research sites primarily in the LTER network, Jaffe et al. (2008) found that the FI was correlated with the carbon-to-nitrogen ratio in the DOM (Figure 9.15), with higher FI

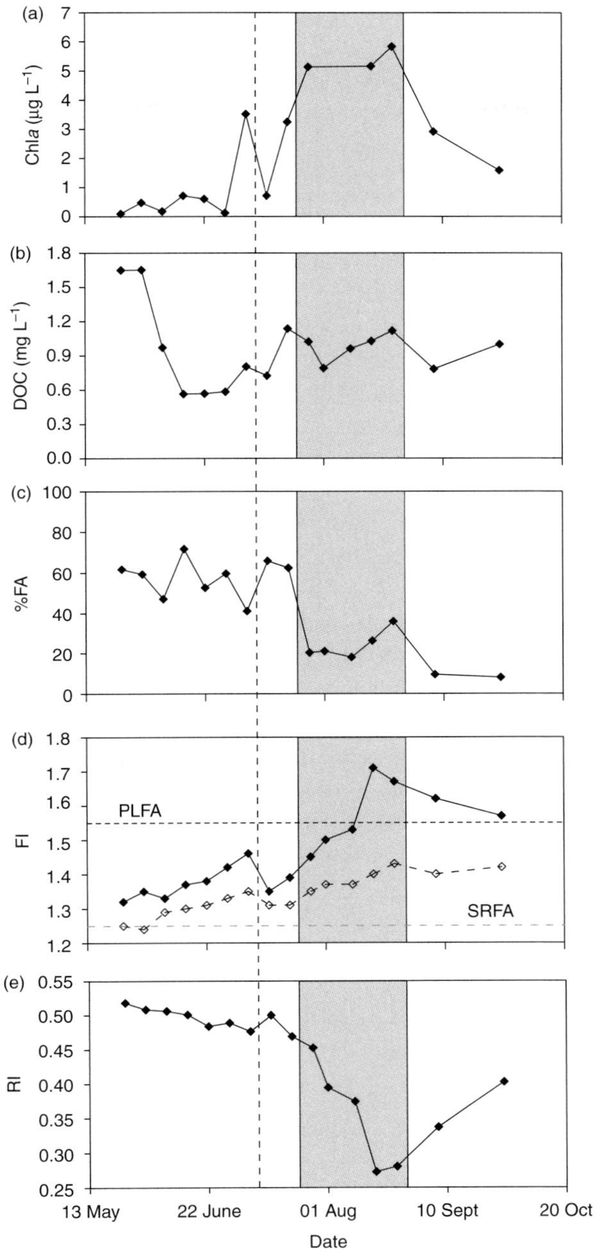

Figure 9.13. Seasonal variation in (a) chlorophyll (b) DOC, (c) percentage fulvic acid, (d) whole water (solid line) and fulvic acid (dashed line) McKnight Fluorescence Index (FI), and (e) whole water redox index (RI) at the outlet to Green Lake 4, an alpine lake in the Colorado Rocky Mountains. The vertical dashed lines indicate the sustained rain event, and the shaded areas represent the timing of the peak in chlorophyll a at the outlet. PLFA represents the FI of the Pony Lake fulvic acid, a microbial end member, and SRFA represents the FI of the Suwannee River fulvic acid, a terrestrial end member. (Adapted from Miller et al., 2009.)

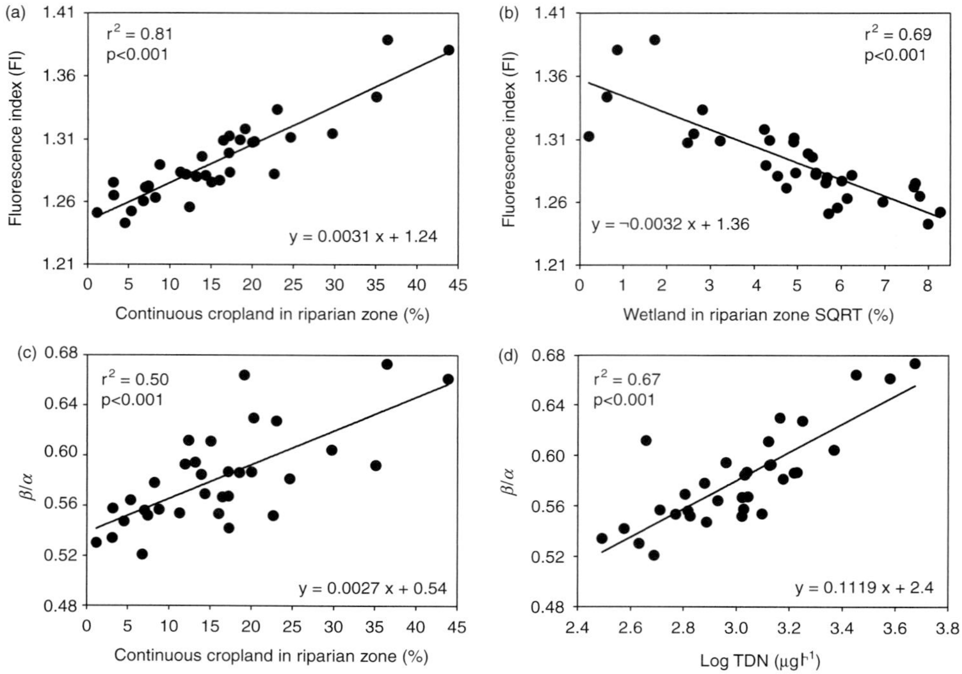

Figure 9.14. Relationship between McKnight et al. (2001) fluorescence index (FI) (a, b) and freshness index (β/α) (c, d) to characteristics of land cover and nutrients for 34 watersheds. FI correlated positively with % of continuous cropland in the riparian zone (a) and negatively with % wetland in the riparian zone (square root transformed) (b) whereas (β/α) correlated positively with both % of continuous cropland (c) and log total dissolved nitrogen (TDN) (d). These results support the ability of fluorescence indices to respond to changes in catchment and stream characteristics. (Adapted from Wilson and Xenopoulos, 2009.)

values being associated with lower C/N ratios. This relationship is consistent with those determined from studying humics isolated from natural waters, with humics derived from microbial sources generally having a lower C:N, reflecting the absence of lignin precursors (McKnight et al., 1997).

9.3.4 Understanding DOM Changes in Estuaries

Huguet et al. (2009) compared the HIX_{EM}, the BIX (the modified β/α ratio of Parlanti et al., 2000) and McKnight et al. (2001) FI when applied to dissolved organic matter in an estuarine environment. The McKnight et al. (2001) FI demonstrated low values, and no trends seasonally or with salinity through this estuarine transect. In contrast, the HIX_{EM} and BIX ratios presented trends with salinity and appeared to provide differing information in an estuarine environment (Figure 9.16). HIX_{EM} and BIX are related in that they look at very

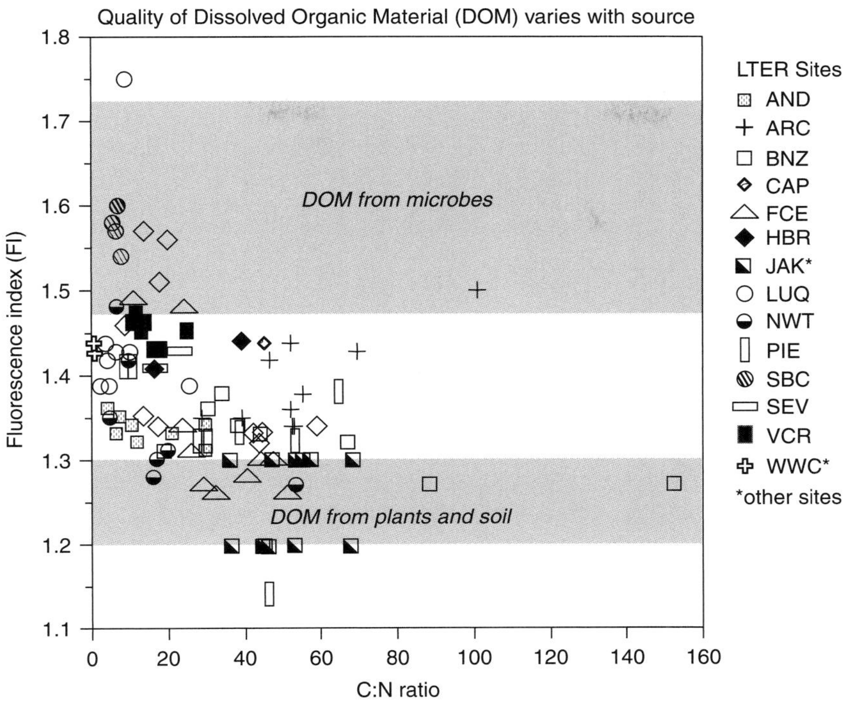

Figure 9.15. Relationship between the McKnight et al. (2001) fluorescence index (FI) and the C/N ratio for DOM samples from a variety of freshwater and marine sites studied in the LTER network and elsewhere. Higher FI DOM, attributed to predominately microbial precursor material, generally had a lower C/N ratio, whereas lower FI DOM, attributed as predominately terrestrial precursor material, had a higher C/N. This also shows the value of using proper measurement and correction techniques so results and indices can be compared between labs and sites. (Adapted from Jaffé et al., 2008.)

similar ratios of emission but with contrasting excitation wavelengths: HIX_{EM} with excitation at 254 nm and BIX at 310 nm. However, the processes affecting the HIX_{EM} index in an estuarine environment are poorly understood; Huguet et al. (2009) interpret HIX_{EM} values as differentiating humic/terrestrial organic matter from recent/autochthonous organic matter. Excitation at 254 nm and emission at 300–345 nm would detect peak T and peak B fluorescence. Emission measured at 435–480 nm would detect peak A fluorescence. So for estuarine samples where peak T and peak B fluorescence is autochthonously sourced, this interpretation is valid. However, where the terrestrial input to the estuary has a high peak T intensity, this simple mixing model would not apply. The extent to which HIX_{EM} directly measures DOM character or is measuring the interactions of peaks A, B, and T with a changing matrix (increasing salinity) and associated chemical effects (flocculation) is yet to be determined. Thus the use of HIX_{EM} for aquatic samples requires careful consideration. In contrast, the BIX ratio, measuring the peak M to peak C intensity ratio, is more

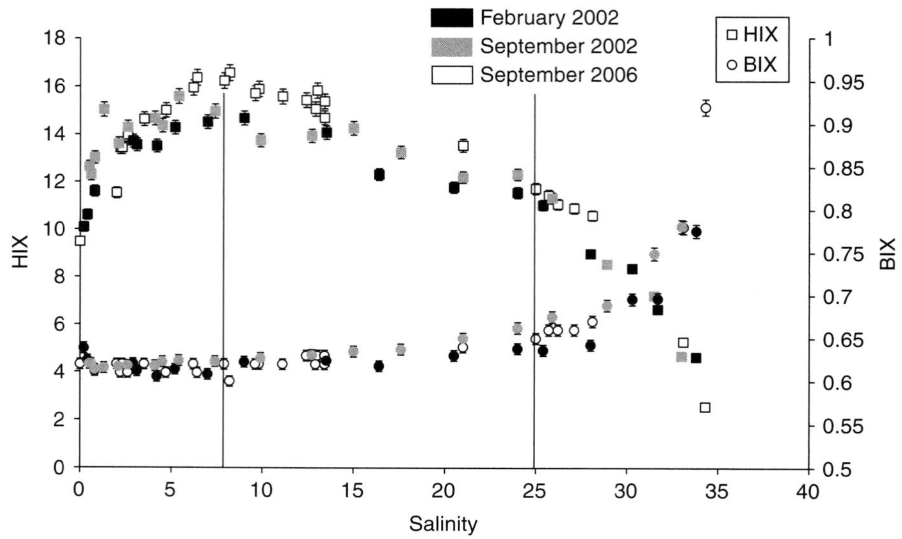

Figure 9.16. Comparison of the BIX and HIX_{EM} for an estuarine transect sampled on three dates. BIX and HIX_{EM} measurements were used to infer regions of different organic matter character, identified by the vertical lines. These differences were attributed to fresh water from the river mixing with saline marine water. (From Huguet et al., 2009.)

likely to reflect the marine (autochthonous and bacterial derived) organic matter and terrestrial (allochthonous) end members. Figure 9.7 demonstrates a trend typical of estuarine mixing for BIX versus salinity, with high BIX values corresponding with highest salinity.

Huguet et al. (2009) suggested that the interpretation of the McKnight et al. (2001) FI is complicated in an estuarine environment because peak C emission wavelength may vary with salinity due to flocculation processes and changes in organic matter molecular weight. The values they reported were much lower than those that have been reported for freshwaters. Jaffé et al. (2004) came to similar conclusions about the difficulty in interpreting the FI when applied to a mangrove environment. However, in a study of chromophoric DOM in a fjord in New Zealand, Gonsoir et al. (2008) also found that the FI had low values in the estuary and that the FI increased with depth in the upper 5 m. These results were interpreted as suggesting that there was an increasing influence of marine-derived DOM at depth, or that changes in the fluorophores occurred across the mixing zone in the fjord.

9.4 Spectroscopic Challenges toward Using the Indices

Fluorescence indices have proven to be powerful tools for analyzing dissolved organic matter as well as organic matter from soils and sediments. However, owing to the many factors that affect fluorescence signatures, it is necessary to pay careful attention to collection of samples and correction of data to properly utilize the indices. This section briefly reviews some common pitfalls and proper spectroscopic techniques for understanding EEMs and applying indices.

9.4.1 Instrument-Specific Effects and Proper EEM Correction

Multiple studies have highlighted the effect of instrument variation on fluorescence DOM analysis. Kalbitz et al. (2001) found that the HIX_{SYN} values determined by two different fluorometers correlated well but recommended using a correction curve from a reference standard in order to compare the entire synchronous scan, suggesting that the index, as a ratio of two points, may be less sensitive to instrument variation. Holbrook et al. (2006) compared various EEM correction methods and their impact on peaks related to HIX_{EM} and FI. They found that for these indices, because they both depend on a single excitation wavelength, emission corrections are the most important to apply, but recommend always collecting in ratio mode (collecting the signal as emission intensity normalized to the lamp intensity) to account for any variations in temperature or lamp intensity. Murphy et al. (2010) compared multiple standard samples measured in 20 different labs using 8 different fluorometer models. They found that correction method generally affected the FI, but by treating all samples identically the variation could be brought down to 8% between labs. This identified a need for a standardized correction procedure to compare EEMs and indices between instruments, labs, and studies. Lawaetz and Stedmon (2009) recommended a procedure for EEM correction and a Raman intensity calibration, resulting in fluorescence intensity in Raman units to better aid comparisons between labs. Finally, Cory et al. (2010) offered a detailed analysis of correction procedures and data comparisons across three common fluorometers. They found that although different instruments, with different signal-to-noise ratios, will have variations, properly applied correction procedures will greatly reduce variation among the FIs.

9.4.2 Concentration Issues and the Inner-Filter Effect

Inner-filter effects refer to the attenuation of light prior to detection by the fluorometer, due either to absorption of excitation light before reaching the fluorescent molecule (the primary inner-filter effect) or absorption of light emitted from the fluorescent molecule before being detected by the fluorometer (the secondary inner-filter effect). To obtain an accurate spectroscopic reading, inner-filter effects must be accounted for in highly absorbing samples either by dilution or application of an inner-filter correction (Mobed et al., 1996; McKnight et al., 2001). The most commonly used inner filter correction is based on a correction for a path length of 0.5 cm in a 1-cm cell (Lakowicz, 2006: 56). However, the application of this inner-filter correction is sufficient only for samples with low enough absorbance for this assumption to adequately represent the attenuation of light, after which dilution is required. Ohno (2002a) demonstrated that, for calculation of HIX_{EM} values (see Section 9.2.2), inner-filter corrections were effective for removing inner-filter effects only up to an absorbance at 254 nm of 0.3 cm^{-1}. Kalbitz et al. (2001) recommended a DOC no higher than 10 mg L^{-1} when comparing samples between instruments and Zsolnay et al. (1999) kept absorbance at 254 below 0.1 cm^{-1} to avoid inner filter effects when developing the humification index. Miller et al. (2010) discussed concentration effects, showing the change in the ex 370 nm peak (used to calculate FI) at A_{254} values of 0.1 cm^{-1}, 0.3 cm^{-1}, and 1.0 cm^{-1}. They demonstrated the unequal impact of concentration across emission intensities and that at high A_{254} values inner-filer corrections do not fully remove concentration effects. Whereas the 0.1 cm^{-1} and 0.3 cm^{-1} absorbance samples were similar, the 1.0 cm^{-1} sample had a noticeably different peak shape.

9.4.3 pH Effect on Fluorescence

The chemical character of the matrix in which the organic matter is dissolved can affect the overall fluorescence of the sample. Laane (1982) identified a change in DOM fluorescence due to pH effects and Miano and Senesi (1992) demonstrated the effect of pH on fulvic and humic acid standards, showing that although the fluorescence emission intensity increases with more acidic pH, it does not increase evenly across the scan. Because fluorescence indices involve the ratio of points or areas in two different regions of the scan this means pH can affect the value of an index. McKnight et al. (2001) collected fluorescence spectra and calculated fluorescence index values (see Section 9.2.4) for solutions of fulvic acids from 7.5 m depth in Lake Fryxell in Antarctica (microbially derived end member) and the Suwannee River in Georgia (terrestrially-derived end member provided by IHSS) at pH 2.0 and pH 7.5, to evaluate the potential for minimizing the effects of metal binding in quenching fluorescence by acidifying natural water samples. At pH 2.0, the fluorescence index of the Lake Fryxell fulvic acid sample was 1.8 and the peak emission location at an excitation wavelength of 370 nm was 442 nm. In contrast, at pH 7.5, the fluorescence index increased to 1.9, and the emission peak location shifted to 448 nm. At pH 2.0, the fluorescence index of the Suwannee River Fulvic acid was 1.3 and the emission peak was at 460 nm. At pH 7.5, the fluorescence index of the Suwannee River fulvic acid increased to 1.4 and the emission peak increased to 461 nm. McKnight et al. (2001) point out that a difference in FI of ± 0.1 may be indicative of a change in source. Thus the differences at pH 7.5 and 2 approach the difference that could represent a biogeochemical difference. Although there is not a specific recommended pH for index calculation, it is important to consistently be aware of pH effects and keep pH values consistent for sample comparison. Chapter 7 includes a more detailed discussion of the effect of pH on DOM fluorescence.

9.5 Conclusions

The literature associated with fluorescence indices shows that over a period of just over a decade, researchers in the fields of soil science, marine science, hydrogeology, and environmental science independently derived a series of fluorescence indices for characterizing organic matter. In general, indices track either the position or intensity of the humic-like fluorescence peak (peak C), or the relative intensity of the humic-like peak versus a microbially derived (or protein-like) peak (peak T and/or M). These indices have been used for a variety of purposes from studying land use patterns to estuary dynamics and can characterize organic matter in filtered water samples, fulvic or humic extracts, or SOM leachates.

Owing to its high sensitivity, fluorescence is an effective technique for identifying subtle changes in DOM character and composition. However, EEMs can be difficult to analyze because of the large amount of information found in them. Fluorescence indices provide a useful tool to isolate specific characteristics of the fluorescence signature that have been correlated to aspects of DOM chemical composition. Several indices capture shifts in the emission wavelengths of the humic-like (peak C) fluorescence which are correlated with the aromaticity, hydrophobicity, extent of humification, source of precursor materials, and electron shuttling ability of humics. These indices include the humification index proposed by Kalbitz et al. (1999), the FI proposed by McKnight et al. (2001), the redox index suggested by Miller et al. (2006), and the indices used to study humic material in stalagmites by Proctor et al. (2000) and Perrette et al. (2005). Other indices focus on the relationship between the humic-like and protein-like peaks in an attempt to compare the relative importance of each component of organic matter. These indices include the freshness index developed by Parlanti et al. (2000) and the peak T/peak C ratio introduced by Baker et al. (2001). The humification index by Zsolnay et al. (1999) utilizes portions from each of these regions. Although each of these indices was developed with a specific goal in mind, many of them have converged to identifying similar chemical changes in DOM that cause shifts in the EEM signature and have proven useful in studies beyond the question they were originally developed to address.

Fluorescence indices are powerful tools for analyzing DOM and tracking variations between samples, but doing so properly requires careful consideration of spectroscopic techniques. It is also important to realize that each index was originally developed for a specific set of samples, and may be useful only in certain environments, such as marine or soil ecosystems. Thus, when choosing to apply a specific index, it is important to ensure the index is relevant to the samples and the question at hand. The use of an established index can help promote synthesis, comparison, and interpretation across studies and ecosystems (Jaffé et al., 2008). Certainly, there may be cases where independent development of a new index may be appropriate for a specific sample set or for use with an in situ probe, and this process could be guided by our understanding as to how previous indices were developed.

Acknowledgments

We thank Michael SanClements, Natalie Mladenov, and Paula Coble, as well as three anonymous reviewers for their help with figures and comments on drafts of this chapter. We also thank Eric Parrish for graphic design work on the figures. While writing this chapter, authors were funded by NSF-0724960, NSF EAR-0738910, and ANT-0839027.

References

Aiken, G., McKnight, D., Harnish, R., and Wershaw, R. (1996), Geochemistry of aquatic humic substances in the Lake Fryxell Basin, Antarctica. *Biogeochemistry*, **34**(3), 157–188.

Akagi, J., Zsolnay, A., and Bastida, F. (2007). Quantity and spectroscopic properties of soil dissolved organic matter (DOM) as a function of soil sample treatments: Air-drying and pre-incubation. *Chemosphere*, **69**, 1040–1046.

American Public Health Association, American Water Works Association, and Water Pollution Control Federation (1965). *Standard Methods for the Examination of Water and Wastewater Including Bottom Sediments and Sludges*. New York: American Public Health Association, Inc.

Baker, A. (2001). Fluorescence excitation emission matrix characterization of some sewage-impacted rivers. *Environ. Sci. Technol.*, **35**(5), 948–953.

Baker, A. (2002). Fluorescence properties of some farm wastes: Implications for water quality monitoring. *Water Res.*, **36**(1), 189–195.

Baker, A. and Genty, D. (1999). Fluorescence wavelength and intensity variations of cave waters. *J. Hydrol.*, **217**(1–2), 19–34.

Baker, A. and Bolton, L. (2000). Speleothem organic acid luminescence intensity ratios: a new palaeoenvironmental proxy. *Cave Karst Sci.*, **27**, 121–124.

Baker, A., Genty, D., and Smart, P. (1998). High-resolution records of soil humification and paleoclimate change from variations in speleothem luminescence excitation and emission wavelengths. *Geology*, **26**(10), 903–906.

Beggs, K.M.H., Summers, R.S., and McKnight, D.M. (2009). Characterizing chlorine oxidation of dissolved organic matter and disinfection by-product formation with fluorescence spectroscopy and parallel factor analysis. *J. Geophys. Res.*, **114**(G4), G04001.

Bu, X., Wang, L., Ma, W., Yu, X., Mcdowell, W.H., and Ruan, H. (2010). Spectroscopic characterization of hot-water extractable organic matter from soils under four different vegetation types along an elevation gradient in the Wuyi Mountains. *Geoderma*, **159**, 139–146.

Cannavo, P., Dudal, Y., Boudenne, J.-L., and Lafolie, F. (2004). Potential for fluorescence spectroscopy to assess the quality of soil water-extracted organic matter. *Soil Sci.*, **169**(10), 688–696.

Chen, Y., Senesi, N., and Schnitzer, N. (1977). Information provided on humic substances by E_4/E_6 ratios. *Soil Sci. Soc. Am. J.*, **41**(2), 352–358.

Coble, P. (1996). Characterization of marine and terrestrial DOM in seawater using excitation-emission matrix spectroscopy. *Mar. Chem.*, **51**(4), 325–346.

Corvasce, M., Zsolnay, A., D'Orazio, V., Lopez, R., and Miano, T. (2006). Characterization of water extractable organic matter in a deep soil profile. *Chemosphere*, **62**(10), 1583–1590.

Cory, R. and McKnight, D. (2005). Fluorescence spectroscopy reveals ubiquitous presence of oxidized and reduced quinones in dissolved organic matter. *Environ. Sci. Technol.*, **39**(21), 8142–8149.

Cory, R.M., McKnight, D.M., Chin, Y.-P., Miller, P., and Jaros, C.L. (2007). Chemical characteristics of fulvic acids from arctic surface waters: Microbial contributions and photochemical transformations. *J. Geophys. Res. Biogeochem.*, **112**(G4), G04S51.

Cory, R.M., Miller, M.P., McKnight, D.M., J Guerard, J.J., and Miller, P.L. (2010). Effect of instrument-specific response on the analysis of fulvic acid fluorescence spectra. *Limnol. Oceanogr. Meth.*, **8**, 67–78.

Ewald, M. and Belin, C. (1987). Fluorescence from photic zone water in the Atlantic Ocean. *Sci. Total Environ.*, **62**, 149–155.

Fellman, J.B., Miller, M.P., Cory, R.M., D'Amore, D.V., and White, D. (2009). Characterizing dissolved organic matter using parafac modeling of fluorescence spectroscopy: A comparison of two models. *Environ. Sci. Technol.*, **43**(16), 6228–6234.

Fellman, J.B., Hood, E., and Spencer, R.G.M. (2010). Fluorescence spectroscopy opens new windows into dissolved organic matter dynamics in freshwater ecosystems: A review. *Limnol. Oceanogr.*, **55**(6), 2452–2462.

Fulton, J.R., McKnight, D.M., Foreman, C.M., Cory, R.M., Stedmon, C., and Blunt, E. (2004). Changes in fulvic acid redox state through the oxycline of a permanently ice-covered Antarctic lake. *Aquat. Sci.*, **66**, 27–46.

Gonsoir, M., Peake, B.M., Cooper, W.J., Jaffe, R., Young, H., Kahn, A.E., and Kowalczuk, P. (2008). Spectral characterization of chromophoric dissolved organic matter (CDOM) in a fjord (Doubtful Sound, New Zealand). *Aquat. Sci.*, **70**, 397–409.

Hassouna, M., Massiani, C., Dudal, Y., Pech, N., and Theraulaz, F. (2010). Changes in water extractable organic matter (WEOM) in a calcareous soil under field conditions with time and soil depth. *Geoderma*, **155**(1–2), 75–85.

Holbrook, R., DeRose, P., Leigh, S., Rukhin, A., and Heckert, N. (2006). Excitation–emission matrix fluorescence spectroscopy for natural organic matter characterization: A quantitative evaluation of calibration and spectral correction procedures. *Appl. Spectrosc.*, **60**(7), 791–799.

Hood, E., McKnight, D., and Williams, M. (2003). Sources and chemical character of dissolved organic carbon across an alpine/subalpine ecotone, Green Lakes Valley, Colorado Front Range, United States. *Water Resour. Res*, **39**(7), 1188–1200.

Hood, E., Williams, M., and McKnight, D. (2005). Sources of dissolved organic matter (DOM) in a Rocky Mountain stream using chemical fractionation and stable isotopes. *Biogeochemistry*, **74**(2), 231–255.

Hudson, N., Baker, A., and Reynolds, D. (2007). Fluorescence analysis of dissolved organic matter in natural, waste and polluted waters – a review. *River Res. Appl.*, **23**(6), 631–649.

Hudson, N., Baker, A., Ward, D, Reynolds, D., Brunsdon, C., Carliell-Marquet, C., and Browning, S. (2008). Can fluorescence spectrometry be used as a surrogate for the Biochemical Oxygen Demand (BOD) test in water quality assessment? An example from South West England. *Sci. Total Environ.*, **391**(1), 149–158.

Huguet, A., Vacher, L., Relexans, S., Saubusse, S., Froidefond, J.M., and Parlanti, E. (2009). Properties of fluorescent dissolved organic matter in the gironde estuary. *Org. Geochem.*, **40**(6), 706–719.

Jaffé, R., Boyer, J., Lu, X., Maie, N., Yang, C., Scully, N., and Mock, S. (2004). Source characterization of dissolved organic matter in a subtropical mangrove-dominated estuary by fluorescence analysis. *Mar. Chem.*, **84**(3–4), 195–210.

Jaffé, R., McKnight, D., Maie, N., Cory, R., McDowell, W., and Campbell, J. (2008). Spatial and temporal variations in DOM composition in ecosystems: The importance of long-term monitoring of optical properties. *J. Geophys. Res.*, **113**, G04032.

Kalbitz, K., and Geyer, W. (2001). Humification indices of water-soluble fulvic acids derived from synchronous fluorescence spectra – effects of spectrometer type and concentration, *J. Plant Nutr. Soil Sci.*, **164**(3), 259–265.

Kalbitz, K., Geyer, W., and Geyer, S. (1999). Spectroscopic properties of dissolved humic substances – a reflection of land use history in a fen area. *Biogeochemistry*, **47**(2), 219–238.

Kalbitz, K., Geyer, S., and Geyer, W. (2000). A comparative characterization of dissolved organic matter by means of original aqueous samples and isolated humic substances. *Chemosphere*, **40**(12), 1305–1312.

Kalbitz, K., Schmerwitz, J., Schwesig, D., and Matzner, E. (2003). Biodegradation of soil-derived dissolved organic matter as related to its properties. *Geoderma*, **113**(3–4), 273–291.

Klapper, L., Mcknight, D.M., Fulton, J.R., Blunt-Harris, E.L., Nevin, K.P., Lovley, D.R., and Hatcher, P.G. (2002). Fulvic acid oxidation state detection using fluorescence spectroscopy. *Environ. Sci. Technol.*, **36**(14), 3170–3175.

Laane, R.W.P.M. (1982). Influence of pH on the fluorescence of dissolved organic matter. *Mar. Chem.*, **11**, 395–401.

Lakowicz, J. (2006). *Principles of Fluorescence Spectroscopy*. New York: Springer Science+Business Media.

Lawaetz, A. and Stedmon, C. (2009). Fluorescence intensity calibration using the Raman scatter peak of water. *Appl. Spectrosc.*, **63**(8), 936–940.

Lovley, D., Coates, J., Blunt-Harris, E., Phillips, E., and Woodward, J. (1996). Humic substances as electron acceptors for microbial respiration. *Nature*, **382**(6590), 445–448.

McKnight, D., Andrews, E., Spulding, S., and Aiken, G. (1994). Aquatic fulvic acids in algal-rich Antarctic ponds. *Limnol. Oceanogr.*, **39**, 1972–1979.

McKnight, D., Boyer, E., Westerhoff, P., Doran, P., Kulbe, T., and Andersen, D. (2001). Spectrofluorometric characterization of dissolved organic matter for indication of precursor organic material and aromaticity. *Limnol. Oceanogr.*, **46**(1), 38–48.

McKnight, D.M., Harnish, R., Wershaw, R.L., Baron, J.S., and Schiff, S. (1997). Chemical characteristics of particulate, colloidal, and dissolved organic matter in Loch Vale watershed, Rocky Mountain National Park. *Biogeochemistry*, **36**, 99–124.

McKnight, D.M., Hood, E., and Klapper, L. (2003). Trace organic moieties of dissolved organic material in natural waters. In S.E.G. Findlay and R.L. Sinsabaugh (Eds.), *Aquatic Ecosystems: Interactivity of Dissolved Organic Matter* (pp. 71–96). San Diego: Elsevier/Academic Press.

Miano, T. and Senesi, N. (1992). Synchronous excitation fluorescence spectroscopy applied to soil humic substances chemistry. *Sci. Total Environ.*, **117**, 41–51.

Miller, M.P., McKnight, D.M., Cory, R.M., Williams, M.W., and Runkel, R.L. (2006). Hyporheic exchange and fulvic acid redox reactions in an alpine stream/wetland ecosystem, Colorado Front Range. *Environ. Sci. Technol.*, **40**(19), 5943–5949.

Miller, M.P., McKnight, D.M., Chapra, S.C., and Williams, M.W. (2009). A model of degradation and production of three pools of dissolved organic matter in an alpine lake. *Limnol. Oceanogr.*, **54**(6), 2213–2227.

Miller, M.P., Simone, B.E., McKnight, D.M., Cory, R.M., Williams, M.W., and Boyer, E.W. (2010). New light on a dark subject: Comment. *Aquat. Sci.*, pp. 1–7.

Mladenov, N., McKnight, D.M., Macko, S.A., Norris, M., Cory, R.M., and Ramberg, L. (2007). Chemical characterization of DOM in channels of a seasonal wetland. *Aquat. Sci.*, **69**(4), 456–471.

Mladenov, N., Huntsman-Mapila, P., Wolski, P., Masamba, W., and McKnight, D. (2008). Dissolved organic matter accumulation, reactivity, and redox state in ground water of a recharge wetland. *Wetlands*, **28**(3), 747–759.

Mladenov, N., López-Ramos, J., McKnight, D., and Reche, I. (2009). Alpine lake optical properties as sentinels of dust deposition and global change. *Limnol. Oceanogr.*, **54**(6), 2386–2400.

Mladenov, N., Zheng, Y., Miller, M.P., Nemergut, D.R., Legg, T., Simone, B., Hageman, C., Rahman, M.M., Ahmed, K.M., and McKnight, D.M. (2010). Dissolved organic matter sources and consequences for iron and arsenic mobilization in Bangladesh aquifers. *Environ. Sci. Technol.*, **44**(1), 123–128.

Mobed, J., Hemmingsen, S., Autry, J., and McGown, L. (1996). Fluorescence characterization of IHSS humic substances: Total luminescence spectra with absorbance correction. *Environ. Sci. Technol.*, **30**(10), 3061–3065.

Murphy, K., Butler, K., Spencer, R., Stedmon, C., Boehme, J., and Aiken, G. (2010). Measurement of dissolved organic matter fluorescence in aquatic environments: An interlaboratory comparison. *Environ. Sci. Technol.*, **44**(24), 9405–9412.

Naden, P.S., Old, G.H., Eliot-Laize, C., Granger, S.J., Hawkins, J.M.B, Bol, R., and Haygarth, P. (2010). Assessment of natural fluorescence as a tracer of diffuse agricultural pollution from slurry spreading on intensely-farmed grasslands. *Water Res*, **44**(6), 1701–1712.

Ohno, T. (2002a). Fluorescence inner-filtering correction for determining the humification index of dissolved organic matter. *Environ. Sci. Technol.*, **36**(4), 742–746.

Ohno, T. (2002b). Response to comment on "fluorescence inner-filtering correction for determining the humification index of dissolved organic matter." *Environ. Sci. Technol.*, **36**(19), 4196.

Ohno, T., Fernandez, I., Hiradate, S., and Sherman, J. (2007). Effects of soil acidification and forest type on water soluble soil organic matter properties. *Geoderma*, **140**(1–2), 176–187.

Parlanti, E., Wörz, K., Geoffroy, L., and Lamotte, M. (2000). Dissolved organic matter fluorescence spectroscopy as a tool to estimate biological activity in a coastal zone submitted to anthropogenic inputs. *Org. Geochem.*, **31**(12), 1765–1781.

Perrette, Y., Delannoy, J., Desmet, M., Lignier, V., and Destombes, J. (2005). Speleothem organic matter content imaging: The use of a fluorescence index to characterise the maximum emission wavelength. *Chem. Geol.*, **214**(3–4), 193–208.

Proctor, C., Baker, A., Barnes, W., and Gilmour, M. (2000). A thousand year speleothem proxy record of North Atlantic climate from Scotland. *Clim. Dyn.*, **16**(10), 815–820.

Scott, D.T., McKnight, D.M., Blunt-Harris, E.L., Kolesar, S.E., and Lovley, D.R. (1998). Quinone moieties act as electron acceptors in the reduction of humic substances by humics-reducing microorganisms. *Environ. Sci. Technol.*, **32**(19), 2984–2989.

Senesi, N., Miano, T., Provenzano, M., and Brunetti, G. (1989). Spectroscopic and compositional comparative characterization of IHSS reference and standard fulvic and humic acids of various origin. *Sci. Total Environ.*, **81**, 143–156.

Senesi, N., Miano, T., Provenzano, M., and Brunetti, G. (1991). Characterization, differentiation, and classification of humic substances by fluorescence spectroscopy. *Soil Sci.*, **152**(4), 259–271.

Spencer, R.G.M., Bolton, L., and Baker, A. (2007). Freeze/thaw and pH effects on freshwater dissolved organic matter fluorescence and absorbance properties from a number of UK locations. *Water Res.*, **41**(13), 2941–2950.

Stewart, A. and Wetzel, R. (1980). Fluorescence: Absorbance ratios – a molecular-weight tracer of dissolved organic matter. *Limnol. Oceanogr.*, **25**(3), 559–564.

Stewart, A. and Wetzel, R. (1981). Asymmetrical relationships between absorbance, fluorescence, and dissolved organic carbon. *Limnol. Oceanogr.*, **26**(3), 590–597.

Trouet, V., Esper, J., Graham, N.E., Baker, A., Scourse, J.D., and Frank, D.C. (2009). Persistent positive North Atlantic oscillation mode dominated the medieval climate anomaly. *Science*, **324**(5923), 78–80.

Weishaar, J.L., Aiken, G.R., Bergamaschi, B.A., Fram, M.S., Fujii, R., and Mopper, K. (2003). Evaluation of specific ultraviolet absorbance as an indicator of the chemical composition and reactivity of dissolved organic carbon. *Environ. Sci. Technol.*, **37**(20), 4702–4708.

Wilson, H.F. and Xenopoulos, M.A. (2009). Effects of agricultural land use on the composition of fluvial dissolved organic matter. *Nature Geosci.*, **2**(1), 37–41.

Zsolnay, A. (2002). Comment on "fluorescence inner-filtering correction for determining the humification index of dissolved organic matter." *Environ. Sci. Technol.*, **36**(19), 4195.

Zsolnay, A., Baigar, E., Jimenez, M., Steinweg, B., and Saccomandi, F. (1999). Differentiating with fluorescence spectroscopy the sources of dissolved organic matter in soils subjected to drying. *Chemosphere*, **38**(1), 45–50.

10

Chemometric Analysis of Organic Matter Fluorescence

KATHLEEN R. MURPHY, RASMUS BRO, AND COLIN A. STEDMON

10.1 Introduction

In multivariate data analysis (MVA), statistical and mathematical modeling techniques are used to analyze data that result from measuring multiple variables on several samples. Important variables influencing the fluorescence characteristics of a dissolved organic matter sample might include, among others, its composition, temperature, pH, and the concentration and nature of any quenching agents that may be present. In Chapter 9, top-down (researcher-driven) univariate techniques for exploring fluorescence data were introduced. Such techniques can reduce the complexity of fluorescence data and aid visualization and interpretation and are rapidly implemented and easy to understand. However, because they require the researcher to identify which features of a data set deserve to be included in models, they presuppose a good understanding of the system being studied. In addition, with this type of approach there is a danger that unanticipated events remain undetected. The aim of *exploratory* analysis is to avoid this, by presenting a model and visualization of the data not focused on prior hypotheses. In fact, the basic idea in exploratory analysis is to use the data analysis to get ideas for hypotheses *from the data*. With advances in technology it is a simple matter these days to acquire comprehensive fluorescence data sets, even in environments that were previously difficult to sample. Thus, data sets can quickly become large or complex, masking subtle but potentially important differences between samples. In cases where there is little existing experience of the sampling environment, methodologies are needed that ensure that in the process of condensing the original data set, all relevant information is retained and detected.

Extracting chemical information from a multivariate chemical data set falls within a field of science known as chemometrics. It uses statistical and mathematical methods combined with chemical and physical insight. Often the analyses are visually driven by the data rather than imposed upon them under a theoretical or statistical framework. Chemometrics has proved particularly useful for exploring and interpreting complex data sets involving large numbers of variables that relate to one another in ways that are poorly understood. A number of reviews of chemometric applications in chemistry and spectroscopy are available (Mobley et al., 1996; Workman et al., 1996; Bro et al., 1997; Bro, 2006; Lavine and Workman, 2010).

A principal objective when analyzing complex data sets is to reduce the dimensionality of the data set in order to separate important features from redundant information and noise. The reduced data set is then simpler to interpret. Thus, an excitation–emission matrix (EEM) data set consisting of many thousands of individual data points (samples × excitation × emission) may be reduced to a fraction of its original size (e.g., samples × intensity at a few wavelength pairs), with greatly increased interpretability and opportunities for visualizing the data graphically, while retaining all of the essential information contained in the original data set. A second principal objective is to detect patterns in the relationships between variables, in order to develop prediction/calibration models for other important parameters that are harder to measure.

This chapter is not intended as a statistics tutorial as much as a broad overview of the available chemometric techniques likely to be of greatest assistance for interpreting fluorescence data. The list is by no means exhaustive – there are literally thousands of different techniques and variations to be discovered, and describing all of them would require many textbooks. Similarly, the algorithms underpinning the chemometric methods discussed here are presented elsewhere (Désiré-Luc Massart et al., 1988; Martens and Næs, 1989; Smilde et al., 2004) and many useful tutorials for understanding and applying various techniques are published (e.g., Geladi and Kowalski, 1986; Thomas, 1994; Bro, 1997; Stedmon and Bro, 2008) or available online (see, e.g., http://www.models.life.ku.dk/). The focus here is on describing how chemometric methods can and have already been put to use for the interpretation of natural organic matter fluorescence.

An example data set is used to demonstrate some of the chemometric techniques discussed in this chapter. The data set is derived from the Horsens catchment, Denmark and consists of fluorescence EEMs and absorbance measured at 375 nm (a375), together with dissolved organic carbon (DOC), and the nutrients total dissolved phosphorus (TDP), total dissolved nitrogen (TDN), dissolved organic phosphorus (DOP = TDP – dissolved inorganic P), dissolved organic nitrogen (DON = TDN – dissolved inorganic N). The locations of sampling sites are illustrated in Figure 10.1. Detailed information about the data set (20 sites, $n = 543$ samples) is presented in Stedmon et al. (2006). Previously, an eight-component model was obtained using PARAllel FACtor analysis (PARAFAC; see later) on a data set consisting of these samples together with more than 600 samples generated during a series of degradation experiments (Stedmon and Markager, 2005a). Analyses presented herein were performed using PLS_toolbox (v. 6.0.1) operated with MATLAB (R2010a).

10.2 Multivariate and Multiway Data Sets

The distinction between multivariate and multiway data is best illustrated by example (Figure 10.2). A simple multivariate data set consists of I samples for which fluorescence intensities at five emission wavelengths ($x_1 - x_5$) at a fixed excitation wavelength have been measured; these data are arranged in a table (Figure 10.2a). Now assume that the experiment had been conducted under four increasing temperatures ($t_1 - t_4$). There are now two possible ways to arrange the new data set, either as a two-way (*multivariate*)

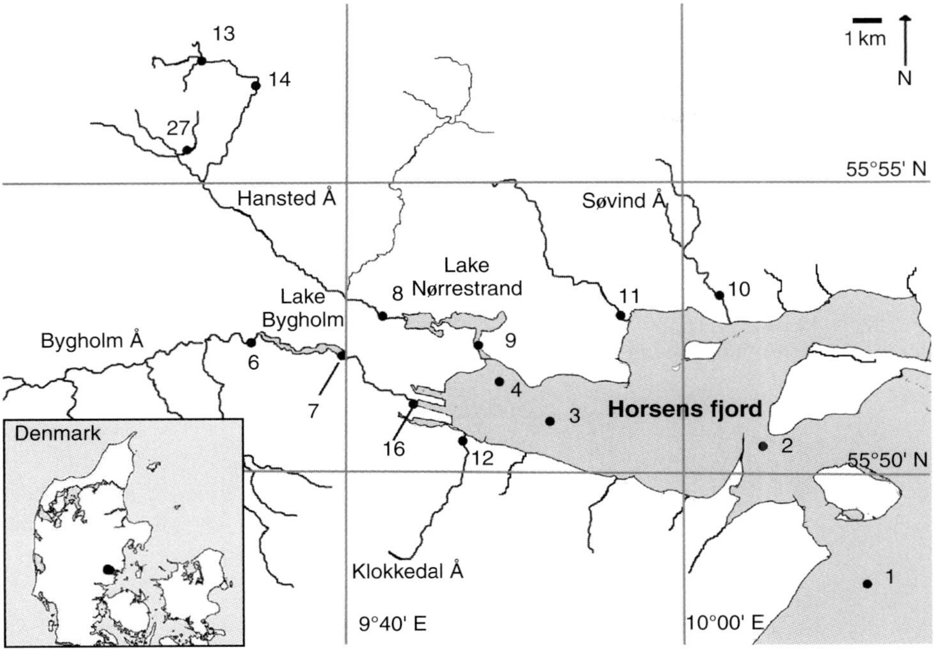

Figure 10.1. Locations of sampling sites in the Horsens catchment data set. (Adapted from Stedmon and Markager, 2005b.)

continuation of the original table (Figure 10.2b) or alternatively, as a three-way (*multiway* in general) data set in which sample, emission wavelength, and temperature form the three right-angled axes of a box (Figure 10.2c). If the entire experiment were to be repeated for a range of excitation wavelengths, a four-way data set would result; and so on for as many modes or ways as there are cross-wise factors in the data set. Notice that for each data set in Figure 10.2, the actual measurements are fluorescence intensities – not temperature, emission, or anything else.

An advantage of multiway techniques is that they retain, and make use of, information about the structure of data sets in more than two modes. For example, in Figure 10.2b, if we take a slice through the box at emission wavelength x_2, we obtain a matrix of samples by temperature that retains the original sequence of variables t_1 to t_4. Conversely, when the data set is "unfolded" as in Figure 10.2c, the temperature sequence is broken, and each temperature in the data set is analyzed without reference to the others. Thus, when a data set that has a genuine three-way underlying structure is unfolded, information implicit in the ordering of variables along the unfolded axes is lost. Some of the models that operate on multiway data also have what is known as the "second-order advantage," a property that is hugely powerful in the analysis of complex data sets (Booksh and Kowalski, 1994). Essentially, this makes it possible to obtain quantitative estimates of chemical analytes

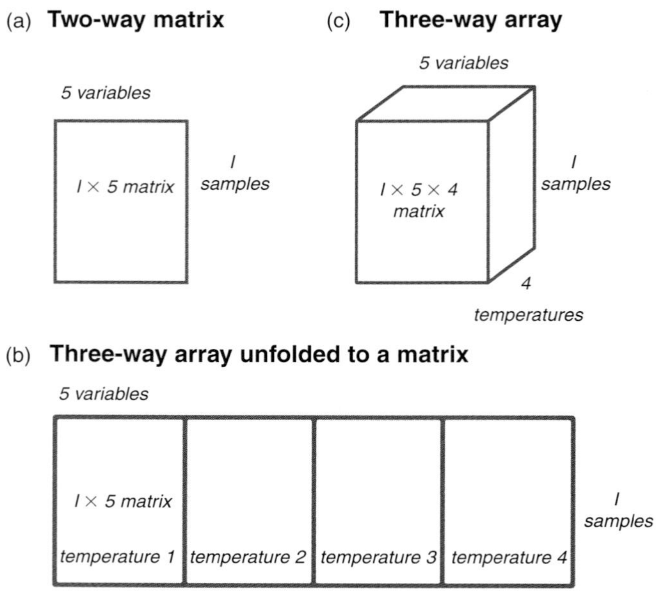

Figure 10.2. A comparison of multivariate and multiway data.

even in a mixture that contains uncalibrated signals associated with chemical interferents (Bro, 2003).

The methods and examples in this chapter are focused primarily on multiway and multivariate analyses of threeway fluorescence EEMs. However, many of the techniques discussed can be applied to other multivariate fluorescence data, such as data sets consisting of spectra obtained by scanning at fixed wavelength, or from synchronous scans. In synchronous scan spectroscopy, scans are performed using a fixed wavelength offset ($\delta\lambda$) between the excitation and emission monochromators, producing profiles of signal strength versus wavelength ($em = ex + \delta\lambda$) with shapes and peak resolutions that depend on $\delta\lambda$ (Miano and Senesi, 1992; Sierra et al., 2005). These synchronous scans can be visualized as diagonal slices through EEMs that intersect with various EEM features according to the value of $\delta\lambda$ (Sierra et al., 2005). It is possible to both analyze a multivariate data set arising from single-offset synchronous scans, and to build a multiway EEM data set by compiling a set of synchronous scan spectra obtained with incrementally increased offsets.

10.3 Preprocessing of Data Matrices and Arrays

Data preprocessing is an important component of successfully implementing multivariate analyses; however, how best to preprocess fluorescence data sets is frequently a point of confusion. Both the type and order of preprocessing steps, and whether these are applied to rows (samples) or columns (variables), can affect the results (Bro and Smilde, 2003).

Information on preprocessing in this chapter is necessarily brief; for broader discussion readers are referred to more comprehensive accounts (Thomas, 1994; Naes et al., 2002; Bro and Smilde, 2003). In general, preprocessing should be based on specific aims such as removing the Rayleigh scattering or giving minor peaks a chance to enter the model. When such practical concerns are guiding the preprocessing, it is often simpler to choose the appropriate tools. An illustration of the role of preprocessing on the principal component analysis (PCA) of fluorescence EEMs is provided later in this chapter.

Wavelength selection is sometimes overlooked in the preprocessing of spectral data. It should be borne in mind that although an instrument may be capable of collecting data across a wide range of excitation and emission wavelengths, such data may be of variable quality and importance. In particular, depending on the type and condition of the spectrophotometer light source, and the characteristics of the sample, data obtained at low excitation wavelengths can have very high uncertainties owing to a combination of factors, including decreasing lamp output, decreasing transmission efficiency of the excitation monochrometer, and increasing light reabsorption by the sample (inner filter effects) (Lakowicz, 2006). For this reason, it is often advisable to exclude fluorescence data obtained at low excitation wavelengths. Alternatively, it should be verified that the inclusion of such data do not skew the results of chemometric analysis. If combining fluorescence data sets collected using more than one fluorometer, intercalibration before analysis is also necessary to ameliorate the effects of instrument biases (Cory et al., 2010; Murphy et al., 2010). Data affected by phenomena unrelated to fluorescence (e.g., Rayleigh and Raman scatter) should always be removed prior to modeling (Andersen and Bro, 2003).

In multivariate matrices in which the variables are not "smoothly" related as they are in spectral data sets, then the preprocessing needs to make sure that a variable is not essentially disregarded just because it has a small scale. For spectral data, a small value usually implies little information, but for discrete data a variable this is not necessarily the case. For example, one variable may be temperature that varies only over a few degrees; however, this is not necessarily less important than another variable measured in weight that varies over thousands of milligrams. To take into account where scales for different variables are not proportional to their importance, analyses on "non-smooth" data are typically performed after first mean centering (subtracting the column average from each column) and scaling (dividing each column by its standard deviation). These steps are often referred to in combination as "auto-scaling" (Thomas, 1994). The centering removes the common features, so that the PCA model focuses on differences *between* the samples. The scaling gives each variable equal weighting in the model, rather than giving greater weighting to variables that naturally exhibit a larger absolute range. In the case of spectra, differences between wavelengths in intensity ranges are chemically meaningful and the auto-scaling of variables (and similar operations operating on the columns of the data set) can distort genuine proportional relationships between wavelengths, especially in data sets that span large concentration ranges. Thus for spectral data, often no preprocessing is needed, although mean centering may be convenient for visualization (Bro and Smilde, 2003).

Although chemometric analysis is often implemented on unscaled fluorescence EEMs without prior preprocessing (Gurden et al., 2001; Bro and Smilde, 2003), this approach may be unsuitable for data sets encompassing large concentration gradients. Samples with higher concentration may exert disproportionately high leverage on the model because models by default concentrate on minimizing differences between high- and low-concentration samples. As a general rule, when data sets encompass large concentration ranges (i.e., varying orders of magnitude) it can be helpful to normalize the area of each EEM to ensure that the modeling is focused on the chemical variations rather than on the magnitude of total signals. This is done by scaling the data in the first (sample) mode to unit norm, that is, dividing by the sum of the squared value of all variables for the sample. Normalization (and other operations affecting rows) should be performed before column operations (such as scaling and mean centering). It is important to note that this normalization or scaling of each sample can also be reversed after estimating the model. That is, scores in the original scale can be obtained by scaling the scores according to the inverse of that which the samples were scaled by.

Preprocessing of multivariate and multiway data sets prior to regression and discriminant analysis follow the general principles outlined earlier with few exceptions. In general, the response matrix (i.e., the data that are to be predicted) should be mean centered because this serves an additional purpose in regression and classification models. By centering both the dependent and independent variables, any possible differences in offsets are removed. Row normalization can be implemented if the priority is to establish a relationship between variables, rather than estimate the magnitude of the response, or to stabilize the impact of differently concentrated samples on models, as previously described. For example, if the calibration model is intended to predict a concentration from data that follow the Beer–Lambert law (e.g., fluorescence), then it is crucial not to normalize as this would cause the loss of concentration information. If, on the other hand, the model is intended to classify samples, then normalization may help the model focus on patterns rather than on concentration-induced variations.

10.4 Exploratory Data Analysis

Most chemometric treatments of DOM fluorescence data to date have been directed towards identifying patterns within data sets (unsupervised pattern recognition or cluster analysis), or deduce the underlying structure of individual EEMs (spectral decomposition) (Table 10.1). These are exploratory techniques in the sense that they are geared toward identifying structures within data sets in order to generate hypotheses about what variables may be important for various purposes (e.g., for classification or prediction), but do not involve hypothesis testing per se. Exploratory data analysis includes methods for analyzing both multivariate and multiway data sets. Cluster analysis is used to sort similar samples into categories, such that two samples with similar measured variables belong to the same group and two samples that have very different measurements belong to different groups. Similarity is measured on the basis of some algorithmically determined distance. Thus

Table 10.1. *Overview of the main available chemometric techniques used for exploring organic matter fluorescence data sets*

Category	Methods	Input data	Goals
Exploratory – visualization and clustering	Cluster analysis, e.g., hierarchical clustering, *k*-means clustering, self organizing maps (SOM)	Multivariate data set consisting of training and validation samples	Assign observations into groups containing similar samples.
	Principal component analysis (PCA)	Multivariate data set consisting of training and validation samples	Reduce dimensionality, explore and visualize linear gradients of variability in the data set, identify clustering among samples
Exploratory – spectral decomposition	Multivariate curve resolution (MCR)	Multivariate data set (e.g., synchronous scan spectra) consisting of training and validation samples	Determine the number, amount and spectral shapes of underlying components in mixtures, explore variability in the data set, visualization, clustering
	PARAFAC, PARAFAC2, Tucker3	Multiway data set consisting of training and validation samples	Determine the number, amounts, and spectral shapes of underlying components in mixtures, explore variability in the dataset, visualization, clustering. PARAFAC allows no spectral shifts, PARAFAC2 allows explicit spectral shifts along one axis, Tucker3 allows implicit spectral shifts along all axes but has no second-order advantage.
Exploratory – time series	Principal filters analysis (PFA)	Time series of fluorescence EEMs	Identify periods of time associated with high fluorescence variability.
Calibration	Principal components regression (PCR), partial least squares regression (PLS), and their multiway versions (N-PCR and N-PLS)	Multivariate or multiway data set consisting of independent and dependent variables. Training and validation samples	Predict dependent variables from independent variables
Classification	PLS discriminant analysis (PLS-DA), soft independent modeling of class analogy (SIMCA), multiway N-PLS-DA	Multivariate or multiway data set, training samples identified by category, test samples	Classify training and test samples into one of several possible categories

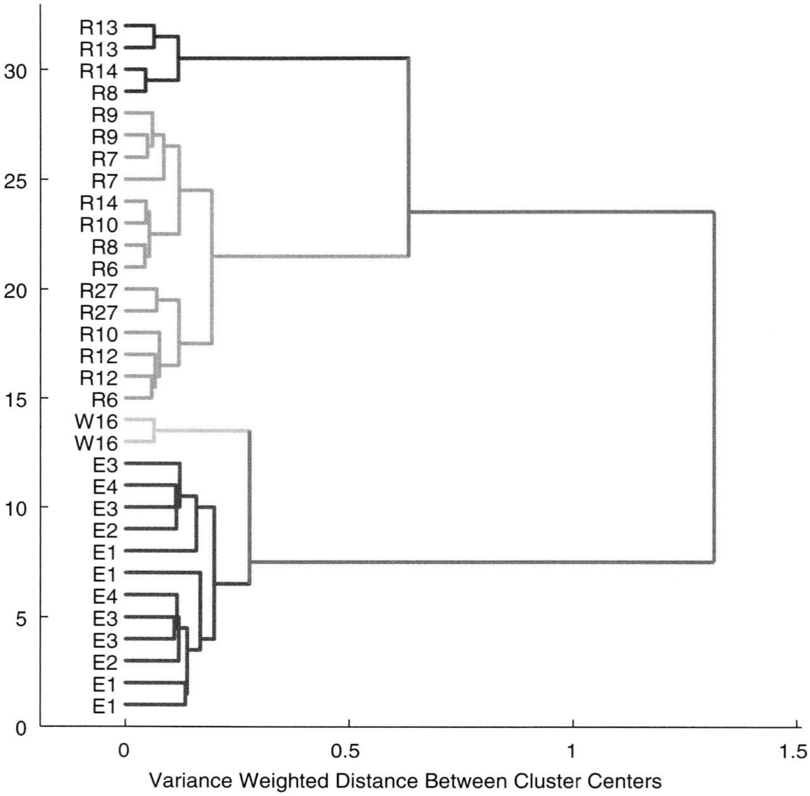

Figure 10.3. Dendrogram of Horsens catchment samples collected in June, 2006. Sample codes indicate Stream (R), Estuary (E), or WTP (W) sites, followed by site number (1–27). Single samples were collected from each site on two occasions 20 days apart, except at E1 and E3, where pairs of replicate samples were collected 20 days apart.

cluster analysis is used to discover structures in data without using prior knowledge of why such should be present. A level of subjectivity in the interpretation of cluster analyses is inevitable. In order that the clusters identified represent chemically distinct groupings rather than happenstance correlations the researcher should establish what degree of clustering is sensible both visually and through validation with additional samples and alternative clustering methods (Bratchell, 1989).

An example of a hierarchical cluster analysis using samples from the Horsens catchment data set is illustrated in Figure 10.3. The data used to produce the figure are the unfolded EEMs preprocessed by normalization followed by mean-centering. For clarity, this analysis is restricted to samples that were collected in June, 2006 ($n = 32$). The graphical output, called a dendrogram, shows similar samples grouped together in a hierarchical fashion. The longer the distance of the connecting line between two samples, the more different they are. The four highest level groups include two clusters of stream samples, a single cluster of estuary samples, and a small cluster of waste water treatment plant (WTP) samples that lie intermediate to the stream and estuary samples. Overall, the dendrogram suggests that in June 2006, there were greater differences among stream sites than between estuary and WTP sites. In addition, the analysis differentiates between the sites located at the head of the Hansted stream system (Figure 10.1) in catchments that are less impacted by agriculture (R13, R14) compared to other sites (Stedmon et al., 2006). Trends in the full data set will be explored in detail in the remainder of the chapter using further chemometric techniques.

A range of clustering and other exploratory techniques have been used to visualize and interpret DOM fluorescence data sets. Jiang (2008) used hierarchical cluster analysis to investigate sources of DOM in the Bohai Sea of China. Nelson (2009) used distance measures, cluster analysis, and multidimensional scaling coupled with classical statistics to examine how similarity in DOM compositions in montane lake chains and their connecting streams related to landscape position and catchment characteristics. Brunsdon and Baker (2002) presented a new tool for exploring and visualizing EEMs, termed principal filters analysis (PFA), that identifies periods of high variability in data sets consisting of fluorescence EEMs represented in a time series. They used this technique to identify three distinct periods with different fluorescence characteristics in the development of a stalagmite during the last 10,000 years (Holocene period).

A type of adaptive artificial neural network called a self-organizing map (SOM) has been used to visualize and identify patterns in fluorescence EEMs from streams, reservoirs, and wastewaters (Bieroza et al., 2009). The technique may be considered a nonlinear extension of PCA in which the loadings of the principal components are generalized from straight lines to curves, which may be useful for data that have a strongly nonlinear underlying distribution. The data set is reduced to a series of low-dimensional maps in which similar objects are clustered close together and dissimilar objects grouped further apart, with distances between objects depicted qualitatively using a graduated color scheme (e.g., the "U-Map" method). Extensions to SOM, such as visualization-induced self- organizing map (ViSOM), have been proposed to simplify visualization relative to traditional SOMs (Yin, 2002).

Data sets consisting of fixed-wavelength scans or fixed-offset synchronous scans, with each sample represented by a vector of data, have been examined using multivariate curve resolution (MCR; Tauler et al., 1995; Antunes and Esteves Da Silva, 2005; Abbas et al., 2006). This technique is also useful for, for example, sets of emission spectra or sets of excitation spectra. The MCR technique attempts to explicitly recover the pure response profiles explaining the chemical variance observed in multivariate matrices, and can therefore provide more physically interpretable results than some other exploratory methods, such as PCA. As in PCA, the MCR model provides scores and loadings, but in MCR the score and loading vectors are not required to be orthogonal. On the contrary, they are often required to be nonnegative. This leads to scores that seem to be estimates of concentrations and loadings that seem to estimate spectra. However, it is important to be careful not to overinterpret the results. MCR suffers from so-called rotational ambiguity, which essentially means that there can be many equally good (and equally good-looking) solutions to a problem. This is in contrast to PCA, for which there is only one single solution. Hence, caution is warranted especially for fluorescence data that have characteristics (e.g., broad overlapping spectra) that increase the difficulty of obtaining identified solutions with MCR (Jaumot and Tauler, 2010).

10.5 Principal Component Analysis

One of the simplest and most often used exploratory methods is principal component analysis (PCA), which identifies the most important uncorrelated variations in a data set, termed principal components. The principal components are defined by a new orthogonal and truncated coordinate system upon which the original data are mapped. PCA is typically used for exploring data to obtain preliminary assessments of the importance of different variables, clustering and classification of objects (samples), and for detecting outliers. The direction accounting for most of the variability in the data set (the "hyperplane of maximum variance") is the first principal component, with each subsequent principal component accounting for the maximum variability in the remainder of the data set once all preceding principal components have been subtracted from it.

PCA is normally considered a nonparametric method that does not rely on any hypotheses about data probability distributions, and provides a unique solution (except sign indeterminacy). It provides the least squares solution for compressing the original set of higher dimensional vectors into a set of lower-dimensional vectors from which the original set can be reconstructed. Typically, PCA is performed on data represented in a table (or matrix) form. PCA can also be performed on three-way EEM data sets that have been unfolded along one dimension so that the rows of the matrix are represented by samples and the columns by unique combinations of one excitation and one emission wavelength, which are each treated as individual variables (see Figure 10.2c). This is sometimes referred to as a Tucker1 or, somewhat misleadingly, a multiway PCA model.

Mathematically, the PCA model decomposes the data matrix into a set of so-called bilinear terms and a residual matrix:

$$X_{ij} = \sum_{f=1}^{F} a_{if} b_{jf} + e_{ij} \quad i = 1,\ldots,I; \quad j = 1,\ldots,J$$

where x_{ij} is the intensity of the ith sample at the jth variable; a_{if} is called a score value and locates each sample along each principal component; and b_{jf} is a loading matrix element describing the contribution of each variable toward each principal component. Finally e_{ij} is the residual error, representing the variability not accounted for by the model.

Typically in a PCA analysis, the first few principal components describe most of the variability in the data set, allowing the transformed data set to be easily visualized as a series scores and loading plots. Score plots depict clustering and separation of objects and the relationships between clusters and the principal components, while loading plots illustrate the projection of the original variables upon each principal axis. Hence, the loadings explain in which part of the spectrum the respective principal components vary. In PCA analyses of unfolded EEMs, loading plots look much like EEMs themselves, with each plot showing the degree that different wavelength regions vary along the direction of the corresponding principal component (Persson and Wedborg, 2001; Boehme et al., 2004). Provided that the number of original variables in the data set is not large, it can also be convenient to view loadings and variables together in a so-called biplot (Gabriel, 1971).

As an illustration of the importance of preprocessing, three alternative preprocessing methods are applied to the Horsens catchment data set prior to PCA decomposition of the spectral data. Preprocessing by mean centering, autoscaling, and normalization followed by mean centering results in the identification of one, two, or three principal components, respectively. With mean centering only it is not possible to obtain a valid multicomponent PCA model – a single component is found describing 98% of the variation in the data set. With auto scaling, the PCA finds two distinctly different phenomena (Figure 10.4a). The first component describes 94% of the variation in the data set and shows a continuum between stream and estuary sites, possibly reflecting concentration as much as chemical differences between samples. The second component describes only 4% of total variation and is due almost entirely to the wastewater samples; a close-up view ignoring site 16 shows little discrimination among other sites along this axis. In contrast, PCA analysis following row normalization and mean centering identifies three principal components, while sites are seen to fall into three distinct clusters representing wastewater samples, estuary samples, and stream samples (Figure 10.4b). Further, less distinct clustering is also apparent between various stream sites. The first two components describe 64.9% and 13.6% of the variation between samples, while the third (not shown) describes approximately 2.5%. The percent variations described by the three different models are not directly comparable because they are generated from differently processed data sets; however, it is apparent that unsuitable preprocessing can reduce the effectiveness of PCA at partitioning variation along secondary and subsequent axes, and in so doing diminish the visualization of multivariate discrimination between samples.

PCA can be performed on fluorescence measurements alone or in combination with other water measurements. It may also be performed on scores obtained from other chemometric models, although there can be redundancy in this approach. It is also important to realize that variables appearing close together on, for example, a PCA score plot are typically not strongly correlated unless they also have high loadings in the corresponding components. Consequently, correlations between variables suggested by a PCA should always be confirmed directly by plotting the variables one against the another (Gabriel, 1971). In previous fluorescence studies, PCA on unfolded EEMs has been used to study DOM fluorescence variability in the oceans. Persson et al. (2001) used PCA to examine the mixing of deep and surface water masses in the Baltic Sea. Boehme et al. (2004) explored seasonal and regional variation in fluorescent DOM in the Gulf of Mexico, determining that 87% of DOM fluorescence variability related to a single PCA component representing terrestrial

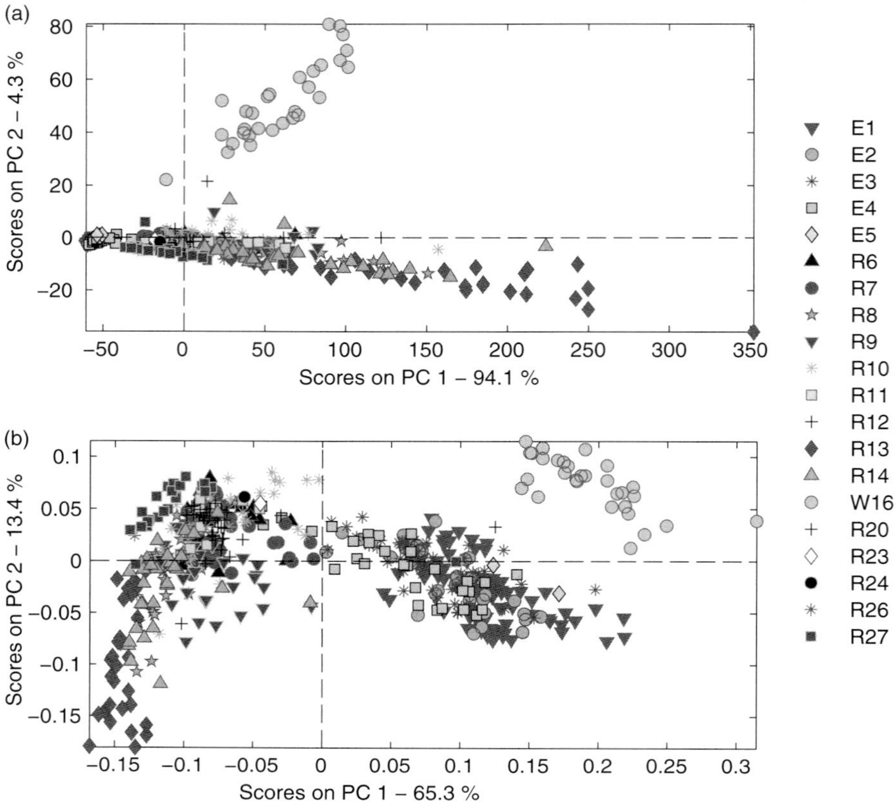

Figure 10.4. Principal components analysis of the Horsens catchment data set showing the effect of different preprocessing methods on model scores. (a) Auto-scaling. (b) Row normalization followed by mean centering. (See Plate 15.)

sources. Wedborg et al. (2007) showed that a humic-like fluorescent component in the Southern Ocean correlated with salinity, density and total inorganic carbon along the first principal axis highlighting its refractory nature, while along the second principal axis, two protein-like fluorescence components correlated strongly with each other and negatively with temperature and nitrate. Chen et al. (2010) used PCA on nine fluorescence components in surface and groundwaters of the Florida coastal Everglades and wetlands to show separation between samples from two surface and two groundwater sources. Terrestrial-humic and protein-like components lay at opposite ends of the first principal axis, while at opposite ends of a second principal axis apparently describing photo- and microbial-degradation were two short-emission humic fluorescence components. PCA has also been used to analyze DOM fluorescence and its interaction with process variables and plant performance in water treatment facilities (Antunes et al., 2007; Lu et al., 2009; Peiris et al., 2010). Typically, published studies involving PCA analyses of CDOM fluorescence data sets from natural and artificial aquatic environments have found the greater part (>80%) of fluorescence variability to be explained by just one or two principal components.

10.6 Parallel Factor Analysis

Parallel factor analysis (PARAFAC, less commonly termed canonical decomposition or CANDECOMP), is a multiway decomposition method originating from the field of psychometrics (Carroll and Chang, 1970; Harshman and Lundy, 1994). Although its utility for analyzing fluorescence data has been known for some years (Appellof and Davidson, 1981), its first applications to DOM fluorescence occurred relatively recently (Søndergaard et al., 2003; Stedmon et al., 2003). PARAFAC is now widely applied in the study of DOM fluorescence, with more than 90 PARAFAC studies of DOM in soils and natural water bodies published in 2005–2011, and a rapidly increasing number appearing in the engineering literature related to monitoring of wastewaters and water treatment systems.

10.7 PARAFAC and Its Properties

The popularity of PARAFAC stems from its ability to mathematically separate the spectra of overlapping fluorescence components. Hence, PARAFAC is similar to MCR mentioned earlier (actually PARAFAC can be considered the three-way version of the two-way MCR) with the important difference being that PARAFAC does not have rotational ambiguity as does MCR. This means that if the model is correct, it *will* give chemically meaningful results. For a mathematical explanation of PARAFAC including tutorials on its application we refer the reader to other references (Bro, 1997; Andersen and Bro, 2003). Briefly stated, PARAFAC of a three-way data set decomposes the data signal into a set of trilinear terms and a residual array:

$$x_{ijk} = \sum_{f=1}^{F} a_{if}\, b_{jf}\, c_{kf} + e_{ijk}$$

$$i = 1, \ldots, I;\ j = 1, \ldots, J\ ;\ k = 1, \ldots, K$$

where x_{ijk} is the intensity of the ith sample at the jth emission value and at the kth excitation value. When successfully modeling an EEM data set, the PARAFAC model allows for a direct chemical interpretation as opposed to the abstract orthogonal components stemming from PCA. For example, the parameter a_{if} is directly proportional to the concentration of the fth analyte of sample i; the vector $\mathbf{b}_f$ with elements b_{jf} is a scaled estimate of the emission spectrum of the fth analyte. Likewise, the vector $\mathbf{c}_f$ with elements c_{kf} is linearly proportional to the specific absorption coefficient (i.e., molar absorptivity) of the fth analyte. Finally, e_{ijk} is the residual representing the variability not accounted for by the model. Although the simple so-called trilinear model is shown in the above equation, it can be extended to higher-order data sets by increasing the number of terms following the summation sign. For example, a four-way data set consisting of EEMs x_{ijkl} replicated in time would require a further multiplicative term d_{lf} to model the fourth (time) dimension.

Three important assumptions for successful PARAFAC models are (1) variability: no two chemical components can have identical excitation or emission spectra or have concentrations that are perfectly correlated (an example could be a set of diluted samples); (2) trilinearity: emission spectra do not vary across excitation wavelengths, excitation spectra do not vary across emission wavelengths, and fluorescence increases approximately linearly with concentration; (3) additivity: fluorescence is due to the linear superposition of a fixed number of components (Bro, 1997). In the case where data consists of a three-way array of EEMs (samples × emission × excitation), PARAFAC has been shown to be capable of recovering accurate spectra and concentrations of known fluorescent materials in mixtures, even in the presence of uncalibrated spectral interferents (Bro, 1997).

Practical issues associated with DOM measurements that can challenge or violate these assumptions include the presence of strongly correlated components with similar spectral properties, inner filter effects at high concentrations, spectral changes due to e.g., varying pH, charge transfer and intersystem crossing, quenching, changes in instrumental set up affecting spectral resolution between samples, Rayleigh and Raman scatter bands and other non-trilinear systematic error structures. The presence of highly correlated fluorophores in a data set is problematic for PARAFAC because it violates the variability assumption. Recently, a new class of factor models (PARALIND) were designed to deal with complications arising from linearly dependent factors (Bro et al., 2009).

10.7.1 Determining the Number of Components

The chemical interpretation of a PARAFAC model relies on the right number of components being specified by the user. When models are underspecified, fewer components are used in the model than there are independently varying fluorescent moieties present at detectable levels. When this occurs, the model may hybridize the spectra of chemically distinct components and produce scores that model fluorescence from multiple unrelated fluorophores. When models are overspecified, more components are present in the model than there are independently varying fluorescent moieties. In this case, two or more PARAFAC components may be used to represent a single fluorophore often in combination with noise, or components may be included that are not chemically meaningful. The presence of highly correlated components in a model can indicate that it is overspecified.

Whether distinct chemical moieties can be separated by PARAFAC depends on a range of factors, including the relative concentrations of the various fluorophores in the sample and the degree of overlap between their spectra, as well as the number of samples and the degree of environmental variability encompassed in the data set. In natural systems where the number of detectable fluorophores is unknown, visualization and diagnostic tools are needed to assess the validity of PARAFAC models and determine the correct number of components (Andersen and Bro, 2003). Most simply, visualization of the model spectra can be used to assess whether its components appear to be reasonable. The spectra of organic matter fluorophores are typically smoothly shaped, with single, often broad emission peaks and either single or multiple excitation peaks. Spectra that look highly irregular, strange or noisy may be distorted – possibly due to the incomplete removal of scatter or other artefacts during preprocessing. Alternatively, the model may be overfitted and such components may represent PARAFAC's attempt to model background noise. Although visualization alone can identify some problems, in most cases additional tools are needed to confirm that visually feasible models are also mathematically robust.

Because the PARAFAC model makes no assumptions about spectral shapes nor the structure of parameters and error terms, if two completely independent models derived from different sets of samples arrive at similar spectral shapes, it provides strong evidence that the spectra represent underlying chemical phenomena. In split-half validation, independent halves of a data set are modeled separately. The model is validated when the same components are found in each half-data set, as this result could not reasonably arise from chance alone (Harshman and Lundy, 1994). When spectrally identical components are uncovered in completely unrelated data sets, as has been reported with increasing frequency (e.g., Stedmon et al., 2007; Murphy et al., 2011; this study), it can be taken as even stronger validation that such PARAFAC components are chemically meaningful.

An additional tool for determining the number of components is the core consistency diagnostic (Bro and Kiers, 2003), which checks the adherence of the data to the trilinear PARAFAC model. Valid PARAFAC models have core consistency close to 100%, unstable models have intermediate core consistencies (around 50%), and invalid models (caused by data are not trilinear, or a model having too many components) have core consistencies that are often near zero or negative. When a sequence of models are developed each with one more component than the previous, the first overspecified model is often identified by a large decrease in core consistency relative to the model with one fewer components.

Unfortunately, diagnostics such as these can at times give ambiguous or contradictory results; for example, models with poor core consistencies can quite often be validated using split half analysis (Murphy et al., 2008; Stedmon and Bro, 2008), or may have better predictive capability than models with fewer components and higher core consistencies (Bosco et al., 2006). The improvement of diagnostics for model selection is an active area of research (Smilde et al., 2004), but it has to be stressed that extensive insight in to the analytical data, the context of the actual problem and the mathematics and statistics behind the modeling is needed to provide scientifically valid results in general. That said, an automated program for calculating PARAFAC models of EEM data is available (Bro and Vidal, 2010). This program takes into account the interdependence of a range of modeling decisions and diagnostics and automatically determines the number of components, possible outliers, and so forth. Although the automated program may be useful, it is imperative to be aware that it is based on certain assumptions and is bound to fail for some data.

10.8 Practical Implementation of PARAFAC

Although the PARAFAC algorithm is designed to search for loadings and scores that produce the least-squares, "best fit" solution, in practice it is possible for the algorithm to converge on local, rather than global minimum residuals. When this happens, an incorrect solution is obtained; or rather; the least squares solution is not obtained at all. To guard against this, it is recommended that models are initialized with random starting conditions, and that it is confirmed that essentially the same solution is obtained from different random starting conditions. Obtaining the same solution means that the sum of the squared residuals from different models should be identical. The loadings from prior PARAFAC models can also be used as "first guesses" To speed up modeling or assist PARAFAC in arriving at a "likely" solution (Bro, 1997), however, this can increase the risk that the model will incorrectly settle on local rather than global minimum residuals.

Unstable models can often be recognized by low core consistencies, or by the fact that the model changes when small numbers of samples are removed from the data set. It is important to identify and remove outliers before modeling so they do not exert undue influence on the model. One technique that is useful in this regard is jack-knifing, a resampling method used to assess the influence and leverage of individual samples within a data set (Riu and Bro, 2003). Residual error plots should be always be examined for evidence of nonrandom structure. Consistent peaks in the residuals suggest that additional components may be needed, whereas peaks and troughs appearing next to each other can indicate an overfitted, or poorly fitted model (Stedmon et al., 2003).

Unstable models can sometimes be improved by applying appropriate constraints during modeling (Andersen and Bro, 2003). For example, it is common in fluorescence applications that concentrations and spectra are constrained to be non-negative. It can also work well to constrain spectra to having no more than a single peak (unimodality). Constraints can assist PARAFAC in arriving at stable, chemically sensible solutions especially for real-world, noisy data sets. However, care has to be taken to ensure that the process does not cover up problems that would be better solved with other approaches, and that important chemical phenomena are not obscured or misrepresented as a result.

10.8.1 PARAFAC Models of Organic Matter

Of all chemometric methods, PARAFAC is currently the one most frequently applied to the analysis of organic matter fluorescence EEMs. Figure 10.5 shows the relationship between sample size and numbers of components identified in 33 PARAFAC models of DOM in natural waters and soils, including only independently derived models by a range of research groups that were published between 2003 and 2010. While there has been a general trend where larger numbers of components are resolved in larger data sets, as many as five components have been identified with as few as 18 samples (Hall and Kenny, 2007). Models with eight or more components were typically derived from data sets that included soil samples (Yamashita et al., 2008; Fellman et al., 2009a, 2009b, 2009c; Chen et al., 2010), with the largest models usually derived from studies that analyzed a diverse range of samples from lakes, streams, soils, wetlands, estuaries and the ocean.

Chemical interpretations of individual PARAFAC components in organic matter fluorescence data sets have ranged from very general (protein-like, humic-like) to somewhat more specific (tryptophan-like, tyrosine-like, quinone-like). Of models created from data sets consisting of more than 100 samples, the vast majority have each resolved PARAFAC

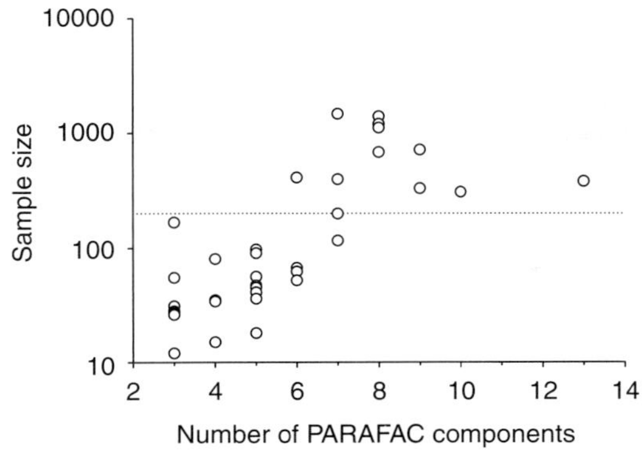

Figure 10.5. The relationship between number of samples and number of components determined by the analysts in 33 PARAFAC models of natural organic matter fluorescence EEMs, published between 2003 and 2010.

components that correspond with the protein-like (T and B), humic-like (A and C) and autochthonous humic (M) peaks described by Coble (1996).

Figure 10.5 suggests there is a relatively low cap on the number of PARAFAC components able to be identified in organic matter data sets. Undoubtedly, the number of fluorophores and spectra in natural samples is much greater. It is not at present clear to what extent the small number of components typically identified is a reflection of undetectably low concentrations of many organic matter fluorophores (i.e., low signal-to-noise ratios), or of combinations of very similar fluorophores being modeled by PARAFAC as if they are single components, or if the low number of components reflect that there are significant deviations from Beer-Lambert's law.

Providing that the assumptions of trilinearity, additivity and variability are met, the chemical interpretation of PARAFAC models is clear; specifically, the equivalence of the PARAFAC loadings and the chemical spectra of the individual underlying fluorophores. However, the chemical interpretation of PARAFAC components in decompositions of DOM data sets is not straightforward, in part because the degree to which these assumptions are reasonable is difficult to assess (e.g., the presence of collinearity in a data set) or else a point of debate (e.g., additivity and trilinearity) (Del Vecchio and Blough, 2004; Boyle et al., 2009). Studies have shown that PARAFAC can provide interpretable if imperfect models, even in the presence of known violations of the assumptions of variability or trilinearity (Bro, 1997; Bro et al., 2009; Murphy et al., 2011), and that the imposition of appropriate model constraints can ameliorate these in some cases. Even so, in natural systems it is difficult to assess the severity of departures from model assumptions, particularly in cases where data sets are large or diverse.

A number of recent studies have projected new data sets on to existing models in order to extract concentrations of "known" PARAFAC components (e.g., Mladenov et al., 2008; Fellman et al., 2009c; Macalady and Walton-Day, 2009; Miller et al., 2009a, 2009b; Miller and McKnight, 2010). In many such studies, judgments of model suitability have been based on the magnitude of error residuals relative to the measured fluorescence signals. A problem with this approach is that the fit of an existing PARAFAC model to a new data set increases *trivially* with the number of components in the model, because irrelevant components can be used to model noise and compensate for poor fit. In fact, a range of outlier diagnostics are needed to verify that existing models provide unbiased representations of new data before using them in a predictive capacity (Rinnan et al., 2007). It should further be noted that the assumptions of PARAFAC may not be sufficiently met by all data sets, depending, for example, on sample concentration (Stedmon and Bro, 2008), stoichiometry (Bro et al., 2009), and interactions between fluorophores (Boyle et al., 2009) or with metal ions, among other factors. For all of these reasons, the interpretation of PARAFAC models of organic matter fluorescence, and particularly their application in a predictive capacity, is at present a developing science necessitating a cautious investigative approach.

10.8.2 PARAFAC Example

PARAFAC modeling of the Horsens catchment data set was performed after normalizing concentrations to unit norm in the first mode and imposing non-negativity constraints on model scores and loadings. The model resolves five components (Figure 10.6) that describe 99.6% of the variability in the EEMs. The model has low core consistency – in fact, core consistency diagnostics alone would indicate that a two component model is most appropriate. However, the five component model is confirmed by modeling independent halves of the data set in a split-half validation (Figure 10.6) and the spectra obtained are regularly shaped and consistent with other published PARAFAC spectra for DOM samples (Figure 10.7). The loadings of the Horsens catchment PARAFAC model, and correlations with similar spectra in previous studies, are provided in the appendices to this chapter.

Identifying the sources of PARAFAC components is also not straightforward, even when aided by comparisons with earlier studies. A difficulty is that the presence of high concentrations of a component near terrestrial sources does not in itself ensure that the component has allochthonous origins. Also, it is generally not possible to distinguish between production and removal of components except in carefully designed experiments (e.g., Stedmon and Markager, 2005b; Stedmon et al., 2007).

In our example, there are $5!/3!2! = 10$ pairs of components that can be plotted in order to explore the relationships between the five PARAFAC components from the Horsens catchment model. Four of these are shown in Figure 10.8 (note log-log scales). Similar to the PCA in Figure 10.4, the PARAFAC analysis detects three clusters of samples. However, due to the additional information provided by the PARAFAC spectra, it is possible to obtain a more direct chemical interpretation of the differences between sites than previously. Component 3 appears to represent a terrestrially derived fraction that is more or less absent in the WTP organic matter (Figure 10.8, first and last panels). Closer analysis shows that this signal was present in the water in conjunction with periods of intensive rain where soil organic matter is also expected to be a more significant fraction of the municipal

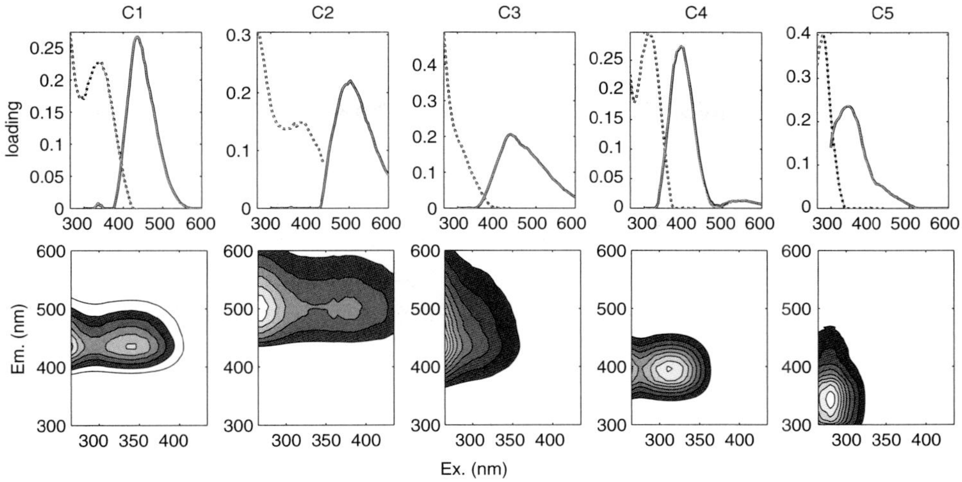

Figure 10.6. PARAFAC decomposition of the Horsens catchment fluorescence dataset. Spectra are shown above contour plots, with spectra for two independently modeled halves of the data set overlaid.

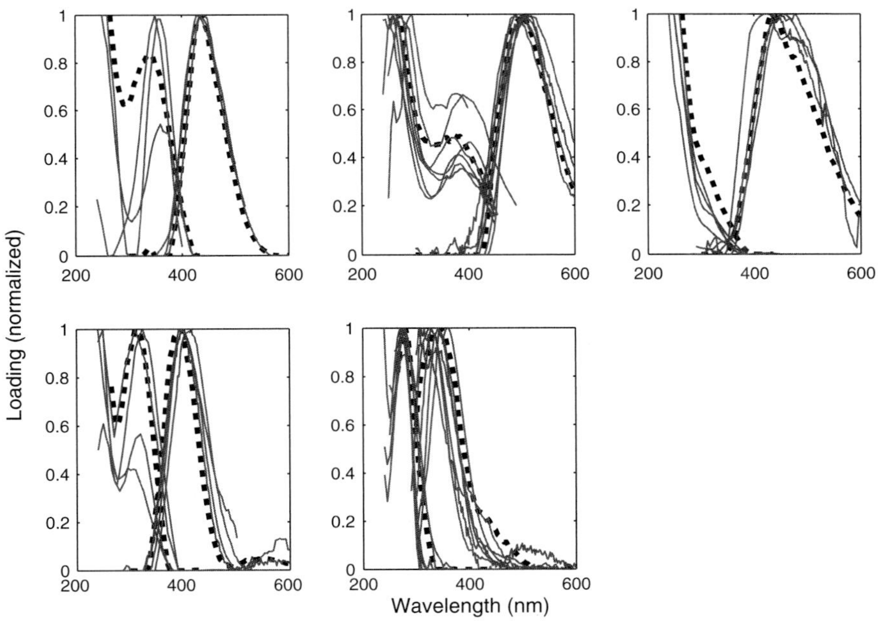

Figure 10.7. PARAFAC spectra in the Horsens catchment model (dotted lines) compared to PARAFAC spectra in 10 published models (thin lines). See the Appendix for inter-model comparisons.

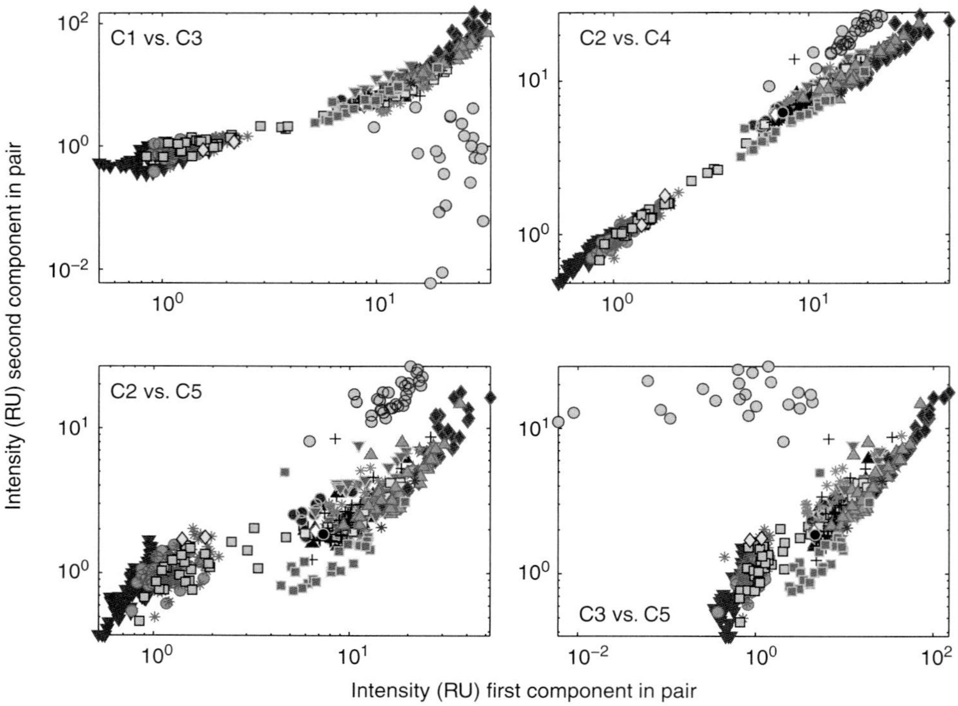

Figure 10.8. Paired comparisons between log-transformed fluorescence intensities of PARAFAC components in the Horsens catchment data set. (See Plate 16.)

waste water (Stedmon, unpublished). The consistent relationship between C2 and C4 for all except the wastewater samples suggests that they represent a common ubiquitous signal from the streams that is traceable into, and sustained in, the estuary. In the plot relating C2 and C5, it is observed that wastewater, streams, and estuary sites form three distinct groups in which wastewater and estuarine samples have a relatively greater proportion of C5, suggesting biological production, relative to the terrestrial signal of C2.

Comparisons between the fluorescent components in the Horsens catchment model and other published PARAFAC spectra for DOM samples are shown in Figure 10.7. Previous studies hypothesize that humic-like C2, which has been identified in many previous models, is derived from terrestrial material (Stedmon et al., 2003; Murphy et al., 2008), possibly as an intermediate formed during photochemical degradation of terrestrial organic matter (Stedmon et al., 2007). Protein-like fluorescence including tryptophan-like C5 is associated with biological production in surface waters (Determann et al., 1994; Determann et al., 1998). Humic-like C1 has an emission spectrum identical to syringealdehyde (produced in the breakdown of lignin), and is associated with waters with high organic matter loadings (Murphy et al., 2011). In this study, C1 dominated samples from forested upstream sites (particularly sites R13 and R14 in the Hansted system), as well as at the WTP.

Components practically identical to humic-like C3 have been identified in at least three previous models (Figure 10.7). In the Horsens catchment data set, the highest abundances of C3 occurred in streams (highest: site 13, lowest: site 27), and the lowest at the WTP site and in the estuary. Previously, Stedmon et al. 2003 sampled the Horsens catchment in 2001 and found that a similar component (component 1) dominated wetland and forested regions. Another similar component identified by Stedmon et al. (2005b, component 1) accumulated in Si- and P-limited mesocosms, where it was produced by microbial degradation and degraded by ultraviolet (UV) and visible light. In the PARAFAC model of Chen et al. (2010), a very similar component (component 2) occurred in surface waters of Florida Bay but not in ground waters. Components almost identical to humic-like C4 have been identified in at least four previous models (Stedmon et al., 2003; Stedmon et al., 2007; Kowalczuk et al., 2009; Murphy et al., 2011), particularly at sites near terrestrial sources. In this study of the Horsens catchment, it was abundant in streams and, in contrast to C3, was also abundant at the WTP site.

10.8.3 PARAFAC PCA Example

In a further example using the Horsens catchment data set, PCA is used to visualize the relationship between the five PARAFAC components, nutrients, DOC, and absorbance (Figure 10.9). The first principal component (62.4%) correlates positively with fluorescence component C2 as well as DOC and a375, but also correlates positively with all other variables. This is evidenced by the first loading plot (Figure 10.9i) in which all variables are placed on the right-most side of zero for the first principal component. Consequently, it appears to be an axis primarily describing variability in quantities of carbon, a phenomenon that is then apparently the main cause of variability overall. The second principal component (16.4%) strongly positively correlates with phosphorus (TDP, DOP) and weakly negatively with DON, and C3, and therefore primarily describes variations in phosphorus. This is seen from the fact that TDP and DOP are placed high on the positive axis of the second component and DON and C3 are placed less extreme on the negative axis (Figure 10.9ii). The third principal component (7.6%) correlates positively with nitrogen and negatively with component C3 (Figure 10.9iii, horizontal axis), and is the first principal component to distinguish C3 from DON. The fourth (6.1%) correlates positively with C5 (Figure 10.9iii, vertical axis).

The most important test of any model, other than to ensure that it is statistically sound, is to determine whether the results are plausible in the context of our understanding of organic matter sources and behavior. The PCA in Figure 10.9 indicates that in the Horsens catchment, fluorescence and carbon (DOC, a375) are strongly correlated, whereas carbon, nitrogen, and phosphorus vary largely independently. This fits with the current understanding of the Horsens catchment and decoupling of C, N, and P supply. The dominant source of organic carbon and CDOM in the catchment is believed to be derived from soils, particularly in the more forested catchments (e.g., R13 in Figure 10.5a), whereas the largest source of P is from agricultural catchments, for example, R10 and R11 (Figure 10.5b) (Stedmon

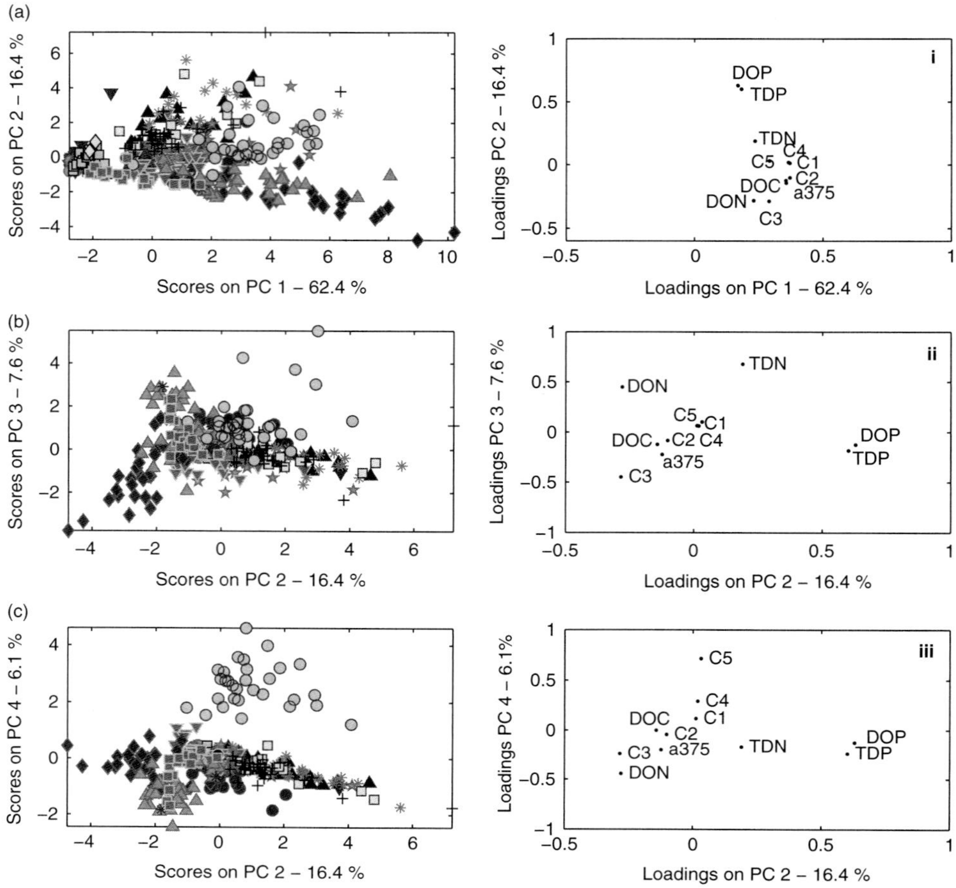

Figure 10.9. Principal components analysis of PARAFAC components and water quality parameters for the Horsens catchment data set. Score plots (a–c) showing variation among sites are on the left, loading plots (i–iii) showing correlations between measured variables are on the right. (See Plate 17.)

et al., 2006). The sampling station on the outflow of a wetland area in the Hansted system (R14) has the highest concentrations of TDN and DON (Fig. 2 in Stedmon et al., 2006). Thus, the PCA components results reveal differences in the supply of organic C, N, and P that appear to be related to land use.

10.8.4 Supervised Learning Techniques

In contrast to exploratory data analysis that mostly uses unsupervised learning, supervised learning involves developing models from data that are paired with a desired set of outcomes, which are used to guide the estimation of the models. In chemometrics, the objective of data analysis often involves classification or prediction. In either case, a basic procedure is followed. A training (or "calibration") data set is collected comprising reference measurements for the properties of interest together with the measurement attributes believed to reflect these properties (in the case of prediction) or the categories corresponding to the samples (in the case of classification). Chemometric models are then used to identify the "best model" of the relationship between the measurement attributes and properties of interest (prediction) or their categories (classification). Performing a validation in which predictions are tested using a new data set (test set validation) or using appropriate subsets of the original data matrix (cross-validation) is critical to ensuring that the model obtained is not overfitting and hence extensively describing random variation (Martens and Næs, 1989). If the validation is successful, the prediction good and if the training set included the full range of conditions to be expected in new samples, then the properties of interest can in future be estimated from the measurement variables without the need to measure them explicitly (except, perhaps, occasionally for checking purposes). Inputs to classification and prediction models can include the output of exploratory data analysis, for example, the principal components determined by PCA or PARAFAC.

10.9 Multivariate Calibration

Calibration aims to develop predictive models that relate properties of interest (that may be difficult or expensive to measure) with more easily measured attributes of the chemical system, for example, spectral data (Thomas, 1994; Gibb et al., 2000). Very often, a linear relationship may be anticipated between the measured data and the variables of interest or else between some (possibly nonlinear) transformation thereof. Although there are many different techniques that could be used, principal components regression (PCR) and partial least squares regression (PLS) are obvious candidates. Both are extensions of the multiple linear regression (MLR) model but utilize different algorithms for calculating regression coefficients and impose different restrictions. They differ from MLR primarily in that they are able to handle highly correlated input variables (such as adjacent wavelengths in fluorescence EEMs), whereas MLR requires that input variables are unique to some extent. Also, PCR and PLS are able to handle the situation of having more variables than samples.

In PCR, the score vectors from a PCA model are used as independent variables in an MLR model for predicting the dependent variable. Whereas in PCR a model is found that best reflects the covariance structure between the predictor variables (the columns of matrix X), a PLS model reflects the covariance structure between the predictor (X) and response (Y) variables. Thus the PLS model is optimized for predicting response. Both PCR and PLS models provide regression coefficients for predicting Y from X. The regression coefficients are different from the regression coefficients of, for example, MLR, because in PCR and PLS, the coefficients are found as a linear combination of the loadings in the model. They are also not directly chemically meaningful, as in PARAFAC. In PCR, the loadings come from PCA so they are orthogonal abstract representations of the most important spectral variation. In PLS, the loadings are similarly abstract, but in this case they reflect the variation in **X** relevant for predicting the dependent variables. The regression coefficients can be interpreted, though, in a similar way as MLR regression coefficients.

A multilinear version of PLS, called N-PLS, has also been developed for building regression models for multiway data (Bro, 1996). The loadings of an N-PLS model of EEMs are analogous to PARAFAC loadings, that is, a single excitation spectrum, a single emission spectrum, and a vector of regression coefficients for each latent variable in the model. Unlike with PARAFAC, however, the latent variables identified by N-PLS are optimized for predicting the response matrix and would not normally represent pure chemical spectra. Hence, the N-PLS loadings are similar in that respect to ordinary two-way PLS loadings.

For the Horsens catchment data set, the correlation between fluorescence and DOC implied by the principal component analysis suggests that it may be possible to predict DOC from fluorescence intensity. Previously, Vasel and Praet (2002) attempted to predict DOC and TOC in a wastewater treatment plant from fluorescence emission scans (300–450 nm) obtained at an excitation wavelength of 280 nm, finding only poor correlations ($R^2 < 0.4$) and low predictive success. Much greater success in predicting DOC from fluorescence ($R^2 > 0.9$) was documented by Marhaba et al. (2003), using a three-component PCR model developed from 69 EEMs (excitation 225–500 nm and emission 231–633 nm) collected from a canal supplying various New Jersey water treatment utilities.

For the current example, the unfolded EEMs are used to predict DOC concentrations in the Horsens catchment using PLS regression. Preprocessing is by mean-centering only because as was described earlier, normalization would remove concentration-related information inherent in the fluorescence data that is relevant to predicting DOC concentrations. Cross-validation indicates that a model with two latent variables has a relatively high correlation coefficient (cross-validated $R^2_{cv} = 0.91$); however, inspection of the model predictions (Figure 10.10) indicates that DOC in the estuary sites are poorly predicted in a combined site model owing to their low concentrations and low influence on the model. A solution in this case is to construct two separate models: one for the river and WTP sites, and another for the estuary sites. Table 10.2 summarizes the PLS results relevant to selecting the number of latent variables (LVs) in each model. When choosing the optimal model, there is a trade-off between selecting a small number of LVs, as increasing the number of latent variables increases the risk of overfit, causing poorer predictions with future data and minimizing cross-validated root-mean-square-error of prediction (RMSECV). Models with lower RMSECV also have higher correlation coefficients, R^2_{cv}. Although models are often selected to have minimum RMSECV, it is not always the case that the model with lowest RMSECV is the best. It has been observed that PLS is "eager-to-please," meaning that it can easily produce over-optimistic assessments of predictive capability, especially in situations where there are many more variables than samples (e.g., in the case of many data sets of unfolded EEMs) or when cross-validation is inadequate (e.g., leave-one-out

Table 10.2. *PLS prediction of DOC concentration from fluorescence intensities in the Horsens catchment, using separate models for the river and wastewater samples (n = 302) and estuary samples (n = 211)*

River and wastewater			Estuary	
nLV	$RMSE_{cv}$	R^2_{cv}	$RMSE_{cv}$	R^2_{cv}
1	120.59	0.80	46.51	0.24
2	**104.11**	**0.85**	36.57	0.52
3	98.25	0.87	**34.51**	**0.58**
4	93.07	0.88	35.53	0.58
5	89.07	0.89	33.21	0.63
6	90.32	0.89	32.49	0.64
7	90.17	0.89	31.48	0.66

Note: The "best model," representing a trade-off between minimum number of latent variables (nLV), minimum $RMSE_{cv}$ and maximum R^2_{cv} is indicated in each case in bold.

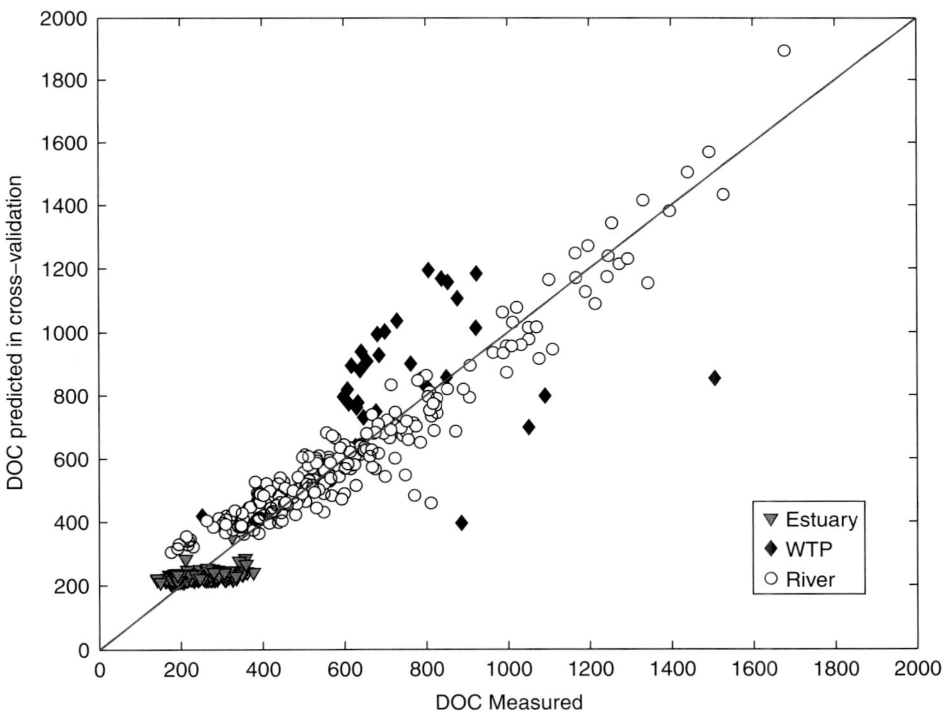

Figure 10.10. PLS prediction of DOC from fluorescence in the Horsens catchment. A single two-latent-variable model incorporating data from all sites produces poor predictions for the estuarine samples.

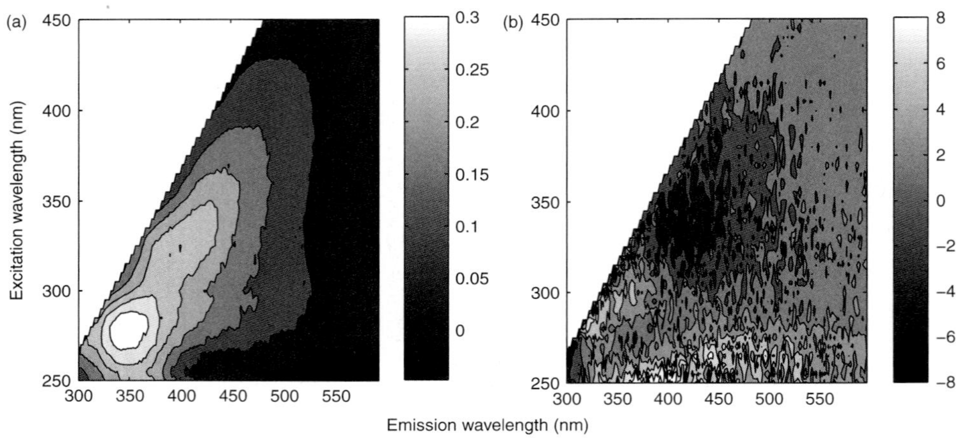

Figure 10.11. Regression coefficients for PLS prediction of DOC from fluorescence in the Horsens catchment. (a) River and WTP model with two latent variables. (b) Estuary model with three latent variables.

cross-validation used with a large data set, or even a small data set if it includes replicate samples) (Westerhuis et al., 2008; Kjeldahla and Bro, 2010). This often leads to the selection of overfitted models that are likely to perform worse when making future predictions than models that have fewer latent variables.

For the river and WTP sites, a five-component model has the lowest RMSECV; however, the two-component model appears to be a better choice, given that relatively small gains in RMSECV (and the correlation coefficient, R^2_{cv}) are obtained by including the last three latent variables. For the estuary sites, a three- component model appears to be sufficient. Again, a seven-component model has lower RMSECV, yet only very small improvements in RMSECV are attained with the addition of many latent variables. A conservative approach is thus to select two latent variables for the river and WTP model, and three for the estuary model.

Plots of regression coefficients might be expected to highlight the EEM regions with greatest influence upon the prediction of DOC concentration (Figure 10.11), however, in the case of spectral data (and non-designed data in general) one must take care not to overinterpret the components (Kjeldahla and Bro, 2010). Figure 10.11 suggests that for the river and WTP model, the T and M peaks are most important, with fluorescence in these regions associated with higher DOC concentrations (positive regression coefficients, Figure 10.11a). In the estuary model, the C peak region has high negative regression coefficients, indicating an inverse relationship with DOC, whereas high positive coefficients are associated with the A-peak region (Figure 10.11b). Whereas the relationship suggested in Figure 10.11a seems plausible, the strong inverse correlation between peak C fluorescence and DOC concentration implied in Figure 10.11b is counterintuitive. In fact, the visual interpretation of this plot may be distorted by overlapping spectral

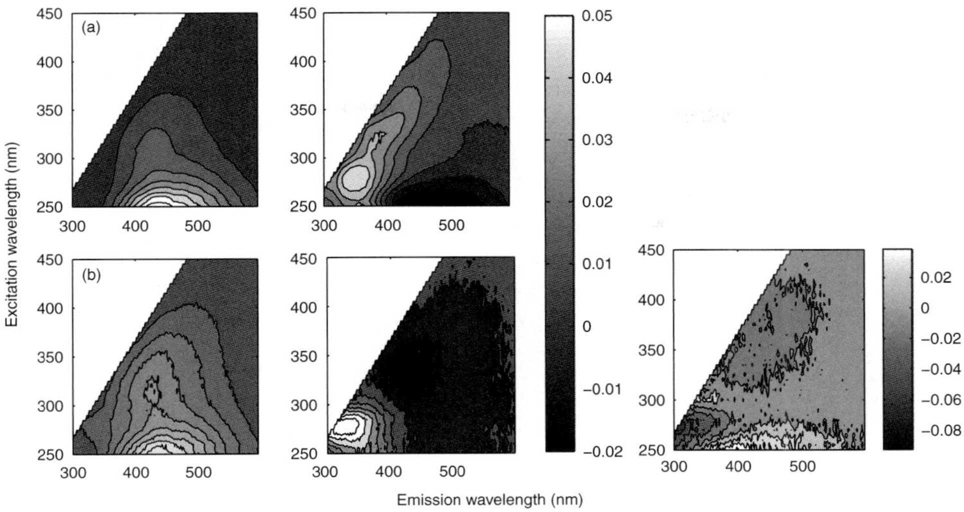

Figure 10.12. Latent variables in the PLS prediction of DOC from fluorescence in the Horsens catchment. (a) River and WTP model with two latent variables. (b) Estuary model with three latent variables.

signals in the latent variables (Figure 10.12), leading to negative regression coefficients being obtained for wavelengths that actually correlate positively to DOC (Kjeldahla and Bro, 2010).

The performance of the PLS regression models can be examined closely in plots of predicted and measured DOC against time, shown in Figure 10.13 for two river sites (R12: agricultural and R13: forested), the WTP site (W16) and an estuarine site (E3). At each site, temporal trends in measured DOC concentrations are also reflected in predicted concentrations. Close correspondence between measured and predicted concentrations is apparent particularly at the river sites (Figure 10.13a, b). Predictions at the wastewater site correspond generally with the DOC data, although not as closely as for the river sites (Figure 10.13c). In the estuary, the pool of nonfluorescent DOM that does not covary with fluorescent DOM might be expected to be larger due to the decoupling of processes that regulate CDOM and DOM production in marine waters (Nelson and Siegel, 2002). Despite this, not only are the overall trends in DOC concentrations at station E3 adequately modeled, but in many cases the differences in predicted DOC concentrations obtained from replicate EEMs is smaller than the difference between replicate DOC samples (Figure 10.13d). Overall, the median absolute errors associated with the DOC predictions at the four sites are 4.9%, 3.3%, 11.6%, and 5.9% at sites R12, R13, W16, and E3 respectively. Thus the error associated with the PLS predictions of DOC concentration is often comparable to, and sometimes better than, the combined (DOC and fluorescence) method measurement errors.

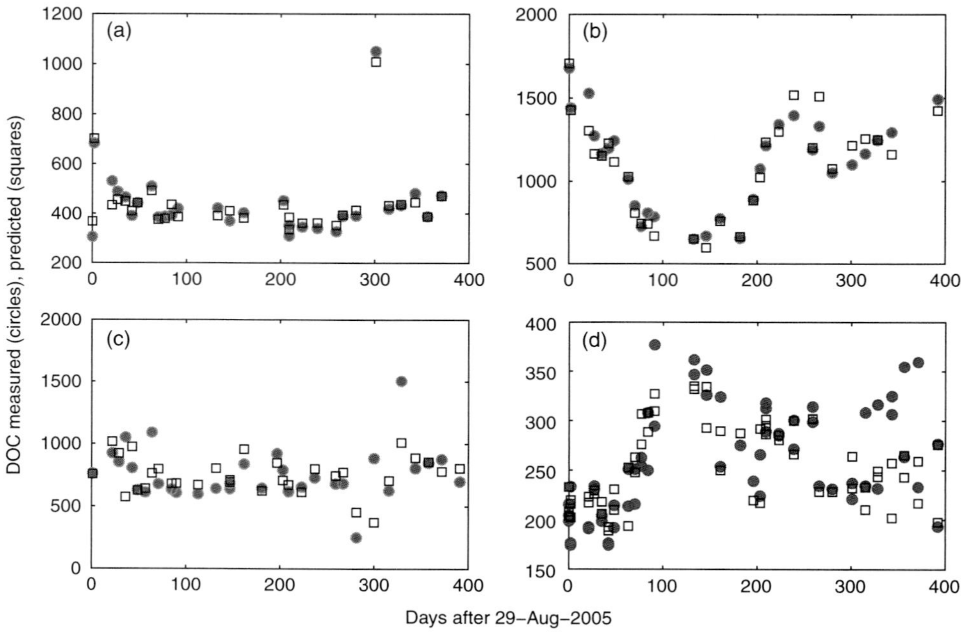

Figure 10.13. PLS prediction of DOC from fluorescence at four sites in the Horsens catchment. (a) agricultural station R12. (b) Forested station R13. (c) WTP station W16. (d) Estuarine station E3. Symbols show measured (circles) and predicted (squares) DOC concentrations over the 13-month study. Replicate samples were collected at station E3 only.

Predictive chemometric techniques, particularly PLS, have frequently been applied to fluorescence data sets in biomedical and industrial applications. Published examples involving DOM data sets are, in contrast, relatively few. Persson and Wedborg (2001) used PLS to predict the pedogenic versus marine origin of fluorescent DOM in the Baltic Sea. Murphy et al. (2009) attempted to predict the origin of marine fluorescence samples (in terms of distance from land) using PLS. Several authors have used fluorescence and PLS or PCR to monitor water quality in wastewater supply systems and in bioreactors (Vasel and Praet, 2002; Marhaba et al., 2003; Morel et al., 2004; Wolf et al., 2007). PLS has also been used to predict a range of chemical and microbiological variables from fluorescence measurements in soils and forest floor leachates (Simonsson et al., 2005; Rinnan and Rinnan, 2007).

10.10 Classification

Often, it is desired to assign samples to preestablished categories or classes. Such classes may describe, for example, differences in the origins of samples in a data set, whether

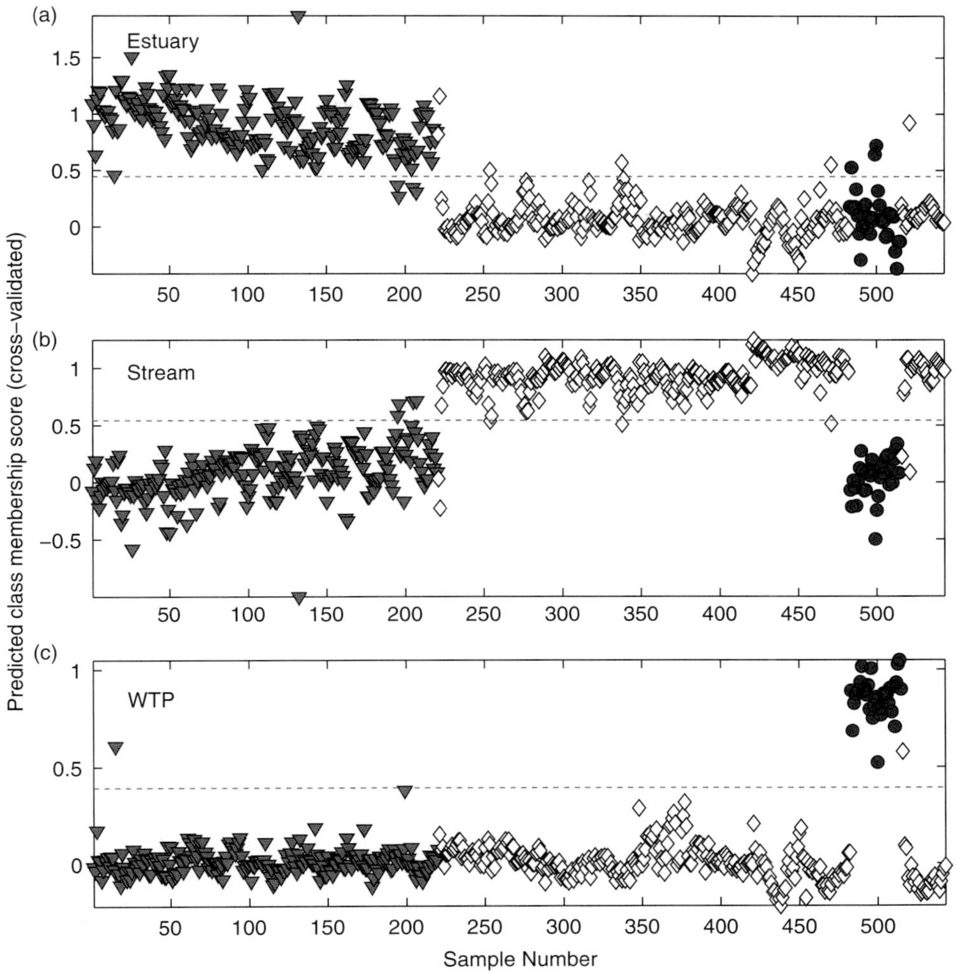

Figure 10.14. Discrimination between estuarine, stream, and WTP samples using PLS-DA, with class membership indicated by symbols plotting above the dotted line. (a) Estuary sites, closed triangles. (b) Stream sites, open diamonds. (c) WTP sites, dark circles.

they have undergone some kind of treatment or another, or whether they are "good" or "bad." A classification model is constructed from a "training" data set in which each sample has a known category. Subsequently, it is tested by classifying samples in a new data set using an algorithmically determined distance statistic to predict whether new samples belong to one category or another. Methods commonly used for this purpose in chemometrics include *k*-nearest neighbor (*k*NN), a supervised classification method based on PCA called soft independent modeling of class analogy (SIMCA), and both linear discriminant analysis (LDA) and discriminant analysis based upon partial least squares regression (PLS-DA).

In Figure 10.14, PLS-DA on the unfolded EEMs (normalized and mean-centered) is used to predict membership of three classes representing estuarine, stream or WTP sites for each of the samples from the Horsens catchment data set. Cross-validation of the PLS-DA model using random subsets of the data indicated that a three component model was appropriate. Unlike in PLS regression, where models are desired to have minimum $RMSE_{cv}$, in PLS-DA models are selected in order to have the lowest possible cross-validated classification error rates (Kjeldahla and Bro, 2010). In the current example, cross-validated classification error rates are 2.2%, 0.2%, and 1.6% for estuary, stream, and WTP sites respectively. In Figure 10.14, the vertical axis shows the predicted class membership (cross-validated) for each sample, plotted in relation to a calculated threshold (dotted line) that distinguishes samples that are assessed as belonging to a particular class (above the line) from samples that are not (below the line). Ideally, a good classification model also shows tight clustering of sites around 1 (class members) or zero (non-class members); in the current example, the absence of tight clustering is presumably a reflection of the continuity between sites in the data set, and the somewhat arbitrary designation of the three classes. In Figure 10.14a, all but four samples from stations E1–E5 are correctly assigned to the estuary class, while a small number of river (n = 3) and WTP (n = 3) samples cannot be distinguished from the estuary class. In Figure 10.14b showing predicted membership of the stream class, similar success rates for classification are observed. In Figure 10.14c, all samples from site 16 are correctly assigned to the WTP class, along with one misclassified sample from the estuary.

In previous studies of DOM fluorescence, a limited range of classification techniques have been applied. Bilal et al. (2010) used classification and regression trees (CART) of DOM fluorescence characteristics to investigate the persistence of farm waste contamination during a biodegradation experiment. Hall and Kenny (2007) used SIMCA coupled to a PARAFAC model to classify port samples according to their harbor of origin along the US east coast, while Hall et al. (2005) used multilinear N-PLS-DA to classify samples by port and river of origin. Overall, however, discrimination techniques have been underutilized in the interpretation of DOM fluorescence data sets, and could play a much larger role in understanding and predicting the behavior of natural organic matter fluorescence in the future.

10.11 Summary

In this chapter, a range of chemometric models for exploring and visualizing CDOM fluorescence data sets and for predicting the relationship between fluorescence and other variables have been introduced. It is apparent that exploratory methods, particularly PARAFAC and PCA, have already been widely implemented. Conversely, calibration models and discriminant analyses have been attempted relatively rarely, yet have considerable potential for improving our understanding of organic matter sources and biogeochemical processes due to their capacity for producing and testing predictive models.

Some technical aspects of developing chemometric models of spectroscopic data have been discussed, particularly the importance of preprocessing data properly to obtain meaningful multivariate models. Further, the importance of implementing cross validation procedures to ensure that especially supervised models are capable of making valid predictions has been discussed.

Most of the methods in this chapter are based on classical statistical estimators such as means and variances, and make predictions based on classical linear regression using least-squares estimates. These approaches have the disadvantage of high sensitivity to atypical observations (outliers) and to small departures from model assumptions. Robust algorithms for computing PCA, PLS, PCR (Verboven and Hubert, 2005), PARAFAC (Engelen et al., 2009), and linear regression (McKean, 2004), employing estimators that are less sensitive to outliers and small departures in model assumptions, are often integrated in modern statistical software packages or available online as free toolboxes for MATLAB.

Several techniques helpful in studying CDOM fluorescence data sets have been demonstrated using a 13-month time series of fluorescence EEMs, DOC, and nutrient measurements obtained from the Horsen's catchment in Denmark. These examples illustrate that both two- and three-way chemometric models can be useful for variable reduction, visualizing sources of variability in DOM fluorescence data sets, and for predicting relationships among future samples. Using PLS regression, DOC concentrations within Horsen's catchment streams were predicted with an accuracy comparable to the error associated with DOC and fluorescence measurements.

Acknowledgments

C. A. Stedmon acknowledges support by the Carlsberg Foundation and the Danish Research Council (grant #272–07–0485). K.R. Murphy acknowledges support under the Australian Research Council's Linkage Projects funding scheme (Project LP0776347)

Appendix

Comparisons between PARAFAC spectra in the 5-component (Horsens) model from this study versus PARAFAC spectra in 10 published studies. Horsens components are numbered from 1 to 5 corresponding to the order of components (top left to bottom right) in Figure 10.7. The number of the corresponding component in the published model is identified in the upper third of the table. Spectral congruence (Tucker, 1951) between similar components in each pair of models is shown separately for the excitation and emission spectra. Matched components are both visually and statistically similar and are overlaid on the Horsens spectra in Figure 10.7.

This study	(Stedmon et al., 2003)	(Stedmon and Markager, 2005a)	(Stedmon and Markager, 2005b)	(Cory and McKnight, 2005)	(Murphy et al., 2006)	(Stedmon et al., 2007)	(Murphy et al., 2008)	(Kowalczuk et al., 2009)	(Chen et al., 2010)	(Murphy et al., 2011)
Component in published study (C_Y) that is similar to a Horsens component (C_H)										
C_H	C_Y									
1		4				1				
2				5	3	3	3	4	5	
3		1	1					2	2	
4			6			2		3		2
5	4			8			7	5		
Spectral congruence between emission spectra										
C_H	Congruence between C_H and C_Y									
1		1				0.99				
2				1	0.99	0.99	1	0.97	0.98	
3		0.98	1					0.98	1	
4			0.94			0.99		0.98		0.94
5	0.98			0.99			0.97	0.97		
Spectral congruence between excitation spectra										
C_H	Congruence between C_H and C_Y									
1		0.92				0.99				
2				0.95	1	0.99	0.98	0.98	0.99	
3		0.99	0.98					0.99	0.99	
4			0.99			0.97		0.99		1
5	0.95			0.96			0.99	1		

References

Abbas, O., Rebufa, C., Dupuy, N., Permanyer, A., Kister, J., and Azevedo, D.A. (2006). Application of chemometric methods to synchronous UV fluorescence spectra of petroleum oils. *Fuel*, **85**, 2653–2661.

Andersen, C.M. and Bro, R. (2003). Practical aspects of PARAFAC modeling of fluorescence excitation – emission data. *J. Chemometr.*, **17**, 200–215.

Antunes, M.C.G. and Esteves Da Silva, J.C.G. (2005). Multivariate curve resolution analysis excitation-emission matrices of fluorescence of humic substances. *Anal. Chim. Acta*, **546**, 52–59.

Antunes, M.C.G, Pereira, C.C.C., and Esteves da Silva, J.C.G. (2007). MCR of the quenching of the EEM of fluorescence of dissolved organic matter by metal ions. *Anal. Chim. Acta*, **595**, 9–18.

Appellof, C.J. and Davidson, E.R. (1981). Strategies for analyzing data from video fluorometric monitoring of liquid-chromatographic effluents. *Anal. Chem.*, **53**, 2053–2056.

Bieroza, M., Baker, A., and Bridgeman, J. (2009). Relating freshwater organic matter fluorescence to organic carbon removal efficiency in drinking water treatment. *Sci. Total Environ.*, **407**, 1765–1774.

Bilal, M., Jaffrezic, A., Dudal, Y., Le Guillou, C., Menasseri, S., and Walter, C. (2010). Discrimination of farm waste contamination by fluorescence spectroscopy coupled with multivariate analysis during a biodegradation study. *J. Agric. Food Chem.*, **58**, 3093–3100.

Boehme, J., Coble, P., Conmy, R., and Stovall-Leonard, A. (2004). Examining CDOM fluorescence variability using principal component analysis: Seasonal and regional modeling of three-dimensional fluorescence in the Gulf of Mexico. *Mar. Chem.*, **89**, 3–14.

Booksh, K.S. and Kowalski, B.R. (1994). Theory of analytical chemistry. *Anal. Chem.*, **66**, A782-A791.

Bosco, M.V., Garrido, M., and Larrechi, M.S. (2006). Determination of phenol in the presence of its principal degradation products in water during a TiO2-photocatalytic degradation process by three-dimensional excitation–emission matrix fluorescence and parallel factor analysis. *Anal. Chim. Acta*, **559**, 240–247.

Boyle, E.S., Guerriero, N., Thiallet, A., Vecchio, R.D., and Blough, N.V. (2009). Optical Properties of humic substances and CDOM: Relation to structure. *Environ. Sci. Technol.*, **43**, 2262–2268.

Bratchell, N. (1989). Cluster analysis. *Chemometr. Intell. Lab. Syst.*, **6**, 105–125.

Bro, R. (1996). Multiway calibration Multilinear PLS. *J. Chemometr.*,**10**, 47–61.

Bro, R. (1997). PARAFAC: Tutorial and applications. *Chemometr. Intell. Lab. Syst.*, **38**, 149–171.

Bro, R. (2003). Multivariate calibration: What is in chemometrics for the analytical chemist? *Anal. Chim. Acta*, **500**, 185–194.

Bro, R. (2006). Review on multiway analysis in chemistry – 2000–2005. *Crit. Rev. Anal. Chem.*, **36**, 279–293.

Bro, R. and Kiers, H.A.L. (2003). A new efficient method for determining the number of components in PARAFAC models. *J. Chemometr.*, **17**, 274–286.

Bro, R. and Smilde, A.K. (2003). Centering and scaling in component analysis. *J. Chemometr.*, **17**, 16–33.

Bro, R., Workman, J.J., Mobley, P.R., and Kowalski, B.R. (1997). Review of chemometrics applied to spectroscopy: 1985–95, Part 3. Multi-way analysis. *Appl. Spectrosc. Rev.*, **32**, 237–261.

Bro, R., Harshman, R.A., Sidiropoulos, N.D., and Lundy, M.E. (2009). Modeling multi-way data with linearly dependent loadings. *J. Chemometr.*, **23**, 324–340.

Bro, R., Vidal, M., and Mizer, E.E. (2010). Automated modeling of fluorescence EEM data. *Chemometr. Intell. Lab. Syst.*, **106**, 86–92.

Brunsdon, C. and Baker, A. (2002). Principal filter analysis for luminescence excitation-emission data. *Geophys. Res. Lett.*, **29**, 2156.

Carroll, J.D. and Chang, J.J. (1970). Analysis of individual differences in multidimensional scaling via an n-way generalization of Eckart-Young decomposition. *Psychometrika*, **35**, 283–319.

Chen, M.L., Price, R.M., Yamashita, Y., and Jaffe, R. (2010). Comparative study of dissolved organic matter from groundwater and surface water in the Florida coastal Everglades using multi-dimensional spectrofluorometry combined with multivariate statistics. *Appl. Geochem.*, **25**, 872–880.

Coble, P.G. (1996). Characterization of marine and terrestrial DOM in seawater using excitation-emission matrix spectroscopy. *Mar. Chem.*, **51**, 325–346.

Cory, R.M. and McKnight, D.M. (2005). Fluorescence spectroscopy reveals ubiquitous presence of oxidized and reduced quinones in dissolved organic matter. *Environ. Sci. Technol.*, **39**, 8142–8149.

Cory, R.M., Miller, M.P., McKnight, D.M., Guerard, J.J., and Miller, P.L. (2010). Effect of instrument-specific response on the analysis of fulvic acid fluorescence spectra. *Limnol. Oceanogr. Meth.*, **8**, 67–78.

Del Vecchio, R. and Blough, N.V. (2004). On the origin of the optical properties of humic substances. *Environ. Sci. Technol.*, **38**, 3885–3891.

Désiré-Luc Massart, B.G.M., Vandeginste, S.N., Deming, Y.M., and Kaufman, L. (1988). *Chemometrics: A Textbook*. Amsterdam: Elsevier.

Determann, S., Reuter, R., Wagner, P., and Willkomm, R. (1994). Fluorescent matter in the early Atlantic-ocean. 1. Method of measurement and near-surface distribution. *Deep-Sea Res. Pt. I*,, **41**, 659–675.

Determann, S., Lobbes, J.M., Reuter, R., and Rullkotter, J. (1998). Ultraviolet fluorescence excitation and emission spectroscopy of marine algae and bacteria. *Mar. Chem.*, **62**, 137–156.

Engelen, S., Frosch, S., and Jorgensen, B.M. (2009). A fully robust PARAFAC method for analyzing fluorescence data. *J. Chemometr.*, **23**, 124–131.

Fellman, J.B., Hood, E., D'Amore, D.V., Edwards, R.T., and White, D. (2009a). Seasonal changes in the chemical quality and biodegradability of dissolved organic matter exported from soils to streams in coastal temperate rainforest watersheds. *Biogeochemistry*, **95**, 277–293.

Fellman, J.B., Hood, E., Edwards, R.T., and Jones, J.B. (2009b). Uptake of allochthonous dissolved organic matter from soil and salmon in coastal temperate rainforest streams. *Ecosystems*, **12**, 747–759.

Fellman, J.B., Miller, M.P., Cory, R.M., D'Amore, D.V., and White, D. (2009c). Characterizing dissolved organic matter using PARAFAC modeling of fluorescence spectroscopy: A comparison of two models. *Environ. Sci. Technol.*, **43**, 6228–6234.

Gabriel, K.R. (1971). Biplot graphic display of matrices with application to principal component analysis. *Biometrika*, **58**, 453–467.

Geladi, P. and Kowalski, B.R. (1986). Partial least-squares regression: A tutorial. *Anal. Chim. Acta*, **185**, 1–17.

Gibb, S.W., Barlow, R.G., Cummings, D.G., Rees, N.W., Trees, C.C., Holligan, P., and Suggett, D. (2000). Surface phytoplankton pigment distributions in the Atlantic Ocean:

An assessment of basin scale variability between 50°N and 50°S. *Prog. Oceanogr.*, **45**, 339–368.

Gurden, S.P., Westerhuis, J.A., Bro, R., and Smilde, A.K. (2001). A comparison of multi-way regression and scaling methods. *Chemometr. Intell. Lab. Syst.*, **59**, 121–136.

Hall, G.J. and Kenny, J.E. (2007). Estuarine water classification using EEM spectroscopy and PARAFAC-SIMCA. *Anal. Chim. Acta*, **581**, 118–124.

Hall, G.J., Clow, K.E., and Kenny, J.E. (2005). Estuarial fingerprinting through multidimensional fluorescence and multivariate analysis. *Environ. Sci. Technol.*, **39**, 7560–7567.

Harshman, R.A. and Lundy, M.E. (1994). Parafac – parallel factor-analysis. *Comput. Stat. Data Anal.*, **18**, 39–72.

Jaumot, J. and Tauler, R. (2010). MCR-BANDS: A user friendly MATLAB program for the evaluation of rotation ambiguities in multivariate curve resolution. *Chemometr. Intell. Lab. Syst.*, **103**, 96–107.

Jiang, F., Lee, F.S-C., Wang, X., and Dai, D. (2008). The application of excitation/emission matrix spectroscopy combined with multivariate analysis for the characterization and source identification of dissolved organic matter in seawater of Bohai Sea, China. *Mar. Chem.*, **110**, 109–119.

Kjeldahla, K. and Bro, R. (2010). Some common misunderstandings in chemometrics. *J. Chemometr.*, **24**, 558–564.

Kowalczuk, P., Durako, M.J., Young, H., Kahn, A.E., Cooper, W.J., and Gonsior, M. (2009). Characterization of dissolved organic matter fluorescence in the South Atlantic Bight with use of PARAFAC model: Interannual variability. *Mar. Chem.*, **113**, 182–196.

Lakowicz, J.R. (2006). *Principles of Fluorescence Spectroscopy*, 3rd ed. New York: Plenum Press.

Lavine, B. and Workman, J. (2010). Chemometrics. *Anal. Chem.*, **82**, 4699–4711.

Lu, F., Chang, C-H., Lee, D.-J., He, P-J., Shao, L-M., and Su, A. (2009). Dissolved organic matter with multi-peak fluorophores in landfill leachate. *Chemosphere*, **74**, 575–582.

Macalady, D.L. and Walton-Day, K. (2009). New light on a dark subject: On the use of fluorescence data to deduce redox states of natural organic matter (NOM). *Aquat. Sci.*, **71**, 135–143.

Marhaba, T.F., Bengraine, K., Pu, Y., and Arago, J. (2003). Spectral fluorescence signatures and partial least squares regression: Model to predict dissolved organic carbon in water. *J. Hazard. Mater.*, **97**, 83–97.

Martens, H. and Næs, T. (1989). *Multivariate Calibration*. Chichester: Wiley & Sons.

McKean, J.W. (2004). Robust analysis of linear models. *Stat. Sci.*, **19**, 562–570.

Miano, T.M. and Senesi, N. (1992). Synchronous excitation fluorescence spectroscopy applied to soil humic substances chemistry. *Sci. Total Environ.*, **117–118**, 41–51.

Miller, M.P. and McKnight, D.M. (2010). Comparison of seasonal changes in fluorescent dissolved organic matter among aquatic lake and stream sites in the Green Lakes Valley. *J. Geophys. Res. Biogeosci.*, **115**. G00F12, doi: 10.1029/2009jg000985.

Miller, M.P., McKnight, D.M., and Chapra, S.C. (2009a). Production of microbially-derived fulvic acid from photolysis of quinone-containing extracellular products of phytoplankton. *Aquat. Sci.*, **71**, 170–178.

Miller, M.P., McKnight, D.M., Chapra, S.C., and Williams, M.W. (2009b). A model of degradation and production of three pools of dissolved organic matter in an alpine lake. *Limnol. Oceanogr.*, **54**, 2213–2227.

Mladenov, N., Huntsman-Mapila, P., Wolski, P., Masarnba, W.R.L, and McKnight, D.M. (2008). Dissolved organic matter accumulation, reactivity, and redox state in ground water of a recharge wetland. *Wetlands*, **28**, 747–759.

Mobley, P.R., Kowalski, B.R., Workman, J.J., and Bro, R. (1996). Review of chemometrics applied to spectroscopy: 1985–95, Part 2. *Appl. Spectrosc. Rev.*, **31**, 347–368.

Morel, E., Santamaria, K., Perrier, M., Guiot, S.R., and Tartakovsky, B. (2004). Application of multi-wavelength fluorometry for on-line monitoring of an anaerobic digestion process. *Water Res.*, **38**, 3287–3296.

Murphy, K.R., Ruiz, G.M., Dunsmuir, W.T.M., and Waite, T.D. (2006). Optimized parameters for fluorescence-based verification of ballast water exchange by ships. *Environ. Sci. Technol.*, **40**, 2357–2362.

Murphy, K.R., Stedmon, C.A., Waite, T.D., and Ruiz, G.M. (2008). Distinguishing between terrestrial and autochthonous organic matter sources in marine environments using fluorescence spectroscopy. *Mar. Chem.*, **108**, 40–58.

Murphy, K.R., Boehme, J.R., Noble, M., Smith, G., and Ruiz, G.M. (2009). Deducing ballast water sources in ships arriving to New Zealand from southeastern Australia. *Mar. Ecol. Prog. Ser.*, **390**, 39–53.

Murphy, K.R., Butler, K.D., Spencer, R.G.M., Stedmon, C.A., Boehme, J.R., and Aiken, G.R. (2010). The measurement of dissolved organic matter fluorescence in aquatic environments: An interlaboratory comparison. *Environ. Sci. Technol.*, **44**, 9405–9412.

Murphy, K.R., Hambly, A., Singh, S., Henderson, R.K., Baker, A., Stuetz, R., and Khan, S.J. (2011). Organic matter fluorescence in municipal water recycling schemes: Towards a unified PARAFAC model. *Environ. Sci. Technol.*, **45**, 2909–2916.

Naes, T., Isaksson, T., Fearn, T., and Davies, T. (2002). *A User Friendly Guide to Multivariate Calibration and Classification*. Chichester: NIR Publications.

Nelson, C.E., Sadro, S., and Melack, J.M. (2009). Contrasting the influences of stream inputs and landscape position on bacterioplankton community structure and dissolved organic matter composition in high- elevation lake chains. *Limnol. Oceanogr.*, **54**, 1292–1305.

Nelson, N.B., and Siegel, D.A. (2002). Chromophoric DOM in the open ocean. In D. A. Hansell and C.A. Carlson (Eds.), *Biogeochemistry of Marine Dissolved Organic Matter* (pp. 547–578). San Diego: Academic Press.

Peiris, R.H., Halle, C., Budman, H., Moresoli, C., Peldszus, S., Huck, P.M., and Legge, R.L. (2010). Identifying fouling events in a membrane-based drinking water treatment process using principal component analysis of fluorescence excitation-emission matrices. *Water Res.*, **44**, 185–194.

Persson, T. and Wedborg, M. (2001). Multivariate evaluation of the fluorescence of aquatic organic matter. *Anal. Chim. Acta*, **434**, 179–192.

Rinnan, R. and Rinnan, A. (2007). Application of near infrared reflectance (NIR) and fluorescence spectroscopy to analysis of microbiological and chemical properties of arctic soil. *Soil Biol. Biochem.*, **39**, 1664–1673.

Rinnan, Å., Riu, J., and Bro, R. (2007). Multi-way prediction in the presence of uncalibrated interferents. *J. Chemometr.*, **21**, 76–86.

Riu, J. and Bro, R. (2003). Jack-knife technique for outlier detection and estimation of standard errors in PARAFAC models. *Chemometr. Intell. Lab. Syst.*, **65**, 35–49.

Sierra, M.M.D., Giovanela, M., Parlanti, E., and Soriano-Sierra, E.J. (2005). Fluorescence fingerprint of fulvic and humic acids from varied origins as viewed by single-scan and excitation/emission matrix techniques. *Chemosphere*, **58**, 715–733.

Simonsson, M., Kaiser, K., Danielsson, R., Andreux, F., and Ranger, J. (2005). Estimating nitrate, dissolved organic carbon and DOC fractions in forest floor leachates using ultraviolet absorbance spectra and multivariate analysis. *Geoderma*, **124**, 157–168.

Smilde, A., Bro, R., and Geladi, P. (2004). *Multi-way Analysis: Applications in the Chemical Sciences*. Chichester: John Wiley & Sons.

Søndergaard, M., Stedmon, C.A., and Borch, N.H. (2003). Fate of terrigenous dissolved organic matter (DOM) in estuaries: Aggregation and bioavailability. *Ophelia*, **57**, 161–176.

Stedmon, C.A. and Markager, S. (2005a). Resolving the variability of dissolved organic matter fluorescence in a temperate estuary and its catchment using PARAFAC analysis. *Limnol. Oceanogr.*, **50**, 686–697.

Stedmon, C.A. and Markager, S. (2005b). Tracing the production and degradation of autochthonous fractions of dissolved organic matter using fluorescence analysis. *Limnol. Oceanogr.*, **50**, 1415–1426.

Stedmon, C.A. and Bro, R. (2008). Characterizing dissolved organic matter fluorescence with parallel factor analysis: A tutorial. *Limnol. Oceanogr. Meth.*, **6**, 572–579.

Stedmon, C.A., Markager, S., and Bro, R. (2003). Tracing dissolved organic matter in aquatic environments using a new approach to fluorescence spectroscopy. *Mar. Chem.*, **82**, 239–254.

Stedmon, C.A., Markager S, Søndergaard, M., Vang, T., Laubel, A., Borch, N.H., and Windelin, A. (2006). Dissolved organic matter (DOM) export to a temperate estuary: Seasonal variations and implications of land use. *Estuaries Coasts*, **29**, 388–400.

Stedmon, C.A., Markager, S., Tranvik, L., Kronberg, L., Slatis, T., and Martinsen, W. (2007). Photochemical production of ammonium and transformation of dissolved organic matter in the Baltic Sea. *Mar. Chem.*, **104**, 227–240.

Tauler, R., Smilde, A., and Kowalski, B. (1995). Selectivity, local rank, 3-way data-analysis and ambiguity in multivariate curve resolution. *J. Chemometr.*, **9**, 31–58.

Thomas, E.V. (1994). A primer on multivariate calibration. *Anal. Chem.*, **66**, 795A-804A.

Tucker, L.R. (1951). *A Method for Synthesis of Factor Analysis Studies* (Personnel Research Section Report No. 984). Washington, DC: Department of the Army.

Vasel, J.L. and Praet, E. (2002). On the use of fluorescence measurements to characterize wastewater. *Water Sci. Technol.*, **45**, 109–116.

Verboven, S. and Hubert, M. (2005). LIBRA: A MATLAB library for robust analysis. *Chemometr. Intell. Lab. Syst.*, **75**, 127–136.

Wedborg, M., Persson, T., and Larsson, T. (2007). On the distribution of UV-blue fluorescent organic matter in the Southern Ocean. *Deep-Sea Res. Pt. I*,**54**, 1957–1971.

Westerhuis, J., Hoefsloot, H., Smit, S., Vis, D., Smilde, A., van Velzen, E., van Duijnhoven, J., and van Dorsten, F. (2008). Assessment of PLSDA cross validation. *Metabolomics*, **4**, 81–89.

Wolf, G., Almeida, J.S., Crespo, J.G., and Reis, M.A.M. (2007). An improved method for two-dimensional fluorescence monitoring of complex bioreactors. *J. Biotechnol.*, 128, 801–812.

Workman, J.J., Mobley, P.R., Kowalski, B.R., and Bro, R. (1996). Review of chemometrics applied to spectroscopy: 1985–95, Part 1. *Appl. Spectrosc. Rev.*, **31**, 73–124.

Yamashita, Y., Jaffe, R., Maie, N., and Tanoue, E. (2008). Assessing the dynamics of dissolved organic matter (DOM) in coastal environments by excitation emission matrix fluorescence and parallel factor analysis (EEM-PARAFAC). *Limnol. Oceanogr.*, **53**, 1900–1908.

Yin, H. (2002). ViSOM – A novel method for multivariate data projection and structure visualization. *IEEE Trans. Neural Networks*, **13**, 237–243.